Materials for Photobiology

Materials for Photobiology

Editors

Antonino Mazzaglia
Angela Scala
Enrico Caruso

MDPI

Basel • Beijing • Wuhan • Barcelona • Belgrade • Novi Sad • Cluj • Manchester

Editors
Antonino Mazzaglia
Institute for the Study of Nanostructured Materials
National Council of Research (CNR-ISMN)
Messina
Italy

Angela Scala
Department of Chemical, Biological, Pharmaceutical and Environmental Sciences (CHIBIOFARAM)
University of Messina
Messina
Italy

Enrico Caruso
Department of Biotechnologies and Life Sciences
University of Insubria
Varese
Italy

Editorial Office
MDPI
St. Alban-Anlage 66
4052 Basel, Switzerland

This is a reprint of articles from the Special Issue published online in the open access journal *International Journal of Molecular Sciences* (ISSN 1422-0067) (available at: www.mdpi.com/journal/ijms/special_issues/photobiology_materials).

For citation purposes, cite each article independently as indicated on the article page online and as indicated below:

Lastname, A.A.; Lastname, B.B. Article Title. *Journal Name* **Year**, *Volume Number*, Page Range.

ISBN 978-3-7258-0696-6 (Hbk)
ISBN 978-3-7258-0695-9 (PDF)
doi.org/10.3390/books978-3-7258-0695-9

Contents

About the Editors

Antonino Mazzaglia

Antonino Mazzaglia is currently Research Director at the National Council of Research-Institute for Nanostructured Materials (CNR-ISMN), Messina (Italy), and responsible for the new research center CNR-ISMN URT at the University of Messina, Dept. ChiBioFarAm. He has been a research scientist since 2001 until 2019 and a senior researcher since 2019 until 2020. He received MIUR Scientific Qualification as a Full Professor in Basic and Inorganic Chemistry, Material Chemistry (since 2017 and renewed in 2023), Pharmaceutical Technology (since 2018), and Physical Chemistry in 2020. He has been affiliated since 2008 with the Dottorato Collegium in Chemical Science at the University of Messina. He attended to European and national consortia by acquiring several grant-funded research projects as a local coordinator, training and industrial projects, and collaborating with national and international institutions and companies. He took part in International Advisory Boards of national, european, and international conferences; he was the Chair of the International Summer School on Cyclodextrin (2013), the XX International Cyclodextrin Symposium (XX ICS) Bridge Virtual Meeting, and XX ICS (13–17 June 2022). He was a visiting scientist and professor at the University of Exeter (UK, 2014) and an invited speaker at several conferences, research centers, and universities in Europe and Italy. He is a member of the editorial board of peer-reviewed journals and a guest editor of *Nanomaterials* and *Int J Mol Sci*. He co-authored about 118 papers, 11 proceedings, 6 book chapters, and one monographic review (H-index 32; 2630 citations; SCOPUS). He was the coauthor of about 235 conference communications, of which 57 were oral contributions (14 invited). His research activity is focused on bio-soft organic/inorganic hybrid materials based on cyclodextrins, porphyrinoids, photosensitizers, polysaccharides, polymers, and carbon nanoplatforms with stimuli-responsiveness properties for therapy, diagnosis, and biosensing.

Angela Scala

Angela Scala is currently an Associate Professor in Organic Chemistry at the University of Messina. Born in 1982, she graduated in Medicinal Chemistry and Technology (110/110 cum laude) in 2006. In 2010, she received the PhD (Doctor Europaeus) in Chemical Sciences. She spent 6 months at Trinity College Dublin (Ireland) as a visiting PhD student for the project "Synthesis and electronic properties of porphyrin arrays as potential photosensitizers for PDT, antenna complexes for solar energy generation, and surface deposition". In 2010, she received a 1-year research grant for the project "Synthesis of small molecule libraries incorporating privileged structures" (PRIN 2008). From 2011 to 2014, she worked at the National Council of Research-Institute for Nanostructured Materials Studies (CNR-ISMN) on the project "Synthesis and characterization of cyclodextrin-based bioactive nanostructures and of small ligands with high cellular affinity" (ESF-EUROCORES EuroBioSAS-ICS). In 2014, she received a scholarship for the project "Synthesis and characterization of nanocapsules of hydrogel/metallic nanoparticles for drug immobilization" (PON HIPPOCRATES). In 2014, she became an Assistant Professor and, in 2021, an Associate Professor in Organic Chemistry at the University of Messina. She supervised PhD, Master, and Bachelor students. She is a member of the editorial board of peer-reviewed journals and a guest editor of some published and incoming issues (*Biomolecules* and *Int J Mol Sci*, MDPI). She is the PI of national projects (PRIN2022 and PRIN2022PNRR), and she is involved in EU-funded and national projects.

Her scientific activity concerns the development of new nanomaterials based on biopolymers, cyclodextrins, carbon nanotubes, graphene, and metal nanoparticles for biological applications and the synthesis of biologically relevant small molecules. She co-authored more than 85 scientific publications and more than 90 contributions at international and national conferences.

Enrico Caruso

Enrico Caruso is currently an Associate Professor in Organic Chemistry at Insubria University. Born in Luino on 6 September 1974, he graduated in Chemistry from the University of Milan in 1998 (110/110). He attended the School of Specialization in Chemical Synthesis at the Polytechnic of Milan (November 1998–October 2000), with positive results (100/100). In December 2000, he won a scholarship at the Department of Structural and Functional Biology (DBSF) of the University of Insubria in Varese. From November 2001 to December 2004, he was employed as a research fellow (DBSF, University of Insubria of Varese) in the Laboratory of Organic Chemistry. Since December 2004, Prof. Enrico Caruso has held the role of Researcher in Organic Chemistry (CHIM06) at the MM.FF.NN Faculty of Sciences. On 1 November 2022, he became an Associate Professor in Organic Chemistry. Now, he is affiliated with the Department of Biotechnology and Life Sciences. His activity in the Organic Chemistry research unit develops along the following lines of research: synthesis of new molecules with photodynamic activity applied to therapy against tumors (PDT) and against bacteria (PACT); extraction of bio-active substances (antibacterial, antiprotozoal, and anticancer) from African plants; and preparation of formulations such as microemulsions and liposomes for the delivery of drugs. Prof. Caruso is the author or co-author of 52 scientific publications in international ISI journals and has presented over 90 contributions (oral presentations and posters) at scientific conferences in the fields of photodynamic therapy, microbiology, and biocatalysis. He has also acted as chairman of several sessions at national conferences and participated in the organization of congresses of the Italian Society of Photobiology.

International Journal of *Molecular Sciences*

MDPI

Editorial

Special Issue "Materials for Photobiology"

Angela Scala [1,*], Enrico Caruso [2] and Antonino Mazzaglia [3,*]

[1] Department of Chemical, Biological, Pharmaceutical and Environmental Sciences, University of Messina, V.le, F. Stagno d'Alcontres, 31, 98166 Messina, Italy

[2] Department of Biotechnologies and Life Sciences, University of Insubria, Via JH Dunant 3, 21100 Varese, Italy; enrico.caruso@uninsubria.it

[3] National Council of Research, Institute for the Study of Nanostructured Materials (CNR-ISMN), URT of Messina c/o Department of Chemical, Biological, Pharmaceutical and Environmental Sciences, University of Messina, V.le, F. Stagno d'Alcontres, 31, 98166 Messina, Italy

* Correspondence: ascala@unime.it or angela.scala@unime.it (A.S.); antonino.mazzaglia@cnr.it (A.M.)

Citation: Scala, A.; Caruso, E.; Mazzaglia, A. Special Issue "Materials for Photobiology". *Int. J. Mol. Sci.* **2024**, *25*, 3209. https://doi.org/10.3390/ijms25063209

Received: 17 February 2024
Revised: 1 March 2024
Accepted: 6 March 2024
Published: 12 March 2024

Photobiology is a challenging research area that aims to explore the interactions between light and living organisms and their biological consequences, with applications in the fields of photomedicine, photo(nano)technology, photosynthesis, and photosensory biology [1]. Nowadays, there is great interest in designing advanced materials with peculiar physicochemical properties and photoresponsive features that, through interactions with light, produce a response which can be exploited for diagnosis and/or therapy [2]. Therefore, the interaction of light with molecules, nanomaterials, cells, and tissues and the subsequent responses elicited represents a multidisciplinary field of research bringing together chemists, physicists, biologists, biochemists, medical specialists, and many others.

In this Special Issue, articles and reviews addressing the latest advances in the use of light-responsive materials for photobiology have been selected for publication, including phototherapeutic biomaterials proposed for light-induced therapies in cancer, microbial infections, atherosclerosis, aesthetic dentistry treatments (e.g., dental bleaching), and photobiomodulation (PBM) for neurological and neuropsychiatric disorder treatment.

In the design of advanced materials aiming to improve photoactivity efficiency, Lu et al. proposed a dual-activated nanoprodrug for combined chemo-photodynamic therapy of breast cancer based on a glutathione (GSH)-responsive BODIPY photosensitizer and a reactive oxygen species (ROS)-responsive thioketal linker connecting BODIPY and the chemotherapeutic agent camptothecin (CPT). After entering the tumor through passive targeting, the prodrug encapsulated with the amphiphilic polymer DSPE-mPEG2000, is activated by the high tumoral concentration of GSH. Light-triggered ROS from activated BODIPY not only induced apoptosis/necrosis of the tumor cells but also cleaved the thioketal linker to release on-demand CPT, achieving combined photodynamic therapy (PDT) and chemotherapy in mouse mammary carcinoma 4T1, human breast cancer MCF-7 cell lines, and also in 4T1 tumor-bearing mice. The IC50 values were 0.50 μM for the 4T1 cells and 0.63 μM for the MCF-7 cells, attesting for the strong photocytotoxicity exhibited against the tumor cells. Moreover, efficient tumor-targeting and tumor-suppressive effects were observed in 4T1 tumor-bearing mice, without toxicity and side effects, which is of great significance for cancer treatment.

With the aim of promoting breast tumor cell death in situ by photosensitization, Díaz et al. proposed FLTX2, a Tamoxifen derivative endowed with antiestrogenic, fluorescent, and photosensitizer properties, as a selective modulator of estrogen receptors (SERM) for the treatment of estrogen receptor (ER)-positive breast cancer. FLTX2, obtained through the covalent binding of tamoxifen as the ER binding core, 7-nitrobenzofurazan (NBD) as the florescent dye, and Rose Bengal (RB) as a source of ROS, showed a strong absorption in the blue spectral range, associated with the NBD moiety, which efficiently transferred the excitation energy to RB through an intramolecular FRET mechanism, generating superoxide anions that induced concentration- and time-dependent MCF7 apoptotic cell death.

Gangemi et al. proposed the synthesis of a novel bichromophoric system consisting of one subunit of curcumin (donor) and one of BODIPY (acceptor) able to emit in the far-red region, offering a large Stokes shift, capable of limiting light scattering processes. The dyad was encapsulated in small-sized mesoporous silica nanoparticles of MCM-41 (50–80 nm) and tested in human fetal osteoblastic cells (hFOB 1.19) and human bone osteosarcoma epithelial cells (U-2 OS), as models of normal and cancer bone cells, respectively. The bichromophoric system maintained a very efficient photoinduced intercomponent energy transfer even within the hybrid silica system, being located within the cytoplasm without losing brightness, confirming their applicability in bioimaging.

The phototoxicity of psoralen with light of different wavelengths (UVC: $\lambda = 254$ nm; UVA: $\lambda = 366$ nm) as well as the effect of ionizing radiation (radioisotope Re-188), the emitted beta radiation of which generates Cherenkov light with a yield of 35 photons per decay, was studied by Hübinger et al. Psoralen itself did not show toxic effects on the plasmid DNA or FaDu human cancer cells. After additional treatment with light, a concentration-dependent increase in single strand breaks (SSBs) was visible due to the photochemical activation of psoralen. Conversely, no additional significant Cherenkov-induced phototoxicity was observed when Re-188 was combined with psoralen.

Won Ahn et al. prepared photoactivatable nanocomplexes for treating atherogenic foam cells by encapsulating hydrophobic chlorin e6 (Ce6) within the triple helix structure of β-glucan (Glu) in aqueous solution, with the aim of overcoming the main limitations of the use of Ce6 as a photosensitizer, i.e., its insolubility in water and low selectivity to target cells or tissues. The Glu/Ce6 nanocomplexes were efficiently internalized into foam cells (due to the specific targeting of the dectin-1 receptor) as compared to normal macrophages and they delivered Ce6 into the cytoplasm. The intracellular uptake increased up to 2.6-fold, compared to free Ce6, leading to enhanced PDT effects. In fact, upon NIR laser irradiation, they generated singlet oxygen, inducing significant in vitro photodynamic effects with membrane damage and cell apoptosis.

A comprehensive summary of the contributions of BODIPY to PDT was provided by Malacarne et al. Despite the few molecules approved for PDT in the clinical setting belonging almost exclusively to the porphyrin family [3], the scientific interest towards other photosensitizers, including BODIPY, is exhibiting tremendous growth. This review focused on a series of structural changes made to BODIPY to favor intersystem crossing and further increase $^1O_2$ production with the final aim of improving cell targeting and/or photoactivity efficiency, pointing out that the absorption and emission features of BODIPY can be modulated by adding suitable substituents to the main chemical structure (e.g., the introduction of heavy atoms, such as bromine and iodine, in the beta positions of the pyrrole ring).

PDT has been shown to be a useful approach not only for cancer treatment but also for treating microbial infection induced by Gram-positive and Gram-negative bacteria, including antibiotic-resistant strains. When the cells being destroyed are microorganisms, this form of therapy is called antimicrobial photodynamic therapy (aPDT) and it shares with "classical" PDT the cooperation of three elements, such as a photosensitizer, light, and oxygen. Interestingly, these components are harmless by themselves but when combined they can lead to the selective destruction of pathogenic cells [4].

The photoinactivation of *Pseudomonas aeruginosa* biofilm by dicationic diaryl-porphyrin was proposed by Orlandi et al. The high binding yield of cationic diaryl-porphyrins (80–100%) could be ascribable to the electrostatic force displayed between negatively charged lipopolysaccharides on the outer layer of the outer membrane and positively charged photosensitizers. Since a mild effect on the formed biofilm was obtained, the results could pave the way for the development of combined strategies to eradicate *P. aeruginosa* biofilms based on dicationic diaryl-porphyrin-mediated aPDT in addition to other antimicrobial approaches.

Higuchi et al. investigated the bactericidal effect of indocyanine green (ICG)-loaded nanospheres coated with chitosan and a diode laser for photodynamic inactivation of a biofilm of *Enterococcus faecalis*, a pathogen of refractory apical periodontitis, an inflammatory lesion causing bone resorption around the apex of teeth. In vitro results showed that antimicrobial photodynamic therapy/photodynamic antimicrobial chemotherapy (aPDT/PACT) can suppress *E. faecalis* in infected root canals with high efficiency (the viable cell counts were reduced by more than 98%) maintaining a temperature rise in the root within a safe range. Morphological observations with SEM confirmed a clear reduction in the biofilm on the dentin block, but the removal was not complete. The authors proposed for future clinical applications, the bactericidal effect should be investigated under changing conditions, such as a smaller interval for the irradiation time and multiple injections of photosensitizer to improve the bactericidal activity.

The influence of incubation time on ortho-toluidine blue (TBO)-mediated aPDT was investigated in vitro against selected *Candida* strains (*C. albicans*, *C. glabrata*, *C. krusei*, *C. parapsilosis*) by Wiench et al. An appropriate incubation time in the aPDT protocol seems to have a great impact on its efficacy, particularly in relation to Candida, due to the size of these cells and the presence of their cell wall. The study pointed out that the most efficient period needed for the uptake of TBO by almost all *Candida* strains was 7–10 min, as confirmed by direct observation by optical microscopy and by evaluation of the efficacy of TBO-mediated aPDT on planktonic cells of these strains.

Amino-functionalized nitrogen-doped graphene quantum dots (amino-N-GQDs) were proposed for aPDT by Kuo et al. as photosensitizers able to generate more ROS than conventional GQDs under 60 s of low-energy (fixed output power: $0.07\ W\cdot cm^{-2}$) excitation exerted by a 670 nm continuous-wave laser. The generated ROS were used to eradicate a multidrug-resistant strain of methicillin-resistant *Staphylococcus aureus* (MRSA), at low energy levels within an extremely short photoexcitation period. Compared with conventional GQDs, the amino-N-GQDs displayed superior optical properties, including stronger absorption and luminescence, a higher quantum yield (0.34), and high stability, contributing to their suitability as contrast probes for biomedical imaging, in addition to their bacteria tracking and localization abilities.

The combination of the use of natural substances with antimicrobial properties and light irradiation at proper light waves, called photodynamic inactivation (PDI), is based on the ability of some natural substances to act as photosensitizers producing bioactive effects under irradiation [5]. Kim et al. demonstrated that fagopyrin F-rich fraction (FFF) separated from *Tartary buckwheat* flower extract when exposed to blue light (BL, 450 nm) produced ROS able to elicit antibacterial photodynamic inactivation (PDI) against *Streptococcus mutans* and its biofilm, which was visually confirmed by confocal laser scanning microscopy (CLSM) and field emission scanning electron microscope (FE-SEM). Interestingly, the PDI effect of FFF against *S. mutans* was similar to curcumin and hypericin and was stronger than riboflavin, although the PDI treatment of FFF was conducted at lower energy fluences of BL and lower concentrations than other photosensitizers.

Some practical applications of PDI are related to food preservation from bacterial contaminants [6]. The development of innovative systems based on natural products and physical methods, such as PDI producing bioactive effects under irradiation, is an emerging and promising research area well reviewed by Munir et al. They explored the antibacterial properties of photoactivated curcumin as a green tool for the preservation of food from bacterial contaminants, since curcumin is a natural antibacterial and effective photosensitizer able to induce photodynamic activation in the visible light range, specifically for blue light.

The great therapeutic potential of photobiomodulation (PBM) therapy in different branches of medicine has been reviewed by Salehpour et al. PBM therapy, namely the application of visible and near-infrared (NIR) light to stimulate cellular processes by changing the biochemical activities of mitochondrial components, is currently applied as a cutting-edge technology in several areas of medicine (i.e., wound healing, dentistry, muscle and tendon repair, dermatology, and neurology). Specifically, this review demonstrated that PBM therapy can serve as a non-invasive neuroprotective strategy for maintaining and optimizing effective brain waste clearance, exerting a neurotherapeutic benefit on glymphatic drainage. The glymphatic drainage system is a waste clearance pathway in the central nervous system devoted to removing toxins and waste metabolites from the brain (i.e., soluble proteins such as amyloid-beta) and its impairment can increase the incidence of neurovascular, neuroinflammatory, and neurodegenerative diseases.

Kim et al. demonstrated that PBM is a promising treatment for neurological and neuropsychiatric disorders, such as autism spectrum disorder (ASD). They used mice exposed to valproic acid (VPA) as a model of ASD to investigate whether PBM treatment during fetal development could attenuate the symptoms of ASD. Their results suggest that the pathological behavioral and histological changes induced by VPA were attenuated by PBM treatment with an 830 nm laser.

One of the molecular mechanisms for the PBM treatment implicates the mitochondrial enzyme, cytochrome C oxidase [7]. With the aim of improving PBM devices and delivery approaches, Amaroli et al. designed a novel handpiece with a flat-top beam profile of irradiation and they compared the beam profile versus a standard handpiece and a fiber probe on mitochondrial activity. They utilized isolated mitochondria and performed treatments at various spots within the beam, namely, the center and left and right edge. The mitochondrial activity was examined by assessing ATP synthesis with the luciferin/luciferase chemiluminescent method as a primary endpoint, while mitochondrial damage was assessed as the secondary endpoint. Their results demonstrated that the novel flat-top beam handpiece enhanced the uniformity of the PBM treatments and can improve the rigor and reproducibility of PBM clinical outcomes.

Since several experimental parameters and factors influence the lifetime of specific fluorophores, different values of fluorescent lifetimes are sometimes found in the literature for the same fluorophore depending on which detection and excitation scheme is used. To clarify this controversy, Kellerer et al. reported a comprehensive and rigorous investigation of parameters influencing fluorescence lifetime imaging microscopy in the frequency domain (FD) and time domain (TD), illustrated by phasor plot analysis. These two most common techniques were implemented in one single microscopy setup and applied to a variety of fluorophores under different conditions of pH, temperature, concentration, and solvent polarity. All the studied parameters fall within two categories (setup-dependent and sample-dependent) and both FD and TD techniques produce reliable and consistent data revealing which of the tested parameters has the strongest influence on the fluorescence lifetime. In addition, their results suggested which technique is most suitable for which research task and how to perform the experiment properly to obtain consistent fluorescence lifetimes.

An in vitro 3D tumor model composed of human melanoma cells and the microalgae *Chlamydomonas reinhardtii*, both seeded into a collagen scaffold, was presented by Holmes et al. as a representative and reproducible model for studying photosynthetic tumor oxygenation. Their results, surprisingly, demonstrated that, although the investigated conditions significantly differ from the optimal microalgae culture settings in key aspects such as temperature, medium composition, and the presence of tumor cells, these conditions do not seem to inhibit the intrinsic ability of the microalgae to provide significant amounts of oxygen in the presence of light. This reproducible and easy-to-use model can be used as a platform for studying the role of other key cell types in the tumor microenvironment and also for studying the antitumoral effects of ROS-dependent photodynamic, chemo-, immuno-, and radiotherapy.

De Plano et al. evaluated the role of capsid rearrangement in engineered phages of M13 in protecting viral DNA and peptide bonds from damage induced by UV-C radiation. The study was performed on two M13 engineered phage clones (P9b and 12III1 phages expressing 9 or 12 additional amino acids in N-terminal end of pVIII capsid protein) and their resistance to environmental stresses, such as UV-C radiation and hydrogen peroxide, was compared to M13 wild-type vector (pC89). Only P9b displayed an increase in resistance against H_2O_2, whereas both clones acquired UV-C resistance due to a reorganization of interactions in the capsid for an increase in H-bond and steric interactions, which in turn depended on the presence of the foreign peptide sequence. These results could help us to understand the molecular mechanisms involved in the stability of new virus variants, useful for developing effective protocols for virus inactivation.

Thacker et al. evaluated the efficacy and cytotoxicity of a dental bleaching gel composed of calcium peroxide (CaO_2) as an active ingredient, visible-light-activating nitrogen-doped titanium dioxide as a photocatalyst, and methylcellulose as the thickener. The study was performed in vitro on bovine teeth stained with coffee and black tea stain solution that were subjected to one minute of visible light irradiation during each bleaching time. The gel demonstrated at neutral pH an efficient bleaching effect with a gradual increase in brightness (ΔL) and color difference (ΔE), without cytotoxicity upon exposure to 3T3 cells. Moreover, the proposed gel allowed for the avoidance of potential side effects usually caused by a highly concentrated hydrogen peroxide-based dental bleaching procedure.

The nineteen articles published in our Special Issue demonstrate the growing interest in the use of light-responsive materials for applications in the field of photobiology and in related biomedical areas in recent years. Materials for photobiology not only were realized by the utilization of novel molecular components for light-responsive processes in living organisms, but also comprised living matter itself in which light-induced processes occur. We hope to supply our readers with some representative and useful snapshots of actual research, perhaps providing the inspiration to push developments in this field a step further. We acknowledge all the contributors of the Special Issue and the Editorial Board of *Int. J. Mol. Sci.* for their support.

Conflicts of Interest: The authors declare no conflicts of interest.

List of Contributors

1. Lu, Z.; Xu, G.; Yang, X.; Liu, S.; Sun, Y.; Chen, L.; Liu, Q.; Liu, J. Dual-Activated Nano-Prodrug for Chemo-Photodynamic Combination Therapy of Breast Cancer. *Int. J. Mol. Sci.* **2022**, *23*, 15656.
2. Díaz, M.; Lobo, F.; Hernández, D.; Amesty, Á.; Valdés-Baizabal, C.; Canerina-Amaro, A.; Mesa-Herrera, F.; Soler, K.; Boto, A.; Marín, R.; et al. FLTX2: A Novel Tamoxifen Derivative Endowed with Antiestrogenic, Fluorescent, and Photosensitizer Properties. *Int. J. Mol. Sci.* **2021**, *22*, 5339.
3. Gangemi, C.M.A.; Salerno, T.M.G.; Barattucci, A.; Cucinotta, F.; Bonaccorsi, P.; Calabrese, G.; Poma, P.; Rizzo, M.G.; Campagna, S.; Puntoriero, F. A Curcumin-BODIPY Dyad and Its Silica Hybrid as NIR Bioimaging Probes. *Int. J. Mol. Sci.* **2022**, *23*, 9542.
4. Hübinger, L.; Runge, R.; Rosenberg, T.; Freudenberg, R.; Kotzerke, J.; Brogsitter, C. Psoralen as a Photosensitizers for Photodynamic Therapy by Means of In Vitro Cherenkov Light. *Int. J. Mol. Sci.* **2022**, 23, 15233.
5. Ahn, J.W.; Kim, J.H.; Park, K. In Vitro Photodynamic Effects of the Inclusion Nanocomplexes of Glucan and Chlorin e6 on Atherogenic Foam Cells. *Int. J. Mol. Sci.* **2021**, *22*, 177.
6. Malacarne, M.C.; Gariboldi, M.B.; Caruso, E. BODIPYs in PDT: A Journey through the Most Interesting Molecules Produced in the Last 10 Years. *Int. J. Mol. Sci.* **2022**, *23*, 10198.
7. Orlandi, V.T.; Martegani, E.; Bolognese, F.; Trivellin, N.; Garzotto, F.; Caruso, E. Photoinactivation of Pseudomonas aeruginosa Biofilm by Dicationic Diaryl-Porphyrin. *Int. J. Mol. Sci.* **2021**, *22*, 6808.
8. Higuchi, N.; Hayashi, J.-i.; Fujita, M.; Iwamura, Y.; Sasaki, Y.; Goto, R.; Ohno, T.; Nishida, E.; Yamamoto, G.; Kikuchi, T.; et al. Photodynamic Inactivation of an Endodontic Bacteria Using Diode Laser and Indocyanine Green-Loaded Nanosphere. *Int. J. Mol. Sci.* **2021**, *22*, 8384.

9. Wiench, R.; Nowicka, J.; Pajączkowska, M.; Kuropka, P.; Skaba, D.; Kruczek-Kazibudzka, A.; Kuśka-Kiełbratowska, A.; Grzech-Leśniak, K. Influence of Incubation Time on Ortho-Toluidine Blue Mediated Antimicrobial Photodynamic Therapy Directed against Selected Candida Strains-An In Vitro Study. *Int. J. Mol. Sci.* **2021**, *22*, 10971.
10. Kuo, W.-S.; Wu, P.-C.; Hung, C.-Y.; Chang, C.-Y.; Wang, J.-Y.; Chen, P.-C.; Hsieh, M.-H.; Lin, S.-H.; Chang, C.-C.; Lin, Y.-S. Nitrogen Functionalities of Amino-Functionalized Nitrogen-Doped Graphene Quantum Dots for Highly Efficient Enhancement of Antimicrobial Therapy to Eliminate Methicillin-Resistant Staphylococcus aureus and Utilization as a Contrast Agent. *Int. J. Mol. Sci.* **2021**, *22*, 9695.
11. Kim, J.; Kim, S.; Lee, K.; Kim, R.H.; Hwang, K.T. Antibacterial Photodynamic Inactivation of Fagopyrin F from Tartary Buckwheat (*Fagopyrum tataricum*) Flower against Streptococcus mutans and Its Biofilm. *Int. J. Mol. Sci.* **2021**, *22*, 6205.
12. Munir, Z.; Banche, G.; Cavallo, L.; Mandras, N.; Roana, J.; Pertusio, R.; Ficiarà, E.; Cavalli, R.; Guiot, C. Exploitation of the Antibacterial Properties of Photoactivated Curcumin as 'Green' Tool for Food Preservation. *Int. J. Mol. Sci.* **2022**, *23*, 2600.
13. Salehpour, F.; Khademi, M.; Bragin, D.E.; DiDuro, J.O. Photobiomodulation Therapy and the Glymphatic System: Promising Applications for Augmenting the Brain Lymphatic Drainage System. *Int. J. Mol. Sci.* **2022**, *23*, 2975.
14. Kim, U.-J.; Hong, N.; Ahn, J.-C. Photobiomodulation Attenuated Cognitive Dysfunction and Neuroinflammation in a Prenatal Valproic Acid-Induced Autism Spectrum Disorder Mouse Model. *Int. J. Mol. Sci.* **2022**, *23*, 16099.
15. Amaroli, A.; Arany, P.; Pasquale, C.; Benedicenti, S.; Bosco, A.; Ravera, S. Improving Consistency of Photobiomodulation Therapy: A Novel Flat-Top Beam Hand-Piece versus Standard Gaussian Probes on Mitochondrial Activity. *Int. J. Mol. Sci.* **2021**, *22*, 7788.
16. Kellerer, T.; Janusch, J.; Freymüller, C.; Rühm, A.; Sroka, R.; Hellerer, T. Comprehensive Investigation of Parameters Influencing Fluorescence Lifetime Imaging Microscopy in Frequency- and Time-Domain Illustrated by Phasor Plot Analysis. *Int. J. Mol. Sci.* **2022**, *23*, 15885.
17. Holmes, C.; Varas, J.; San Martín, S.; Egaña, J.T. Towards an In Vitro 3D Model for Photosynthetic Cancer Treatment: A Study of Microalgae and Tumor Cell Interactions. *Int. J. Mol. Sci.* **2022**, *23*, 13550.
18. De Plano, L.M.; Franco, D.; Rizzo, M.G.; Zammuto, V.; Gugliandolo, C.; Silipigni, L.; Torrisi, L.; Guglielmino, S.P.P. Role of Phage Capsid in the Resistance to UV-C Radiations. *Int. J. Mol. Sci.* **2021**, *22*, 3408.
19. Thacker, M.; Chen, Y.-N.; Lin, C.-P.; Lin, F.-H. Nitrogen-Doped Titanium Dioxide Mixed with Calcium Peroxide and Methylcellulose for Dental Bleaching under Visible Light Activation. *Int. J. Mol. Sci.* **2021**, *22*, 3759.

References

1. Leister, D. Photobiology: Introduction, overview and challenges. *Front. Photobiol.* **2023**, *1*, 1253330. [CrossRef]
2. Di Martino, M.; Sessa, L.; Diana, R.; Piotto, S.; Concilio, S. Recent Progress in Photoresponsive Biomaterials. *Molecules* **2023**, *28*, 3712. [CrossRef] [PubMed]
3. Yang, F.; Xu, M.; Chen, X.; Luo, Y. Spotlight on porphyrins: Classifications, mechanisms and medical applications. *Biomed. Pharmacother.* **2023**, *164*, 114933. [CrossRef] [PubMed]
4. Piksa, M.; Lian, C.; Samuel, I.C.; Pawlik, K.J.; Samuel, D.W.; Matczyszyn, K. The role of the light source in antimicrobial photodynamic therapy. *Chem. Soc. Rev.* **2023**, *52*, 1697–1722. [CrossRef] [PubMed]
5. Pucelik, B.; Dąbrowski, J.M. Photodynamic inactivation (PDI) as a promising alternative to current pharmaceuticals for the treatment of resistant microorganisms. *Adv. Inorg. Chem.* **2022**, *79*, 65–103.

6. Cossu, M.; Ledda, L.; Cossu, A. Emerging trends in the photodynamic inactivation (PDI) applied to the food decontamination. *Food Res. Int.* **2021**, *144*, 110358. [CrossRef]
7. Liebert, A.; Capon, W.; Pang, V.; Vila, D.; Bicknell, B.; McLachlan, C.; Kiat, H. Photophysical Mechanisms of Photobiomodulation Therapy as Precision Medicine. *Biomedicines* **2023**, *11*, 237. [CrossRef] [PubMed]

International Journal of
Molecular Sciences

MDPI

Article

Photobiomodulation Attenuated Cognitive Dysfunction and Neuroinflammation in a Prenatal Valproic Acid-Induced Autism Spectrum Disorder Mouse Model

Ui-Jin Kim [1], Namgue Hong [2,*] and Jin-Chul Ahn [3,*]

1 Department of Medical Laser, Graduate School, College of Medicine, Dankook University, Cheonan 31116, Republic of Korea
2 Medical Laser Research Center, College of Medicine, Dankook University, Cheonan 31116, Republic of Korea
3 Department of Biomedical Science, College of Medicine, Dankook University, Cheonan 31116, Republic of Korea
* Correspondence: hnk210@dankook.ac.kr (N.H.); jcahn@dankook.ac.kr (J.-C.A.)

Abstract: Autism spectrum disorder (ASD) is a neurodevelopmental condition characterized by social communication and interaction disorders, as well as repetitive and restrictive behaviors. To date, no effective treatment strategies have been identified. However, photobiomodulation (PBM) is emerging as a promising treatment for neurological and neuropsychiatric disorders. We used mice exposed to valproic acid (VPA) as a model of ASD and found that pathological behavioral and histological changes that may have been induced by VPA were attenuated by PBM treatment. Pregnant mice that had been exposed to VPA were treated with PBM three times. Thereafter, we evaluated the offspring for developmental disorders, motor function, hyperactivity, repetitive behaviors, and cognitive impairment. PBM attenuated many of the pathological behaviors observed in the VPA-induced ASD mouse model. In addition, pathophysiological analyses confirmed that the increase in activated microglia and astrocytes observed in the VPA-induced ASD mouse model was attenuated by PBM treatment. This suggests that PBM can counteract the behavioral changes caused by neuroinflammation in ASD. Therefore, our data show that PBM has therapeutic potential and may reduce the prevalence of neurodevelopmental disorders such as ASD.

Keywords: autism spectrum disorder; neuroinflammation; photobiomodulation; valproic acid; cognitive function

Citation: Kim, U.-J.; Hong, N.; Ahn, J.-C. Photobiomodulation Attenuated Cognitive Dysfunction and Neuroinflammation in a Prenatal Valproic Acid-Induced Autism Spectrum Disorder Mouse Model. *Int. J. Mol. Sci.* **2022**, *23*, 16099. https://doi.org/10.3390/ijms232416099

Academic Editor: Zsuzsanna Helyes

Received: 14 November 2022
Accepted: 15 December 2022
Published: 17 December 2022

1. Introduction

Autism spectrum disorder (ASD) is a neurodevelopmental disorder characterized by impaired social communication, as well as restrictive and repetitive behaviors [1]. ASD is also closely linked with neurodegenerative diseases such as schizophrenia, Alzheimer's disease, Parkinson's disease, and dementia [2,3]. A report in 2020 from the Centers for Disease Control and Prevention estimated that 1 in 44 children had been diagnosed with ASD. Approximately four times as many males as females are diagnosed with ASD [4]. The cause of autism is not clear, but the most widely accepted explanation is that it is a complex neural developmental disorder characterized by abnormalities in the brain network [5]. According to recent studies, one of the most common risk factors for ASD is ongoing neuroinflammation in various brain regions [6,7]. Recent studies have implicated non-genetic factors in the development of ASD, including exposure to antiepileptic drugs, viral infections, autoimmune diseases, pathogenic substances, and valproic acid (VPA) during pregnancy [8].

VPA is an antiepileptic drug also used to treat bipolar disorder [9]. However, VPA is associated with side effects when administered during pregnancy, including an increased risk of congenital malformation and delayed cognitive development of offspring. Some studies have also suggested that treatment with VPA during pregnancy is associated with an increased risk of ASD, attention deficit disorders, hyperactivity, and fetal valproate syndrome [10].

Photobiomodulation (PBM) uses lasers or light-emitting diodes to apply red or near-infrared light to the brain, thereby improving the metabolic capacity of neurons and stimulating anti-inflammatory responses, antioxidants, neurogenesis, and synaptogenesis [11–14]. Unlike drug treatments, PBM is noninvasive and has few side effects [15]. PBM improved cognitive rigidity, behavior, attention span, and sleep quality in children with ASD [5]. Recent studies have shown that 635 nm wavelength reduces ASD-related hypersensitivity, among other symptoms [16]. However, the therapeutic mechanism by which PBM influences ASD remains unclear.

We investigated the potential neuroprotective and developmental effects of 830 nm wavelength laser in an ASD model. We identified behavioral changes in the ASD model, which differed according to PBM treatment. We investigated the importance of the developmental stage and performed behavioral tests to assess social interactions, motor function, repetitive behavior, and cognitive function. The neuroinflammatory response occurs due to activated microglia and astrocytes in neurological diseases [17]. According to previous studies, activated microglia and astrocytes are increased in ASD [7,18,19]. PBM is attracting attention as an effective treatment method for neuroinflammation [20]. Therefore, it suggests that PBM can effectively control the neuroinflammation that occurs in ASD. To understand the underlying causes of the changes in behavior, we studied brain samples and found that histological changes in the brain, the neuroinflammatory response, correlated with the pathological findings of behavioral tests.

2. Results

2.1. Effect of Laser Treatment on VPA-Induced Developmental Abnormalities

Prenatal VPA exposure caused developmental abnormalities in the pups. We found that 830 nm laser treatment attenuated the developmental abnormalities observed in mice exposed to VPA. The effect of 830 nm laser treatment on developmental abnormalities was assessed using body weight measurements, eye opening, and the righting reflex test. Figure 1 (panel 1a,b) shows that VPA decreases body size and weight. Figure 1d shows late eye-opening scores for the VPA group compared to the vehicle group, and improved eye-opening scores for the VPA + 830 nm laser group compared to the VPA group. The total number of live births did not differ significantly between mice exposed to VPA and the vehicle (Figure 1c). The VPA group exhibited late eye-opening, and the differences in eye-opening scores between the vehicle and VPA groups were significant at P14–P17. Eye-opening scores in the VPA + 830 nm laser group improved, with significant differences seen at P14, P16, and P17 (Figure 1d). These data indicate that VPA induced a significant developmental delay in mice, and that the 830 nm laser treatment attenuated this developmental delay. The difference in righting reflex times between the vehicle and VPA groups also highlights the developmental delay in the VPA group, with significant differences seen at P10 and 11. The righting reflex time for the VPA + 830 nm laser group improved, and was significantly different at P11 (Figure 1e). Therefore, VPA induced a significant developmental delay in mice, which the 830 nm laser treatment attenuated.

2.2. Laser Treatment Improved Motor Function in Mice Exposed to VPA

Prenatal VPA exposure impaired motor function. We found that 830 nm laser treatment improved motor function in mice exposed to VPA. The effect of 830 nm laser treatment on motor function was assessed using a negative geotaxis test (Figure 2a) and a hanging wire test (Figure 2b). The negative geotaxis test revealed a significant delay for the VPA group compared with the vehicle group. However, there was a significant improvement in the VPA + 830 nm laser group compared with the VPA group. These data show that VPA significantly impairs motor function in pups, and that the 830 nm laser treatment improved motor function. The hanging wire test revealed a nonsignificant delay for the VPA group compared with the vehicle group. There was also an improvement in the VPA + 830 nm laser group. Therefore, VPA impaired motor function in the pups, but the 830 nm laser treatment improved motor function.

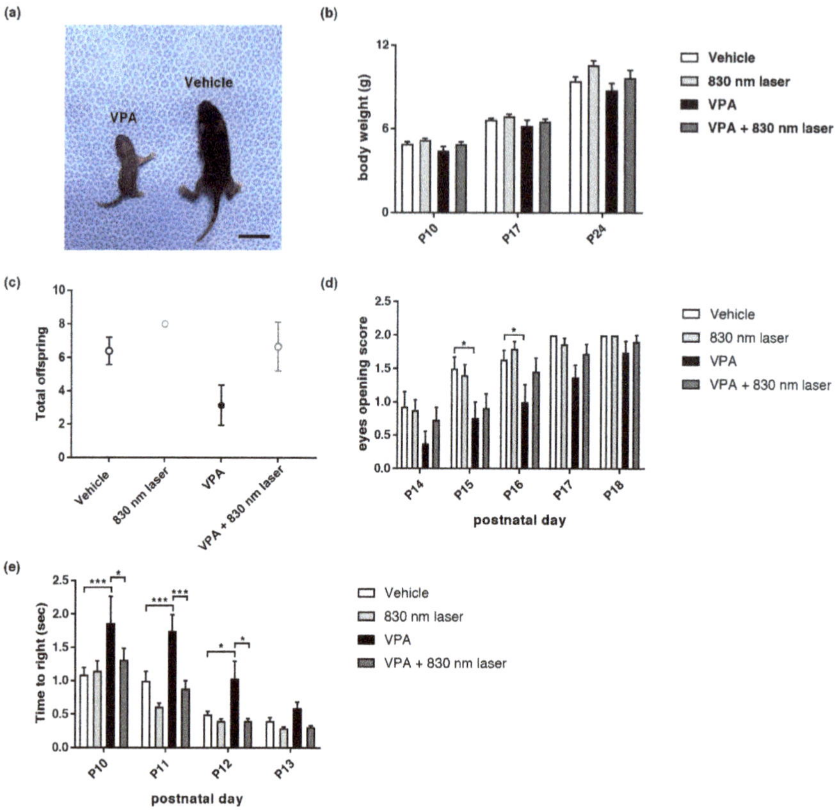

Figure 1. Mice exposed to valproic acid (VPA) in utero exhibited delayed development. (**a**) A pup exposed to VPA was compared to one exposed to the vehicle (postnatal day 9). Scale bar: 1 cm. (**b**) Body weight of mice exposed to VPA (in utero). (**c**) The number of live pups from each pregnancy on postnatal day 10. (**d**) Developmental delays were assessed based on the eye-opening test and (**e**) righting reflex. Eye opening was scored from postnatal days 14 to 18 (score: 0 = closed eyes, 1 = one eye open, 2 = both eyes open). The righting reflex was measured every day from P10 to 13. Data are means ± standard error of the mean (SEM) (vehicle: n = 14,830 nm laser: n = 15, VPA: n = 8, VPA + 830 nm laser: n = 11). * $p < 0.05$ and *** $p < 0.001$ compared to the vehicle group. * $p < 0.05$ and *** $p < 0.001$ compared to the VPA group.

2.3. *Laser Treatment Attenuated Social Cognitive Dysfunction in Mice Exposed to VPA*

Impairment of social interaction is one of the core symptoms of ASD, and the three-chamber test is frequently used to investigate social interactions. Mice exposed to VPA exhibited reduced social interaction, i.e., spent little time in the stranger chamber. The 830 nm laser treatment improved social interaction in mice exposed to VPA. The effect of 830 nm laser treatment on social interaction was assessed using the three-chamber test (Figure 3a–c). Mice exposed to VPA exhibited defects in social interaction, spending less time in the stranger than empty chambers compared to vehicle group mice. However, mice in the VPA + 830 nm laser group spent more time in the stranger chamber than mice in the VPA group, indicating improved social interaction.

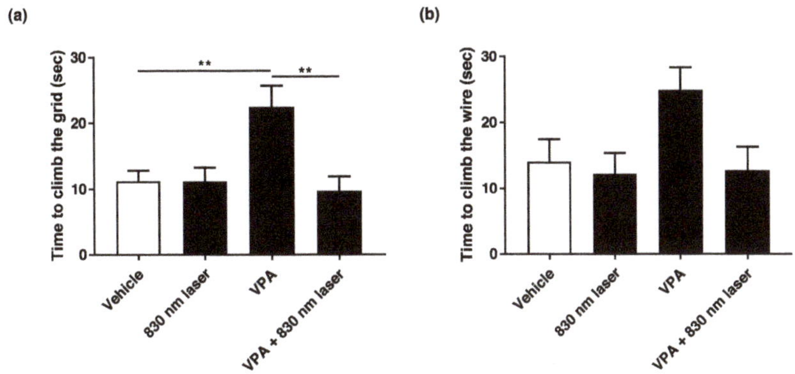

Figure 2. Effect of laser treatment on the motor function of mice exposed to VPA in utero. (**a**) The effect of 830 nm laser treatment on mice exposed to VPA (in utero) was assessed using the negative geotaxis test and (**b**) hanging wire test. Mice were placed on the slope with their heads facing downward. The righting reflex was measured as the time taken for mice to reach the top of the slope. In the hanging wire test, the time elapsed between the holding of the wire with both forelimbs and climbing with the hind limbs was measured. Data are means ± SEM (vehicle: n = 14,830 nm laser: n = 15, VPA: n = 8, VPA + 830 nm laser: n = 11). ** $p < 0.01$ compared to the group exposed to VPA.

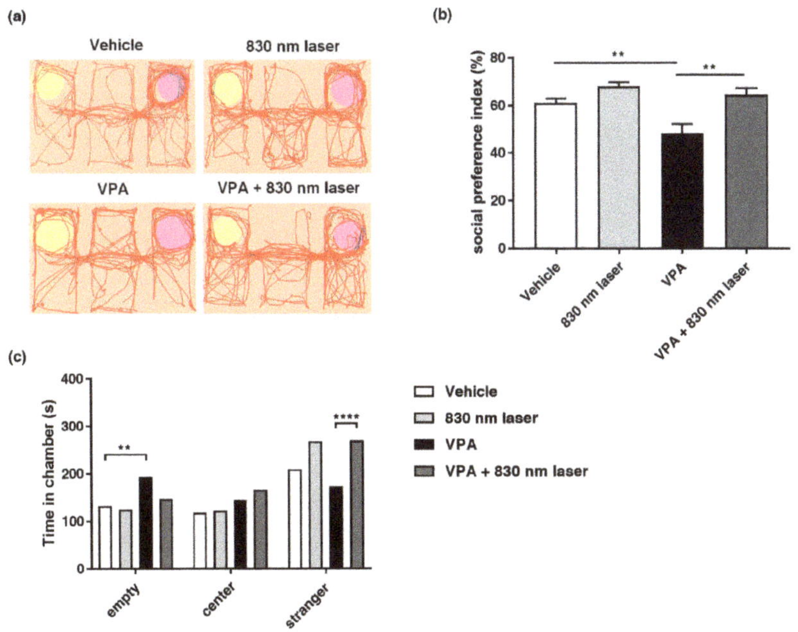

Figure 3. Effect of laser treatment on the social interactions of mice exposed to VPA in utero. (**a**) The effect of 830 nm laser treatment on mice exposed to VPA (in utero) was assessed using the three-chamber test. (**b**) Social preference index (%) = time spent in the stranger chamber divided by the total time spent in all chambers. (**c**) The bar graph indicates that mice spent time in each room, empty, center, and room with a stranger. Data are means ± SEM (vehicle: n = 9, 830 nm laser: n = 9, VPA: n = 6, VPA + 830 nm laser: n = 7). ** $p < 0.01$ compared to the vehicle group. **** $p < 0.0001$ compared to the group exposed to VPA.

2.4. Laser Treatment Attenuated Impairment in Repetitive Behavior of Mice Exposed to VPA

As with social cognitive dysfunction, impairment in repetitive behavior is one of the core symptoms of ASD. Therefore, we assessed the effect of 830 nm laser treatment on repetitive behavior using the Y-maze (Figure 4a,b) and cylinder test (Figure 4c). The number of forelimb contacts as each mouse reared against the wall of the cylinder was quantified, and there was no significant difference between the VPA and vehicle groups. Although there was an improvement in the VPA + 830 nm laser group, the difference between the VPA + 830 nm laser and VPA groups was not significant. Mice exposed to VPA exhibited impairment in repetitive behavior, and showed a decrease in spontaneous alternation compared to vehicle group mice. Mice in the VPA + 830 nm laser group showed an increase in spontaneous alternation compared to the VPA group mice. These data suggest that the laser treatment attenuated impairment in repetitive behavior of mice exposed to VPA.

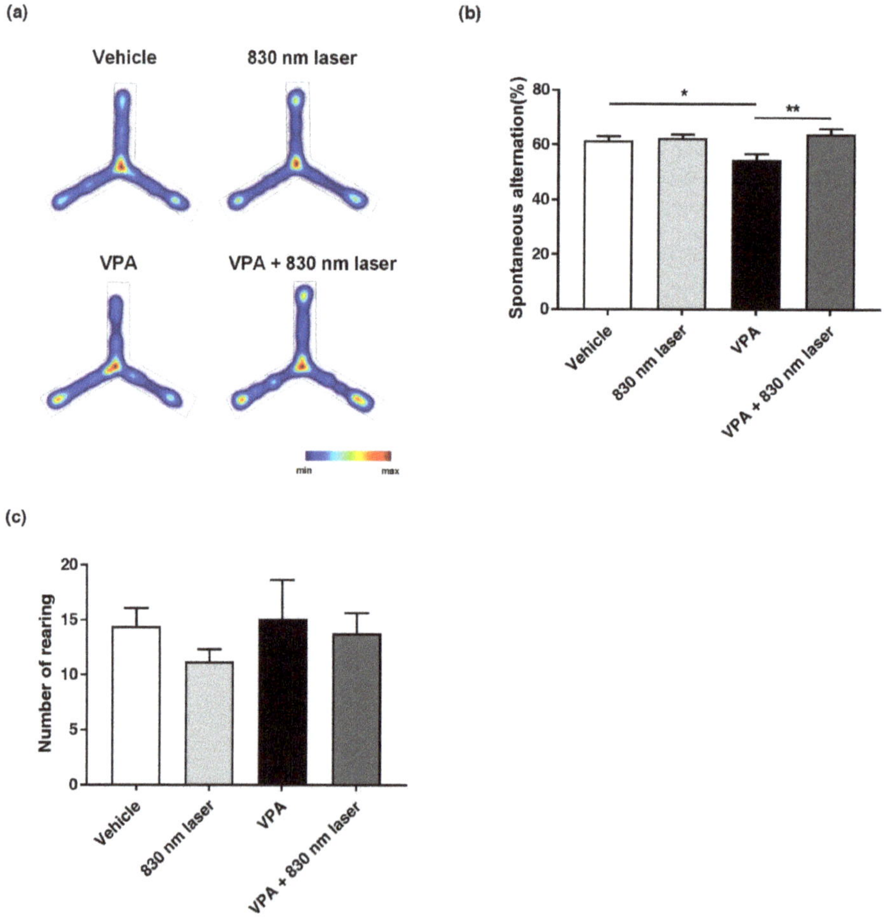

Figure 4. Effect of laser treatment on the repetitive behavior of mice exposed to VPA in utero. The effect of 830 nm laser treatment on mice exposed to VPA (in utero) was assessed using (**a**,**b**) the Y-maze and (**c**) the cylinder rearing test. Spontaneous alternation (%) = total alternations/(total arm entries—2) × 100. Data are means ± SEM (vehicle: n = 14,830 nm laser: n = 15, VPA: n = 8, VPA + 830 nm laser: n = 11). * $p < 0.05$ compared to the vehicle group. * $p < 0.05$ and ** $p < 0.01$ compared to the group exposed to VPA.

2.5. Laser Treatment Attenuated Hyperactivity in Mice Exposed to VPA

Attention deficit disorder and hyperactivity are common manifestations of ASD [21]. Mice exposed to VPA exhibit hyperactivity. The effect of 830 nm laser treatment on hyperactivity was assessed using the open field test (Figure 5a–d). Compared to mice in the vehicle group, mice exposed to VPA exhibited hyperactivity, reflected in the total distance traveled, number of crossings, and time spent in motion. Mice in the VPA + 830 nm laser group exhibited decreased hyperactivity compared to those in the VPA group.

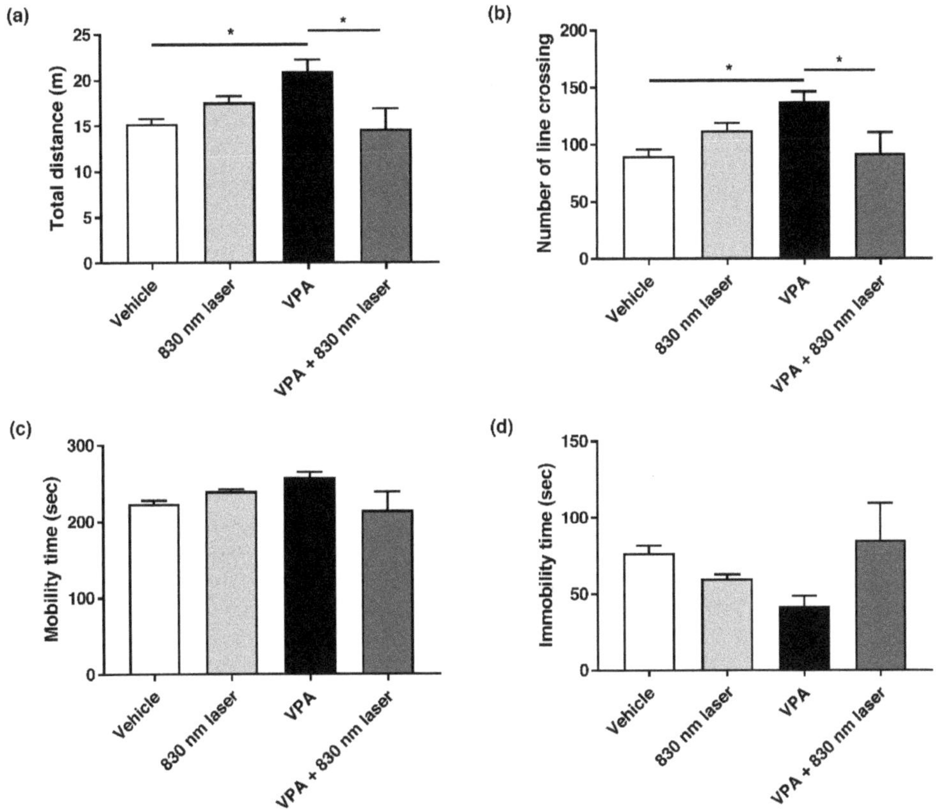

Figure 5. Effect of laser treatment on the hyperactivity of mice exposed to VPA in utero. The effect of 830 nm laser treatment on mice exposed to VPA (in utero) was assessed using (**a–d**) the open field test. Data are means ± SEM (vehicle: n = 14,830 nm laser: n = 15, VPA: n = 8, VPA + 830 nm laser: n = 11). * $p < 0.05$ compared to the vehicle group. * $p < 0.05$ compared to the group exposed to VPA.

2.6. Laser Treatment Improved Cognitive Function in Mice Exposed to VPA

VPA impairs the spatial memory of adults and children. The mice exposed to VPA in this study exhibited decreased cognitive function. The effect of 830 nm laser treatment on cognitive function was assessed using the novel object recognition test (Figure 6a,b) and Morris water maze (Figure 6c,d). Mice exposed to VPA exhibited cognitive function deficits, spending less time with novel objects compared to the vehicle group mice. Mice in the VPA + 830 nm laser group exhibited improved cognitive function, spending more time with novel objects compared to VPA group mice. In the Morris water maze, mice exposed to VPA exhibited defective cognitive function, taking longer to reach the platform than vehicle group mice. Mice in the VPA + 830 nm laser group exhibited improved cognitive function, taking less time to reach the platform than VPA group mice.

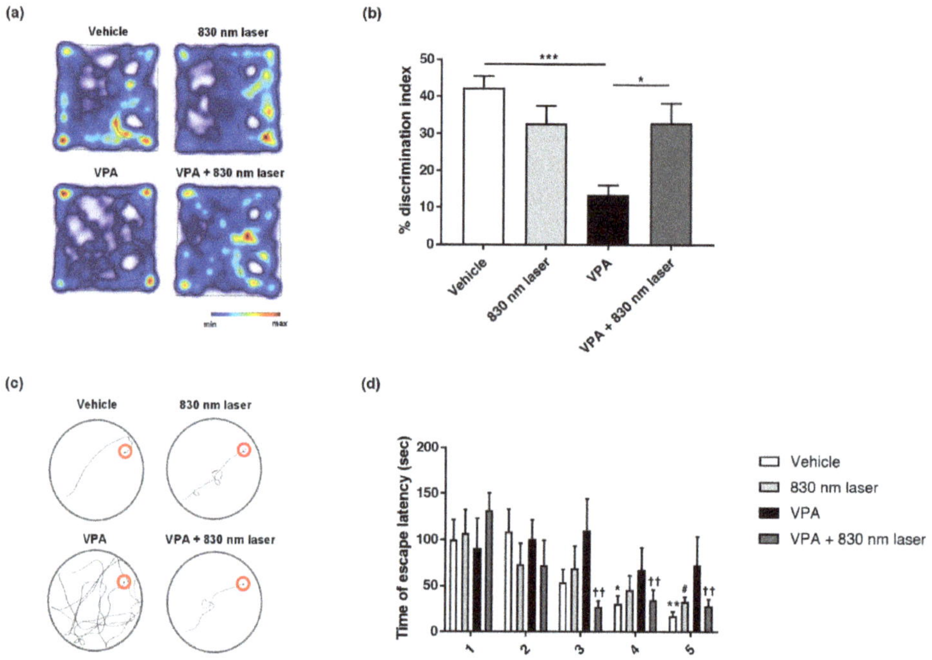

Figure 6. Effect of laser treatment on the cognitive function of mice exposed to VPA in utero. The effect of 830 nm laser treatment on cognitive dysfunction in mice exposed to VPA (in utero) was assessed using (**a**,**b**) the novel object recognition test and (**c**,**d**) Morris water maze. Discrimination index (%) = time spent on the novel object divided by the total time spent exploring both objects. (**b**) The data show significant differences compared to the group exposed to VPA (vehicle: $n = 9$, 830 nm laser: $n = 9$, VPA: $n = 6$, VPA + 830 nm laser: $n = 7$). *** $p < 0.001$ compared to the vehicle group. * $p < 0.05$ compared to the VPA+ 830 nm laser group. (**d**) The data show significant differences compared to the group exposed to VPA (vehicle: $n = 9$, 830 nm laser: $n = 9$, VPA: $n = 6$, VPA + 830 nm laser: $n = 7$). * $p < 0.05$ and ** $p < 0.01$ compared to the vehicle on day 1. # $p < 0.05$ compared to the 830 nm laser group on day 1. †† $p < 0.01$ compared to the VPA + 830 nm laser group on day 1.

2.7. Laser Treatment Decreased GFAP Expression in the PFC and Hippocampus of Mice Exposed to VPA

The hippocampus plays an important role in cognitive function [22]. The prelimbic area of the medial PFC (mPFC) is involved in decision-making, cognitive function, attention, and motor activities [23,24]. Because astrocytes are involved in learning and the regulation of long-term memory [25], we performed immunohistochemical analyses of the hippocampus (Figure 7a) and mPFC (Figure 8a) using an antibody for GFAP. Quantitative analyses revealed significant statistical differences in the number of $GFAP^+$ astrocytes in the hippocampus and PFC, not only for the vehicle group, but also the VPA + 830 nm laser group compared with the VPA group. The number of $GFAP^+$ astrocytes in the CA1, CA3, and hilus regions of the hippocampus was significantly decreased in the VPA + 830 nm laser group compared with the VPA group (Figure 7b–d). The number of $GFAP^+$ astrocytes in the mPFC also significantly decreased in the VPA + 830 nm laser group compared with the VPA group (Figure 8b). Western blotting was used to quantify the expression of GFAP (Figures 7e,f and 8c,d). The full-length blots are presented in Figures S1 and S2.

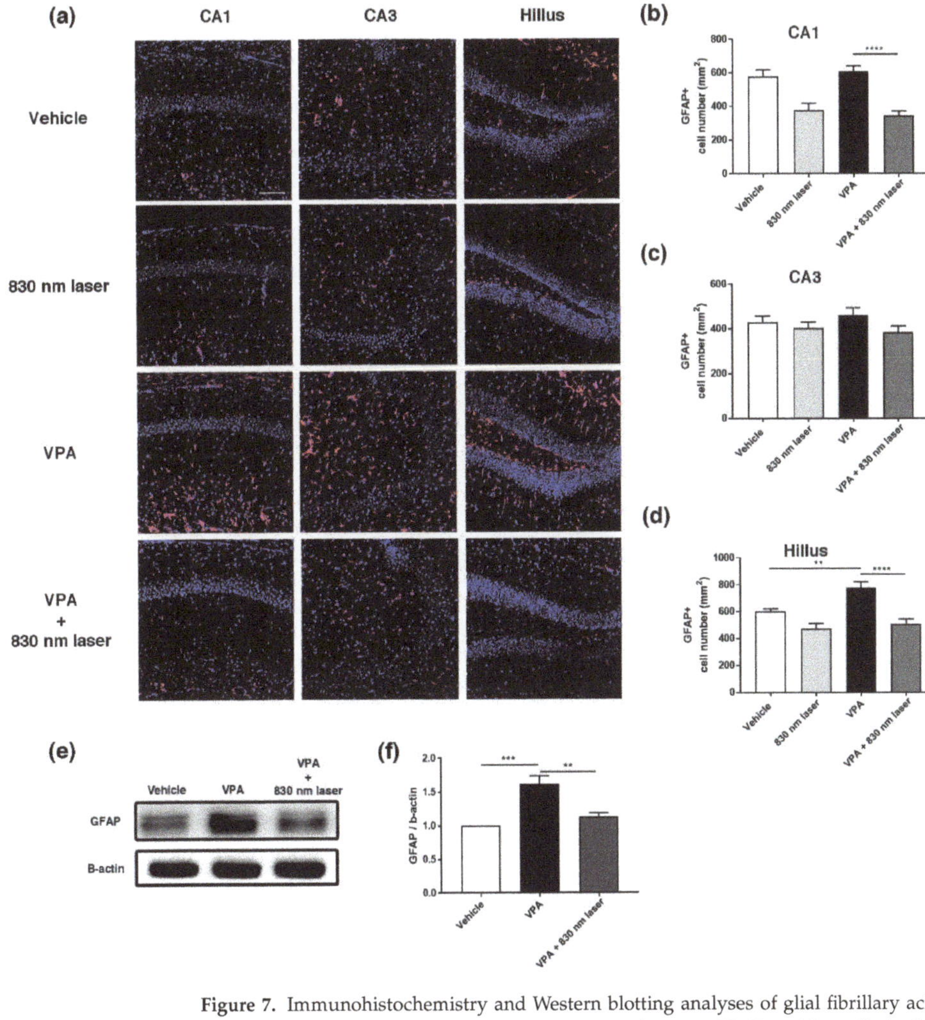

Figure 7. Immunohistochemistry and Western blotting analyses of glial fibrillary acidic protein (GFAP) expression in the hippocampus. (**a**) Image showing GFAP-positive cells (red) in the CA1, CA3, and hilus regions of the hippocampus. (**b–d**) Quantification of GFAP-positive cells in the CA1, CA3, and hilus regions of the hippocampus. (**e**) Image of a Western blot. Full-length blots/gels are presented in Figure S1. (**f**) Average relative intensity values for the Western blots. The bar graph summarizes changes in the GFAP-positive cells. Nuclei were stained using 4′,6-diamidino-2-phenylindole (DAPI). Scale bar: 100 μm. Data are means ± SEM. (vehicle: $n = 3$, 830 nm laser: $n = 3$, VPA: $n = 3$, VPA + 830 nm laser: $n = 3$). ** $p < 0.01$ compared to the vehicle group. *** $p < 0.001$ and **** $p < 0.0001$ compared to the group exposed to VPA.

2.8. Laser Treatment Decreased Iba1 Expression in the Hippocampus of Mice Exposed to VPA

Microglia are macrophages activated when the central nervous system is damaged by infection or disease [26]. Increasing evidence indicates that microglial cell activation and dysfunction can have a profound effect on neurodevelopment, leading to neurodevelopmental disorders including autism [27]. In a previous study, RNA sequencing revealed a close association between genes involved in ASD and glial cell activation, and genes involved in immune and inflammatory responses [28]. Because microglial activation is involved in developmental disorders [29], we performed immunohistochemical analyses of the hippocampus (Figure 9a) and mPFC using an antibody for Iba1. Quantitative analyses revealed significant differences in the number of Iba1$^+$ microglia cells in the hippocampus, not only for the vehicle group, but also for the VPA + 830 nm laser group compared with the VPA group. The number of Iba1$^+$ microglia in the CA1, CA3, and hilus regions of the hippocampus (but not the mPFC) was also significantly decreased in the VPA + 830 nm laser group compared with the VPA group (Figure 9b–d). Western blotting was used to quantify the expression of Iba1 (Figures 9e,f and S1).

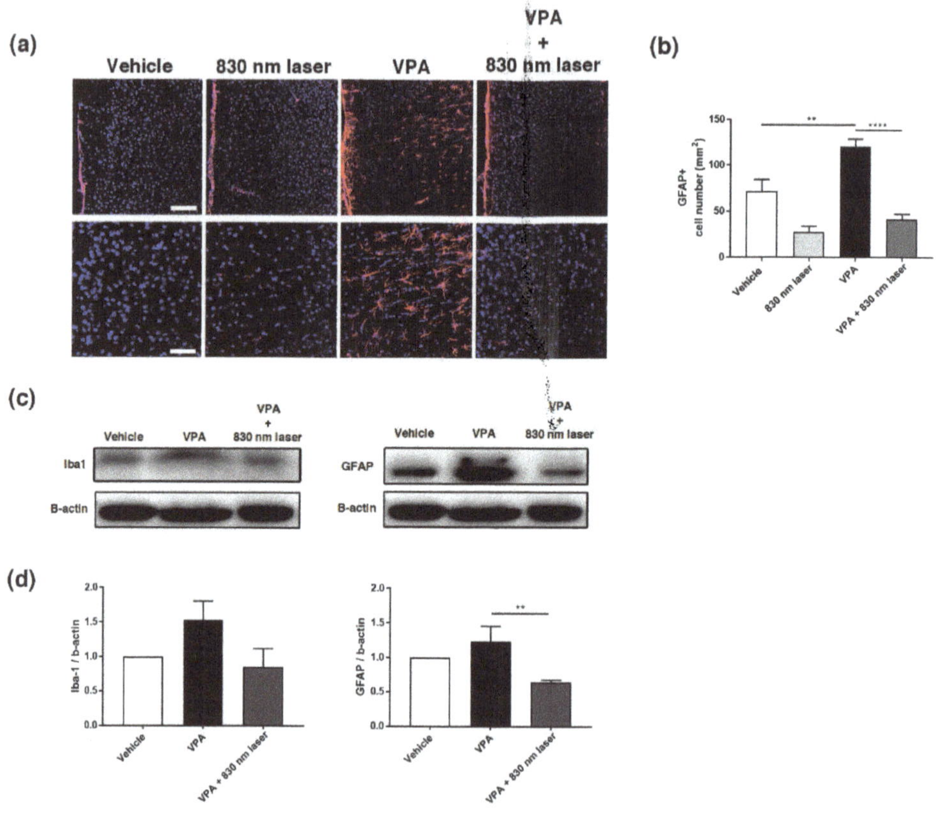

Figure 8. Immunohistochemistry and Western blotting analyses of GFAP expression in the medial prefrontal cortex (mPFC). (**a**) Image showing GFAP-positive cells (red) in the mPFC. (**b**) Quantification of GFAP-positive cells in the mPFC. (**c**) Image of a Western blot. Full-length blots/gels are presented in Figure S2. (**d**) Average relative intensity values for the Western blots. The bar graph summarizes changes in the GFAP-positive cells. Scale bars: 100 and 50 μm. Data are means ± SEM (vehicle: n = 3, 830 nm laser: n = 3, VPA: n = 3, VPA + 830 nm laser: n = 3). ** $p < 0.01$ compared to the vehicle group. **** $p < 0.0001$ compared to the group exposed to VPA.

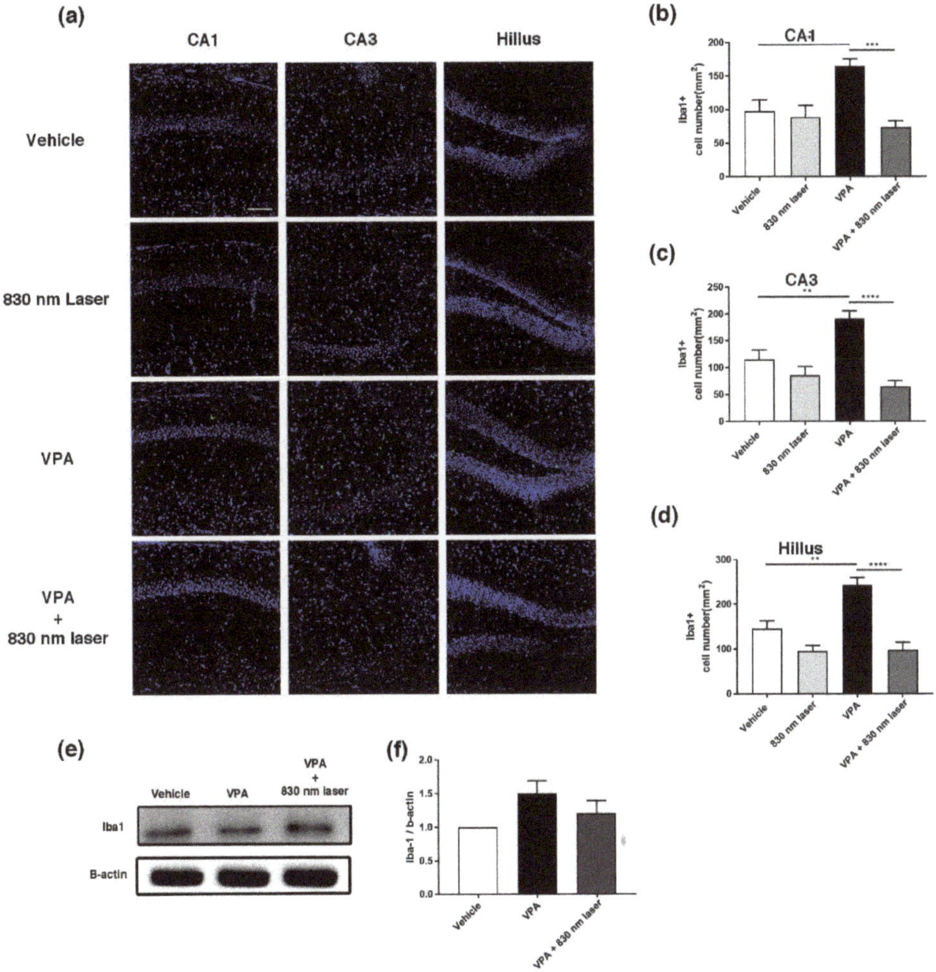

Figure 9. Immunohistochemistry and Western blotting analyses of ionized calcium-binding adapter molecule 1 (Iba1) expression in the hippocampus. (**a**) Image showing Iba1-positive cells (green) in the CA1, CA3, and hilus regions of the hippocampus. (**b–d**) Quantification of Iba1-positive cells in the CA1, CA3, and hilus regions of the hippocampus. (**e**) Image of a Western blot. Full-length blots/gels are presented in Figure S1. (**f**) Average relative intensity values for the Western blots. The bar graph summarizes changes in the Iba1-positive cells. Nuclei were stained using DAPI. Scale bar: 100 μm. Data are means ± SEM. (vehicle: $n = 3$, 830 nm laser: $n = 3$, VPA: $n = 3$, VPA + 830 nm laser: $n = 3$). ** $p < 0.01$ compared to the vehicle group. *** $p < 0.001$ and **** $p < 0.0001$ compared to the group exposed to VPA.

3. Discussion

Prenatal exposure of mice to VPA can be used to mimic the pathogenesis of ASD and provide important insight into the morphological and behavioral characteristics of the disease [30]. However, the mechanism by which VPA generates the ASD phenotype remains unclear. In mice, neurogenesis occurs from E12 until the late embryonic developmental stage [31], and exposure to VPA disrupts normal neurogenesis [32]. Intrauterine exposure to VPA may cause ASD symptoms such as developmental abnormalities, social and communication disorders, and restrictive behaviors [33–35]. Especially, it was reported that the intraperitoneal injection to maternal of VPA 400-600mg/kg on 12 embryonic days could have a symptom related to autism such as social impairments, cognitive rigidity, and repetitive behaviors [36]. We generated an ASD model by exposing mice to VPA at E12 and confirmed the symptoms of ASD using behavioral tests. The battery of behavior tests was performed in the least stressing and challenging order to avoid potential training and learning effects [37]. Our data suggest that VPA exposure inhibits neurogenesis and triggers the onset of ASD and behavioral disorders.

In this study, we investigated whether PBM treatment during fetal development could attenuate the symptoms of ASD in a mouse model. PBM treatment with an 830 nm laser attenuated the developmental and behavioral abnormalities observed in a VPA-induced ASD mouse model.

PBM has been confirmed to improve memory decline, amyloid plaques, tau hyperphosphorylation, neurodegeneration, spinal damage, and synapse loss due to the neuroprotective effect of 808nm PBM in Alzheimer's disease, which shows cognitive deficits among various neurological diseases [38]. In addition to this, NIR shows several neuroprotective effects in animal models with various neurological diseases such as brain stroke and Parkinson's disease [39,40]. Dysregulation of Neurogenesis is caused by the neuroinflammatory response [41], and PBM can promote neurogenesis by stimulating neural progenitor cells [42,43]. Because 830 nm PBM has the lowest absorption rate of water and hemoglobin in tissue, it has better tissue penetration in the deeper range and can reduce nerve damage [44–46]. This was expected to reduce neuroinflammation due to VPA using the 830 nm wavelength. Our results confirmed through GFAP and Iba1 that 830 nm laser-activated astrocytes and microglia were reduced. Therefore, PBM may counteract intrauterine exposure to VPA at E12.5 and prevent the inhibition of neurogenesis.

As well as developmental abnormalities, we also observed various behavioral changes in our VPA-induced ASD mouse model. Reduced motor function is often characteristic of ASD [37]. Using the wire maneuver and negative geotaxis tests, we confirmed that the VPA-induced ASD mouse model exhibited a decrease in motor function that was relieved by PBM. However, we did not observe corresponding neuronal changes in the motor cortex, which is responsible for motor function.

The three-chamber and Y-maze test results demonstrated deficiencies in social interaction, as well as repetitive and restrictive behaviors, in our VPA-induced ASD mouse model, which are all symptoms of ASD. In addition, these symptoms were attenuated by PBM treatment. Our histological studies also revealed corresponding neurological changes in the mPFC region of the brain, which is particularly important for social cognition and behavior [47].

Children with autism often exhibit cognitive deficiencies [48]. Our VPA-induced ASD mouse model exhibited cognitive deficiencies that were attenuated by PBM. Our histological studies also revealed corresponding changes in the hippocampus, which may affect cognitive function [22]. Previous studies have shown that exposure to VPA can alter oxidation status and the expression of proinflammatory genes [49]. Microglia and astrocytes are frequently activated in neurodegenerative diseases [50,51], and have also been found in postmortem brain tissue from patients diagnosed with ASD [7,18,19].

Glial fibrillary acidic protein (GFAP) is a structural marker protein of astrocytes, and ionized calcium-binding adapter molecule 1 (Iba1) is mainly used as an activation marker of microglia [52,53]. In response to CNS damage, the increase in activated glial cells causes cytokine activation, resulting in a gliosis reaction affecting nerve cells, resulting in a neuroinflammatory response [54]. Therefore, we confirmed the activation of astrocytes and microglia through the expression of GFAP and Iba1 in the ASD group exposed to VPA, and demonstrated that neuroinflammation was reduced through the 830 nm laser reducing their expression.

In summary, PBM may decrease neuroinflammation and improve cognitive function in mice exposed to VPA by deactivating microglia and astrocytes. Our results suggest that deficiencies in cognitive function associated with ASD are attenuated when PBM reduces the activity of microglia and astrocytes in the hippocampus. Therefore, this study demon-strates that PBM treatment may benefit patients with ASD and related neurodegenerative diseases.

4. Methods and Materials

4.1. Animals

C57/BL6 mice were purchased from Daehan Biolink Co., Ltd. (Samseong-myeon, Republic of Korea) and maintained in the Dankook University Laser Research Center (Yongin, Republic of Korea) under a 12 h light/dark cycle on an autonomic nervous system diet. All experiments were conducted in compliance with the NIH Guidelines for the Care and Use of Laboratory Animals and approved by the Institutional Animal Care and Use Committee (IACUC; DKU-20-036) of Dankook University. Both sexes were assayed.

4.2. Prenatal VPA Mouse Model for ASD

Experiments were performed on 8-week-old adult male and female mice. C57/BL6 mice were housed together overnight to obtain pregnant mice for the prenatal model. After overnight mating, a vaginal plug was applied to pregnant female mice on embryonic day 0 (E0) [55]. On embryonic day 12 (E12), pregnant mice received a single intraperitoneal injection of 600 mg/kg VPA (P4543; Sigma-Aldrich, St. Louis, MO, USA) dissolved in saline. Control female mice were treated with the vehicle (i.e., saline solution) only. Then, prenatal PBM mice were placed in a decapicone (MDC200; Braintree Scientific, Inc., Braintree, MA, USA) and treated three times with a laser (830 nm, 70 mW/cm^2, 10 min, Figure 10a–c, Table 1) The laser setup was based on previous work [56]. After weaning (P21), the male and female mice in each group were kept in different cages. The number of pups in each group at P10 was as follows: vehicle, n = 14; 830 nm laser, n = 15; VPA, n = 8; and VPA + 830 nm laser, n = 11. Both males and females were used in behavioral experiments, but females were not included in the data because they did not show behavioral changes (Figure S3).

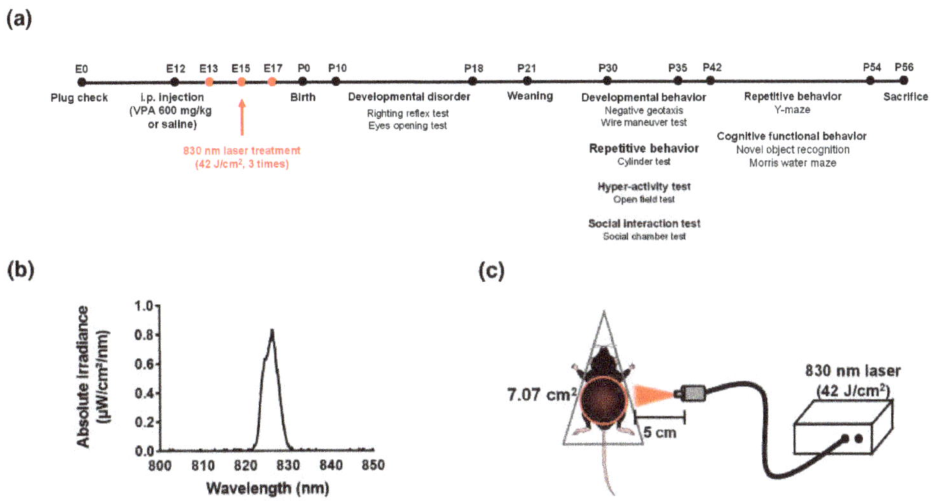

Figure 10. The diode laser. (**a**) Timetable. (**b**) Laser spectrum at a wavelength of 830 nm. (**c**) Schematic diagram showing laser irradiation of the abdomen (diameter: 7.07 cm^2).

4.3. Behavioral Tests

The expression of the ASD-like phenotype and effects of PBM treatment were evaluated by behavioral tests performed during the period P8–54 (Figure 10), between 10:00 am and 6:00 pm on the test days. Developmental disorder, SHIRPA, hyperactivity, social interaction, and cognitive behavior tests were completed. Mice were transferred from the housing cage to the experimental chamber 30 min before each test, to habituate to the test environment.

Table 1. Specifications for laser parameters.

Parameter	Value
Wavelength	830 nm
Operating mode	Continuous
Distance	5 cm
Beam shape	Circular
Spot size	Diameter = 7.07 cm^2
Power density (mW/cm^2)	70 mW/cm^2
Exposure duration (min)	10 min
Energy density (J/cm^2)	42 J/cm^2
Application technique	Without skin contact on the abdomen
Number and frequency of treatment sessions	3 sessions, once for 2 days

4.3.1. Eye Opening Test

Eye opening was observed once daily from P14 to P18 and scored as follows: 0 = both eyes closed, 1 = one eye open, and 2 = both eyes open [37].

4.3.2. Righting Reflex Test

The righting reflex test measured the pup's ability to regain its footing from the supine position [57]. At P10–13, the pups were placed on their backs on a flat surface and held in that position for 5 s. The time it took for each pup to return to the prone position after being released was recorded.

4.3.3. Bodyweight

Weight was measured on P10, P17, and P24.

4.3.4. Negative Geotaxis

Negative geotaxis was measured by placing each pup face down along a 45° incline. The time taken to climb after turning 180° was recorded [58]. If the mouse did not climb within 30 s, the maximum time was recorded.

4.3.5. Wire Maneuver Test

A wire maneuver test was used to assess whole-body force [59]. A metal wire was maintained horizontally 20 cm above a thick layer of soft bedding. Each mouse was allowed to grasp the wire by its forepaws. The time taken for each mouse to climb onto the wire was recorded. If the mouse did not raise its hind paws within 3 min, the maximum time was recorded.

4.3.6. Three-Chamber Test

The sociability test involved a rectangular box (60 cm × 40 cm × 20 cm) divided into three chambers with holes that allowed movement between the chambers. Each test was started after 10 min habituation and lasted for 10 min. During the test, a new "stranger" mouse was placed in the holding cage of one chamber. The time spent in each chamber with/without the stranger mouse was measured. All recorded videos were analyzed using EthoVision XT software (version 15.0; Noldus, Wageningen, The Netherlands). The social preference index was calculated as follows: Social preference index (%) = Time in stranger chamber/(Time in empty chambers + Time in stranger chamber) × 100 [60].

4.3.7. Y-Maze

The Y-maze spontaneous alternation test is used to assess repetitive behavior and spatial working memory [61]. All mice were placed in the same arm of the maze and allowed to explore it for 5 min. The maze was cleaned with 70% ethanol after each animal was tested. All recorded videos were analyzed by Ethovision XT software (version 15.0; Noldus, Wageningen, The Netherlands) in terms of the number of alternating entries. Spontaneous alternation was calculated as follows: Spontaneous alternation (%) = Total alternations/(Total arm entries—2) × 100 [33].

4.3.8. Cylinder Test

The cylinder test involved placing each mouse in a transparent cylindrical chamber for 3 min. This test evaluates locomotor asymmetry by measuring the number of forelimb contacts as the mouse rears against the wall of the cylinder [62].

4.3.9. Open Field Test

The open field test was performed as described previously [63] to assess hyperactivity at P35. Test mice were placed in the center of a cube-shaped open field arena (50 cm each side) and allowed to explore freely for 5 min. The arena was cleaned with 70% ethanol after each animal was tested. All recorded videos were analyzed by Ethovision XT software (version 15.0; Noldus, Wageningen, The Netherlands) in terms of the total distance covered and the number of times a line was crossed, as well as the total time spent in motion/stationary.

4.3.10. Novel Object Recognition Test

The novel object recognition test used the same apparatus as the open field test [64]. The test consisted of three phases: habituation, training, and testing. The habituation phase lasted for 5 min; the animals were exposed to the arena and then returned to the home cage while the arena was cleaned. During the training phase, each animal was placed in the arena with two identical objects located in opposite corners. The animal was returned to the home cage for 10 min to ensure that its memory was being tested. During the test phase, a familiar object and unfamiliar object were placed in the arena. Exploratory activity was monitored for 10 min in each test. Preference for the novel object was defined as the ratio of time spent with the novel object compared to that spent with the familiar object. All recorded videos were analyzed by Ethovision XT software (version 15.0; Noldus, Wageningen, The Netherlands) and the time spent sniffing each object was measured. The discrimination index was calculated as follows: Discrimination index (%) = Time spent on novel object/Total time spent exploring both objects [65].

4.3.11. Morris Water Maze

The Morris water maze apparatus was a circular metal box (diameter: 100 cm) with distinct visual cues placed at the quadrant points [66]. White nontoxic paint was applied to hide the platform below the water. The water was maintained at 20–25 °C. Each test lasted 180 s, or until the mouse found the platform. A sequence of five trials was performed over 5 days. All recorded videos were analyzed by Ethovision XT software (version 15.0; Noldus, Wageningen, The Netherlands).

4.4. Immunohistochemistry

Mouse brains were fixed in 4% paraformaldehyde overnight at 4 °C, and then dehydrated overnight at 4 °C by incubation in a series of sucrose solutions (10%, 20%, and 30%). Whole brains were then embedded in optimal cutting temperature compound and cryosectioned into 18 mm slices using a cryostat (Leica, Wetzlar, Germany). For immunohistochemical analyses, frozen sections were washed with 0.1 M phosphate-buffered saline (PBS) before permeabilization with 0.5% Triton X-100 for 5 min at room temperature. Sections were blocked with 5% bovine serum albumin (BSA) in 0.3% Triton X-100 in 1× PBS for 1 h 30 min at room temperature. Next, sections were incubated overnight with primary antibodies against the following in 1% BSA and 0.3% Triton X-100: glial fibrillary acidic protein (GFAP; mouse, 1:500; MAB360; Sigma-Aldrich, Burlington, MA, USA) or ionized calcium-binding adapter molecule 1 (Iba1; rabbit, 1:250; PA5-27436; Thermo Fisher, Waltham, MA, USA). Finally, sections were incubated with secondary antibody (Alexa Fluor 488- or 555-conjugated antibody) for 1 h at room temperature before being washed with 1× PBS and mounted in a solution containing 4′,6-diamidino-2-phenylindole (DAPI).

4.5. Western Blot Analysis

The hippocampus and prefrontal cortex (PFC) were dissected from male pups in all groups (P56). Tissue was homogenized in ice-cold radio-immune precipitation assay buffer with protease inhibitor, followed by protein isolation and quantification in a detergent-compatible assay. Equal concentrations of protein (30 μg) were loaded onto 15% sodium dodecyl sulfate-polyacrylamide gel electrophoresis gels and then transferred to polyvinylidene difluoride membranes. The membranes were blocked with 5% BSA for 1 h and then incubated overnight at 4 °C with the following primary antibodies: anti-GFAP (1:1000; MAB360; Sigma-Aldrich, Burlington, MA, USA), anti-Iba1 (1:500; PA5-27436; Thermo Fisher, Waltham, MA, USA), or anti-β-actin (1:5000). Then, the membranes were washed with 1× Tris buffered saline with Tween-20 (TBST) buffer (5 min, three times) and incubated with horseradish peroxidase-bound secondary antibody for 1 h. After washing with TBST, Western blot images were acquired using a Chemi Doc system (Bio-Rad, Hercules, CA, USA). Images were analyzed using ImageJ software (version 1.5; National Institutes of Health, Bethesda, MD, USA).

4.6. Statistics

Data are reported as means ± standard error of the mean (SEM) and analyses were performed using GraphPad Prism software (ver. 7.0; GraphPad Software, Inc., San Diego, CA, USA). Data were analyzed using one-way analysis of variance (ANOVA) or two-way ANOVA followed by the Bonferroni test. A p-value < 0.05 was considered statistically significant.

Supplementary Materials: The supporting information can be downloaded at: https://www.mdpi.com/article/10.3390/ijms232416099/s1.

Author Contributions: U.-J.K. contributed to in vivo studies, and writing of the original draft; N.H. contributed to conceptualization, and review and editing of the manuscript; J.-C.A. supervised the whole work, and administered the project. All authors have read and agreed to the published version of the manuscript.

Funding: This research was supported by the Basic Science Research Program through the National Research Foundation of Korea (NRF), funded by the Ministry of Education (NRF-2020R1A6A1A03043283). National Research Facilities and Equipment Center (NFEC) grant funded by the Korea government (Ministry of Education) (No. 2019R1A6C1010033).

Institutional Review Board Statement: All animal experiments were performed in compliance with the National Institutes of Health guidelines for animal research. All animal experiments were approved by IACUC of the Dankook University (IACUC; DKU-19-017). Extensive efforts were made to ensure minimal suffering of the animals used during the study.

Informed Consent Statement: Not applicable.

Data Availability Statement: The data that support the findings of this study are available from the corresponding author upon reasonable request.

Acknowledgments: The English in this document has been checked by at least two professional editors, both native speakers of English. For a certificate, please see: http://www.textcheck.com/certificate/VZL5yR, and the certificate accessed on 31 July 2022.

Conflicts of Interest: The authors declare that they have no known competing financial interests or personal relationships that could have appeared to influence the work reported in this paper.

References

1. Edition, F. Diagnostic and statistical manual of mental disorders. *Am. Psychiatr. Assoc.* **2013**, *21*, 591–643.
2. Jęśko, H.; Cieślik, M.; Gromadzka, G.; Adamczyk, A. Dysfunctional proteins in neuropsychiatric disorders: From neurodegeneration to autism spectrum disorders. *Neurochem. Int.* **2020**, *141*, 104853. [CrossRef] [PubMed]
3. Kern, J.K.; Geier, D.A.; Sykes, L.K.; Geier, M.R. Evidence of neurodegeneration in autism spectrum disorder. *Transl. Neurodegener.* **2013**, *2*, 17. [CrossRef] [PubMed]
4. Nag, H.E.; Nordgren, A.; Anderlid, B.-M.; Nærland, T. Reversed gender ratio of autism spectrum disorder in Smith-Magenis syndrome. *Mol. Autism* **2018**, *9*, 1–9. [CrossRef]
5. Pallanti, S.; Di Ponzio, M.; Grassi, E.; Vannini, G.; Cauli, G. Transcranial Photobiomodulation for the Treatment of Children with Autism Spectrum Disorder (ASD): A Retrospective Study. *Children* **2022**, *9*, 755. [CrossRef]
6. Depino, A.M. Peripheral and central inflammation in autism spectrum disorders. *Mol. Cell. Neurosci.* **2013**, *53*, 69–76. [CrossRef]
7. Vargas, D.L.; Nascimbene, C.; Krishnan, C.; Zimmerman, A.W.; Pardo, C.A. Neuroglial activation and neuroinflammation in the brain of patients with autism. *Ann. Neurol.* **2004**, *57*, 67–81. [CrossRef]
8. Sato, A.; Kotajima-Murakami, H.; Tanaka, M.; Katoh, Y.; Ikeda, K. Influence of Prenatal Drug Exposure, Maternal Inflammation, and Parental Aging on the Development of Autism Spectrum Disorder. *Front. Psychiatry* **2022**, *13*, 821455. [CrossRef]
9. Nevitt, S.J.; Sudell, M.; Cividini, S.; Marson, A.G.; Smith, C.T. Antiepileptic drug monotherapy for epilepsy: A network meta-analysis of individual participant data. *Cochrane Database Syst. Rev.* **2022**, *4*, CD20114122. [CrossRef]
10. Clayton-Smith, J.; Bromley, R.; Dean, J.; Journel, H.; Odent, S.; Wood, A.; Williams, J.; Cuthbert, V.; Hackett, L.; Aslam, N.; et al. Diagnosis and management of individuals with Fetal Valproate Spectrum Disorder; a consensus statement from the European Reference Network for Congenital Malformations and Intellectual Disability. *Orphanet J. Rare Dis.* **2019**, *14*, 180. [CrossRef]
11. Salehpour, F.; Mahmoudi, J.; Kamari, F.; Sadigh-Eteghad, S.; Rasta, S.H.; Hamblin, M.R. Brain Photobiomodulation Therapy: A Narrative Review. *Mol. Neurobiol.* **2018**, *55*, 6601–6636. [CrossRef] [PubMed]
12. Chen, Q.; Wu, J.; Dong, X.; Yin, H.; Shi, X.; Su, S.; Che, B.; Li, Y.; Yang, J. Gut flora-targeted photobiomodulation therapy improves senile dementia in an Aß-induced Alzheimer's disease animal model. *J. Photochem. Photobiol. B Biol.* **2021**, *216*, 112152. [CrossRef] [PubMed]
13. De la Torre, J.C. Treating cognitive impairment with transcranial low level laser therapy. *J. Photochem. Photobiol. B Biol.* **2017**, *168*, 149–155. [CrossRef] [PubMed]
14. Montazeri, K.; Farhadi, M.; Fekrazad, R.; Akbarnejad, Z.; Chaibakhsh, S.; Mahmoudian, S. Transcranial photobiomodulation in the management of brain disorders. *J. Photochem. Photobiol. B Biol.* **2021**, *221*, 112207. [CrossRef]
15. Hamblin, M.R. Could Photobiomodulation Treat Autism Spectrum Disorder? *Photobiomodul. Photomed. Laser Surg.* **2022**, *40*, 367–369. [CrossRef]
16. Leisman, G.; Machado, C.; Machado, Y.; Chinchilla-Acosta, M. Effects of Low-Level Laser Therapy in Autism Spectrum Disorder. *Adv. Exp. Med. Biol.* **2018**, *1116*, 111–130. [CrossRef]
17. Kwon, H.S.; Koh, S.H. Neuroinflammation in neurodegenerative disorders: The roles of microglia and astrocytes. *Transl Neurodegener* **2020**, *9*, 42. [CrossRef]
18. Morgan, J.T.; Chana, G.; Pardo, C.A.; Achim, C.; Semendeferi, K.; Buckwalter, J.; Courchesne, E.; Everall, I.P. Microglial activation and increased microglial density observed in the dorsolateral prefrontal cortex in autism. *Biol Psychiatry* **2010**, *68*, 368–376. [CrossRef]
19. Pardo, C.A.; Vargas, D.L.; Zimmerman, A.W. Immunity, neuroglia and neuroinflammation in autism. *Int. Rev. Psychiatry* **2005**, *17*, 485–495. [CrossRef]
20. Cardoso, F.D.S.; Salehpour, F.; Coimbra, N.C.; Gonzalez-Lima, F.; Gomes da Silva, S. Photobiomodulation for the treatment of neuroinflammation: A systematic review of controlled laboratory animal studies. *Front. Neurosci.* **2022**, *16*, 1006031. [CrossRef]
21. Mabunga, D.F.N.; Gonzales, E.L.T.; Kim, J.-W.; Kim, K.C.; Shin, K.C.K.A.C.Y. Exploring the Validity of Valproic Acid Animal Model of Autism. *Exp. Neurobiol.* **2015**, *24*, 285–300. [CrossRef] [PubMed]
22. Sweatt, J.D. Hippocampal function in cognition. *Psychopharmacology* **2004**, *174*, 99–110. [CrossRef] [PubMed]
23. Vertes, R.P. Differential projections of the infralimbic and prelimbic cortex in the rat. *Synapse* **2003**, *51*, 32–58. [CrossRef] [PubMed]
24. Liang, L.-Y.; Shewokis, P.A.; Getchell, N. Brain Activation in the Prefrontal Cortex during Motor and Cognitive Tasks in Adults. *J. Behav. Brain Sci.* **2016**, *6*, 463–474. [CrossRef]
25. Haydon, P.G.; Nedergaard, M. How Do Astrocytes Participate in Neural Plasticity? Cold Spring Harb. *Perspect. Biol.* **2014**, *7*, a020438. [CrossRef]
26. Casano, A.M.; Peri, F. Microglia: Multitasking Specialists of the Brain. *Dev. Cell* **2015**, *32*, 469–477. [CrossRef]
27. Takano, T. Role of Microglia in Autism: Recent Advances. *Dev. Neurosci.* **2015**, *37*, 195–202. [CrossRef]
28. Voineagu, I.; Wang, X.; Johnston, P.; Lowe, J.K.; Tian, Y.; Horvath, S.; Mill, J.; Cantor, R.M.; Blencowe, B.J.; Geschwind, D.H. Transcriptomic analysis of autistic brain reveals convergent molecular pathology. *Nature* **2011**, *474*, 380–384. [CrossRef]
29. de Baumont, A.; Maschietto, M.; Lima, L.; Carraro, D.M.; Olivieri, E.H.; Fiorini, A.; Barreta, L.A.N.; Palha, J.A.; Belmonte-De-Abreu, P.; Filho, C.A.M.; et al. Innate immune response is differentially dysregulated between bipolar disease and schizophrenia. *Schizophr. Res.* **2015**, *161*, 215–221. [CrossRef]

30. Schneider, T.; Roman, A.; Basta-Kaim, A.; Kubera, M.; Budziszewska, B.; Schneider, K.; Przewłocki, R. Gender-specific behavioral and immunological alterations in an animal model of autism induced by prenatal exposure to valproic acid. *Psychoneuroendocrinology* **2008**, *33*, 728–740. [CrossRef]
31. Bond, A.M.; Berg, D.A.; Lee, S.; Garcia-Epelboim, A.S.; Adusumilli, V.S.; Ming, G.-L.; Song, H. Differential Timing and Coordination of Neurogenesis and Astrogenesis in Developing Mouse Hippocampal Subregions. *Brain Sci.* **2020**, *10*, 909. [CrossRef] [PubMed]
32. Zhao, H.; Wang, Q.; Yan, T.; Zhang, Y.; Xu, H.-J.; Yu, H.-P.; Tu, Z.; Guo, X.; Jiang, Y.-H.; Li, X.-J.; et al. Maternal valproic acid exposure leads to neurogenesis defects and autism-like behaviors in non-human primates. *Transl. Psychiatry* **2019**, *9*, 267. [CrossRef] [PubMed]
33. Campolongo, M.; Kazlauskas, N.; Falasco, G.; Urrutia, L.; Salgueiro, N.; Höcht, C.; Depino, A.M. Sociability deficits after prenatal exposure to valproic acid are rescued by early social enrichment. *Mol. Autism* **2018**, *9*, 36. [CrossRef] [PubMed]
34. Yang, E.-J.; Ahn, S.; Lee, K.; Mahmood, U.; Kim, H.-S. Correction: Early Behavioral Abnormalities and Perinatal Alterations of PTEN/AKT Pathway in Valproic Acid Autism Model Mice. *PLoS ONE* **2016**, *11*, e0157202. [CrossRef] [PubMed]
35. Schneider, T.; Przewłocki, R. Behavioral Alterations in Rats Prenatally Exposed to Valproic Acid: Animal Model of Autism. *Neuropsychopharmacology* **2004**, *30*, 80–89. [CrossRef]
36. Chaliha, D.; Albrecht, M.; Vaccarezza, M.; Takechi, R.; Lam, V.; Al-Salami, H.; Mamo, J. A Systematic Review of the Valproic-Acid-Induced Rodent Model of Autism. *Dev. Neurosci.* **2020**, *42*, 12–48. [CrossRef]
37. Al Sagheer, T.; Haida, O.; Balbous, A.; Francheteau, M.; Matas, E.; Fernagut, P.-O.; Jaber, M. Motor Impairments Correlate with Social Deficits and Restricted Neuronal Loss in an Environmental Model of Autism. *Int. J. Neuropsychopharmacol.* **2018**, *21*, 871–882. [CrossRef]
38. Yang, L.; Wu, C.; Parker, E.; Li, Y.; Dong, Y.; Tucker, L.; Brann, D.W.; Lin, H.W.; Zhang, Q. Non-invasive photobiomodulation treatment in an Alzheimer Disease-like transgenic rat model. *Theranostics* **2022**, *12*, 2205–2231. [CrossRef]
39. Vogel, D.D.S.; Ortiz-Villatoro, N.N.; Araújo, N.S.; Marques, M.J.G.; Aimbire, F.; Scorza, F.A.; Scorza, C.A.; Albertini, R. Transcranial low-level laser therapy in an in vivo model of stroke: Relevance to the brain infarct, microglia activation and neuroinflammation. *J. Biophotonics* **2021**, *14*, e202000500. [CrossRef]
40. Foo, A.S.C.; Soong, T.W.; Yeo, T.T.; Lim, K.L. Mitochondrial Dysfunction and Parkinson's Disease-Near-Infrared Photobiomodulation as a Potential Therapeutic Strategy. *Front. Aging Neurosci.* **2020**, *12*, 89. [CrossRef] [PubMed]
41. Fan, L.W.; Pang, Y. Dysregulation of neurogenesis by neuroinflammation: Key differences in neurodevelopmental and neurological disorders. *Neural Regen. Res.* **2017**, *12*, 366–371. [CrossRef] [PubMed]
42. Yang, L.; Tucker, D.; Dong, Y.; Wu, C.; Lu, Y.; Li, Y.; Zhang, J.; Liu, T.C.; Zhang, Q. Photobiomodulation therapy promotes neurogenesis by improving post-stroke local microenvironment and stimulating neuroprogenitor cells. *Exp. Neurol.* **2018**, *299*, 86–96. [CrossRef]
43. Hong, N.; Kang, G.W.; Park, J.O.; Chung, P.S.; Lee, M.Y.; Ahn, J.C. Photobiomodulation regulates adult neurogenesis in the hippocampus in a status epilepticus animal model. *Sci. Rep.* **2022**, *12*, 15246. [CrossRef] [PubMed]
44. Chung, H.; Dai, T.; Sharma, S.K.; Huang, Y.Y.; Carroll, J.D.; Hamblin, M.R. The nuts and bolts of low-level laser (light) therapy. *Ann. Biomed. Eng.* **2012**, *40*, 516–533. [CrossRef] [PubMed]
45. Tsai, C.M.; Chang, S.F.; Li, C.C.; Chang, H. Transcranial photobiomodulation (808 nm) attenuates pentylenetetrazole-induced seizures by suppressing hippocampal neuroinflammation, astrogliosis, and microgliosis in peripubertal rats. *Neurophotonics* **2022**, *9*, 015006. [CrossRef] [PubMed]
46. Tucker, L.D.; Lu, Y.; Dong, Y.; Yang, L.; Li, Y.; Zhao, N.; Zhang, Q. Photobiomodulation Therapy Attenuates Hypoxic-Ischemic Injury in a Neonatal Rat Model. *J. Mol. Neurosci.* **2018**, *65*, 514–526. [CrossRef] [PubMed]
47. Maliske, L.; Kanske, P. The Social Connectome—Moving Toward Complexity in the Study of Brain Networks and Their Interactions in Social Cognitive and Affective Neuroscience. *Front. Psychiatry* **2022**, *13*, 845492. [CrossRef]
48. Vig, S.; Jedrysek, E. Autistic features in young children with significant cognitive impairment: Autism or mental retardation? *J. Autism Dev. Disord.* **1999**, *29*, 235–248. [CrossRef]
49. Gąssowska-Dobrowolska, M.; Cieślik, M.; Czapski, G.A.; Jęśko, H.; Frontczak-Baniewicz, M.; Gewartowska, M.; Dominiak, A.; Polowy, R.; Filipkowski, R.K.; Babiec, L.; et al. Prenatal Exposure to Valproic Acid Affects Microglia and Synaptic Ultrastructure in a Brain-Region-Specific Manner in Young-Adult Male Rats: Relevance to Autism Spectrum Disorders. *Int. J. Mol. Sci.* **2020**, *21*, 3576. [CrossRef]
50. Carson, M.J.; Bilousova, T.V.; Puntambekar, S.S.; Melchior, B.; Doose, J.M.; Ethell, I.M. A rose by any other name? The potential consequences of microglial heterogeneity during CNS health and disease. *Neurotherapeutics* **2007**, *4*, 571–579. [CrossRef]
51. Edmonson, C.; Ziats, M.N.; Rennert, O.M. Altered glial marker expression in autistic post-mortem prefrontal cortex and cerebellum. *Mol. Autism* **2014**, *5*, 3. [CrossRef] [PubMed]
52. Eng, L.F. Glial fibrillary acidic protein (GFAP): The major protein of glial intermediate filaments in differentiated astrocytes. *J. Neuroimmunol.* **1985**, *8*, 203–214. [CrossRef]
53. Ohsawa, K.; Imai, Y.; Sasaki, Y.; Kohsaka, S. Microglia/macrophage-specific protein Iba1 binds to fimbrin and enhances its actin-bundling activity. *J. Neurochem.* **2004**, *88*, 844–856. [CrossRef] [PubMed]
54. Czlonkowska, A.; Kurkowska-Jastrzebska, I. Inflammation and gliosis in neurological diseases–clinical implications. *J. Neuroimmunol.* **2011**, *231*, 78–85. [CrossRef] [PubMed]

55. Behringer, R.; Gertsenstein, M.; Nagy, K.V.; Nagy, A. Selecting Female Mice in Estrus and Checking Plugs. *Cold Spring Harb. Protoc.* **2016**, *8*, pdb-prot092387. [CrossRef] [PubMed]
56. Yang, L.; Dong, Y.; Wu, C.; Youngblood, H.; Li, Y.; Zong, X.; Li, L.; Xu, T.; Zhang, Q. Effects of prenatal photobiomodulation treatment on neonatal hypoxic ischemia in rat offspring. *Theranostics* **2021**, *11*, 1269–1294. [CrossRef]
57. Soria-Ortiz, M.B.; Reyes-Ortega, P.; Martínez-Torres, A.; Reyes-Haro, D. A Functional Signature in the Developing Cerebellum: Evidence From a Preclinical Model of Autism. *Front. Cell Dev. Biol.* **2021**, *9*, 727079. [CrossRef]
58. Wang, R.; Tan, J.; Guo, J.; Zheng, Y.; Han, Q.; So, K.-F.; Yu, J.; Zhang, L. Aberrant Development and Synaptic Transmission of Cerebellar Cortex in a VPA Induced Mouse Autism Model. *Front. Cell. Neurosci.* **2018**, *12*, 500. [CrossRef]
59. Dorchies, O.M.; Reutenauer-Patte, J.; Dahmane, E.; Ismail, H.M.; Petermann, O.; Patthey-Vuadens, O.; Comyn, S.A.; Gayi, E.; Piacenza, T.; Handa, R.J.; et al. The Anticancer Drug Tamoxifen Counteracts the Pathology in a Mouse Model of Duchenne Muscular Dystrophy. *Am. J. Pathol.* **2013**, *182*, 485–504. [CrossRef]
60. Nygaard, K.R.; Maloney, S.E.; Dougherty, J.D. Erroneous inference based on a lack of preference within one group: Autism, mice, and the social approach task. *Autism Res.* **2019**, *12*, 1171–1183. [CrossRef]
61. Markram, K.; Rinaldi, T.; La Mendola, D.; Sandi, C.; Markram, H. Abnormal Fear Conditioning and Amygdala Processing in an Animal Model of Autism. *Neuropsychopharmacology* **2007**, *33*, 901–912. [CrossRef]
62. Fleming, S.M.; Ekhator, O.R.; Ghisays, V. Assessment of Sensorimotor Function in Mouse Models of Parkinson's Disease. *J. Vis. Exp.* **2013**, *17*, e50303. [CrossRef]
63. Kim, J.-W.; Seung, H.; Kwon, K.J.; Ko, M.J.; Lee, E.J.; Oh, H.A.; Choi, C.S.; Kim, K.C.; Gonzales, E.L.; You, J.S.; et al. Subchronic Treatment of Donepezil Rescues Impaired Social, Hyperactive, and Stereotypic Behavior in Valproic Acid-Induced Animal Model of Autism. *PLoS ONE* **2014**, *9*, e104927. [CrossRef]
64. Takuma, K.; Hara, Y.; Kataoka, S.; Kawanai, T.; Maeda, Y.; Watanabe, R.; Takano, E.; Hayata-Takano, A.; Hashimoto, H.; Ago, Y.; et al. Chronic treatment with valproic acid or sodium butyrate attenuates novel object recognition deficits and hippocampal dendritic spine loss in a mouse model of autism. *Pharmacol. Biochem. Behav.* **2014**, *126*, 43–49. [CrossRef] [PubMed]
65. Zou, M.; Liu, Y.; Xie, S.; Wang, L.; Li, D.; Li, L.; Wang, F.; Zhang, Y.; Xia, W.; Sun, C.; et al. Alterations of the endocannabinoid system and its therapeutic potential in autism spectrum disorder. *Open Biol.* **2021**, *11*, 200306. [CrossRef] [PubMed]
66. Bromley-Brits, K.; Deng, Y.; Song, W. Morris Water Maze Test for Learning and Memory Deficits in Alzheimer's Disease Model Mice. *J. Vis. Exp.* **2011**, *53*, e2920. [CrossRef] [PubMed]

International Journal of *Molecular Sciences*

MDPI

Article

Comprehensive Investigation of Parameters Influencing Fluorescence Lifetime Imaging Microscopy in Frequency- and Time-Domain Illustrated by Phasor Plot Analysis

Thomas Kellerer [1,2], Janko Janusch [1,3,4], Christian Freymüller [3,4], Adrian Rühm [3,4], Ronald Sroka [3,4] and Thomas Hellerer [1,*]

1 Multiphoton Imaging Lab, Munich University of Applied Sciences, 80335 Munich, Germany
2 Faculty of Physics, Soft Condensed Matter, Ludwig-Maximilians-University, 80539 Munich, Germany
3 Laser-Forschungslabor, LIFE Center, Department of Urology, University Hospital, Ludwig-Maximilians-University, 82152 Planegg, Germany
4 Department of Urology, University Hospital, Ludwig-Maximilians-University, 81377 Munich, Germany
* Correspondence: hellerer@hm.edu

Abstract: Having access to fluorescence lifetime, researchers can reveal in-depth details about the microenvironment as well as the physico-chemical state of the molecule under investigation. However, the high number of influencing factors might be an explanation for the strongly deviating values of fluorescent lifetimes for the same fluorophore reported in the literature. This could be the reason for the impression that inconsistent results are obtained depending on which detection and excitation scheme is used. To clarify this controversy, the two most common techniques for measuring fluorescence lifetimes in the time-domain and in the frequency-domain were implemented in one single microscopy setup and applied to a variety of fluorophores under different environmental conditions such as pH-value, temperature, solvent polarity, etc., along with distinct state forms that depend, for example, on the concentration. From a vast amount of measurement results, both setup- and sample-dependent parameters were extracted and represented using a single display form, the phasor-plot. The measurements showed consistent results between the two techniques and revealed which of the tested parameters has the strongest influence on the fluorescence lifetime. In addition, quantitative guidance as to which technique is most suitable for which research task and how to perform the experiment properly to obtain consistent fluorescence lifetimes is discussed.

Keywords: fluorescence lifetime imaging microscopy; FLIM; fluorescence microscopy; two-photon microscopy; phasor-plot; image analysis; time-domain; frequency-domain; bioimaging

Citation: Kellerer, T.; Janusch, J.; Freymüller, C.; Rühm, A.; Sroka, R.; Hellerer, T. Comprehensive Investigation of Parameters Influencing Fluorescence Lifetime Imaging Microscopy in Frequency- and Time-Domain Illustrated by Phasor Plot Analysis. *Int. J. Mol. Sci.* **2022**, *23*, 15885. https://doi.org/10.3390/ijms232415885

Academic Editors: Antonino Mazzaglia, Angela Scala and Enrico Caruso

Received: 16 November 2022
Accepted: 9 December 2022
Published: 14 December 2022

1. Introduction

Modern microscopy excels at uncovering the morphology of living samples [1]. Today, advanced techniques, such as stimulated Raman scattering (SRS) [2], fluorescence correlation spectroscopy (FCS) [3,4] and fluorescence lifetime imaging microscopy (FLIM) [5], give additional access to photo-physical, chemical and biological information, to mention just three types [6]. Furthermore, this information can be quantified, e.g., by measuring the local concentration of a molecular species in an area as small as one femtoliter [7]. Unfortunately, the majority of microscopy methods rely on signal intensities, which are dependent on the experimental setup, such as the excitation/detection efficiencies, or susceptible to unwanted phenomena, such as photo-bleaching of the markers used [8]. Therefore, only relative changes are commonly quantified. FLIM is an exception here because the measured lifetime is not dependent on the kind of excitation, e.g., via one-photon (1P) or two-photon (2P) absorption or on matching the laser wavelength to the excitation profile of the sample. Furthermore, different kinds of detection, e.g., single photon counting with a point-detector or a camera-based widefield approach, deliver the same results. This paper refers to this desirable feature as the FLIM-advantage. Why then is the lifetime of a specific fluorophore not the same under all circumstances? Consulting the literature, one finds differing values that lead to the impression that FLIM may not be a reliable method. As a benchmark, two examples are represented: the reported lifetime of rhodamine B ranges from 1.74 to 3.13 ns [9–12], or the lifetime of rose bengal ranges from 0.095 to 2.4 ns [13–16]. Despite these huge differences, there is no evidence that the authors of the published data made any mistake in their measurements or conclusions. Therefore, the aim of this paper is to clarify how these discrepancies arise and what circumstances lead to differing lifetime values. On the other hand, it will be shown that time domain (TD) and frequency domain (FD) FLIM lead to the same results although they employ the different setup parameters mentioned above. Unfortunately, they are not directly comparable because the corresponding data analysis is different. Therefore, the data of both methods are transformed to retrieve a so-called phasor plot, which makes the comparison an easy task. The reason for the superficial discrepancy in lifetimes is due to certain parameters that influence the sample directly, such as concentration, pH-value, solvent polarity, and temperature, etc. This fact will be demonstrated in our comprehensive investigation. On the other hand, if these parameters are under the control of the researcher, they can use this to their advantage and utilize the fluorophore as the smallest possible reporter inside a living cell.

2. Results

All parameters that were studied in this rigorous investigation fall within two categories—setup-dependent and sample-dependent. The former determines the accuracy and range in which the lifetime can be obtained. The latter includes all parameters related to the microenvironment and the photo-physical properties of the fluorophore that affect its lifetime.

2.1. Setup Dependent Parameters

Starting with the different measurement setups, interdependent advantages and disadvantages of each were worked out. While the TD setup stands out with its diffraction-limited resolution, deep tissue penetration and optical sectioning possibilities, the needed detection time must be long enough to gather sufficient photon statistics. If a high resolution is not needed, the fast image acquisition of the FD setups provides the greatest advantage.

2.1.1. Wavelength

An often-formulated question is whether the kind of absorption process, e.g., 1P- or 2P-absorption, or the excitation wavelength, affect the fluorescence lifetime [12]. Therefore, the setups chosen in this series differ in the way the excitation takes place. For the TD setup, two ultrashort pulsed lasers with center wavelengths at 780 and 1034 nm were utilized, whereas the FD setup employed two wavelengths at 405 and 445 nm. To compare the lifetimes resulting from two-photon (TD FLIM) and one-photon (FD FLIM) absorption, a single dye with suitable spectral properties was used (Figure 1A). Rose bengal dissolved in ethanol at a concentration of 10^{-2} M met this requirement. The lifetimes determined using all four wavelengths, as well as one-photon and two-photon absorption, gave consistent results of around 0.85 ns (Figure 1B). A p-value test yielded a result of 0.046, which met the requirement for the measurement to be considered statistically significant. To also show the mono-exponential behavior within the FD, a p-value between the phase and demodulation lifetime was determined (Table 1). The mono-exponential behavior in the TD was verified by the residuals of the single exponential fit.

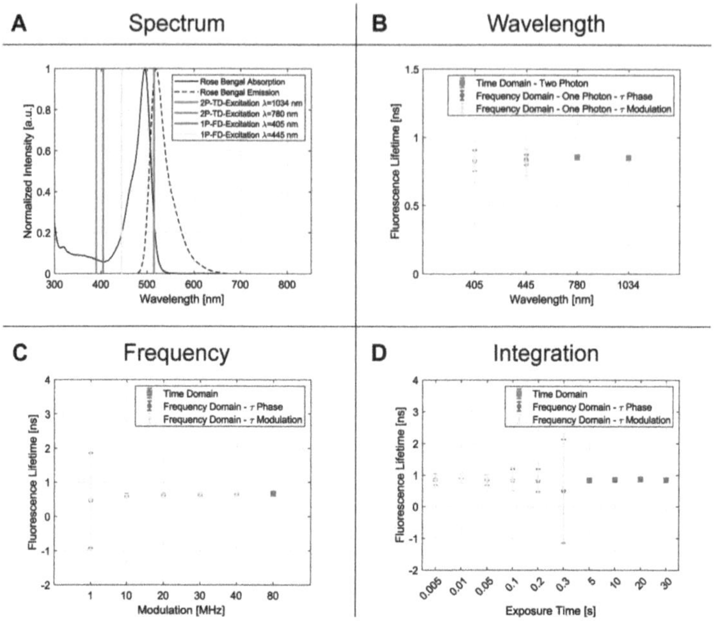

Figure 1. In (**A**), the absorption and emission spectrum of rose bengal and the individual excitation wavelengths are shown. The blue (405 nm) and cyan line (445 nm) represent the FD Wavelengths. For TD FLIM, the two-photon excitation wavelengths are shown in pink (two-photon excitation at 390 nm) and red (two-photon excitation at 515 nm). Although the laser lines are only slightly within the absorption spectrum, a sufficiently high fluorescence can be generated. In (**B**), the obtained fluorescence lifetimes for the wavelength dependency measurements were presented. In blue (phase lifetime) and cyan (demodulation lifetime), the results for the FD are shown, and in red the results for the TD are shown. In section (**C**), the influence of the repetition rate (TD) and the modulation frequency (FD) are shown. For the FD, the tangential relationship leads to inaccurate results for small modulation frequencies for lifetimes in the range of 1–10 ns. As the last parameter, the time constant for the different measurement methods is illustrated in (**D**). For the FD FLIM, high exposure times lead to pixel saturation and therefore creates bigger standard deviations. For the TD, the correlation is inverse. For longer integration times, a smaller error bar is detected due to the larger number of photons that form the decay statistic.

2.1.2. Modulation Frequency

The modulation frequency, with which the sample is excited, plays an important role; hence, it defines in which range the fluorescence lifetime can be detected. For the TD, the limiting factor is the repetition rate of the laser source. All photons detected during an excitation cycle can be accurately assigned to a specific excitation-pulse and thus counted for the statistics. If the time between pulses increases (i.e., the repetition rate decreases), longer lifetimes can be measured and vice versa.

Because of the tangent function in Equation (17), it can be derived that a smaller modulation frequency must be selected for longer lifetimes. If this criterion is not met, the measurement results are more scattered. The reason for this could be explained easily by the phase lifetime. While the frequency increases, the working line becomes steeper due to the stretched tangent function. The phase determination is, therefore, less accurate and results in scattered lifetimes with bigger errors, as can be seen in Figure 1C and Table 1.

Table 1. Single and overall *p*-values for each frequency for the phase- and demodulation-lifetime of the FD technique, when measuring rose bengal.

	Modulation Frequency					
	1 MHz	**10 MHz**	**20 MHz**	**30 MHz**	**40 MHz**	**Overall *p*-Value**
τ_P [ns]	0.468	0.599	0.613	0.628	0.645	0.048
σ_P [ns]	1.387	0.056	0.053	0.04	0.052	
τ_M [ns]	0.501	0.589	0.602	0.631	0.645	0.046
σ_M [ns]	1.499	0.055	0.055	0.072	0.071	
Single *p*-Values	0.063	0.034	0.021	0.027	0	

For statistical evaluation, the *p*-value analysis for all phase- and demodulation-lifetimes was performed. The overall *p*-value (Table 1, last column) is the statistical proof that, for the different frequencies, the same lifetime is detected, while the single *p*-value (Table 1 bottom row) represents the consistent results between the phase- and demodulation-lifetimes. Due to the obviously too-large standard deviation for the frequency of 1 MHz, the corresponding raw data were left out for the calculation. This resulted in an overall *p*-value of 0.048 for the phase lifetime and 0.046 for the demodulation lifetime in a frequency range of 10 to 40 MHz. For the TD, a fluorescence lifetime of 0.633 ns was obtained for a frequency of 80 MHz. This value coincides with those from the FD.

The experimental results imply that the lifetime is not dependent on the modulation frequency as long as the repetition rate of the pulsed laser in TD or the modulation frequency of the laser in FD are adequately chosen.

2.1.3. Integration Time

The time period for how long the fluorescence signal is collected is given by the exposure time of the camera, whereas in TD, it is the time interval over which the laser focus remains at each sample point to gather sufficient events for building the histogram. For short integration times down to 0.3 s, only FD measurements are presented, whereas for longer times, TD measurements only up to 30 s are shown. Here, the expectation would be that the longer the integration time, the more accurate the determination of the lifetime due to increased statistics. This is confirmed for TD but not in the same manner for FD, as can be clearly seen in Figure 1D, with increasing error bars for FD measurements. Here, a saturation effect caused by the limited dynamic range of the camera plays an important role, which will be discussed later in detail.

Again, the *p*-value is used to make statistical statements. Here, the raw data at 300 ms exposure time in the FD were not taken into account due to the high standard deviation caused by pixel saturation. For the other measurements, the overall *p*-value was 0.010 (phase lifetime) and 0.038 (modulation lifetime). The *p*-value of the TD analysis was 0.024 and thus statistically significant. The comparison of the FD measurement series and their individual lifetime components (single *p*-values) can be seen in Table 2.

On the one hand, the longer the integration time in TD, the more events are collected, resulting in more accurate lifetimes due to better statistics. On the other hand, for the camera-based lifetime measurement, the exposure time can not be extended to arbitrary values because pixel saturation counteracts the improvement in accuracy due to the limited dynamic range of 14 bit in our case. If the exposure time is too long, the saturation distorts the measurement and causes larger error bars. To circumvent the saturation effect, several camera images with shorter integration times should be accumulated to improve the accuracy more effectively.

Table 2. Single and overall *p*-values for each exposure time for the phase- and demodulation-lifetime of the FD setup, as well as the exposure time of the TD setup for measurements on rose bengal.

			Integration Time (TD)				
	5 s	10 s	20 s	30 s			Overall *p*-Value
τ [ns]	0.842	0.859	0.875	0.856			0.024
σ [ns]	0.020	0.015	0.013	0.011			
			Exposure Time (FD)				
	5 ms	10 ms	50 ms	100 ms	200 ms	300 ms	Overall *p*-Value
τ_P [ns]	0.852	0.897	0.852	0.847	0.832	0.495	0.010
σ_P [ns]	0.181	0.166	0.180	0.348	0.368	1.636	
τ_M [ns]	0.832	0.895	0.842	0.877	0.904	0.595	0.038
σ_M [ns]	0.190	0.169	0.200	0.377	0.480	1.712	
Single *p*-Values	0.001	0.001	0.020	0.042	0.049	0.053	

2.2. *Sample Dependent Parameter*

Because the setup-dependent parameters have shown consistent results for TD and FD, the parameters regarding the fluorophore itself and its microenvironment were investigated with the following setup parameters fixed: FD excitation of 405 nm and 40 MHz, while for TD, the excitation wavelength was set to 780 nm (2P excitation at 390 nm) and 80 MHz. The values for the exposure or integration time are set individually depending on the fluorescence intensity. For all coming figures, only the phase lifetime is presented because up to this point, the consistency of phase- and demodulation-lifetimes is convincingly demonstrated. In addition to the *p*-values for the fluorophore-dependent parameters, *t*-tests were carried out to check if linear correlations are statistically significant [17].

2.2.1. Concentration

Although fluorescence is an intra-molecule relaxation process, some research groups were able to experimentally demonstrate a dependency on the molecular concentration, which suggests an inter-molecular influence. From moderate to high concentrations, an increase in the lifetime [12] was noticeable, while for even higher concentrations, drastic lifetime reductions occurred [6]. Inter-molecular interactions may cause these effects, which are, for example, re-absorption, self-quenching or other energy transfer mechanisms. The spectral overlap between the absorption and emission spectrum determines how strongly these mechanisms have an impact on the measured lifetime.

For the investigation of these effects, two dyes were chosen, which were measured in a concentration range starting from 10^{-1} M down to 10^{-5} M in the TD and the FD. The first used dye is lucifer yellow dissolved in ethanol, which has a small spectral overlap of its excitation and emission spectra. Fluorescein (dissolved in water) is used as the counterpart, which has a very large spectral overlap and should show a strong lifetime change in concentration-dependent measurements (Figure 2A).

As can be derived from Table 3, the series of measurements for lucifer yellow exhibits an approximately constant lifetime of around (10.289 ± 0.371) ns for the FD over the concentration range of 10^{-5} to 10^{-4} M. The TD shows an increase from 8.837 to 11.476 ns presumably due to a re-absorption process. For high concentrations beginning from 10^{-3} M, a slight decrease in TD and FD lifetime measurements can be detected that may result from energy transfer mechanisms. For a concentration of 10^{-1} M, the FD-results showed an average lifetime of 9.685 ns. The small effects of re-absorption and energy transfer for lucifer yellow can be seen in its spectrum in Figure 2B. Within the limits of the measurement accuracy, an agreement of the TD with the FD method could be shown. Furthermore, the *p*-values (Table 3) imply congruous experimental data. The phasor plot illustrates the consistent lifetime, which is characterized by the point cloud along the semicircle (Figure 2).

Table 3. TD and FD fluorescence lifetimes and p-values obtained for the concentration series for lucifer yellow and fluorescein.

			Lucifer Yellow		
Concentration	**0.00001 M**	**0.0001 M**	**0.001 M**	**0.01 M**	**0.1 M**
τ_{TD} [ns]	8.837	11.476	11.087	10.265	9.907
σ_{TD} [ns]	1.027	0.565	0.221	0.053	0.049
τ_{FD} [ns]	10.532	10.646	10.294	10.288	9.685
σ_{FD} [ns]	1.020	0.647	0.506	0.267	0.298
Single p-Value	0.052	0.499	0.045	0.042	0.049
			Fluorescein		
Concentration	**0.00001 M**	**0.0001 M**	**0.001 M**	**0.01 M**	**0.1 M**
τ_{TD} [ns]	4.015	4.049	4.212	3.176	0.209
σ_{TD} [ns]	0.064	0.025	0.018	0.028	0.002
τ_{FD} [ns]	3.356	3.623	4.336	3.520	0.380
σ_{FD} [ns]	0.203	0.115	0.075	0.045	0.106
Single p-Value	0.062	0.051	0.034	0.021	0.039

The second investigated fluorophore is fluorescein, whose spectral overlap between the absorption and emission spectra is large compared to lucifer yellow, as can be seen in Figure 2B. The TD-derived lifetimes show an increase in concentrations of 10^{-5} M to 10^{-3} M, mainly through the re-absorption process (Table 3). If the concentration is increased further (up to 10^{-1} M), a strong reduction in the lifetime to a value of 0.209 ns can be detected caused by self-quenching effects [18]. Compared to FD, the rise of the lifetime for concentrations of 10^{-5} M to 10^{-3} M is larger and possible due to the different absorption processes. The self-quenching caused a reduction to a value of 0.380 ns for 10^{-1} M, which is illustrated in Figure 2B.

The visual representation of the lifetimes by the phasor plot illustrates the self-quenching effect, which is characterized by an elongated point cloud distribution along the universal circle. By comparing the phasor plot of the TD and the FD, a slightly different distribution can be observed, which is also obvious by looking at the p-values in Table 3.

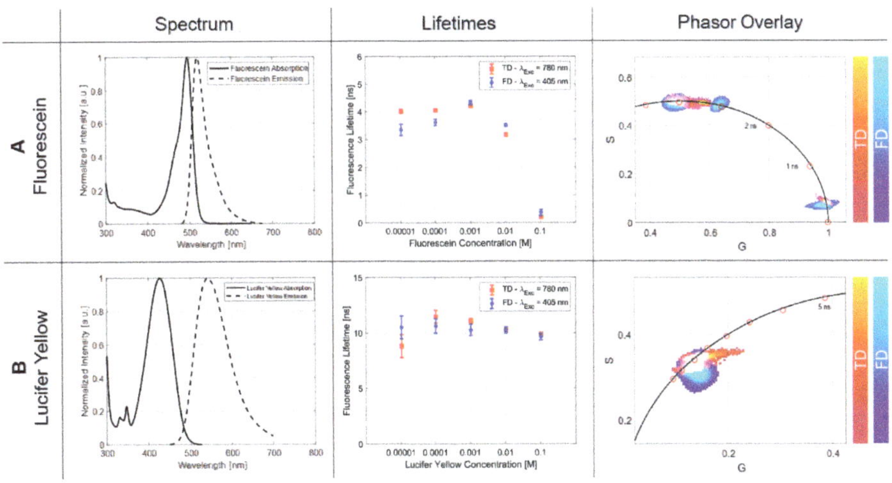

Figure 2. In this figure, the spectral properties of the two fluorophores used, fluorescein and lucifer yellow, are shown in (**A**) to check if the concentration could lead to an influencing parameter while measuring lifetime values, the spectral overlap must be observed. Fluorescein shows a relatively big spectral overlap compared to lucifer yellow. With the graphs in (**B**), the lifetimes of the TD are shown in red and the ones of the FD are in blue. For fluorescein, the lifetime change for high concentrations can be seen clearly, whereas the results of lucifer yellow stay nearly the same. The last column presents this relation in the phasor plot for the TD, as well as for the FD data.

2.2.2. Solvent Polarity

The fluorescence lifetime of a fluorophore changes based on the surrounding medium due to induced conformational changes [19]. For a variety of applications, these media must be chosen accordingly.

The extent of the solvent influence was measured by using both FLIM setups on rose bengal. Figure 3A visualizes the results for the fluorescence lifetimes in relation to the used media. The main influencing factor of the solvent is its polarity, therefore, a direct correlation can be seen (Figure 3B). An increase in the polarity results in a reduction in the lifetime and vice versa.

The graphical presentation (Figure 3C) of the lifetimes measured by TD as well as FD shows consistent results within the standard deviation. The lifetimes with the respective *p*-values for each solvent are given in Table 4.

Table 4. Calculated *p*-values, detected lifetimes and relative polarity values for different solvents.

Figure Declaration	Solvent	Relative Polarity	τ_{TD} [ns]	σ_{TD} [ns]	τ_{FD} [ns]	σ_{FD} [ns]	Single *p*-Value
a	Water	1	0.127	0.014	0.093	0.052	0.050
b	Methanol	0.762	0.645	0.012	0.605	0.052	0.045
c	Ethanol	0.654	0.852	0.011	0.883	0.062	0.039
d	Aceton	0.355	2.725	0.014	2.819	0.214	0.021

The phasor plot shows the individual point clouds of the different measurements for the corresponding media. Due to the linear relationship between the lifetime and the polarity change in the solvent, it can be stated that the polarity increases along the semicircle.

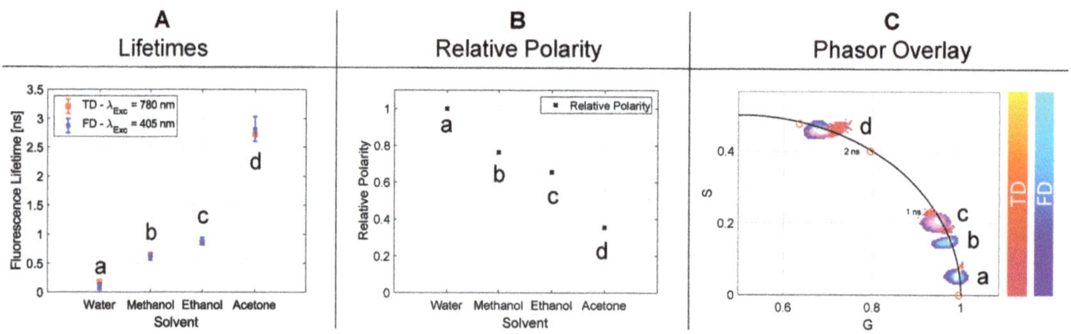

Figure 3. In section (**A**), the relative polarity of each individual solvent is illustrated. Compared to the obtained lifetimes in (**B**), an indirectly proportional behavior to the polarity is noticeable. Section (**C**) represents the overlayed phasor plots of the TD and FD data.

2.2.3. Temperature

Although there are well-known temperature influences on the fluorescence lifetime, which are treated later in the discussion section, this investigation uses fluorescein as an example to show the opposite. Measuring fluorescein (10^{-3} M in ethanol), the lifetime curve for different temperatures in the range of 21–42 °C is shown in Figure 4A and all individual lifetimes in Table 5. Using linear regression, a slope of 0.0021 ns/°C in the TD and a slope of 0.0015 ns/°C for the FD technique can be obtained. For most FLIM setups, accuracy in the two- to three-digit picosecond range is possible. The fluorescence lifetime change resulting from the temperature variation for fluorescein can thus be neglected. Furthermore, the standard deviation of each measurement showed no strong variation.

A *t*-test was used to check the consistency of the slopes determined and their intercept for TD and FD. A significance value of $T_{Slope} = 0.00313$ was determined for the slope and $T_{Axis} = 0.00689$ for the axis intercept. Both values are thus below the significance level of 0.05 and are consistent.

The phasor plot analysis illustrates the mono-exponential behavior of the temperature dependency measurement due to the point cloud on the semicircle (Figure 4A). Since the values change only by a few picoseconds per measurement, all measurements together result in a uniform distribution.

Table 5. P-values and obtained lifetimes for the temperature series of fluorescein.

	Fluorescein								
Temperature [°C]	**23**	**25**	**27**	**29**	**31**	**33**	**35**	**37**	**39**
τ_{TD} [ns]	3.857	3.813	3.830	3.769	3.789	3.783	3.781	3.796	3.763
σ_{TD} [ns]	0.110	0.107	0.113	0.112	0.115	0.112	0.111	0.111	0.0122
τ_{FD} [ns]	3.859	3.833	3.834	3.832	3.828	3.823	3.815	3.816	3.816
σ_{FD} [ns]	0.110	0.110	0.120	0.130	0.130	0.120	0.110	0.120	0.120
Single *p*-Values	0.016	0.038	0.018	0.041	0.038	0.038	0.034	0.021	0.040

2.2.4. pH-Value

The pH-value of a fluorophore solution is of central importance during a fluorescence lifetime measurement. Based on the chemical composition of the dye, different spectral property changes can be obtained related to the pH-value. Therefore, a pH series of a 10^{-4} M concentration of fluorescein in water was measured in the TD and FD. The adjusted pH-values cover a range from 3 to 12. The detected lifetimes show a plateau for strongly acidic and alkaline pH-ranges, whereas the transition between them shows an approximately linear dependence. The resulting sigmoid function showed an overall lifetime change of ca. 1.5 ns for the adjusted pH range (Figure 4B and Table 6). The linear section from pH 6 to 9 was fitted and a *t*-test was performed. The slope of the TD was (0.336 ± 0.022) ns and for the FD (0.348 ± 0.028) ns. The *t*-test for the slope has a value of $T_{Slope} = 0.039$ and for the axis intercept $T_{Axis} = 0.021$.

Table 6. Measured lifetimes for the pH-series in the TD and FD for fluorescein.

	Fluorescein								
pH-Value	**3**	**4.5**	**7.5**	**7.8**	**8**	**8.2**	**9.3**	**11.2**	**12**
τ_{TD} [ns]	3.107	3.330	3.944	3.945	4.144	4.154	4.345	4.424	4.514
σ_{TD} [ns]	0.084	0.084	0.076	0.053	0.038	0.038	0.038	0.039	0.042
pH-Value	**3**	**4.5**	**6.5**	**7**	**7.5**	**8**	**8.5**	**10.5**	**12**
τ_{FD} [ns]	3.104	3.303	3.565	3.832	3.885	3.985	4.327	4.445	4.615
σ_{FD} [ns]	0.108	0.127	0.146	0.189	0.134	0.162	0.226	0.168	0.155

Furthermore, for the pH dependence, comparable results within the respective standard deviations could be retrieved for TD and FD. With the aid of the phasor plot, the lifetime change can be easily tracked and, with the appropriate calibration, also mapped to the corresponding pH level of the microenvironment (Figure 4B).

2.2.5. Quencher

Similar to the fluorescence lifetime dependency on the microenvironment (solvent polarity, temperature, pH), there are also other possibilities to intentionally quench a dye and thus obtain a shortening of the lifetime. For example, this can be utilized to mimic pathologically relevant coenzymes such as nicotinamide adenine dinucleotide (NADH) or flavin adenine dinucleotide (FAD) in optical phantoms used for clinical studies [20]. Here, coumarin 1 and coumarin 6 have similar spectra to the coenzymes but different lifetimes. To fix this issue and to fine-tune the desired lifetime of the phantom, quenching materials such as 4-hydroxy TEMPO are applied.

For the demonstration of the artificial quenching effect, this system is used as well. The starting solution was a coumarin 1 and a coumarin 6 mixture in ethanol with a concentration of $2 \cdot 10^{-4}$ M. The initial measurements showed a consistent lifetime for coumarin 1 of $\tau_{TD} = (3.239 \pm 0.017)$ ns in TD and $\tau_{FD} = (3.268 \pm 0.097)$ ns in FD (Figure 5A), and for coumarin 6 of $\tau_{TD} = (2.663 \pm 0.013)$ ns and $\tau_{FD} = (2.615 \pm 0.118)$ ns, respectively

(Figure 5B). Shortened lifetimes could be produced by adding different concentrations of 4-hydroxy TEMPO, which are shown in Figure 5.

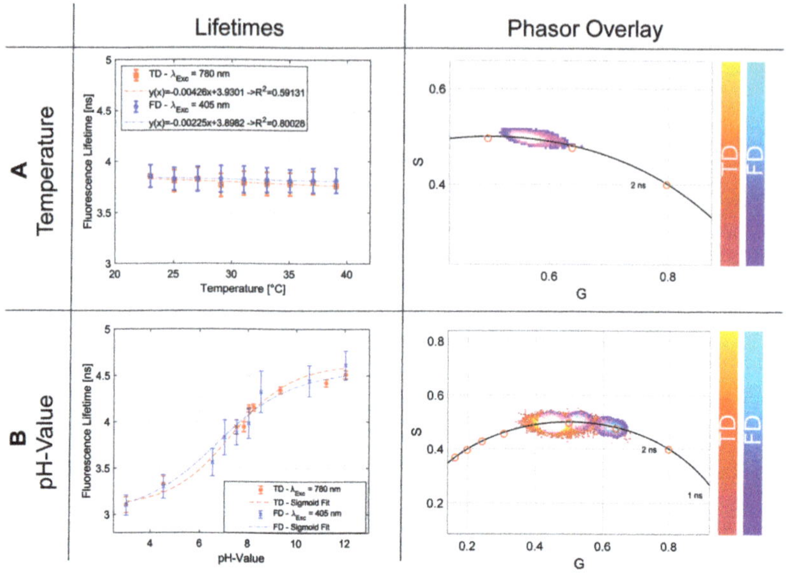

Figure 4. (**A**) shows the obtained results for the temperature dependency between 22–39 °C and the slope of the fitted linear regression for fluorescein in ethanol. The overall lifetimes for the TD and FD are also represented in the phasor plot where a single point cloud indicates no significant lifetime change. In (**B**) the pH dependency for values between 3 and 12 are shown, which follow a sigmoid function.

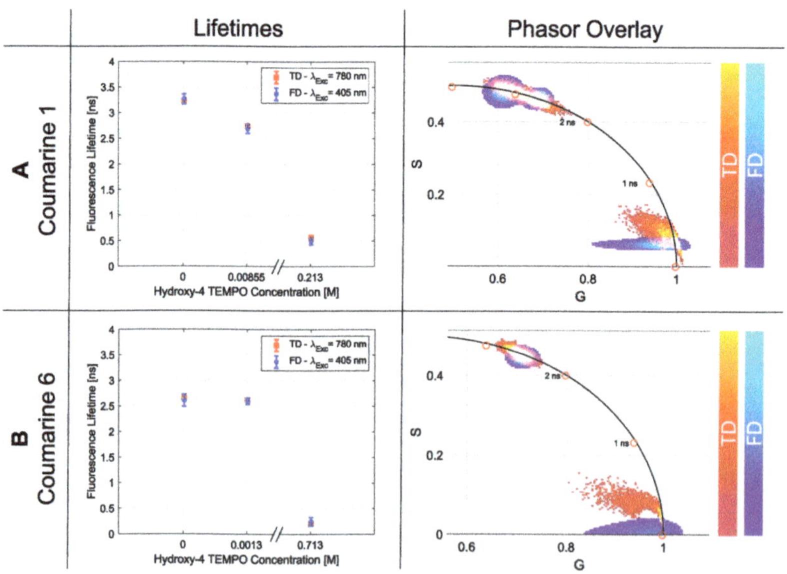

Figure 5. Results of the fluorescence lifetimes and phasor plot for the quenching series of coumarine 1 are illustrated in (**A**,**B**) for coumarine 6. With the help of 4-hydroxy TEMPO, it is possible to decrease the fluorescence lifetime to a specific value.

The phasor plot illustrates the shortening of the lifetime. The obtained values together with the p-values are listed in Table 7.

Table 7. Results for the quenching experiments for coumarine 1 and coumarine 6.

	Coumarine 1			Coumarine 6		
4-hydroxy TEMPO	**0 M**	**0.00855 M**	**0.234 M**	**0 M**	**0.0013 M**	**0.713 M**
τ_{TD} [ns]	3.239	2.745	0.569	2.663	2.617	0.209
σ_{TD} [ns]	0.017	0.016	0.011	0.013	0.014	0.002
τ_{FD} [ns]	3.268	2.689	0.492	2.615	2.599	0.25
σ_{FD} [ns]	0.097	0.086	0.063	0.118	0.060	0.082
Single p-Value	0.031	0.041	0.050	0.022	0.044	0.061

3. Discussion

By comparing TD- with FD-FLIM data of various fluorophores in different microenvironmental conditions and measured with different excitation and detection schemes, a statistical proof could be given that both techniques deliver equal lifetime values and become visible when using the phasor plot evaluation.

3.1. Setup Dependent Parameter

Specifically, experiments were performed by changing system parameters such as one- and two-photon excitation at various wavelengths, detection of many photons with a camera or single photons with a point detector, and using different modulation frequencies and exposure times (FD-FLIM) or repetition rates and acquisition times (TD-FLIM), respectively. All results retrieved with either technique were consistent, as demonstrated by statistical analysis. This remarkable outcome is based on the FLIM-advantage, which is extensively discussed in Section 4.1 of this manuscript.

3.1.1. Wavelength

The results showed that the lifetime data are independent with respect to the wavelength and to the kind of absorption process (1P- or 2P-excitation). The Einstein model for a two-level system gives an accurate explanation for both observations. Here, the fluorescence intensity resembles the population in the excited state, whereas the lifetime corresponds to the inverse radiative transition rate described by the so-called Einstein coefficient [21]. In this picture, the absorption process leading to the population in the excited state depends not only on the transition rate but also on the intensity of the excitation light source and how well it matches the photo-physical properties of the molecule. This is worth considering because, for excitation with two photons, one would expect to double the wavelength of the laser to reach the same excitation state as with one photon. However, it was demonstrated already in 1995 by Watt W. Webb and Chris Xu that 2P absorption spectra are most likely blue shifted to the 1P spectra because the cross sections of excited singlet states are not the same for 1P and 2P excitation [22]. This leads to different excitation probabilities for the various wavelengths and, thus, different fluorescence intensities. Interestingly, the radiative transition to the ground state seems to always start from the same excited singlet state. Therefore, the transition rate responsible for the lifetime is not affected by the kind of excitation process giving rise to the FLIM-advantage (see Section 4.1).

3.1.2. Modulation Frequency

For both techniques, the modulation frequency defines the range in which the fluorescence lifetime can be measured. While the FD laser can be modulated internally, the TD laser is limited by its repetition rate. Therefore, only the FD-FLIM setup was tested for different modulation frequencies showing a consistent lifetime in the range of 10–40 MHz. While the modulation frequency decreases, a slightly larger standard deviation can be noticed that results from the tangent function in Equation (10).

3.1.3. Integration Time

For different integration times in both techniques, an overall consistent lifetime for rose bengal in ethanol could be achieved (Figure 1D and Table 2). Whereas the TD-FLIM setup benefits from a longer integration time and statistics, the FD-FLIM setup suffers from overexposure and pixel saturation. This context can also be seen in the standard deviations listed in Table 2.

3.2. Sample Dependent Parameter

More interestingly are the sample-dependent parameters because they are responsible for differing lifetimes if comparing experiments performed under circumstances that are not always openly reported. One important goal of this extensive investigation was to demonstrate as many influencing parameters as possible to raise awareness among FLIM microscopists for taking full control over their experiment.

3.2.1. Concentration

The fact that the fluorescence lifetime is subject to a statistically independent process leads to the assumption that the concentration of a dye has no influence on this parameter. Even the fluorescence of only a small number of fluorophores is sufficient to record a decay curve if the integration/exposure time is set accordingly long.

Here, several effects cause variations in lifetime, as can be seen in Table 3 and Figure 2. For example, the apparently longer lifetimes can be explained by the effect of re-absorption: one molecule relaxes back to the ground state emitting a photon that is subsequently re-absorbed by another molecule lifting it to the excited state [23]. In contrast to the true lifetime τ, which is not affected, the detected lifetime τ_{Det} is prolonged because many photons may reach the detector only after this secondary step. Parameters that influence this process are the quantum yield Φ, as well as the concentration-dependent spectral overlap $J(\lambda,C)$ of the fluorophore (Equation (1)).

$$\tau_{Det} = \frac{\tau}{1 - J(\lambda, C) \cdot \Phi} \quad (1)$$

The difference in the 1P- to 2P-absorption spectra, absorption cross section and, further, the quantum yield suggest different τ_{Det} for 1P- and 2P-absorptions. This translates in this experiment to FD for 1P- and to TD for 2P-absorption.

Further, energy transfer mechanisms such as the self-quench effect or the Förster Resonance Energy Transfer (FRET) are responsible for the lifetime decrease in highly concentrated samples [6,24]. Here, a more efficient energy transfer occurs, where, for example, an excited molecule partially transfers its energy to a molecule in the ground state. This leads to the fact that only very fast transitions can undergo the radiative relaxation path of emission, which leads to a shortened detectable lifetime [25]. Again, the probability of these effects is based on several spectral and geometrical properties of the sample. The most important factor is the spectral overlap $J(\lambda, C)$ between the absorption $E(\lambda)$ and the emission $F(\lambda)$ spectrum (Equation (2)). This context determines whether a concentration-dependent lifetime change can be detected experimentally or not. The higher the overlap, the larger the possible self-quench effect.

$$J(\lambda, C) = \int E(\lambda) \cdot F(\lambda) \cdot \lambda^4 \, d\lambda \quad (2)$$

Fortunately, consistent lifetimes are most likely for the widely used working dilutions of around 10^{-6} M. However, for time-resolved measurements with high or changing concentrations, these effects have to be kept in mind.

3.2.2. Solvent Polarity

In Section 2.2.2, the fluorescent lifetimes of rose bengal were related to the solvent used. The graphic representation in Figure 3A showed that the increasing lifetime could be correlated to a decrease in the solvent polarity (Figure 3B). The decisive parameter for this behavior will probably be the so-called solvatochromism [26]. It describes the spectral changes of a fluorophore while interacting with solvent molecules. A decreasing solvent polarity is understood as a negative solvatochromism, which is expressed by a higher energy difference between the ground and excited state. This change causes a hypsochromic shift, thus the blue shift of the absorption spectrum. The contrary case is a bathochromic shift and, therefore, has a lower energy difference. In addition to these spectral properties, parameters such as dipole moment, dielectric constant, refractive index and others are also changed, which are related to the observed fluorescence lifetime [27,28]. Again, time-resolved microscopy can be used to determine the environment more precisely and, in this case, to make statements about the polarity of a solvent [29].

3.2.3. Temperature

When changing the temperature, the non-radiative rate k_{NR}, also known as the quenching rate, is affected. With the help of the Stokes–Einstein–Debye relationship, the quenching rate can be expressed with a dependence on the temperature and the viscosity [6,30]. Looking at Equation (3), it is obvious that an increase in the temperature T is related to a decrease in the detected fluorescence lifetime τ_{Det}:

$$\tau_{Det} = \frac{1}{\frac{1}{\tau} + \frac{k_B T}{4\pi r^3 \eta}} \tag{3}$$

In Equation (3), next to the Boltzmann constant k_B, the radius r and the viscosity η are also included. How strongly the temperature affects the detected lifetime τ_{Det} depends on the structure of the fluorophore. While some of them only show a small change, such as fluorescein in this case, other specially designed dyes show a strong and well-defined lifetime change that is used as temperature sensors [31,32].

3.2.4. pH-Value

The pH-induced fluorescence lifetime change was also demonstrated by the group of Seungrag Lee, showing its importance in cancer research in modern days [33]. A chemical restructuring due to the additional uptake of H^+ ions with increasing pH-value results in altered photo-physical properties. In addition to this, there are also specially designed dyes that use the FRET mechanism to detect pH level changes by time-resolved methods [34].

3.2.5. Quencher

With quenching reagents such as 4-hydroxy TEMPO, it is possible to actively shorten the fluorescence lifetime due to a quenching process. The two most common quenching mechanisms are the following: The first one is the so-called static quenching process. In this situation, the quenching molecule, together with the fluorophore, creates a non-fluorescent complex. The outcome is a reduced concentration of fluorescent molecules leading to a weaker intensity. The radiative transition rate remains the same, and therefore, no fluorescence lifetime change can be noticed [35]. The second mechanism is the dynamic quenching effect. In this situation, a collision between the quenching molecule and the excited fluorophore takes place. Due to this collision, the fluorophore converts to the ground state without emitting a photon and therefore shortening the fluorescence lifetime [35]. The concentration of the quencher determines the extent to which the lifetime is shortened. This makes it possible to set almost any time constant desired. For different tasks, it is therefore possible to mimic defined biological components, for example, NADH and FAD by coumarine 1 and coumarine 6 [20]. With the help of the Stern-Volmer representation, it is possible to distinguish wherever a dynamic or static quenching process takes place [35]. For the sake of completeness, it has to be mentioned that next to the two above-mentioned quenching mechanisms, the FRET process [36] and proton/electron transfer reactions could possibly create a shortened lifetime as well [37].

3.3. Technique Comparison

Nevertheless, there are some differences between both techniques that are necessary to mention here. First, the image acquisition is either serial in TD or parallel in FD. In TD, the laser is focused and scanned over the sample. This takes more time than imaging the entire field of view at once with a camera, such as in FD. Therefore, the FD approach is faster compared to TD. On the other hand, two-photon excitation is only possible with focused light because of the high intensities needed for the nonlinear optical process to occur. The associated longer wavelengths lie in the near infrared, which is related to an optical window for biological tissue [38]. This enables deeper optical penetration depths by the excitation laser light in highly scattering media. Another advantage of two-photon over one-photon excitation is the optical sectioning capability, i.e., the sample can be virtually sliced like in confocal microscopy [7,39]. In conclusion, the TD technique plays out its potential the most in multiphoton microscopy, where an expensive ultrashort pulsed laser is required. On the other hand, if acquisition speed is of utmost importance, the FD technique clearly wins the race.

In addition to the acquisition, the data post processing and representation is complementary. While for the TD, no prior knowledge of the detected lifetime is needed, for the FD, more accurate values can be achieved if the modulation frequency of the excitation light is adjusted accordingly. Further, the extraction of the lifetime in the TD is complicated by the fact that dependent on the sample system, a mono- or multi-exponential fit has to be calculated. However, the interpretation is relatively easy, while for each exponential fit, a single lifetime and amplitude are calculated. In the FD, the raw data result in two lifetime components that are difficult to understand and illustrate in a concordant form.

With the help of the phasor plot, a graphical representation can be created that overcomes the above-mentioned problems. While the fit algorithm in the TD is omitted, the two components in the FD are compressed to a single vector. The phasor plot also helps to understand the lifetime distribution in an intuitive way. By comparing these graphs in biological contexts, changes in fluorophore ratios, quenching effects or other events can be easily observed and interpreted [40,41].

4. Materials and Methods

4.1. Theoretical Background

In order to understand the FLIM-advantage, the underlying quantum process is described briefly. The excitation photon elevates the sample molecules to an excited electronic state where they remain for a stochastic time period before they return to the electronic ground state either by radiative relaxation (k_R) under emission of one photon per molecule or by non-radiative relaxation (k_{NR}) through one or more so-called dark channels. The overall decay rate k is given by Equation (4):

$$k = k_R + k_{NR} \tag{4}$$

Using this definition, the decay of the population N of the excited state is governed by the following differential equation:

$$\frac{dN}{dt} = -k \cdot N \tag{5}$$

The solution to this is a single-exponential function:

$$N(t) = N_0 \cdot \exp(-k \cdot t) \tag{6}$$

The initial population N_0 is dependent on the excitation and, therefore, susceptible to the kind of excitation (1P or 2P) and the efficiency thereof, which is, besides other influences, defined by the excitation wavelength. The signal intensity corresponds to the number of emitted photons and is thus the negative rate described in Equation (5), that is $k \cdot N(t)$. The product reflects not only the decay rate k but also the amount of excited molecules

N—the better the excitation, the stronger the signal. However, if bleaching occurs, the initial population N_0 diminishes, and the signal drops accordingly.

On the other hand, FLIM only determines the decay rate k defined in Equation (4) and does not share this flaw of intensity measurements. However, it should be noted here that, though FLIM detects photons associated with the radiative channel k_R, it also reflects the non-radiative channels k_{NR} and, therefore, the overall rate k. Therefore, one must distinguish the detected lifetime $\tau_{Det} = \frac{1}{k}$ from the so-called true lifetime $\tau = \frac{1}{k_R}$. This is the reason for the differing lifetimes reported in the literature: experiments performed under different circumstances incorporate different dark channels that lead to the above-mentioned discrepancies in τ_{Det}. What makes it even more complicated is the fact that for a variety of experimental scenarios, Equation (6) may be written as a series of exponential functions accounting for several fluorescent components present in the sample. For convenience, this series shall be written in terms of the signal intensity $I(t)$ instead of the population $N(t)$:

$$I(t) = \sum_{n=1}^{N} I_n \cdot \exp\left(-\frac{t}{\tau_n}\right) \tag{7}$$

4.1.1. Time-Domain (TD)

In TD-FLIM, a time-correlated single photon counting (TCSPC) device measures the stochastic time period between excitation and emission of the fluorescent molecule, which determines the arrival time of each measured photon. Although advanced versions such as rapidFLIM [42] increased the detection rate considerably, the latter is still limited because the detection of single photons is a prerequisite of this technique. For data analysis, sufficient events are required for making the subsequent histogram of the arrival times statistically significant. Furthermore, the origin of each photon is localized by scanning the sample point-wise in a serial fashion. Therefore, statistics based on single-photon detection combined with serial point-scanning of the sample make the acquisition time long compared to frequency domain FLIM.

In this experiment, all samples were made with dye solutions containing neither structures nor different components to increase the validity of the results. Therefore, the measured histograms were fitted with a single-exponential reconvolution fit (Equation (8)), where the instrument response function (IRF) is convolved ($\otimes$) with the intensity decay curve including a background B.

$$FIT(t) = IRF \otimes I_0 \cdot \exp\left(-\frac{t}{\tau}\right) + B \tag{8}$$

The samples were scanned over a range of 200 times 200 pixels to average over the many molecules of the solutions. As expected, the recorded lifetimes of all pixels showed a statistically normal distribution. For calculating the expectation value and the standard deviation of the measurements, a Gaussian fit function was used.

4.1.2. Frequency Domain (FD)

In FD-FLIM, consecutive widefield images are taken in a rapid sequence with a camera to extract the lifetime for each pixel in parallel. Using specially developed sensors, these sequences can be recorded very fast and efficiently in a homodyne or heterodyne fashion [11]. Compared to conventional image sensors, each pixel of this sensor has two charge collection sites, including a switch that enables fast recording of two consecutive images. The switch is synchronized to the modulation of the excitation laser so that each image corresponds to half a period of the modulation. Please see [43,44] for a detailed description.

The excitation $E(t)$ of the sample is performed using a sinusoidal modulated light source and can be expressed as follows:

$$E(t) = A_{Exc} + B_{Exc} \cdot \sin(2\pi \cdot f_{Mod} \cdot t + \phi_{Exc}) \tag{9}$$

The parameter A_{Exc} describes the offset, B_{Exc} the amplitude and ϕ_{Exc} the phase of the excitation. With previous knowledge of the lifetime range τ to be measured, one can choose the optimal modulation frequency f_{Mod} in the experiment according to Equation (10):

$$f_{Mod} = \frac{1}{2\pi \cdot \tau} \tag{10}$$

In analogy to a forced oscillation, the generated fluorescence $F(t)$ is delayed, as well as demodulated (Equation (11)):

$$F(t) = C_{Em} + D_{Em} \cdot \sin(2\pi \cdot f_{Mod} \cdot t + \phi_{Em}) \tag{11}$$

The modulation depths M of the excitation $E(t)$, as well as of the fluorescence emission $F(t)$, are defined in Equations (12) and (13):

$$M_{Exc} = \frac{B_{Exc}}{A_{Exc}} \tag{12}$$

$$M_{Em} = \frac{D_{Em}}{C_{Em}} \tag{13}$$

The transfer function between excitation $E(t)$ and fluorescence $F(t)$ shows a demodulation (Equation (14)) and a phase shift (Equation (15)), which are directly related to the photo-physical properties of the sample.

$$M = \frac{M_{Em}}{M_{Exc}} \tag{14}$$

$$\phi = \phi_{Em} - \phi_{Exc} \tag{15}$$

Finally, with definitions made in Equations (14) and (15), two lifetimes τ_M and τ_P can be derived in the FD approach:

$$\tau_M = \frac{\sqrt{1-M^2}}{2\pi \cdot f_{Mod} \cdot M} \tag{16}$$

$$\tau_P = \frac{\tan(\phi)}{2\pi \cdot f_{Mod}} \tag{17}$$

For single-exponential systems, both lifetimes τ_M and τ_P are the same. However, if the sample contains several fluorescent components with lifetimes lying in the range between τ_1 and τ_2, the demodulation lifetime τ_M differs from the phase lifetime τ_P [45]. Here, no direct assignment to the individual lifetimes of the components can be made. In this case, the single lifetime τ in Equation (10) should be substituted with the expression $\sqrt{\tau_1 \cdot \tau_2}$ for choosing the best possible modulation frequency when performing the experiment.

The FD camera (pco.FLIM, pco AG, Germany) used in the experiments detects the two components τ_M and τ_P with a homodyne detection scheme [43,46]. In contrast to TD-FLIM, an intensity-modulated diode laser was used for widefield illumination that excited the samples via one-photon absorption. Analogous to the TD evaluation, all detected pixel lifetimes are collected in a histogram and evaluated using a normal distribution fit (Figure 6B).

4.1.3. Phasor Plot Approach

Assuming that the emitted fluorescence of a fluorophore has a linear time-invariant behavior, it is possible to measure the lifetime in the TD as well as in the FD [46] without loss of information. For both approaches, however, the lifetime is not measured directly but has to be extracted from the data by subsequent analysis. Especially in the TD, this often leads to erroneous results depending on the selected fit function. Using the FD technique instead, the statistical fit is omitted, but the fact that two lifetimes are calculated complicates the interpretation of the results if the sample contains not just one but several components.

The strength of the phasor plot lies in the consistent representation of lifetimes measured with either technique. Furthermore, the fit algorithm needed in TD becomes obsolete, and the ambiguity present in FD is avoided [47,48]. For this expedient representation, the respective pixel information is transformed into a vector with a defined magnitude and angle. Because the pixel information differs in FD- and TD-FLIM, the two vector components G and S are calculated in different ways. In the FD, the transformation is based on the demodulation $M_{x,y}$ and the phase shift $\phi_{x,y}$ for each pixel (x, y):

$$G_{FD}(x,y) = M_{x,y} \cdot \cos(\phi_{x,y}) \tag{18}$$

$$S_{FD}(x,y) = M_{x,y} \cdot \sin(\phi_{x,y}) \tag{19}$$

For data collected in the TD, the decay curve $I(t)$ for each pixel (x, y) must be Fourier-transformed and normalized, as described by Equations (20) and (21):

$$G_{TD}(x,y) = \frac{\int_0^T I_{x,y}(t) \cdot \cos(2\pi \cdot f_{Rep} \cdot t) \cdot dt}{\int_0^T I_{x,y}(t) \cdot dt} \tag{20}$$

$$S_{TD}(x,y) = \frac{\int_0^T I_{x,y}(t) \cdot \sin(2\pi \cdot f_{Rep} \cdot t) \cdot dt}{\int_0^T I_{x,y}(t) \cdot dt} \tag{21}$$

The tips of the vectors, when plotted, are located within a semicircle, also called the universal circle. This graphical representation allows a quick and detailed overview of the lifetime distribution. If the vector point ends on the universal circle, it represents a single-exponential lifetime. However, if the vector points are inside the circle, this indicates that multiple fluorescent components have been detected (Figure 6C). Using this plot, it is possible to graphically represent the correspondence of TD to FD data. When plotting the vectors of all pixels in a single phasor plot, one can check the accuracy of the measurement by comparing the sizes and densities of the respective clouds. With this representation, further analysis methods such as image segmentation can be used, where only a defined portion of the phasor plot is reproduced in the intensity image [49]. This eliminates the need for shape-based segmentations for some applications and could be a way to label images for artificial-intelligence-based algorithms [50]. Furthermore, an innovative approach to metabolic imaging could be achieved with the help of the phasor plot [51]. For example, in the field of bacteria research, different metabolic states induced by their environment could be differentiated by applying segmentation to the phasor plot representation [52].

For comparing the obtained results of lifetime data and their different influence factors, a self-made Matlab (R2020a, MathWorks, USA) program calculates the phasor plot from the raw data detected either by the TCSPC device or the FD camera. The phasor plot depiction is then represented according to an additive color overlay. For the FD- and TD-results, a separate color map was created, which generates a corresponding blended color when superimposed (here pink to white).

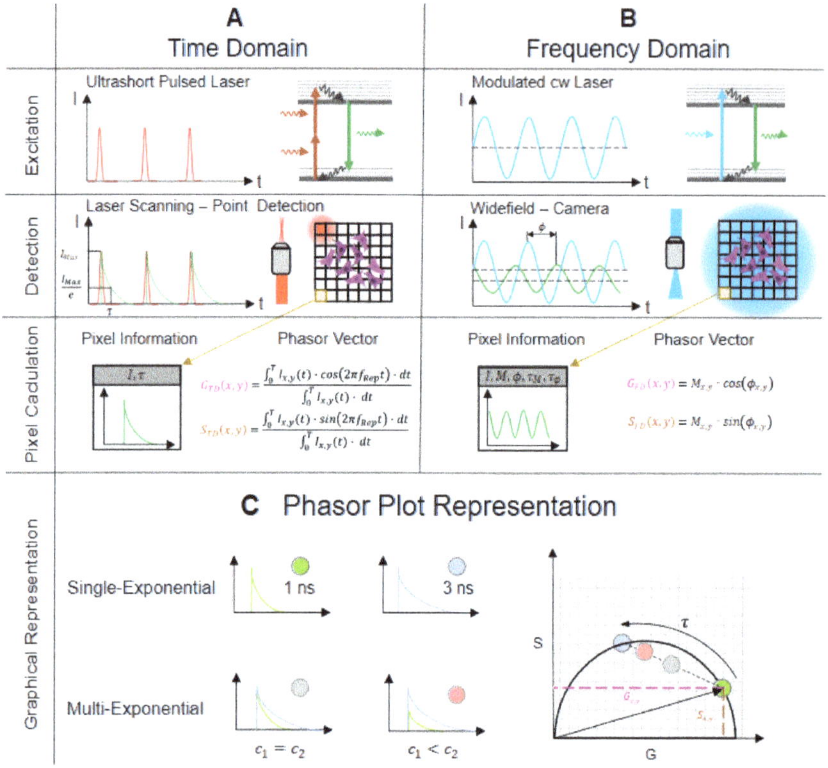

Figure 6. This figure shows both methods of obtaining the fluorescence lifetime in the (**A**) time-domain as well as in the frequency-domain (**B**). In the time-domain (**A**), a TCSPC device acts as a stopwatch and registers the time differences between the excitation (top row) and the arrival of the emitted fluorescence photon detected by a single photon detector (middle row). In the frequency-domain (**B**), the lifetime information is calculated by comparing the phase and amplitude between the modulated excitation (top row) and the modulated fluorescence of the sample (middle row). The obtained values in TD- and FD-FLIM (bottom row) are correlated by the Fourier transformation. (**C**) shows a way to represent both measurement results in a single plot, the so-called phasor plot. Each lifetime of a pixel is represented by a vector with the components G and S. Single exponential decays are located on the semicircle while multi-exponential decays are located within the semicircle. The position of the point cloud along the connection line between the two dyes indicates the concentration ratio c between them.

4.2. Experimental Setup

In order to obtain fluorescence lifetime data of different fluorophores, a custom-built two-photon excited fluorescence microscope (TPEFM) was used. The optical setup is built around a Nikon Eclipse Ti2 body, which was modified with various 3D-printed parts to make it suitable for a broad variety of multiphoton microscopy techniques as well as for the combination of TD- and FD-FLIM (Figure 7).

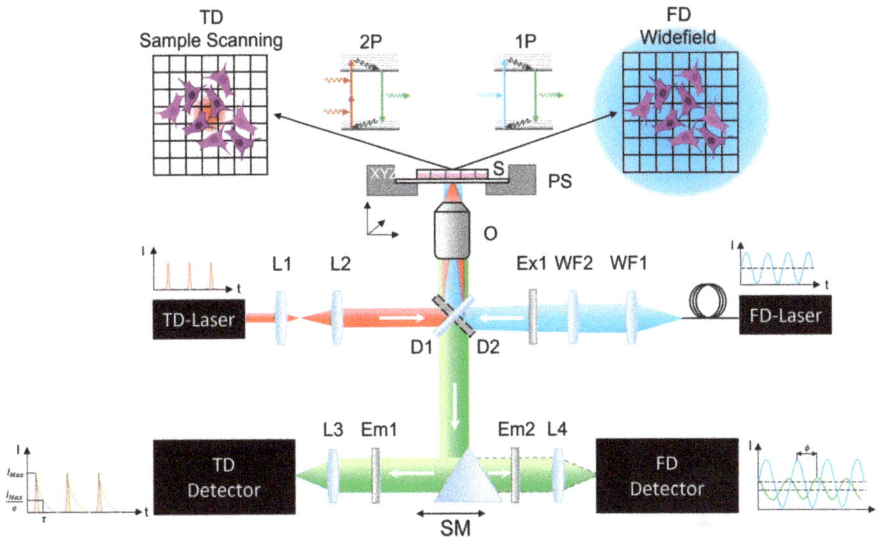

Figure 7. Schematic drawing of the optical setup. The excitation beam path for the TD setup is indicated in red. The fluorescence (green) for each pixel is detected with a single photon detector by switching the selection mirror **SM** to the right position. The widefield excitation beam path for the FD setup is shown in blue. Because one-photon excitation is used in FD and two-photon excitation is used in TD, the dichroic mirror **D1** differs from **D2**, thus excluding a simultaneous measurement. Instead, individual filter cubes, one containing the excitation filter **Ex1** and dichroic mirror **D1** and the other containing **D2**, are placed in the beam path depending on the measurement. L: lens, WF: widefield lens, Ex: excitation filter, D: dichroic mirror, S: sample, PS: piezo stage, Em: emission filter, SM: selection mirror, 1P: one-photon excitation, 2P: two-photon excitation, O: objective lens.

4.2.1. Time-Domain Experiment

Optical Setup: For the TD measurements, a sample scanning approach was realized (Figure 7). An ultrashort pulsed fiber laser (FemtoFiber Dichro Design, TOPTICA Photonics AG, Gräfelfing, Germany)—referred to as **TD Laser**—emitting pulses of 95 fs duration at two wavelengths of $\lambda_1 = 780$ nm and $\lambda_2 = 1034$ nm was used for excitation. Both pulse trains with a repetition rate of 80 MHz have a maximum laser power of 100 mW each. Next, a Kepler-telescope consisting of two lenses **L1** ($f_1 = 20$ mm) and **L2** ($f_2 = 100$ mm) with a 5× magnification was integrated. It was used to overfill the back aperture of the objective, thus focusing the light to the diffraction limit. The beam enters the microscope body through the rear input port and is reflected by a dichroic mirror **D1** (F38-749, AHF Analysentechnik, Tübingen, Germany) towards the objective. This optical component is located in a filter cube. The water immersion objective (CFI Apo MRD77200, Nikon, Japan) has a 20× magnification and a numerical aperture of 0.95. A piezo stage **PS** (Nano-LPS, Mad City Labs, Madison, WI, USA) moves the sample **S** along a defined path while the laser focus stays fixed. The generated fluorescence is then collected through the same objective and passes the dichroic mirror. A selection mirror **SM** reflects the fluorescence light either to the **TD Detector** or the **FD Camera**. For TD measurements, the fluorescence is filtered by a bandpass filter **Em1** (F39-653, AHF Analysentechnik, Germany) at (550 ± 25) nm before a lens **L3** ($f = 7.5$ mm) focuses it onto a single photon detector (MPD-APD 50, Micro Photon Devices Srl, Bolzano, Italy).

Electronic Signal Path: A TCSPC card (Multiharp 150, PicoQuant, Berlin, Germany) was used to detect the laser reference signal, as well as the signal of the single photon detector to measure the time difference between them. Care was taken that each detected photon was assigned correctly to the corresponding excitation laser pulse, which is recorded by a photodiode. For this purpose, the photodiode signal is fed into an electronic pulse delay device (Picosecond Delayer, Micro Photon Devices Srl, Italy). This can shift the electrical signal with picosecond accuracy to compensate for time delays caused by the optical and electronic setup when recording the signals. The frame- and line-trigger required for image acquisition are generated by the control unit of the piezo stage.

Control Software: Two control programs were used for image acquisition. The Symphotime software (SymPhoTime64 Version: v2.6.5544, PicoQuant, Germany) manages data acquisition and raw data fitting. A self-made program (LabView Version: 17.0, National Instruments, Austin, TX, USA) is used to control the piezostage for sample scanning and generates the required line-start, line-end and frame trigger signals.

4.2.2. Frequency-Domain Setup

Optical Setup: For lifetime detection in the FD, a widefield approach is required (Figure 7). Thus, the moving piezostage **PS** with which the sample is moved in the TD was omitted. Two separate laser diodes (LuxX Series, Omicron, Germany) were used, offering an internal modulation capability to generate sinusoidal radiation at wavelengths of 405 and 445 nm with 100 mW laser power for each **FD Laser**. A liquid light guide directs the light to a widefield coupling located at the rear of the microscope containing appropriate lenses to collimate (**WF1**) and to focus (**WF2**) the emerging light into the back focal plane of the objective. A clean-up filter **Ex1** (405 nm: F49-406; 445 nm: F39-448, AHF Analysentechnik, Germany) in the beam path blocks all other wavelength components but the laser wavelength itself. To image the sample, the TD Objective **O** was reused. The laser light was separated by a dichroic mirror **D2** (405 nm: F48-425; 445 nm: F38-470, AHF Analysentechnik, Germany). For the elimination of any laser radiation or ambient light, an appropriate fluorescence filter **Em2** (405 nm: F47-424; 445 nm: F76-460, AHF Analysentechnik, Germany) was used. The selection mirror **SM** was aligned in order to focus the fluorescence signal via a lens **L4** onto an FD FLIM camera (pco.FLIM, pco AG, Kelheim, Germany).

Signal Path: The signal path includes two BNC cables between the camera and the laser. These cables were used to synchronize the homodyne detection method within the camera and the modulation signal of the laser. The data transfer between the control computer and the pco.FLIM camera was done via a USB 3.0 interface.

Control Software: The NIS Elements software (V5.0, Nikon, Japan) serves as control software. In addition to all control parameters for the lifetime detection, it is also used for analyzing the raw data. For calibration, a reference measurement with a sample of known lifetime (Starna Green, Starna Scientific Limited, Ilford, UK) has to be performed at the start of each measurement campaign.

4.3. Sample Preparation

To cover a wide range of dyes, a sample selection was made on the basis of spectral data to be measurable both with the TD and the FD systems. Among common fluorophores, those were selected with regard to their chemical properties to react, for example, strongly to a pH change such as fluorescein [53]. All sample fluorophores are listed in Table 8. For microscopic examination, a volume of 30 µL was placed in a sterile ibidi μ-Slide VI0.4 (80606, ibidi GmbH, Gräfelfing, Germany) and subsequently measured.

Table 8. Sample fluorophores with the respective concentration.

Number	Fluorophore	Distributor	Article No.	Molar Weight (g/mol)	Concentration (M)
1	Rose Bengal	Sigma Aldrich	330000	1017.64	$10^{-1}, 10^{-2}, 10^{-3}, 10^{-4}$
2	Lucifer Yellow	Sigma Aldrich	LO144	521.57	$10^{-1}, 10^{-2}, 10^{-3}, 10^{-4}, 8\cdot10^{-5}$
3	Fluorescein	Sigma Aldrich	46955	332.31	$10^{-1}, 10^{-2}, 10^{-3}, 10^{-4}$
4	Coumarin 1	Sigma Aldrich	D87759	231.29	$10^{-1}, 10^{-2}, 10^{-3}, 10^{-4}$
5	Coumarin 6	Sigma Aldrich	546283	350.43	$10^{-1}, 10^{-2}, 10^{-3}, 10^{-4}$

Concentration: All sample fluorophores here were in salt form. Using the substance-specific property of the molar weight M and the desired amount of salt for a specific concentration c, the final volume V was calculated according to Equation (22). For concentration series, a stock solution of V = 1.5 µL with a concentration of $c = 0.1$ mol/L was produced that was afterward diluted to the desired concentration value.

$$m_{Fluorophore} = V \cdot c \cdot M \quad (22)$$

Solvent Polarity: To investigate whether the polarity has an influence on the fluorescence lifetime, the dyes were dissolved in different solvents (Table 9).

Table 9. Solvent used for the different types of fluorophores.

Solvent	Distributor	Article No.	Fluorophores
Water	Merck	1.15333	1, 2, 3
Methanol	Merck	1.06002	1, 2, 3, 4, 5
Ethanol	Merck	1.0098	1, 2
Aceton	Sigma Aldrich	270725-1L	1, 2

Microenvironment: As variable environmental parameters, the pH level and the temperature of the solvent were varied. The pH level adjustments were carried out on a dye solution of 20 ml that was diluted with HCl or NaOH, respectively, to achieve pH-values in the range between 3 and 12. The exact pH-value was checked with a pH-meter (CyberScan ph 100, Eutech InstrumentsPte Lte, Singapore) measuring device before and after each lifetime measurement.

A stage top incubation system (10720, ibidi GmbH, Germany) was used for controlled temperature changes. The temperature range to be covered was between 21 and 42 °C. For temperature verification, a chamber of the µ-Slide filled with the fluorophore solution under investigation was equipped with a temperature probe. For each measurement, a 10 min settling period was ensured to avoid possible temperature fluctuation during the measurement.

Quencher: Quencher experiments were made with 4-hydroxy TEMPO (176141, Sigma Aldrich Chemie GmbH, Darmstadt, Germany) in combination with coumarine 1 and coumarine 6 in two different concentrations. For coumarine 1, quencher concentration of 0.00855 and 0.234 M were used and for coumarine 6, quencher concentrations of 0.0013 and 0.713 M.

4.4. Statistics

To meet statistical standards, both TD and FD experiments were repeated seven times for each measurement. To confirm the reproducibility of the series, a p-value was calculated [54]. Therefore, a binomial distribution of τ and its error was assumed. If the p-value lies under a threshold of 0.05, it is statistically proven that the lifetimes that are obtained under different conditions are consistent. These conditions may be the application of a specific setup such as TD and FD or changing specific parameters such as temperature, pH level and so forth. For the graphical presentation of the measurement results, the mean values and their averaged standard deviation are displayed next to the phasor plot.

5. Conclusions

In this work, several experimental parameters are demonstrated to have an influence on the lifetime of specific fluorophores. This sensitivity can be used to the advantage of the experimenter: monitoring the lifetime will reveal how much a parameter changes during the measurement. On the other hand, having these findings in mind, one has to be careful to have control over all influencing parameters to make the results comparable and consistent with other measurements [55]. Therefore, it would be beneficial in future publications to also name the solvent polarity, the temperature and the pH-value next to the fluorophore. If a large spectral overlap between the absorption and emission spectrum of the fluorophore exists, its concentration should also be mentioned.

The statistical analysis of all measurements proved without exception that the results of the TD setup are consistent with the ones obtained with the FD setup. Therefore, it can be concluded that both setups produce reliable data if all influencing parameters are under the control of the experimenter. Both techniques have the advantage in common of not relying on the way the fluorophore is excited. Therefore, their independence from the excitation laser wavelength or from the number of photons absorbed to transfer the fluorophore to its excited state was demonstrated.

While the discussion section explains which fundamental process lies behind each influencing parameter, it is possible to use this information and create fluorophores that function as biological sensors. For example, fluorescein is used in clinical studies to image the pH-value and infer cancer in breast tissue [33]. Further, special dyes were developed that can monitor the intracellular viscosity through the fluorescence lifetime [56]. Other fluorophores give the possibility to detect intracellular temperature changes [31]. Each experiment can thus benefit from selecting the right dye for the particular parameter that has to be determined.

Author Contributions: T.H. supervised the project. T.K. and J.J. made the experimental work along with the evaluation, validation and interpretation. C.F. helped with the sample preparation. T.K., J.J. and T.H. wrote the manuscript. T.H., R.S., A.R., J.J., T.K. and C.F. made reviews of the manuscript. All authors have read and agreed to the published version of the manuscript.

Funding: This research was funded by the BMBF project "OMOXI" (FKZ 13N14507), the FFG Bridge project "LITEScope" and the Bavarian Academic Forum (BayWISS)—Doctoral Consortium "Health Research".

Institutional Review Board Statement: Not applicable.

Informed Consent Statement: Not applicable.

Data Availability Statement: The data presented in this study are available in the article.

Acknowledgments: The authors would like to thank Lukas Krainer and Stefanie Kiderlen of the company Prospective Instruments and all their members for the possibility of using their fluorescent lifetime devices and their outstanding cooperation.

Conflicts of Interest: The authors declare no conflict of interest.

References

1. Ettinger, A.; Wittmann, T. Fluorescence live cell imaging. *Methods Cell Biol.* **2014**, *123*, 77–94. [PubMed]
2. Freudiger, C.W.; Min, W.; Saar, B.G.; Lu, S.; Holtom, G.R.; He, C.; Tsai, J.C.; Kang, J.X.; Xie, X.S. Label-free biomedical imaging with high sensitivity by stimulated Raman scattering microscopy. *Science* **2008**, *322*, 1857–1861. [CrossRef] [PubMed]
3. Ries, J.; Schwille, P. Fluorescence correlation spectroscopy. *BioEssays* **2012**, *34*, 361–368. [CrossRef] [PubMed]
4. Elson, E.L.; Magde, D. Fluorescence correlation spectroscopy. I. Conceptual basis and theory. *Biopolym. Orig. Res. Biomol.* **1974**, *13*, 1–27. [CrossRef]
5. Gratton, E.; Breusegem, S.; Sutin, J.D.B.; Ruan, Q.; Barry, N.P. Fluorescence lifetime imaging for the two-photon microscope: Time-domain and frequency-domain methods. *J. Biomed. Opt.* **2003**, *8*, 381–390. [CrossRef]
6. Berezin, M.Y.; Achilefu, S. Fluorescence lifetime measurements and biological imaging. *Chem. Rev.* **2010**, *110*, 2641–2684. [CrossRef]
7. Rubart, M. Two-photon microscopy of cells and tissue. *Circ. Res.* **2004**, *95*, 1154–1166. [CrossRef]
8. Icha, J.; Weber, M.; Waters, J.C.; Norden, C. Phototoxicity in live fluorescence microscopy, and how to avoid it. *BioEssays* **2017**, *39*, 1700003. [CrossRef]
9. Boens, N.; Qin, W.; Basarić, N.; Hofkens, J.; Ameloot, M.; Pouget, J.; Lefevre, J.P.; Valeur, B.; Gratton, E.; VandeVen, M.; et al. Fluorescence lifetime standards for time and frequency domain fluorescence spectroscopy. *Anal. Chem.* **2007**, *79*, 2137–2149. [CrossRef]
10. Casey, K.G.; Quitevis, E.L. Effect of solvent polarity on nonradiative processes in xanthene dyes: Rhodamine B in normal alcohols. *J. Phys. Chem.* **1988**, *92*, 6590–6594. [CrossRef]
11. Gadella Jr, T.W.; Jovin, T.M.; Clegg, R.M. Fluorescence lifetime imaging microscopy (FLIM): Spatial resolution of microstructures on the nanosecond time scale. *Biophys. Chem.* **1993**, *48*, 221–239. [CrossRef]

12. Kristoffersen, A.S.; Erga, S.R.; Hamre, B.; Frette, Ø. Testing fluorescence lifetime standards using two-photon excitation and time-domain instrumentation: Rhodamine B, coumarin 6 and lucifer yellow. *J. Fluoresc.* **2014**, *24*, 1015–1024. [CrossRef] [PubMed]
13. Fleming, G.; Knight, A.; Morris, J.; Morrison, R.; Robinson, G. Picosecond fluorescence studies of xanthene dyes. *J. Am. Chem. Soc.* **1977**, *99*, 4306–4311. [CrossRef]
14. Periasamy, A.; Wodnicki, P.; Wang, X.F.; Kwon, S.; Gordon, G.W.; Herman, B. Time-resolved fluorescence lifetime imaging microscopy using a picosecond pulsed tunable dye laser system. *Rev. Sci. Instruments* **1996**, *67*, 3722–3731. [CrossRef]
15. Krishnan, R.V.; Masuda, A.; Centonze, V.F.E.; Herman, B.A. Quantitative imaging of protein–protein interactions by multiphoton fluorescence lifetime imaging microscopy using a streak camera. *J. Biomed. Opt.* **2003**, *8*, 362–367. [CrossRef]
16. Lakowicz, J. *Principles of Fluorescence Spectroscopy*, 2nd ed.; Plenum: New York, NY, USA, 1999.
17. Andrade, J.; Estévez-Pérez, M. Statistical comparison of the slopes of two regression lines: A tutorial. *Anal. Chim. Acta* **2014**, *838*, 1–12. [CrossRef]
18. Glasgow, B.J. Fluorescence lifetime imaging microscopy reveals quenching of fluorescein within corneal epithelium. *Exp. Eye Res.* **2016**, *147*, 12–19. [CrossRef]
19. Love, L.C.; Upton, L.M.; Ritter, A.W. Solvent effects on fluorescence spectra decay times and quantum yields of atabrine and its homologs. *Anal. Chem.* **1978**, *50*, 2059–2064. [CrossRef]
20. Freymüller, C.; Kalinina, S.; Rück, A.; Sroka, R.; Rühm, A. Quenched coumarin derivatives as fluorescence lifetime phantoms for NADH and FAD. *J. Biophoton.* **2021**, *14*, e202100024. [CrossRef]
21. Hilborn, R.C. Einstein coefficients, cross sections, f values, dipole moments, and all that. *Am. J. Phys.* **1982**, *50*, 982–986. [CrossRef]
22. Xu, C.; Webb, W.W. Measurement of two-photon excitation cross sections of molecular fluorophores with data from 690 to 1050 nm. *JOSA B* **1996**, *13*, 481–491. [CrossRef]
23. Shibasaki, Y.; Suenobu, T.; Nakagawa, T.; Katoh, R. Effect of reabsorption of fluorescence on transient absorption measurements. *Spectrochim. Acta Part A Mol. Biomol. Spectrosc.* **2019**, *220*, 117127. [CrossRef] [PubMed]
24. Ishikawa-Ankerhold, H.C.; Ankerhold, R.; Drummen, G.P. Advanced fluorescence microscopy techniques—Frap, Flip, Flap, Fret and flim. *Molecules* **2012**, *17*, 4047–4132. [CrossRef] [PubMed]
25. Wallrabe, H.; Periasamy, A. Imaging protein molecules using FRET and FLIM microscopy. *Curr. Opin. Biotechnol.* **2005**, *16*, 19–27. [CrossRef] [PubMed]
26. Loving, G.S.; Sainlos, M.; Imperiali, B. Monitoring protein interactions and dynamics with solvatochromic fluorophores. *Trends Biotechnol.* **2010**, *28*, 73–83. [CrossRef]
27. Berezin, M.Y.; Lee, H.; Akers, W.; Achilefu, S. Near infrared dyes as lifetime solvatochromic probes for micropolarity measurements of biological systems. *Biophys. J.* **2007**, *93*, 2892–2899. [CrossRef] [PubMed]
28. Spange, S.; Reuter, A.; Vilsmeier, E. On the determination of polarity parameters of silica by means of solvatochromic probe dyes. *Colloid Polym. Sci.* **1996**, *274*, 59–69. [CrossRef]
29. Xiao, H.; Li, P.; Tang, B. Recent progresses in fluorescent probes for detection of polarity. *Coord. Chem. Rev.* **2021**, *427*, 213582. [CrossRef]
30. Jue, T. *Fundamental Concepts in Biophysics*; Springer: Cham, Switzerland, 2009; Volume 1.
31. Okabe, K.; Inada, N.; Gota, C.; Harada, Y.; Funatsu, T.; Uchiyama, S. Intracellular temperature mapping with a fluorescent polymeric thermometer and fluorescence lifetime imaging microscopy. *Nat. Commun.* **2012**, *3*, 705. [CrossRef]
32. Zhang, H.; Huang, H.; He, S.; Zeng, H.; Pralle, A. Monodisperse magnetofluorescent nanoplatforms for local heating and temperature sensing. *Nanoscale* **2014**, *6*, 13463–13469. [CrossRef]
33. Lee, J.; Kim, B.; Park, B.; Won, Y.; Kim, S.Y.; Lee, S. Real-time cancer diagnosis of breast cancer using fluorescence lifetime endoscopy based on the pH. *Sci. Rep.* **2021**, *11*, 16864. [CrossRef] [PubMed]
34. Burgstaller, S.; Bischof, H.; Gensch, T.; Stryeck, S.; Gottschalk, B.; Ramadani-Muja, J.; Eroglu, E.; Rost, R.; Balfanz, S.; Baumann, A.; et al. pH-Lemon, a fluorescent protein-based pH reporter for acidic compartments. *ACS Sensors* **2019**, *4*, 883–891. [CrossRef] [PubMed]
35. Albrecht, C. *Joseph R. Lakowicz: Principles of Fluorescence Spectroscopy*; Springer: Cham, Switzerland, 2008.
36. Förster, T. Zwischenmolekulare energiewanderung und fluoreszenz. *Ann. Der. Phys.* **1948**, *437*, 55–75. [CrossRef]
37. Doose, S.; Neuweiler, H.; Sauer, M. Fluorescence quenching by photoinduced electron transfer: A reporter for conformational dynamics of macromolecules. *ChemPhysChem* **2009**, *10*, 1389–1398. [CrossRef] [PubMed]
38. Anderson, R.R.; Parrish, J.A. The optics of human skin. *J. Investig. Dermatol.* **1981**, *77*, 13–19. [CrossRef]
39. Conchello, J.A.; Lichtman, J.W. Optical sectioning microscopy. *Nat. Methods* **2005**, *2*, 920–931. [CrossRef] [PubMed]
40. Lou, J.; Scipioni, L.; Wright, B.K.; Bartolec, T.K.; Zhang, J.; Masamsetti, V.P.; Gaus, K.; Gratton, E.; Cesare, A.J.; Hinde, E. Phasor histone FLIM-FRET microscopy quantifies spatiotemporal rearrangement of chromatin architecture during the DNA damage response. *Proc. Natl. Acad. Sci. USA* **2019**, *116*, 7323–7332. [CrossRef]
41. Ranjit, S.; Datta, R.; Dvornikov, A.; Gratton, E. Multicomponent analysis of phasor plot in a single pixel to calculate changes of metabolic trajectory in biological systems. *J. Phys. Chem. A* **2019**, *123*, 9865–9873. [CrossRef]
42. Orthaus-Mueller, S.; Kraemer, B.; Dowler, R.; Devaux, A.; Tannert, A.; Roehlicke, T.; Wahl, M.; Rahn, H.J.; Erdmann, R. RapidFLIM: The new and innovative method for ultra fast flim imaging. *PicoQuant Appl. Note* **2017**, *2017*, 1–8.
43. Chen, H.; Holst, G.; Gratton, E. Modulated CMOS camera for fluorescence lifetime microscopy. *Microsc. Res. Tech.* **2015**, *78*, 1075–1081. [CrossRef]

44. Esposito, A.; Oggier, T.; Gerritsen, H.; Lustenberger, F.; Wouters, F. All-solid-state lock-in imaging for wide-field fluorescence lifetime sensing. *Opt. Express* **2005**, *13*, 9812–9821. [CrossRef] [PubMed]
45. Verveer, P.J.; Squire, A.; Bastiaens, P.I. Global analysis of fluorescence lifetime imaging microscopy data. *Biophys. J.* **2000**, *78*, 2127–2137. [CrossRef] [PubMed]
46. Franke, R.; Holst, G.A. Frequency-domain fluorescence lifetime imaging system (pco. flim) based on a in-pixel dual tap control CMOS image sensor. *Imaging Manip. Anal. Biomol. Cells Tissues* **2015**, *9328*, 241–259.
47. Ranjit, S.; Malacrida, L.; Jameson, D.M.; Gratton, E. Fit-free analyss of fluorescence lifetime imaging data using the phasor approach. *Nat. Protoc.* **2018**, *13*, 1979–2004. [CrossRef]
48. Sun, Y.; Liao, S.C. *The Ultimate Phasor Plot and Beyond*; ISS Inc.: San Antonio, TX, USA, 2014.
49. Zhang, Y.; Hato, T.; Dagher, P.C.; Nichols, E.L.; Smith, C.J.; Dunn, K.W.; Howard, S.S. Automatic segmentation of intravital fluorescence microscopy images by K-means clustering of FLIM phasors. *Opt. Lett.* **2019**, *44*, 3928–3931. [CrossRef]
50. Mannam, V.; Zhang, Y.; Yuan, X.; Ravasio, C.; Howard, S.S. Machine learning for faster and smarter fluorescence lifetime imaging microscopy. *J. Phys. Photon.* **2020**, *2*, 042005. [CrossRef]
51. Datta, R.; Heylman, C.; George, S.C.; Gratton, E. Label-free imaging of metabolism and oxidative stress in human induced pluripotent stem cell-derived cardiomyocytes. *Biomed. Opt. Express* **2016**, *7*, 1690–1701. [CrossRef]
52. Bhattacharjee, A.; Datta, R.; Gratton, E.; Hochbaum, A.I. Metabolic fingerprinting of bacteria by fluorescence lifetime imaging microscopy. *Sci. Rep.* **2017**, *7*, 3743. [CrossRef]
53. Martin, M.M.; Lindqvist, L. The pH dependence of fluorescein fluorescence. *J. Lumin.* **1975**, *10*, 381–390. [CrossRef]
54. Krzywinski, M.; Altman, N. Significance, *p* values and *t*-tests. *Nat. Methods* **2013**, *10*, 1041–1042. [CrossRef]
55. Montero Llopis, P.; Senft, R.A.; Ross-Elliott, T.J.; Stephansky, R.; Keeley, D.P.; Koshar, P.; Marqués, G.; Gao, Y.S.; Carlson, B.R.; Pengo, T.; et al. Best practices and tools for reporting reproducible fluorescence microscopy methods. *Nat. Methods* **2021**, *18*, 1463–1476. [CrossRef] [PubMed]
56. Peng, X.; Yang, Z.; Wang, J.; Fan, J.; He, Y.; Song, F.; Wang, B.; Sun, S.; Qu, J.; Qi, J.; et al. Fluorescence ratiometry and fluorescence lifetime imaging: Using a single molecular sensor for dual mode imaging of cellular viscosity. *J. Am. Chem. Soc.* **2011**, *133*, 6626–6635. [CrossRef] [PubMed]

International Journal of
Molecular Sciences

MDPI

Article

Dual-Activated Nano-Prodrug for Chemo-Photodynamic Combination Therapy of Breast Cancer

Ziyao Lu [1,2,†], Gan Xu [3,†], Xiaozhen Yang [3], Shijia Liu [1], Yang Sun [1,4], Li Chen [1,2], Qinying Liu [1,*] and Jianyong Liu [3,5,6,*]

1 Fujian Provincial Key Laboratory of Tumor Biotherapy, Clinical Oncology School of Fujian Medical University, Fujian Cancer Hospital, Fuzhou 350014, China
2 Fujian Provincial Key Laboratory of Medical Instrument and Pharmaceutical Technology, College of Biological Science and Technology, Fuzhou University, Fuzhou 350108, China
3 National & Local Joint Biomedical Engineering Research Center on Photodynamic Technologies, College of Chemistry, Fuzhou University, Fuzhou 350108, China
4 Department of Gynecology, Clinical Oncology School of Fujian Medical University, Fujian Cancer Hospital, Fuzhou 350014, China
5 Key Laboratory of Molecule Synthesis and Function Discovery, Fujian Province University, College of Chemistry, Fuzhou University, Fuzhou 350108, China
6 State Key Laboratory of Photocatalysis on Energy and Environment, College of Chemistry, Fuzhou University, Fuzhou 350108, China
* Correspondence: liuqy@fjmu.edu.cn (Q.L.); lkw82@fzu.edu.cn (J.L.)
† These authors contributed equally to this work.

Abstract: Herein, we developed a dual-activated prodrug, BTC, that contains three functional components: a glutathione (GSH)-responsive BODIPY-based photosensitizer with a photoinduced electron transfer (PET) effect between BODIPY and the 2,4-dinitrobenzenesulfonate (DNBS) group, and an ROS-responsive thioketal linker connecting BODIPY and the chemotherapeutic agent camptothecin (CPT). Interestingly, CPT displayed low toxicity because the active site of CPT was modified by the BODIPY-based macrocycle. Additionally, BTC was encapsulated with the amphiphilic polymer DSPE-mPEG$_{2000}$ to improve drug solubility and tumor selectivity. The resulting nano-prodrug passively targeted tumor cells through enhanced permeability and retention (EPR) effects, and then the photosensitizing ability of the BODIPY dye was restored by removing the DNBS group with the high concentration of GSH in tumor cells. Light-triggered ROS from activated BODIPY can not only induce apoptosis or necrosis of tumor cells but also sever the thioketal linker to release CPT, achieving the combination treatment of selective photodynamic therapy and chemotherapy. The antitumor activity of the prodrug has been demonstrated in mouse mammary carcinoma 4T1 and human breast cancer MCF-7 cell lines and 4T1 tumor-bearing mice.

Keywords: nano-prodrug; photodynamic therapy; reactive oxygen species; activation; combination therapy

Citation: Lu, Z.; Xu, G.; Yang, X.; Liu, S.; Sun, Y.; Chen, L.; Liu, Q.; Liu, J. Dual-Activated Nano-Prodrug for Chemo-Photodynamic Combination Therapy of Breast Cancer. *Int. J. Mol. Sci.* **2022**, *23*, 15656. https://doi.org/10.3390/ijms232415656

Academic Editors: Antonino Mazzaglia, Angela Scala and Enrico Caruso

Received: 20 October 2022
Accepted: 7 December 2022
Published: 10 December 2022

1. Introduction

Cancer has become one of the most important causes of death in the past few decades due to invasion and metastasis [1,2]. Although chemotherapy has always been a powerful tool to induce apoptosis or necrosis of cancer cells [3,4], it has limited effects due to its low selectivity and serious systemic toxicity [5]. Therefore, there is an urgent need to develop antitumor drugs with low toxicity and high efficiency.

Photodynamic therapy (PDT) is regarded as one of the most effective treatments for cancer due to its small trauma [6], poor toxicity and high spatiotemporal selectivity [7,8]. PDT contains three critical components, including a photosensitizer (PS), light with a specific wavelength and oxygen to generate reactive oxygen species (ROS), particularly singlet oxygen, to induce cell apoptosis or necrosis [9]. Pioneering studies have confirmed that the combination of PDT and chemotherapy is a promising treatment mode, which not only reduces the dose of the most toxic component but also overcomes single-mode limitations [10,11]. However, most encapsulation or conjugation of PSs and chemotherapeutic agents typically result in drug leakage and low drug toxicity [12,13]. An effective method has been developed to overcome the above shortcomings and improve drug biological activity. Specifically, chemotherapeutic agents are activated by internal (such as hypoxia, high concentrations of GSH, etc.) or external (such as temperature, ROS, etc.) stimuli to release on-demand drugs [14,15]. Youngjae You et al. linked phthalocyanine to PTX using a ROS-sensitive linker. ROS produced after illumination can be used to kill cancer cells as well as cut off the linker to release PTX [16,17]. However, prodrugs can still cause phototoxicity and side effects to normal cells due to their poor selectivity and irregular tumor margins [18,19]. Thus, more practical PSs that only work in tumor tissue through activation by the tumor environment have emerged.

Compared with the normal cell microenvironment, tumor cells have some inimitable pathological signals, such as vascular abnormalities, high concentrations of GSH, acidosis and hypoxia [20–23]. The PSs lose their photoactivity by connecting to quenchable groups through chemical bonds sensitive to the tumor microenvironment. In the tumor microenvironment, the chemical bonds break, the quenchable effect disappears and the photoactivity of PSs recovers [24–26]. There are many types of intermolecular quenching mechanisms, such as photoinduced electron transfer (PET) [27,28], forster resonance energy transfer (FRET) [29,30] and internal charge transfer (ICT) [31,32]. Akkaya et al. designed a GSH-activated BODIPY based on the PET effect between the 2,4-dinitrobenzenesulfonate (DNBS) group and the parent nucleus of BODIPY [33]. Zheng et al. designed an MMP-7-responsive PS in which porphyrin and a quenching agent (BHQ3) could cause the FRET effect [34]. Zhao et al. designed an activated PS based on the ICT effect between the DNBS group and iodinated BODIPY [35]. Although environmentally responsive PSs avoid damage to normal tissues, nontargeting agents still lead to low PDT efficiency [36,37]. To solve the above drawback, specific chemical bonds are usually added to PSs to connect polypeptides, vitamins, antibodies and small molecules, which increase tumor targeting [38–42]. Alternatively, taking advantage of the large space between vascular endothelial cells at the tumor, nanoparticles loaded into PSs also enter the tumor tissue by virtue of the enhanced permeability and retention (EPR) effect [43–46].

In this study, we designed a prodrug that combines passive targeting and stepwise activation. First, the fluorescence of BODIPY was quenched due to the PET effect existing between BODIPY and the DNBS group. In an environment of high concentration GSH, the sulfhydryl group of GSH and the nitro sulfonate structure undergo a nucleophilic substitution reaction which leaves the 2,4-dinitrobenzenesulfonyl moiety, the PET effect is inhibited and the fluorescence of BODIPY is restored. Then, CPT is attached to BODIPY by means of a ROS-sensitive linker. The fluorescence of CPT is quenched by BODIPY derivatives, accounting for FRET, and the low toxicity of CPT is displayed because the active site of CPT is modified by a BODIPY-based macrocycle. Meanwhile, nanoparticles are coated with the amphiphilic polymer DSPE-mPEG$_{2000}$ to improve drug solubility and tumor selectivity. As demonstrated in Scheme 1, after entering the tumor through passive targeting, the prodrug is activated by a high concentration of GSH in the tumor, and ROS is produced under light stimulation to damage tumor cells. Meanwhile, ROS ruptures the TK linker between BODIPY and CPT, releasing CPT to achieve the combination treatment of selective PDT and chemotherapy.

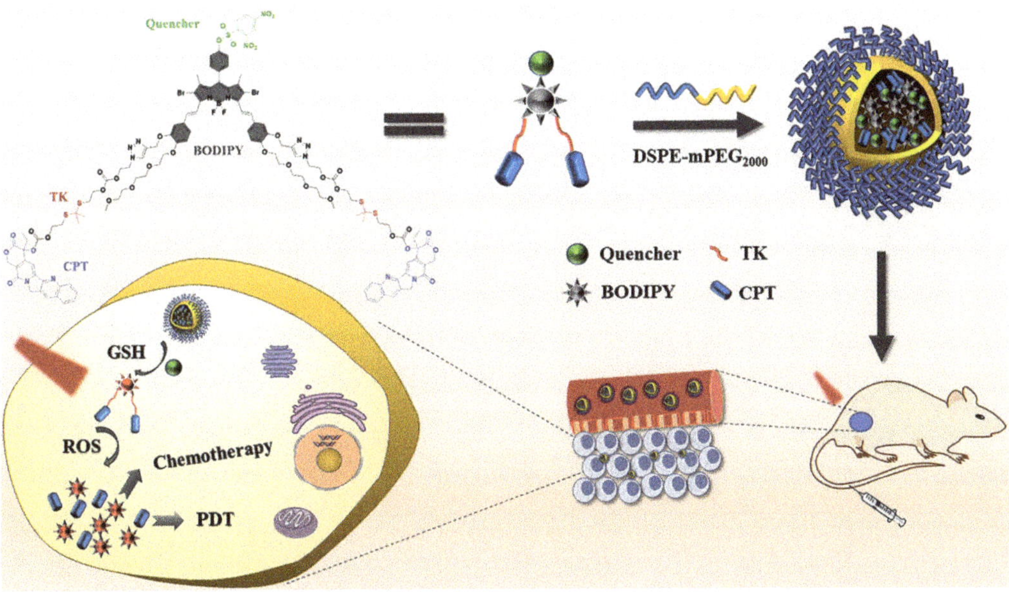

Scheme 1. The chemical structure and activation mechanism of nano-prodrug BTC NPs.

2. Results and Discussion

2.1. Molecular Design and Synthesis

In this paper, we rationally designed GSH and a ROS-responsive prodrug that consists of the DNBS group, BODIPY-based PS, a ROS-cleavable thioketal linker and CPT. As is known, the BODIPY-based fluorescent dye is a hopeful fluorescent chromophore, which is mainly due to its strong absorption in the visible and near infrared region (NIR), and robust photo-stability. Interestingly, its spectral properties can be significantly tuned by reasonable chemical modification. Scheme 2 depicts the synthetic route for the BTC. First, the distyryl-substituted BODIPY **1** was modified with the 3,4-dinitrobenzenesulfonyl chloride to produce GSH-responsive BODIPY **2**. Owing to photo-induced electron transfer (PET), GSH-responsive BODIPY **2** had a relatively weak fluorescence emission and a red-shift absorption, which is beneficial to the PDT of deep tissues. Then, BODIPY **2** was clicked with covalent CPT **3a** containing a ROS-sensitive linkage to attain BTC. For comparison, the reference compounds with ROS-insensitive carbon chains, BCC was also prepared in a similar procedure. In brief, BTC and BCC were nontoxic to healthy cells with or without light. In contrast, BCC only restored phototoxicity in tumor cells with a laser. However, BTC restored phototoxicity and chemotoxicity. The synthesis process and characterization in the Supporting Information demonstrate that BTC and BCC were successfully obtained (Figures S12–S17).

2.2. Photophysical and Photochemical Properties

2.2.1. Photophysical Properties

The electronic absorption spectra of BTC, BCC, BODIPY **2** and CPT were measured in dimethylsulfoxide (DMSO). As shown in Figures S1 and S2, BTC, BCC and BODIPY **2** showed a strong Q band with a maximum absorption wavelength of 683 nm. It was found that the maximum absorption wavelength of BODIPY **2** was redshifted by 10 nm compared with that of BODIPY **1** (Table S1), which may have been due to the strong electron-absorbing effect of introducing the DNBS group. In addition, the absorption of BTC and BCC were the superposition of CPT and BODIPY **2**, which indicated that the introduction of CPT negligibly affected the electron absorption of BTC and BCC (Figure S1). When BODIPY **1** was selectively excited at 610 nm, BODIPY **1** exhibited considerable fluorescence emission. The fluorescence quantum yield (ΦF) of BODIPY **1** was determined to be 0.319 relative to unsubstituted zinc (II) phthalocyanine (ZnPc). However, BODIPY **2**, BCC and BTC were virtually nonfluorescent. This change was attributed to the conjugation with the DNBS group. Their fluorescence quantum yields were estimated to be 0.081, 0.076, 0.078, respectively (Figure 1a and Table S1). This result demonstrated that the DNBS moiety could effectively reduce the fluorescence emission.

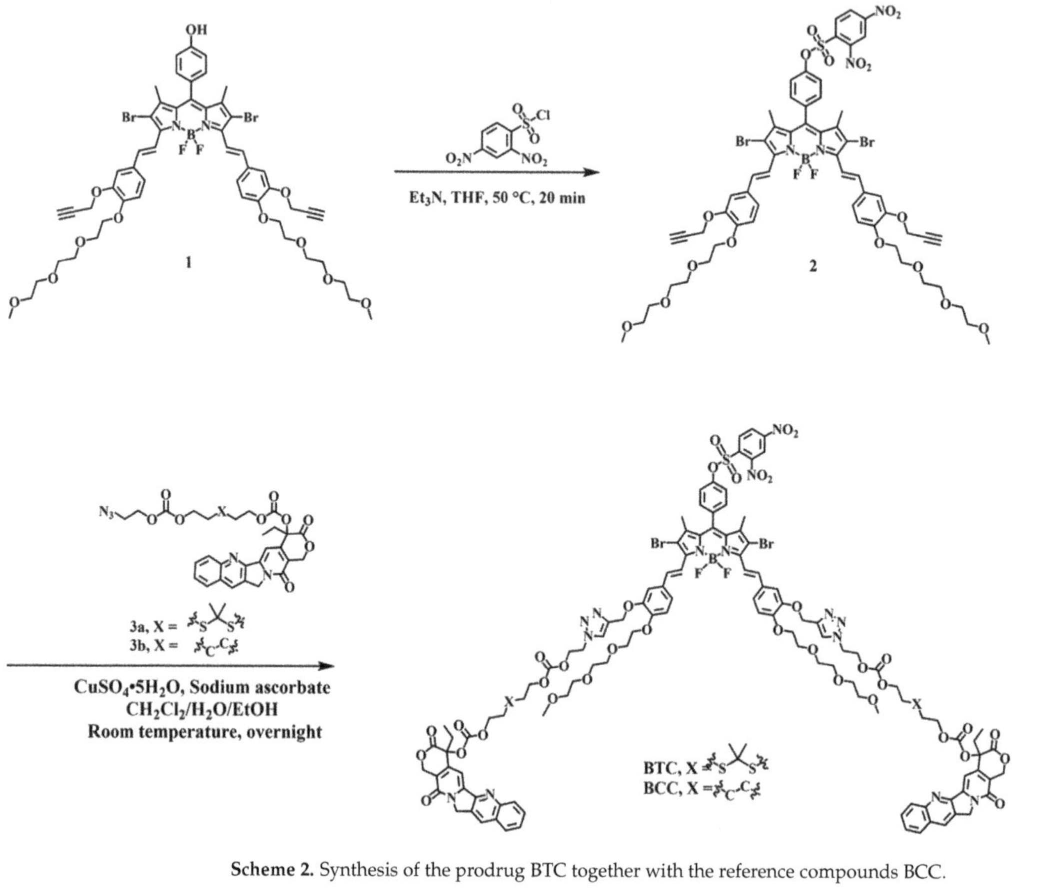

Scheme 2. Synthesis of the prodrug BTC together with the reference compounds BCC.

2.2.2. GSH-Responsive Behavior

The intracellular concentration of GSH is approximately one thousand times that of extracellular GSH. Here, we used concentrations of 2.5 μM and 2.5 mM GSH to mimic the inside and outside of the tumor cells, respectively. As can be seen from Figures 1b and S3a,c, the fluorescence emission intensity of BCC and BTC at 718 nm showed no significant change after two hours of incubation with 2.5 μM GSH. However, when they were incubated with 2.5 mM GSH (Figures 1b and S3b,d), the fluorescence emission intensity steadily increased with the extension of incubation time and finally stabilized, and the fluorescence intensity of BTC and BCC at 718 nm increased remarkably, by approximately 4-fold, compared with incubating with 2.5 μM GSH for 2 h, showing that the DNBS group was susceptible to thiol-mediated cleavage in a millimolar concentration of GSH, and the recovery degree of PS fluorescence was positively correlated with the concentration of GSH and incubation time. This also means that BTC and BCC can remain in a quenched state in the blood circulation; however, once they enter the tumor cells, they can be activated rapidly by GSH inside the tumor cells.

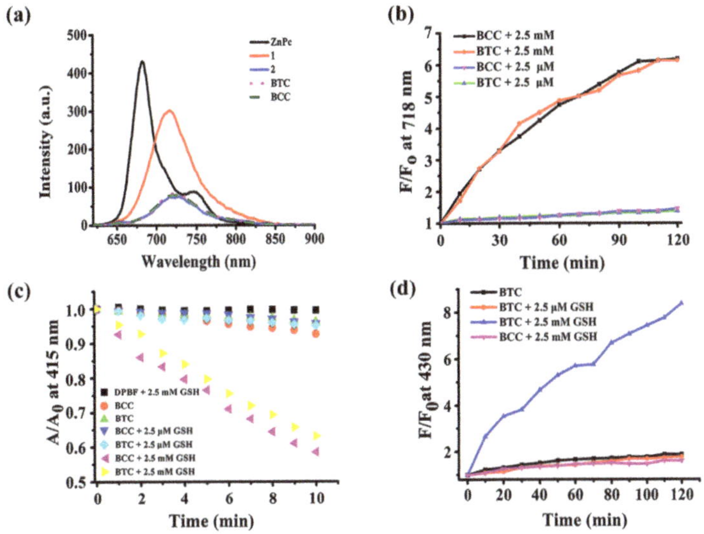

Figure 1. (**a**) Fluorescence emission spectra of ZnPc, **1**, **2**, BCC, and BTC with absorbances between 0.04–0.05 at 610 nm in DMSO (λex = 610 nm). (**b**) Changes in fluorescence intensity of BTC and BCC at 718 nm with culture time in DMSO/PBS (3:1, *v*/*v*) after addition of GSH (2.5 μM or 2.5 mM). (**c**) Comparison of the degradation rates of DPBF as monitored by the decrease in electronic absorbance at 415 nm with irradiation (660 nm, 1 mW/cm^2) time in DMSO/PBS (3:1, *v*/*v*) under various conditions. (**d**) Changes in fluorescence intensity of BTC and BCC at 430 nm with irradiation (660 nm, 30 mW/cm^2) time in DMSO/PBS (3:1, *v*/*v*) after treatment with GSH (2.5 μM or 2.5 mM) for 2 h.

As expected, the singlet oxygen ($^1O_2$) generation efficiency of BTC and BCC also responded similarly. The trapping agent of singlet oxygen, 1,3-diphenyliso-benzofuran (DPBF), is the best choice for singlet oxygen detection due to its high stability and sensitivity. When the $^1O_2$ reacts with DPBF in solution, the conjugated structure of DPBF will be destroyed and the absorption of DPBF at 415 nm will be reduced. Therefore, the $^1O_2$ production capacity of the compound was indirectly reflected by observing the decrease rate of DPBF absorption at 415 nm. As depicted in Figure S4b,e, there were no considerable absorption changes for DPBF at 415 nm in the presence of BTC and BCC under light (660 nm, 1 mW/cm^2). The results indicated that both BTC and BCC did not effectively generate singlet oxygen. Under 2.5 μM GSH, there was essentially unchanged absorbance of DPBF at 415 nm under irradiation at 660 nm (Figures 1c and S4c,f), which indicated that BTC and BCC could not effectively generate singlet oxygen after coculturing for 120 min with 2.5 μM GSH. However, after incubating these two compounds with 2.5 mM GSH, the absorbance of DPBF at 415 nm decreased remarkably (Figures 1c and S4d,g), indicating that these two compounds obviously boosted the photo-degradation of DPBF. All the above findings revealed that the intramolecular PET process in BTC and BCC disappeared with the high concentration of GSH, increasing the fluorescence emission and singlet oxygen generation efficiency.

2.2.3. ROS-Triggered CPT Release

The first step was to confirm the intramolecular FRET process in BTC and BCC. The fluorescence of BTC or BCC was compared with that of CPT under the same conditions. Upon excitation at 370 nm, the fluorescence of CPT was strong at 430 nm (Figure S5). By contrast, we found that the fluorescence intensity of BTC and BCC decreased dramatically at 430 nm, and a new fluorescence peak appeared at 724 nm (Figure S5). This phenomenon accounted for the FRET effect between BODIPY units and CPT, such that the fluorescence of CPT was quenched by BODIPY units. The next step was to evaluate the CPT moiety release in the prodrug BTC by monitoring the fluorescence changes at 430 nm under different conditions. According to Figures 1d and S6a, after the treatment of the solution of BTC with laser light (660 nm, 30 mW/cm^2), the fluorescence intensity of CPT hardly changed, indicating the CPT moiety was barely released. After pretreating with 2.5 μM of GSH for 120 min, the fluorescence emission of CPT moiety from BTC was insignificant under laser irradiation (660 nm, 30 mW/cm^2) (Figure S6b). By contrast, after treatment with 2.5 mM GSH for 120 min (Figures 1d and S6c), the fluorescence of CPT in the solution increased significantly and reached a plateau under light (660 nm, 30 mW/cm^2). Due to the fluorescence of the CPT moiety in BTC being in the "off" state through the FRET process, the increased fluorescence intensity at 430 nm was attributed to the released CPT from BTC. Furthermore, the fluorescence of CPT could not enhance the reference BCC upon irradiation after the treatment with 2.5 mM GSH (Figures 1d and S6d).

2.3. Preparation and Characterization of BTC NPs and BCC NPs

Since BTC and BCC are water-insoluble, we wrapped BTC and BCC into soluble nanoparticles by using the amphiphilic polymer DSPE-mPEG$_{2000}$, which can increase drug solubility but can also improve tumor targeting through the EPR effect. The hydrodynamic diameters of BTC NPs and BCC NPs were 99.7 nm and 112.5 nm, respectively, and the polydispersity index (PDI) of BTC NPs and BCC NPs were approximately 0.2 (Figures 2a and S7a), showing a promising distribution in solution. Transmission electron microscopy (TEM) results showed that the BTC NPs and BCC NPs had uniform and monodispersed spherical shapes. Additionally, the loading content and loading efficiency of BTC (BCC) in the BTC NPs (BCC NPs) were 1.5% (1.3%), and 85% (80%), respectively. The BTC NPs and the BCC NPs also showed superior stability in an aqueous solution, and no precipitation or significant changes in average size were observed within 7 days (Figures 2b and S7b), which guaranteed their potential application in vivo. We found that the maximum absorption wavelength of the BTC NPs and BCC NPs was lower than those of BTC and BCC, which confirmed the BTC NPs and BCC NPs have aggregation (Figures 2c and S7c). Upon excitation at 640 nm, both NPs emitted a negligible fluorescence (Figures 2d and S7d), which indicated that aggregation favored quenching the fluorescence.

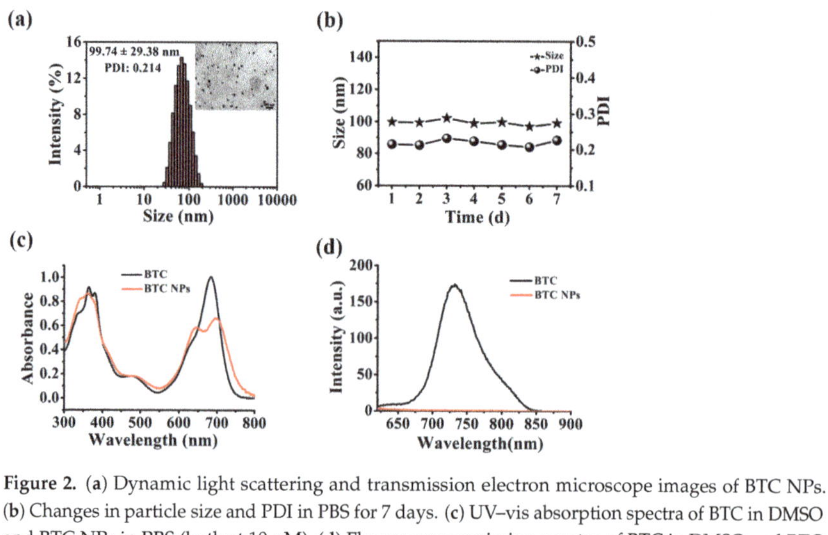

Figure 2. (**a**) Dynamic light scattering and transmission electron microscope images of BTC NPs. (**b**) Changes in particle size and PDI in PBS for 7 days. (**c**) UV–vis absorption spectra of BTC in DMSO and BTC NPs in PBS (both at 10 μM). (**d**) Fluorescence emission spectra of BTC in DMSO and BTC NPs in PBS (both at 10 μM).

2.4. In Vitro Assessments

2.4.1. GSH-Responsive Intracellular Fluorescence

The activation effect of BTC NPs in mouse mammary carcinoma 4T1 and human breast cancer MCF-7 cell lines was also monitored to estimate the intracellular fluorescence of BDP units via confocal laser scanning microscopy (CLSM). According to Figures 3a,c and S8a,c, both the BTC NPs and BCC NPs showed bright intracellular fluorescence in treated 4T1 and MCF-7 cells, indicating that the nanoparticles could be activated by GSH in tumor cells. In contrast, the intracellular fluorescence intensity of the BTC NPs and BCC NPs decreased significantly after the tumor cells were pretreated with L-buthionine sulfoximine (BSO) to deplete intracellular GSH [47], which indicated that GSH inside the tumor cells should be the key to activating the fluorescence emission of the BTC NPs and BCC NPs.

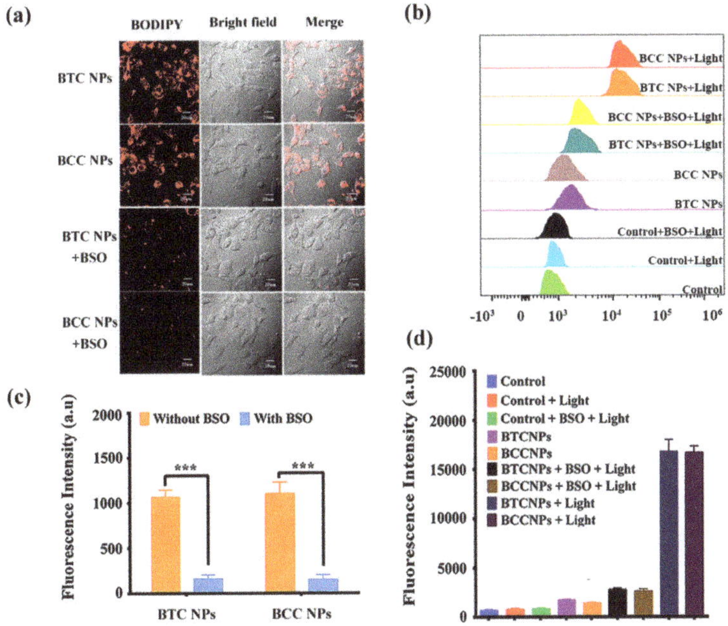

Figure 3. (**a**) CLSM images of 4T1 cells after incubation with the BTC NPs and BCC NPs (5 μM) for 24 h, or cells pretreated with BSO (10 mM) for 12 h. (**b**) ROS evaluation of BTC NPs and BCC NPs in 4T1 cells under light irradiation at 660 nm (20 mW/cm^2, 5 min) by flow cytometry. (**c**) Quantitative results of (**a**) (*** $p < 0.001$). (**d**) Quantitative results of (**b**).

2.4.2. Intracellular ROS Level

Since ROS produced by PSs can directly kill tumor cells, the ROS generation ability of PSs in tumor cells is extremely important. Here, the release of CPT was also controlled by the ROS. Next, we investigated ROS generation in 4T1 and MCF-7 cells using 2,7-dichlorofluorescein yellow diacetate (DCFH-DA) as the ROS indicator. After DCFH-DA enters cells through diffusion, DCFH is rapidly oxidized by intracellular ROS and then generates DCF with strong fluorescence, which can be detected by flow cytometry. There was negligible fluorescence in 4T1 (Figure 3b,d) and MCF-7 cells (Figure S8b,d) treated with PBS with or without light at 660 nm, indicating that the content of endogenous ROS in cells was relatively low. When the BTC NPs- and BCC NPs-treated cells were stimulated with a laser at 660 nm, the fluorescence was markedly enhanced and was 5-fold higher than that without laser stimulation. The results implied that nanoparticles could be activated by intracellular GSH and produced large amounts of ROS under light. We also investigated whether ROS could still be produced by the nanoparticles when intracellular

GSH was depleted. As the results showed, when cells were pretreated with BSO followed by the BTC NPs and BCC NPs, the intracellular fluorescence of DCF was significantly weakened, demonstrating that BSO reduced the intracellular concentration of GSH, leading the prodrug to inactivate.

2.4.3. Intracellular CPT Release

We further evaluated the release efficiency of CPT from the BTC NPs triggered by light at the cellular level by CLSM. After treatment with BTC NPs for 24 h, there was only negligible blue fluorescence in the CPT channel in 4T1 (Figure 4a,c,d) and MCF-7 cells (Figure S9a–c). However, the fluorescence of CPT was significantly enhanced when illuminated at 660 nm for 2 min, demonstrating the release of CPT from the BTC NPs under light. Moreover, regardless of the light, the fluorescence of CPT from the BCC NPs was very minimal due to the absence of an ROS-cleavable linker. Notably, when the cells were treated with the BTC NPs or BCC NPs, the red fluorescence in the BODIPY channel of the laser group was similar to that of the no-laser group.

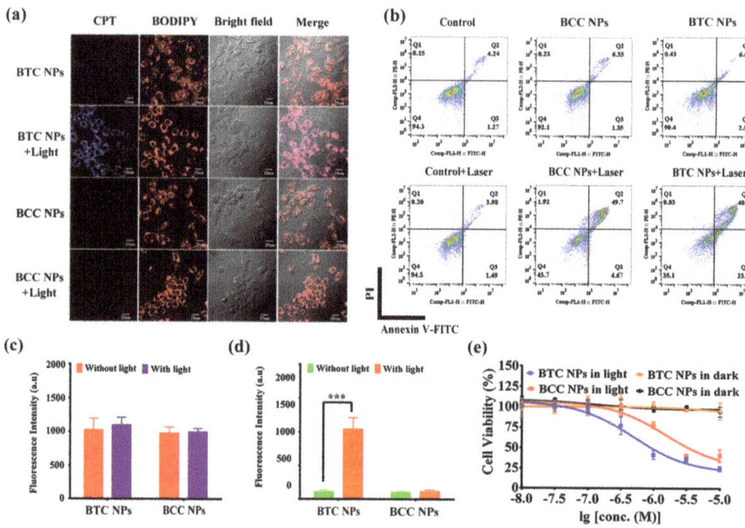

Figure 4. (**a**) CLSM images of 4T1 cells treated with BTC NPs (or BCC NPs) (5 μM) in the presence/absence of light illumination (660 nm, 20 mW/cm^2, 2 min). (**b**) Flow cytometry analysis for 4T1 cells incubated with BTC NPs (or BCC NPs) (5 μM) with or without light (20 mW/cm^2, 5 min). (**c**) Quantitative results of (**a**) for average fluorescence intensity of BODIPY moiety in 4T1 cells. (**d**) Quantitative results of (**a**) for average fluorescence intensity of CPT moiety in 4T1 cells. (*** $p < 0.001$). (**e**) Cytotoxic effects of BTC NPs (or BCC NPs) against 4T1 cells in the absence and presence of light (660 nm, 20 mW/cm^2, 5 min); Data are expressed as mean ± SD of three independent experiments, each performed in sextuplicate.

2.4.4. Cytotoxicity Assays

To explore the killing activity of nanoparticles on tumor cells directly, the cytotoxicities of the BTC NPs and BCC NPs were investigated in 4T1 and MCF-7 cells by methyl thiazolyl tetrazolium (MTT) assays. In Figures 4e and S9d, the BTC NPs and BCC NPs exhibited negligible dark toxicity, even at a concentration of 10 μM. The results indicated the cytotoxicity of CPT was largely inhibited by conjugating with BODIPY-based PSs. However, following laser exposure, the half maximal inhibitory concentrations (IC50) values for the BTC NPs were decreased to 0.50 μM for 4T1 cells and 0.63 μM for MCF-7 cells, which were lower than the BCC NPs (1.6 and 1.5 μM, respectively). The cytotoxicity differences between the BTC NPs and the BCC NPs were probably because the BTC NPs killed the tumor cells by ROS and ROS-triggered release of CPT in a combination treatment of PDT and chemotherapy, while the BCC NPs damaged the tumor cells only by ROS via PDT treatment. Furthermore, we detected the cell apoptosis induced by the BTC NPs and the BCC NPs by an Annexin V-FITC/PI Apoptosis Kit. As shown in Figure 4b, both the BTC NPs and BCC NPs induced obvious apoptosis under light conditions, but the apoptosis rate induced by the BTC NPs was higher than that induced by the BCC NPs, which suggested that the BTC NPs could kill tumor cells more efficiently through the combination of PDT and chemotherapy.

2.5. In Vivo Studies

The biodistribution of the BTC NPs in tumor tissues was evaluated in 4T1 tumor-bearing BALB/c mice. Because BODIPY-based PSs could emit near-infrared (NIR) fluorescence, the time-dependent accumulation process of the BTC NPs was directly observed by fluorescence molecular tomography (FMT). As shown in Figure 5a, there was a fluorescence signal for mice after tail vein injection with the BCC NPs and BTC NPs for 4 h. The fluorescence signal increased first and then decreased, and the maximum accumulation time of BCC NPs and BTC NPs was 12 h. This was because of the EPR effect of the nano-prodrug which caused pronounced tumor enrichment of BCC NPs and BTC NPs; the nano-prodrug was further activated by high concentrations of GSH in the tumor tissues. Furthermore, BCC and BTC displayed weak fluorescence signals at the tumor sites (Figure S10a), which was due to poor tumor accumulation. We further collected tumor tissue and other main organs to analyze the prodrug distribution at the maximum accumulation time and found that the concentrations of the BCC NPs and BTC NPs in tumor tissue were much higher than that in other main organs (Figure 5b), suggesting that the nano-prodrug could target tumor tissues. Additionally, similar results for the fluorescence image were observed in BTC and BCC (Figure S10b).

Furthermore, we investigated the antitumor effect of the BTC NPs and BCC NPs with or without a light trigger in vivo. Thirty female BALB/c mice bearing 4T1 tumors were blindly separated into six groups treated with saline, saline plus light, BCC NPs, BCC NPs plus light, BTC NPs, and BTC NPs plus light. As shown in Figure 6a, tumors grew rapidly in mice treated with saline with or without laser irradiation. A similar tumor growth tendency was displayed for mice treated with the BCC NPs or the BTC NPs but without light, showing negligible dark toxicity. However, both BTC NPs plus light and BCC NPs plus light exhibited satisfactory antitumor effects. More interestingly, BTC NPs plus light displayed better antitumor efficacy than BCC NPs plus light, revealing the combined therapeutic effect of photodynamic therapy and chemotherapy. The tumor weight (Figure 6b) and photographs (Figure 6d) also confirmed the above results. We further examined the antitumor effect of the BTC NPs by analyzing hematoxylin and eosin (H&E)-stained tumor tissues. As shown in Figure 6e, there was almost no cell necrosis or apoptosis in the group treated with saline, saline plus light, BCC NPs or BTC NPs, while a large number of shrinking and nuclear pyknosis cells appeared in the other two groups of BCC NPs plus light and BTC NPs plus light. Moreover, tumor cell proliferation seemed to be inhibited significantly in the BTC NPs plus light group.

Finally, the safety of the BTC NPs and BCC NPs were evaluated by observing the body weight change of mice under treatment and analyzing H&E-stained main organs (heart, liver, spleen, lung and kidney). The body weight of mice after BTC NPs plus light and BCC NPs plus light treatment remained stable (Figure 6c). Additionally, there were no pathological changes in the main H&E-stained organs in the BTC NPs plus light and BCC NPs plus light group (Figure S11), suggesting that no obvious toxicity or side effects were observed. Overall, the above results proved that BTC NPs had superior biocompatibility and outstanding antitumor effects with light.

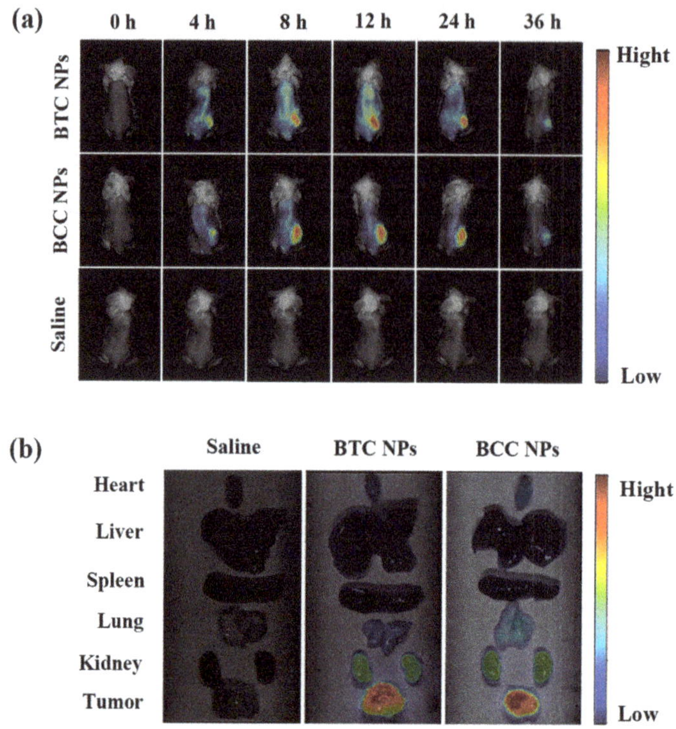

Figure 5. (**a**) Fluorescence images of 4T1 tumor-bearing mice at different time points after treatment with BTC NPs, BCC NPs and saline. (**b**) The Ex vivo fluorescence images of the different organs and tumors at 12 h after intravenous injection with BTC NPs (or BCC NPs).

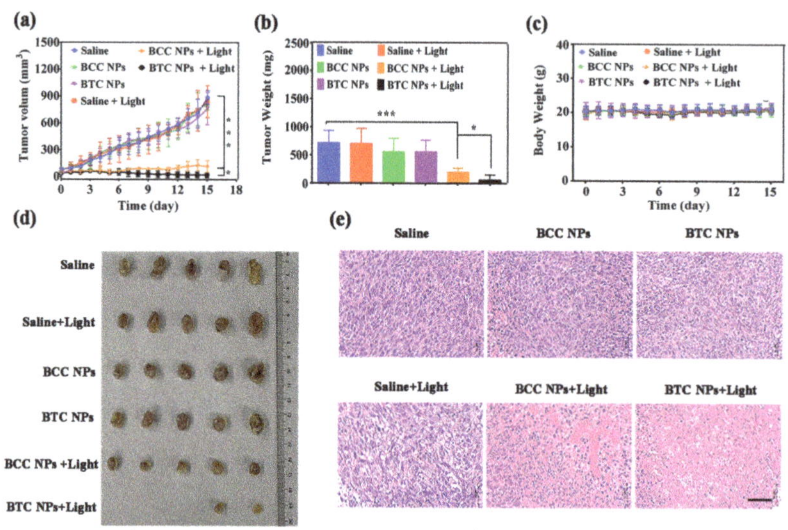

Figure 6. Antitumor effects on BALB/c mice with 4T1 tumors, (**a**) Tumor growth curves of BALB/c mice. (**b**) Weight of the tumors, (**c**) body weight of mice, (**d**) photographs of the tumors. (**e**) H&E staining images of tumors. (* $p < 0.05$, *** $p < 0.001$).

3. Materials and Methods

3.1. General

The purification of solvent, instrumentation, photophysical and photochemical investigations are described in the Supporting Information. Compounds **1**, **2**, **3a** and **3b** were prepared, as previously reported [48].

3.1.1. Synthesis of BTC

A solution of **2** (0.16 g, 0.12 mmol) and **3a** (21 g, 31 μmol) in dichloromethane (6 mL) was added to a mixture of $CuSO_4 \cdot 5H_2O$ (15 mg, 60 μmol) and sodium ascorbate (30 mg, 0.15 mmol) in water (0.5 mL). Then, 0.5 mL of ethanol was added. The resulting mixture was stirred overnight under N_2, poured into saturated sodium chloride and extracted with dichloromethane twice (2 × 15 mL). The crude product was further purified by column chromatography on silica gel using dichloromethane/MeOH (50/1, *v*/*v*) as the eluent to produce BTC as a dark green solid. (39 mg, 65%). ^{1}H NMR (400 MHz, $CDCl_3$): δ = 8.71 (s, 1 H), 8.55 (d, J = 8.4 Hz, 1 H), 8.37 (s, 2 H), 8.25 (d, J = 8.8 Hz, 1 H), 8.18 (d, J = 8.8 Hz, 2 H), 8.00 (d, J =16.4 Hz, 2 H), 7.90 (d, J = 8. 4 Hz, 2 H), 7.88 (s, 2 H), 7.80 (t, J = 7.6 Hz, 2 H), 7.63 (t, J = 7.6 Hz, 2 H), 7.52 (d, J = 16.4 Hz, 2 H), 7.44 (d, J = 8.0 Hz, 2 H), 7.39 (d, J = 7.6 Hz, 2 H), 7.31 (s, 2 H), 7.20 (d, J = 8.4 Hz, 4 H), 6.90 (d, J = 8.4 Hz, 2 H), 5.67 (d, J = 17.2 Hz, 2 H), 5.36 (d, J = 16.8 Hz, 2 H), 5.28 (s, 4 H), 5.24 (s, 4 H), 4.57 (t, J = 4.8 Hz, 4 H), 4.46 (t, J = 5.2 Hz, 4 H), 4.27–4.20 (m, 8 H), 4.15 (t, J = 6.8 Hz, 4 H), 3.89 (t, J = 4.8 Hz, 4 H), 3.74 (t, J = 4.4 Hz, 4 H), 3.67–3.62 (m, 8 H), 3.52 (t, J = 4.0 Hz, 4 H), 3.34 (s, 6 H), 2.85 (t, J = 6.8 Hz, 4 H), 2.78 (t, J = 6.8 Hz, 4 H), 2.29–2.09 (m, 4 H), 1.50 (d, J = 9.2 Hz, 12 H), 1.34 (s, 6 H), 0.98 (t, J = 7.2 Hz, 6 H). ^{13}C NMR (100.6 MHz): δ 167.33, 157.21, 154.26, 153.44, 152.26, 151.07, 150.89, 149.58, 148.93, 148.76, 148.32, 148.06, 146.52, 145.62, 144.07, 140.38, 138.88, 136.10, 135.00, 133.94, 132.96, 131.86, 131.32, 130.79, 130.67, 130.11, 129.52, 128.45, 128.29, 128.17, 128.02, 126.63, 124,27, 123,27, 122.42, 120.63, 120.11, 116.17, 114.76, 113,99, 110.56, 95.90, 77.98, 71.89, 70.76, 70.61, 70.47, 69.57, 68.59, 67.51, 67.06, 67.04, 65.73, 63.53, 58.97, 56.57, 50.03, 48.88, 31.83, 30.87, 30.85, 28.78, 28.74, 13.87, 7.67. HRMS-ESI (m/z): $[M+H]^+$ calced for $C_{121}H_{125}BBr_2F_2N_{14}O_{35}S_5$, 2704.5549, found 2704.5558.

3.1.2. Synthesis of BCC

Similarly, **2** (0.16 g, 0.12 mmol) was treated with **3b** (21 g, 31 μmol), $CuSO_4 \cdot 5H_2O$ (15 mg, 60 μmol) and sodium ascorbate (30 mg, 0.15 mmol) to produce BCC as a dark green solid. (39 mg, 71%). ^{1}H NMR (400 MHz, $CDCl_3$): δ = 8.70 (s, 1 H), 8.53 (d, J = 8.4 Hz, 1 H), 8.38 (s, 2 H), 8.21 (d, J = 10.0 Hz, 1 H), 8.19 (d, J = 8.8 Hz, 2 H), 8.00 (d, J =16.4 Hz, 2 H), 7.91 (d, J = 8.4 Hz, 2 H), 7.88 (s, 2 H), 7.80 (t, J = 7.6 Hz, 2 H), 7.64 (t, J = 7.6 Hz, 2 H), 7.51 (d, J = 16.8 Hz, 2 H), 7.43 (d, J = 8.0 Hz, 2 H), 7.37 (d, J = 8.0 Hz, 2 H), 7.32 (s, 2 H), 7.21 (d, J = 6.4 Hz, 4 H), 6.92 (d, J = 8.8 Hz, 2 H), 5.67 (d, J = 17.2 Hz, 2 H), 5.37 (d, J = 17.2 Hz, 2 H), 5.29 (s, 4 H), 5.25 (s, 4 H), 4.60 (t, J = 5.2 Hz, 4 H), 4.47 (t, J = 5.2 Hz, 4 H), 4.22 (t, J = 4.8 Hz, 4 H), 4.12–4.04 (m, 4 H), 4.00 (t, J = 6.8 Hz, 4 H), 3.90 (t, J = 4.8 Hz, 4 H), 3.75 (t, J = 4.8 Hz, 4 H), 3.66 (t, J = 4.8 Hz, 4 H), 3.63 (t, J = 4.8 Hz, 4 H), 3.52 (t, J = 4.4 Hz, 4 H), 3.35 (s, 6 H), 2.29–2.11 (m, 4 H), 1.66–1.54 (m, 8 H), 1.37–1.33 (m, 8 H), 1.28 (s, 6 H),0.99 (t, J = 7.6 Hz, 6 H). ^{13}CNMR (100.6 MHz): δ 167.47, 157.23, 155.26, 154.54, 153.77, 152.24, 151.06, 150.86, 149.58, 148.94, 148.74, 148.35, 148.05, 146.42, 145.78, 144.11, 138.90, 136.08, 135.00, 133.92, 132.97, 131.84, 131.35, 130.76, 130.71, 130.13, 129.47, 128.51, 128.28, 128.18, 128.04, 126.60, 124.20, 123.27, 122.42, 120.61, 120.16, 116.19, 114.74, 113.98, 110.55, 95.95, 77.68, 71.88, 70.75, 70.60, 70.46, 69.56, 68.89, 68.58, 68.33, 67.02, 65.46, 63.54, 58.96, 50.00, 48.95, 31.86, 28.36, 28.32, 25.18, 13.85, 7.65. HRMS-ESI (m/z): $[M+H]^+$ calced for $C_{119}H_{121}BBr_2F_2N_{14}O_{35}S$, 2548.6353, found 2548.6357.

3.2. Photophysics and Photochemistry Investigations

3.2.1. GSH-Responsive Fluorescence Emission

GSH was dissolved in deionized water to obtain 0.5 M stock solution. BTC and BCC were dissolved in DMSO to produce 2 mM stock solutions. The samples of BTC (or BCC)

(2.5 μM) with various concentrations of GSH (0 μM, 2.5 μM, 2.5 mM) in a mixture of DMSO and PBS (*v*/*v*, 3:1) were then obtained. There were four groups: (1) BTC + 2.5 μM GSH; (2) BTC + 2.5 mM; (3) BCC + 2.5 μM GSH; and (4) BTC + 2.5 mM. The fluorescence emission spectra of the sample solutions were determined continuously for 2 h under excitation at 640 nm.

3.2.2. Intramolecular FRET Process Evaluation

CPT, **2**, BTC and BCC were dissolved in DMSO to produce a 1 mM solution, which was then diluted to 2 μM with a mixture of DMSO and PBS (*v*/*v*, 3:1). The fluorescence emission spectrum of these solutions was recorded at 380–900 nm (λ_{ex} = 370 nm).

3.2.3. Photoinduced CPT Release

BTC and BCC were dissolved in DMSO to produce 2 mM stock solutions. GSH was dissolved in deionized water to obtain 0.5 M stock solution. BTC (2.5 μM) and GSH (0 μM, 2.5 μM and 2.5 mM) were prepared in a mixture of DMSO and PBS (*v*/*v*, 3:1) and cocultured at room temperature for 2 h. BCC was directly incubated with GSH (2.5 mM) at room temperature for 2 h. Then, the sample solution was irradiated by a 660 nm laser with a power of 30 mW/cm^2 for 120 min. The release of CPT was recorded by monitoring the fluorescence changes at 430 nm (λ_{ex} = 370 nm) with illumination time.

3.3. Preparation and Characterization of BTC NPs and BCC NPs

3.3.1. Preparation of BTC NPs and BCC NPs

In short, 25 mg DSPE-mPEG$_{2000}$ was dissolved in 10 mL water and ultrasonicated for 5 min at room temperature. A solution of BTC (or BCC) (1 mM) in 100 μL DMSO was added to the DSPE-mPEG$_{2000}$ aqueous solution by dropwise addition and stirred overnight at room temperature to stabilize and create uniform nanoparticles.

3.3.2. Characterization of BTC NPs and BCC NPs

The hydrodynamic diameters of the BTC NPs and BCC NPs were recorded by dynamic light scanning (DLS). The morphology of the BTC NPs and BCC NPs were determined by transmission electron microscopy (TEM). The stability of the BTC NPs and BCC NPs were recorded by detecting the changes in size through DLS over 7 days in phosphate-buffered saline (PBS) solution (10 μM, pH 7.4) at 37 °C.

3.4. In Vitro Studies

3.4.1. GSH-Responsive Intracellular Fluorescence

Buthionine sulfoximine (BSO) can inhibit GSH synthesis in cells. The 4T1 and MCF-7 cells (1×10^5) in 1 mL of Dulbecco's modified Eagle's medium (DMEM) were seeded and cultured in a confocal dish and incubated overnight. Then, the cells were continued to incubate with BSO (10 mM) in 1 mL DMEM medium for 12 h. Free DMEM without BSO was added to another group. After removing the old medium and rinsing three times with PBS, 1 mL DMEM medium bearing the BTC NPs and BCC NPs (5 μM) was added and incubated for 24 h. The cells were washed three times with PBS, serum-free medium was added to the dishes and intracellular fluorescence images were obtained by confocal laser scanning microscopy (CLSM). The BODIPY unit was excited at 633 nm and collected at 650–800 nm.

3.4.2. Intracellular ROS Measurement

Intracellular ROS levels of the BTC NPs and BCC NPs were detected by a DCFH-DA probe. First, 4T1 and MCF-7 cells (1×10^5) were seeded in a 12-well plate and incubated overnight at 37 °C with 5% CO_2. Then, the cells were incubated with BSO (10 mM) in 1 mL DMEM medium for 12 h. Free DMEM medium without BSO was added to another group, after removing the old DMEM medium and rinsing three times with PBS, 1 mL DMEM medium containing BTC NPs (or BCC NPs) (2 μM) was added and fresh DMEM was added as a control. After incubation for 24 h, the old DMEM medium was removed, and the plates were rinsed thrice with PBS. The 1 mL DMEM medium bearing DCFH-DA (10 μM) was added to each well and incubated for another 1 h. After washing three times with PBS, fresh DMEM without serum and phenol red was added and the cells were placed under 660 nm laser irradiation at 20 mW/cm^2 for 5 min. Moreover, flow cytometry was used to collect quantitative data on the intracellular ROS level.

3.4.3. Intracellular CPT Release

First, 4T1 and MCF-7 cells (1 × 10^5) in 1 mL of DMEM were seeded and cultured in a confocal dish and incubated overnight. After washing three times with PBS, 1 mL DMEM medium of the BTC NPs (or BCC NPs) (5 μM) was added to the dish and incubated for 24 h. Meanwhile, the old medium was removed and rinsed three times with PBS. Then, 1 mL serum-free medium was added to the dishes. Subsequently, the dishes were irradiated with LED irradiation (660 nm, 20 mW/cm^2) for 2 min, incubated for an additional 15 min and finally imaged by CLSM. The CPT was excited at 405 nm and detected at 425–475 nm and the BODIPY unit was excited at 633 nm and collected at 650–800 nm.

3.4.4. Photocytotoxicity Assay

MTT can detect the cytotoxicity of each compound. In short, 5000 4T1 and MCF-7 cells in 100 μL were seeded into 96-well plates and cultured overnight. Thereafter, the old medium was removed, and the cells were cultured with free medium containing BTC NPs (or BCC NPs) at different concentrations. After 24 h of incubation, the cells were washed three times with PBS and the new DMEM medium was added. The cells were irradiated with or without a 660 nm LED lamp (20 mW/cm^2) for 5 min and then cultured for another 12 h. Afterward, 10 μL of MTT (5 mg/mL) was added to each well and cultured for 4 h. Finally, the media was discarded and 100 μL of DMSO was added to each well. The optical density (OD) was measured at 570 nm by a microplate reader. The larger the OD value is, the stronger the cell activity. Cell viability = (OD $_{sample}$/OD $_{control}$) × 100%.

3.4.5. Cell Apoptosis Assay

The 4T1 and MCF-7 cells were seeded into 6 wells at a density of 1 × 10^5 cells per well and incubated overnight. Then, BTC NPs (or BCC NPs) (5 μM) was added for an additional 24 h. The cells were treated with a 660 nm LED lamp at 20 mW/cm^2 for 5 min and incubated for another 12 h. Meanwhile, the cells were collected, washed three times with PBS, and stained with PI (5 μL) and Annexin V-FITC (15 μL) solutions for 15 min based on the guidelines of the apoptosis kit (Beyotime Biotechnology, C1062L, Shanghai, China). Finally, the cells were placed on ice and detected by flow cytometry.

3.5. *In Vivo Studies*

All BALB/c mice (female, 20–25 g) were purchased from the Wushi Laboratory Animal Services Centre (Fuzhou, China). All experiments were performed according to the Institutional Animal Care and Treatment Committee of Fuzhou University. The healthy 4T1 cells were digested, centrifuged, washed twice with PBS and then collected in centrifugal tubes. Thereafter, 4T1 cells were subcutaneously injected at a density of 2 × 10^6 cells (100 μL) per mouse on the right side of the back. After the average tumor size reached a certain volume (approximately 80 mm^3), all animal experiments were performed.

3.5.1. In Vivo Fluorescence Imaging Study

A saline solution of BTC NPs, BCC NPs, BTC and BCC (2 mg/kg) were prepared, and 25 mice were randomly assigned to five groups (BCC NPs, BTC NPs, BCC, BTC and control). Each mouse was injected with 100 μL of solution via the tail vein. The changes in drug fluorescence in mice were observed by using the PerkinElmer FMT2500LX imaging system at different time points (0, 4, 6, 12, 24, 36) after injection. At the time point of maximum concentration of drugs in tumor tissues, the experimental mice were euthanized, and the heart, liver, spleen, lung and kidney were removed. A small animal imager was used to compare the drug distribution in different tissues of each group.

3.5.2. In Vivo Antitumor Assay

Thirty mice were randomly assigned to six groups: (1) saline; (2) saline plus laser; (3) BCC NPs; (4) BCC NPs plus laser; (5) BTC NPs; and (6) BTC NPs plus laser. Each group was injected with 100 μL solution via the tail vein. Then, the mice in the (2), (4) and (6) groups were irradiated with a 660 nm laser at a power of 300 mW/cm^2 for 10 min. After light treatment, the mice were weighed, and the tumor volume was measured at regular times every day for 15 days.

After treatment, a mouse was randomly selected from each group and killed by the cervical dislocation method. The tumor, heart, liver, spleen, lung and kidney were removed and immersed in 4% paraformaldehyde solution for fixation for 24 h. Paraffin-wrapped sections were prepared for H&E staining. The histological morphology of each tissue section was observed and photographed under a light microscope.

4. Conclusions

In conclusion, we designed and synthesized a GSH-activatable and ROS-responsive nano-prodrug, BTC, to attain the target PDT–chemo treatment for breast cancer. The BTC NPs were dormant without chemotoxicity and phototoxicity in healthy tissues. After the BTC NPs entered the 4T1 and MCF-7 tumor cells, the DNBS group was removed and the BODIPY derivative was activated by the high concentration of GSH. Subsequently, BTC NPs generated ROS after irradiation, and then the ROS cleaved the ROS-cleavable TK linkage to realize on-demand CPT release. The IC_{50} values for the BTC NPs were 0.50 μM toward 4T1 cells and 0.63 μM against MCF-7 cells, showing strong photocytotoxicities to tumor cells. The BTC NPs have been proven to have efficient tumor-targeting and tumor-suppressive effects without obvious toxicity and side effects in 4T1 tumor-bearing mice, which is of great significance for cancer treatment.

Supplementary Materials: The supporting information can be downloaded at: https://www.mdpi.com/article/10.3390/ijms232415656/s1.

Author Contributions: J.L., Q.L., L.C. and Z.L. designed the experiments; Z.L. and G.X. performed the experiments and prepared the manuscript; X.Y., S.L. and Y.S. participated in part of the experiments; J.L. and Q.L. revised the paper and supervised the whole process. All authors have read and agreed to the published version of the manuscript.

Funding: This study was funded by the National Natural Science Foundation of China (81873045, 21471033), The Natural Science Foundation of Fujian Province, China (2020J011118, 2020J011115, 2021J01608, 2020J01440), Fujian Province Health Care Young and Middle-aged Backbone Talents Training Project (2020GGA015), Startup Fund for Scientific Research, Fujian Medical University (2020QH1225), Excellent Young Talents Training Plan of Fujian Cancer Hospital (2020YNYQ01), Fujian Provincial Clinical Research Center for Cancer Radiotherapy and Immunotherapy (2020Y2012), The Fuzhou University Testing Fund of Precious Apparatus (No. 2022T041).

Institutional Review Board Statement: Not applicable.

Informed Consent Statement: Not applicable.

Data Availability Statement: Not applicable.

Conflicts of Interest: The authors declare that they have no known competing financial interests or personal relationships that could have appeared to influence the work reported in this paper.

References

1. Michael, M.D.; Louise, L.J. The Role of Inflammation in Progression of Breast Cancer: Friend or Foe? (Review). *Int. J. Oncol.* **2015**, *47*, 797–805. [CrossRef]
2. Cai, Y.; Wang, B.; Li, B.; Huang, X.; Guo, H.; Liu, Y.; Chen, B.; Zhao, S.; Wu, S.; Li, W.; et al. Collection on Reports of Molecules Linked to Epithelial-Mesenchymal Transition in the Process of Treating Metastasizing Cancer: A Narrative Review. *Ann. Transl. Med.* **2021**, *9*, 946. [CrossRef] [PubMed]
3. Sakaue-Sawano, A.; Kobayashi, T.; Ohtawa, K.; Miyawaki, A. Drug-induced Cell Cycle Modulation Leading to Cell-cycle Arrest, Nuclear Mis-segregation, or Endoreplication. *BMC Cell Biol.* **2011**, *12*, 2. [CrossRef] [PubMed]
4. Su, Z.; Dong, S.; Zhao, S.; Liu, K.; Tan, Y.; Jiang, X.; Assaraf, Y.; Qin, B.; Chen, Z.; Zou, C. Novel Nanomedicines to Overcome Cancer Multidrug Resistance. *Drug Resist. Updates* **2021**, *58*, 100777. [CrossRef] [PubMed]
5. Chudy, M.; Tokarska, K.; Jastrzębska, E.; Bułka, M.; Drozdek, S.; Lamch, Ł.; Wilk, K.; Brzózka, Z. Lab-on-a-chip Systems for Photodynamic Therapy Investigations. *Biosens. Bioelectron.* **2018**, *101*, 37–51. [CrossRef] [PubMed]
6. Cheng, X.; Gao, J.; Ding, Y.; Lu, Y.; Wei, Q.; Cui, D.; Fan, J.; Li, X.; Zhu, E.; Lu, Y.; et al. Multi-Functional Liposome: A Powerful Theranostic Nano-Platform Enhancing Photodynamic Therapy. *Adv. Sci.* **2021**, *8*, 2100876. [CrossRef]
7. Wan, Y.; Fu, L.; Li, C.; Lin, J.; Huang, P. Conquering the Hypoxia Limitation for Photodynamic Therapy. *Adv. Mater.* **2021**, *33*, 2103978. [CrossRef]
8. Zhuang, Z.; Dai, J.; Yu, M.; Li, J.; Shen, P.; Hu, R.; Lou, X.; Zhao, Z.; Tang, B. Type I Photosensitizers Based on Phosphindole Oxide for Photodynamic Therapy: Apoptosis and Autophagy Induced by Endoplasmic Reticulum Stress. *Chem. Sci.* **2020**, *11*, 3405–3417. [CrossRef]
9. Zhang, L.; Ding, D. Recent Advances of Transition Ir(III) Complexes as Photosensitizers for Improved Photodynamic Therapy. *VIEW* **2021**, *2*, 20200179. [CrossRef]
10. Majerník, M.; Jendželovský, R.; Vargová, J.; Jendželovská, Z.; Fedoročko, P. Multifunctional Nanoplatforms as A Novel Effective Approach in Photodynamic Therapy and Chemotherapy, to Overcome Multidrug Resistance in Cancer. *Pharmaceutics* **2022**, *14*, 1075. [CrossRef]
11. Alves, S.; Calori, I.; Tedesco, A. Photosensitizer-based Metal-organic Frameworks for Highly Effective Photodynamic Therapy. *Mater. Sci. Eng. C Mater. Biol. Appl.* **2021**, *131*, 112514. [CrossRef]
12. Nishiyama, N.; Morimoto, Y.; Jang, W.; Kataoka, K. Design and Development of Dendrimer Photosensitizer-incorporated Polymeric Micelles for Enhanced Photodynamic Therapy. *Adv. Drug Deliv. Rev.* **2009**, *61*, 327–338. [CrossRef]
13. Brown, J.M. Tumor Microenvironment and the Response to Anticancer Therapy. *Cancer Biol. Ther.* **2002**, *1*, 453–458. [CrossRef]
14. Wiese, M.; Stefan, S.M. The A-B-C of Small-molecule ABC Transport Protein Modulators: From Inhibition to Activation—A Case Study of Multidrug Resistance-associated Protein 1 (ABCC1). *Med. Res. Rev.* **2019**, *39*, 2031–2081. [CrossRef]
15. Hao, Y.; Chen, Y.W.; He, X.; Yu, Y.; Han, R.; Li, Y.; Yang, C.; Hu, D.; Qian, Z. Polymeric Nanoparticles with ROS-responsive Prodrug and Platinum Nanozyme for Enhanced Chemophotodynamic Therapy of Colon Cancer. *Adv. Sci.* **2020**, *7*, 2001853. [CrossRef]
16. Liu, L.; Qiu, W.; Li, B.; Zhang, C.; Sun, L.; Wan, S.; Rong, L.; Zhang, X. A Red Light Activatable Multifunctional Prodrug for Image-guided Photodynamic Therapy and Cascaded Chemotherapy. *Adv. Funct. Mater.* **2016**, *26*, 6257–6269. [CrossRef]
17. Yang, C.; Wen, F.; Du, Y.; Luo, M.; Lu, Y.; Liu, Y.; Lin, H. A Hypoxia-Activated Prodrug Conjugated with a BODIPY-BasedPhotothermal Agent for Imaging-Guided Chemo-PhotothermalCombination Therapy. *ACS Appl. Mater.* **2022**, *14*, 40546–40558. [CrossRef]
18. Tam, L.K.B.; Yu, L.; Wong, R.C.H.; Fong, W.P.; Ng, D.K.P.; Lo, P.C. Dual Cathepsin and Glutathione-Activated Dimeric and Trimeric Phthalocyanine-Based Photodynamic Molecular Beacons for Targeted Photodynamic Therapy. *J. Med. Chem.* **2021**, *64*, 17455–17467. [CrossRef]
19. Yu, Z.; Wang, H.; Chen, Z.; Dong, X.; Zhao, W.; Shi, Y.; Zhu, Q. Discovery of an Amino Acid-Modified Near-Infrared Aza-BODIPY Photosensitizer as an Immune Initiator for Potent Photodynamic Therapy in Melanoma. *J. Med. Chem.* **2022**, *65*, 3616–3631. [CrossRef]
20. Chu, D.; Dong, X.; Zhao, Q.; Gu, J.; Wang, Z. Photosensitization Priming of Tumor Microenvironments Improves Delivery of Nanotherapeutics via Neutrophil Infiltration. *Adv. Mater.* **2017**, *29*, 1701021. [CrossRef]
21. Wang, X.; Wang, X.; Jin, S.; Muhammad, N.; Guo, Z. Stimuli-Responsive Therapeutic Metallodrugs. *Chem. Rev.* **2019**, *119*, 1138–1192. [CrossRef] [PubMed]
22. Gasser, S.; Lim, L.H.K.; Cheung, F.S.G. The Role of the Tumour Microenvironment in Immunotherapy. *Endocr. Relat. Cancer* **2017**, *24*, T283–T295. [CrossRef] [PubMed]
23. Huai, Y.; Hossen, M.N.; Wilhelm, S.; Bhattacharya, R.; Mukherjee, P. Nanoparticle Interactions with the Tumor Microenvironment. *Bioconjug. Chem.* **2019**, *30*, 2247–2263. [CrossRef] [PubMed]
24. Ha, S.; Zou, Y.; Fong, W.; Ng, D.K.P. Multifunctional Molecular Therapeutic Agent for Targeted and Controlled Dual Chemo-and Photodynamic Therapy. *J. Med. Chem.* **2020**, *63*, 8512–8523. [CrossRef]
25. Ihsanullah, K.M.; Kumar, B.N.; Zhao, Y.; Muhammad, H.; Liu, Y.; Wang, L.; Liu, H.; Jiang, W. Stepwise-Activatable Hypoxia Triggered Nanocarrier-Based Photodynamic Therapy for Effective Synergistic Bioreductive Chemotherapy. *Biomaterials* **2020**, *245*, 119982. [CrossRef] [PubMed]

26. Pan, W.; Tan, Y.; Meng, W.; Huang, N.; Zhao, Y.; Yu, Z.; Huang, Z.; Zhang, W.; Sun, B.; Chen, J. Microenvironment-Driven Sequential Ferroptosis, Photodynamic Therapy, and Chemotherapy for Targeted Breast Cancer Therapy by a Cancer-Cell-Membrane-Coated Nanoscale Metal-Organic Framework. *Biomaterials* **2022**, *282*, 121449. [CrossRef]
27. James, N.S.; Joshi, P.; Ohulchanskyy, T.Y.; Chen, Y.; Tabaczynski, W.; Durrani, F.; Shibata, M.; Pandey, R.K. Photosensitizer (PS)-Cyanine Dye (CD) Conjugates: Impact of the Linkers Joining the PS and CD Moieties and Their Orientation in Tumor-Uptake and Photodynamic Therapy (PDT). *Eur. J. Med. Chem.* **2016**, *122*, 770–785. [CrossRef]
28. Wu, X.; Wu, W.; Cui, X.; Zhao, J.; Wu, M. Preparation of Bodipy–ferrocene Dyads and Modulation of the Singlet/Triplet Excited State of Bodipy via Electron Transfer and Triplet Energy Transfer. *J. Mater. Chem. C Mater.* **2016**, *4*, 2843–2853. [CrossRef]
29. Bio, M.; Rajaputra, P.; You, Y. Photodynamic Therapy via FRET Following Bioorthogonal Click Reaction in Cancer Cells. *Bioorg. Med. Chem. Lett.* **2016**, *26*, 145–148. [CrossRef]
30. Yang, G.; Sun, X.; Liu, J.; Feng, L.; Liu, Z. Light-Responsive, Singlet-Oxygen-Triggered On-Demand Drug Release from Photosensitizer-Doped Mesoporous Silica Nanorods for Cancer Combination Therapy. *Adv. Funct. Mater.* **2016**, *26*, 4722–4732. [CrossRef]
31. Sasan, S.; Chopra, T.; Gupta, A.; Tsering, D.; Kapoor, K.K.; Parkesh, R. Fluorescence "turn-off" and Colorimetric Sensor for Fe^{2+}, Fe^{3+}, and Cu^{2+} Ions Based on a 2,5,7-Triarylimidazopyridine Scaffold. *ACS Omega* **2022**, *7*, 11114–11125. [CrossRef]
32. Thapa, P.; Li, M.; Bio, M.; Rajaputra, P.; Nkepang, G.; Sun, Y.; Woo, S.; You, Y. Far-Red Light-Activatable Prodrug of Paclitaxel for the Combined Effects of Photodynamic Therapy and Site-Specific Paclitaxel Chemotherapy. *J. Med. Chem.* **2016**, *59*, 3204–3214. [CrossRef]
33. Turan, I.S.; Cakmak, F.P.; Yildirim, D.C.; Cetin-Atalay, R.; Akkaya, E.U. Near-IR Absorbing BODIPY Derivatives as Glutathione-Activated Photosensitizers for Selective Photodynamic Action. *Chemistry* **2014**, *20*, 16088–16092. [CrossRef]
34. Zheng, G.; Chen, J.; Stefflova, K.; Jarvi, M.; Li, H.; Wilson, B.C. Photodynamic Molecular Beacon as an Activatable Photosensitizer Based on Protease-Controlled Singlet Oxygen Quenching and Activation. *Proc. Natl. Acad. Sci. USA* **2007**, *104*, 8989–8994. [CrossRef]
35. Huang, L.; Yang, W.; Zhao, J. Switching of the Triplet Excited State of Styryl 2,6-Diiodo-Bodipy and its Application in Acid-Activatable Singlet Oxygen Photosensitizing. *J. Org. Chem.* **2014**, *79*, 10240–10255. [CrossRef]
36. Deng, C.; Liu, Y.; Zhou, F.; Wu, M.; Zhang, Q.; Yi, D.; Yuan, W.; Wang, Y. Engineering of Dendritic Mesoporous Silica Nanoparticles for Efficient Delivery of Water-Insoluble Paclitaxel in Cancer Therapy. *J. Colloid Interface Sci.* **2021**, *593*, 424–433. [CrossRef]
37. Harmatys, K.M.; Overchuk, M.; Chen, J.; Ding, L.; Chen, Y.; Pomper, M.G.; Zheng, G. Tuning Pharmacokinetics to Improve Tumor Accumulation of a Prostate-Specific Membrane Antigen-Targeted Phototheranostic Agent. *Bioconjug. Chem.* **2018**, *29*, 3746–3756. [CrossRef]
38. Gulzar, A.; Xu, J.; Wang, C.; He, F.; Yang, D.; Gai, S.; Yang, P.; Lin, J.; Jin, D.; Xing, B. Tumour Microenvironment Responsive Nanoconstructs for Cancer Theranostic. *Nano Today* **2019**, *26*, 16–56. [CrossRef]
39. Li, M.; Cui, X.; Wei, F.; Wang, Z.; Han, X. Red Blood Cell Membrane-Coated Biomimetic Upconversion Nanoarchitectures for Synergistic Chemo-photodynamic Therapy. *New J. Chem.* **2021**, *45*, 22269–22279. [CrossRef]
40. Meng, X.; Wang, J.; Zhou, J.; Tian, Q.; Qie, B.; Zhou, G.; Duan, W.; Zhu, Y. Tumor Cell Membrane-Based Peptide Delivery System Targeting the Tumor Microenvironment for Cancer Immunotherapy and Diagnosis. *Acta Biomater.* **2021**, *127*, 266–275. [CrossRef]
41. Fu, X.; Yin, W.; Shi, D.; Yang, Y.; He, S.; Hai, J.; Hou, Z.; Fan, Z.; Zhang, D. Shuttle-shape Carrier-Free Platinum-Coordinated Nanoreactors with O2 Self-supply and ROS Augment for Enhanced Phototherapy of Hypoxic Tumor. *ACS Appl. Mater.* **2021**, *13*, 32690–32702. [CrossRef]
42. Zhang, H.; Lin, H.; Su, D.; Yang, D.; Liu, J. Enzyme-activated Multifunctional Prodrug Combining Site-specific Chemotherapy with Light-triggered Photodynamic therapy. *Mol. Pharm.* **2022**, *19*, 630–641. [CrossRef] [PubMed]
43. Chu, D.; Dong, X.; Shi, X.; Zhang, C.; Wang, Z. Neutrophil-based Drug Delivery Systems. *Adv. Mater.* **2018**, *30*, 1706245. [CrossRef] [PubMed]
44. Chen, Y.; Xu, Z.; Zhu, D.; Tao, X.; Gao, Y.; Zhu, H.; Mao, Z.; Ling, J. Gold Nanoparticles Coated with Polysarcosine Brushes to Enhance their Colloidal Stability and Circulation Time in vivo. *J. Colloid Interface Sci.* **2016**, *483*, 201–210. [CrossRef] [PubMed]
45. Wang, X.; Luo, D.; Basilion, J.P. Photodynamic Therapy: Targeting Cancer Biomarkers for the Treatment of Cancers. *Cancers* **2021**, *13*, 2992. [CrossRef]
46. Chen, Y.; Zhang, X.; Cheng, D.; Zhang, Y.; Liu, Y.; Ji, L.; Guo, R.; Chen, H.; Ren, X.; Chen, Z.; et al. Near-infrared Laser-triggered in Situ Dimorphic transformation of BF2-azadipyrromethene Nanoaggregates for Enhanced Solid Tumor Penetration. *ACS Nano* **2020**, *14*, 3640–3650. [CrossRef]
47. Zhu, J.; Jiao, A.; Li, Q.; Lv, X.; Wang, X.; Song, X.; Li, B.; Zhang, Y.; Dong, X.C. Mitochondrial Ca^{2+}-overloading by Oxygen/Glutathione Depletion-Boosted Photodynamic Therapy Based on a $CaCO_3$ Nanoplatform for Tumor Synergistic Therapy. *Acta Biomater.* **2022**, *137*, 252–261. [CrossRef]
48. Xu, G.; Zhang, H.; Li, X.; Yang, D.; Liu, J. Red light Triggered Photodynamic-Chemo Combination Therapy Using a Prodrug Caged by Photosensitizer. *Eur. J. Med. Chem.* **2021**, *2015*, 113251. [CrossRef]

International Journal of
Molecular Sciences

MDPI

Article

Psoralen as a Photosensitizers for Photodynamic Therapy by Means of In Vitro Cherenkov Light

Lisa Hübinger *, Roswitha Runge, Tobias Rosenberg, Robert Freudenberg, Jörg Kotzerke * and Claudia Brogsitter

Department of Nuclear Medicine, University Hospital Carl Gustav Carus, Technische Universität Dresden, 01307 Dresden, Germany
* Correspondence: lisa.huebinger@ukdd.de (L.H.); joerg.kotzerke@uniklinikum-dresden.de (J.K.); Tel.: +49-351-458-11616 (L.H.); +49-351-458-4160 (J.K.)

Abstract: Possible enhancements of DNA damage with light of different wavelengths and ionizing radiation (Rhenium-188—a high energy beta emitter (Re-188)) on plasmid DNA and FaDu cells via psoralen were investigated. The biophysical experimental setup could also be used to investigate additional DNA damage due to photodynamic effects, resulting from Cherenkov light. Conformational changes of plasmid DNA due to DNA damage were detected and quantified by gel electrophoresis and fluorescent staining. The clonogene survival of the FaDu cells was analyzed with colony formation assays. Dimethyl sulfoxide was chosen as a chemical modulator, and Re-188 was used to evaluate the radiotoxicity and light (UVC: λ = 254 nm and UVA: λ = 366 nm) to determine the phototoxicity. Psoralen did not show chemotoxic effects on the plasmid DNA or FaDu cells. After additional treatment with light (only 366 nm—not seen with 254 nm), a concentration-dependent increase in single strand breaks (SSBs) was visible, resulting in a decrease in the survival fraction due to the photochemical activation of psoralen. Whilst UVC light was phototoxic, UVA light did not conclude in DNA strand breaks. Re-188 showed typical radiotoxic effects with SSBs, double strand breaks, and an overall reduced cell survival for both the plasmid DNA and FaDu cells. While psoralen and UVA light showed an increased toxicity on plasmid DNA and human cancer cells, Re-188, in combination with psoralen, did not provoke additional DNA damage via Cherenkov light.

Keywords: psoralen; Re-188; plasmid DNA; FaDu cells; Cherenkov light

Citation: Hübinger, L.; Runge, R.; Rosenberg, T.; Freudenberg, R.; Kotzerke, J.; Brogsitter, C. Psoralen as a Photosensitizers for Photodynamic Therapy by Means of In Vitro Cherenkov Light. *Int. J. Mol. Sci.* **2022**, 23, 15233. https://doi.org/10.3390/ijms232315233

Academic Editors: Antonino Mazzaglia, Angela Scala and Enrico Caruso

Received: 4 November 2022
Accepted: 1 December 2022
Published: 3 December 2022

1. Introduction

UV light is established for therapeutic purposes for numerous skin diseases. UVB light is especially used for treating psoriasis and UVA light can be performed when treating atopic eczema [1–3]. Since light can activate primarily non-toxic substances to phototoxic ones, it is a so-called photosensitizer. The typical procedure during photodynamic therapy (PDT) activates a photosensitizer via the radiation of visible light. The therapeutic effect is either performed directly by effects on the DNA or indirectly by the formation of singlet oxygen or reactive oxygen species such as hydrogen peroxide or superoxide anions. The small penetration depth of light in the visible and the UV scope, however, limits the applicability of PDT to surface lesions and does not allow for the treatment of solid tumors inside the body. These limitations could be overcome by Cherenkov light (CL), which can be produced by diagnostically or therapeutically used radioisotopes (e.g., Ga-68, Re-188, or Y-90) [4–6]. CL is already used diagnostically, for example, in visual imaging [3,7–9]. This features a new therapeutic approach: depositing phototoxic substances inside the tumor and activating these via CL in addition to internal or external irradiation. Thereby, a local combination of radiotoxicity and phototoxicity induces a more profound treatment [10,11]. This vision of PDT is promising; however, the mechanism is not yet illuminated completely. Recent research follows up with different methods and studies that confirm this vision. Yoon et al. [12] used trioxsalen (a psoralen derivative) and irradiated melanoma and breast cancer cells. The CL was created by external radiation, producing CL in a solid water block.

This work investigated this idea in a nuclear medicine setting. In doing so, the technique postulated by Yoon et al. was combined with the precision of nuclear medicine directly targeting tumor volume, even deep inside the patient. Our proof-of-principle study expands on the research by Yoon et al. by implementing CL via the radioisotope Re-188 in comparison to external radiation and investigated the photoactivation of psoralen by the Re-188-produced CL. Not only this new method, but also an additional setting with protein-free plasmid DNA, were tested for performance and suitability. The results were compared with the phototoxicity of a defined wavelength scope.

This simple setup lays the groundwork for follow-up studies approaching PDT enhancement with CL, radioisotopes, and psoralen, along with demonstrating fundamental methods in photoactivation and the measuring of CL in a nuclear medicine approach.

2. Material and Methods

2.1. Plasmid DNA and Cell Culture

2.1.1. Plasmid DNA

A pUC19 plasmid with 2686 base pairs with a molar mass of 1.75×10^6 Dalton was used (New England Biolabs, Ipswich, UK). The DNA stock solution was adjusted to a concentration of 0.1 µg/µL via a TE buffer (10 mM Tris-HCl, 1 mM EDTA, pH 7.5). All plasmid DNA solutions contained ≥95% supercoiled plasmid DNA. To obtain linear marker plasmid DNA, pUC19 was enzymatically treated with BamHI (Invitrogen, Karlsruhe, Germany).

2.1.2. FaDu Cells

FaDu cells are squamous cell carcinoma cells of the pharynx. They were retained via biopsy in 1968 and have since grown as a monolayer [13]. The sub cell line $FaDu_{DD}$ has been used in radiobiological experiments since the 1980s and possesses a doubling time of about 18 h in the exponential phase [14]. FaDu cells (ATCC® HTB-43™) originate from an undifferentiated human squamous cell carcinoma. In our experiments, the sub cell line $FaDu_{DD}$, kindly provided by the Department of Radiotherapy and Radiation Oncology, Medical Faculty, Technical University Dresden, was used. The cells were maintained in Dulbecco's minimum Essential medium (DMEM) containing 2% Hepes buffer, 1% of non-essential amino acids, 1% sodium pyruvate, and 10% fetal calf serum.

2.2. Psoralen and DMSO

Psoralen (CAS: 66-97-7, purity ≥99%, Sigma Aldrich, Darmstadt, Germany) is the basic compound of the linear furanocoumarins and shows photosensitive properties. At first, a non-covalent bond is formed with the DNA. Due to UV irradiation, a covalent bond between psoralen and a pyrimidine base (preferentially thymine) is formed via cycloaddition [15]. With further UV irradiation, more interactions between the psoralen monoadduct and the pyrimidine base can occur, so that crosslinks between DNA strands are formed. Furthermore, psoralens react with other cellular structures such as proteins or lipids. The general absorption maximum of psoralen lies between 320 nm and 400 nm, while the psoralen used in this publication has its maximum at 355 nm [16]. Dimethyl sulfoxide (DMSO, CAS: 67-85-5, purity = 99.9%, Sigma Aldrich, Darmstadt, Germany) was used as a radical scavenger in a final concentration of 0.2 M.

2.3. Light and Radioactivity

The UV irradiation took place with a UV lamp (Type 022.9230, LAMAG, Berlin, Germany) in the UVC region ($\lambda = 254$ nm) and in the UVA region ($\lambda = 366$ nm) of the spectrum. The irradiation times varied between 1 and 12 min (254 nm) and 10 to 120 min (366 nm). For a 5 cm distance between the probes and the light source, the spectro-radiometric measurements showed an irradiance of 20.5 W/m^2 and 14.0 W/m^2 for $\lambda = 366$ nm and $\lambda = 254$ nm, respectively [5].

Re-188 was extracted from a tungsten/rhenium generator (Isotope Technologies Garching GmbH, Garching, Germany). The high-energy beta particles of Re-188 (maximum beta energy $E_{\beta,max}$ = 2.12 MeV, mean energy 765 keV, and maximum penetration depth of 1.05 cm in tissue) can be used therapeutically, whereas its gamma radiation (energy 155 keV, intensity 15.8%) is used for imaging.

During its decay, Cherenkov light with a yield of 35 photons per decay event is produced [4]. Its physical half-life is 16.98 h and Re-188 has an average linear energy transfer (LET) of 0.19 keV/µm.

The irradiation geometry used in Monte Carlo simulations for plasmid DNA in a 20 µL volume resulted in 7.7 Gy/(1 MBq * 1 h) and 120 Gy/(1 MBq * 24 h), respectively. Thus, cells in a 6-well plate with a 2 mL volume receive 2 Gy/(1.37 MBq * 24 h).

2.4. Agarose Gel Electrophoresis, Colony Formation Assay and Cherenkov Light

2.4.1. Agarose Gel Electrophoresis

Plasmid probes with each 200 ng DNA (0.1 µg/µL) were incubated with varying volumes of chemical noxa or radionuclide solutions in 1.5 mL micro tubes (Eppendorf, Hamburg; Germany) in a total volume of 20 µL. After treating the plasmid probes, 10 µL of each DNA solution was mixed with 1.25 µL loading buffer (Invitrogen, Karlsruhe, Germany).

The probes were pipetted into the pocket of a 1.4% agarose gel in Tris-Acetat-EDTA buffer (TAE, Sigma, Darmstadt, Germany). The electrophoresis (Bio-Rad Laboratories GmbH, Munich, Germany) took place on ice with a voltage of 4 V/cm over 120 min. Due to differences in the mobility of the plasmid conformations, supercoiled (SC), open circular (OC), and linear (L) conformations can be distinguished.

The agarose gel was stained with an ethidium bromide solution (0.5 µg/mL) and the plasmid DNA was detected with a UV transilluminator (DIANA III Digital Imaging System, Straubenhardt, Germany). The fluorescence intensities of the DNA bands were quantified with the analysis software Fiji [17] and interpreted as undamaged native plasmid DNA (SC conformation), DNA single strand breaks (OC form) as well as DNA double strand breaks (L conformation).

2.4.2. CL Verification

CL was verified in the same system that the electrophoresis used (Bio-Rad Laboratories GmbH, Munich, Germany). The included charge-coupled device camera detects the CL produced by Re-188, since it can detect light in the wavelength region of the Cherenkov spectrum. Each verification image shows the intensity of light recorded by the camera over 10 min. The ImageLab software provides the opportunity to analyze different areas and evaluates the integrated volume light intensity. However, this analysis is not quantitative and only works as a general verification and holds for comparisons between different setups.

2.4.3. Colony Formation Assay

To determine the clonogene survival, a colony formation assay was used [18]. After 24 h, the cells were stripped, an aliquot for each dose point (or UV irradiation time) of the cell suspension was taken for the colony formation test in T25 culture flasks, and finally placed in an incubator for 9 days. To stop colony formation, the cells were fixed with ethanol 80% v/v and stained with crystal violet. The counting of the colonies was performed on a microscope (magnification 25×). The plating efficiency and the survival rate (or survival fraction) was calculated for both the irradiated and unirradiated cells [18].

2.5. Statistics

All results are shown as the average as well as the SEM (standard error of the mean) of the plasmid DNA and the pooled standard deviation (SD_{pooled}) of the cell experiments based on three independent tests (each experimental setup was determined as a triplicate). To prove the statistical significance, a Student's t-test was used. A difference between two independent samples was seen as significant if the probability of error was $p \leq 0.05$. The statistical analysis was conducted with MS Office Excel.

3. Results

3.1. Plasmid DNA and Psoralen with UVA

Preliminary tests excluded that the incubation time (1 h or 24 h) of the plasmid DNA with psoralen without exposure to light influences the DNA integrity. Thus, small time differences in the sample processing can be neglected.

The effect of increasing concentrations of psoralen with or without UVA exposure can be seen in Figure 1. The running behavior depends on the conformational change in each lane, thus supercoiled (SC), linear (L) and open circular (OC) conformations can be distinguished. Even though non-UVA-activated psoralen in the highest concentration did not show chemotoxicity, there was a change in the running behavior after light exposure, which can be explained by the uncoiling of the SC DNA and the formation of psoralen DNA monoadducts [19].

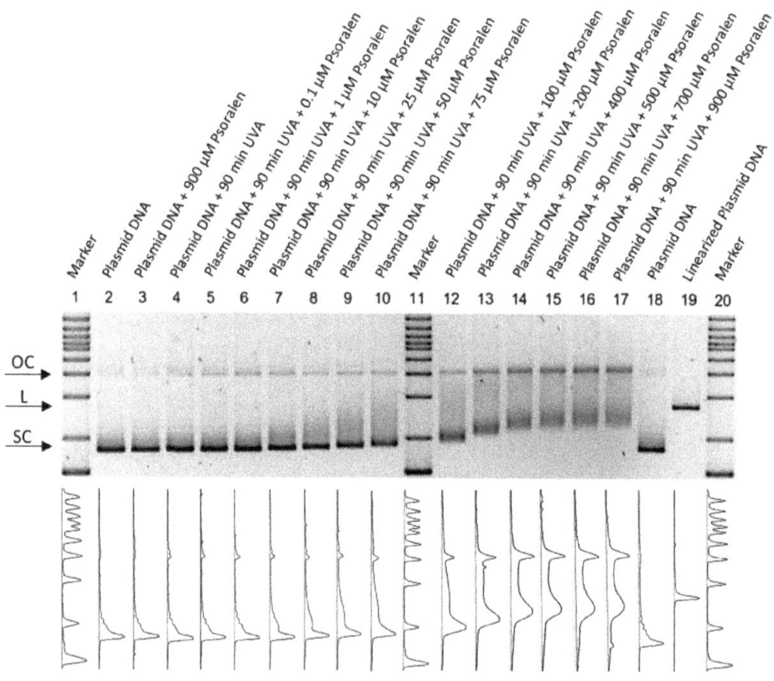

Figure 1. Fluorescent measurement of an agarose gel for combining psoralen with UV-366 nm. The aperture detects the intensity spectra (**bottom**) of the fluorescent gels and displays them visually (**top**). The figure is inverted, thus showing a dark color for high intensities. Since this measurement is not quantitative, the intensity scale only shows arbitrary units. Markers are placed in lanes 1, 11, and 20. Lane 2 and 18 show untreated plasmid DNA, whilst lane 19 contains enzymatically linearized plasmid as the control. In lanes 4–10 and 12–17, different psoralen concentrations (0, 0.1, 1, 10, 25, 50, 75, 100, 200, 400, 500, 700, 900 µM) were irradiated with UV-366 nm for 90 min. Each lane was normalized and shows the running behavior of the different setups. A higher concentration of psoralen showed a higher amount of open circular conformations (OCs). Lane 3 demonstrates non-UV-activated psoralen (900 µM).

A similar effect was visible for a constant concentration of psoralen and an increasing integral dose of UVA light achieved by longer irradiation times (Figure 2). When looking at the case of 30 min UV treatment and psoralen concentrations in the range of 100 µM to 500 µM, Figure 2 nicely shows that the fraction of SC conformations quickly dropped to zero, and thus the OC conformations increased.

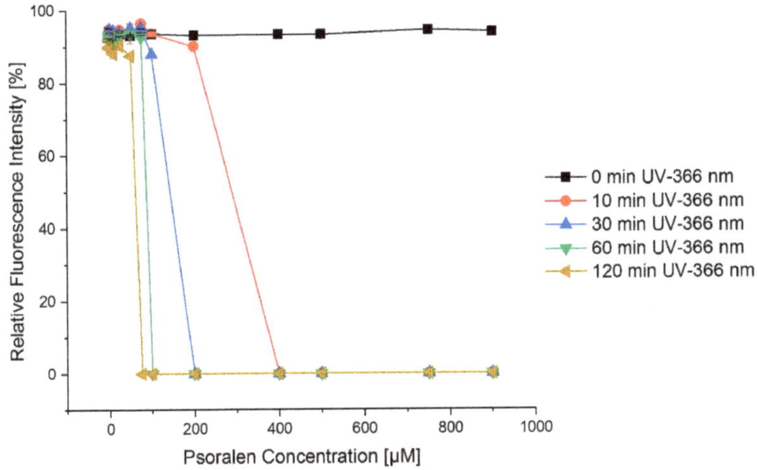

Figure 2. Exemplary concentration–effect relations for the combination of psoralen and varying irradiation times of UV-366 nm for the SC conformation. The longer the irradiation with UV-366 nm, the more intense the fractions of the open circular confirmation as a consequence of the SC conformation quickly dropping to zero. The error bars show the SEM.

Figure 3a depicts the dependency on the time of irradiation for UVC (254 nm). However, this effect could not be increased for varying concentrations of psoralen (Figure 3b).

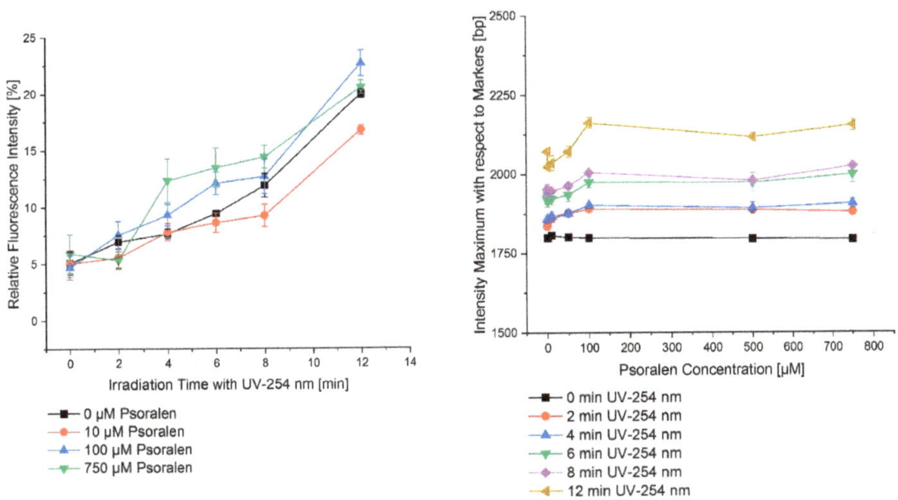

(a) (b)

Figure 3. UV irradiation time–effect relations for the combination of psoralen with UV-254 nm. (**a**) OC fraction as a function of irradiation time, (**b**) OC fraction as a function of psoralen concentration. The effects are displayed as changes in the migration distances of DNA in relation to the markers bp. To increase the irradiation time, the effect as in relative fluorescence intensity also increased. However, this effect was nearly concentration independent. The error bars show the SEM.

Psoralen was dissolved in DMSO (1% $v/v \equiv 0.14$ M DMSO) to obtain a stock solution of 1000 µM psoralen. To further investigate the additional effects of DMSO on the damage of DNA, the light exposure was repeated with and without DMSO (0.2 M) concentrations with varying concentrations of psoralen. However, in either way, there was no visible effect (Figure 4).

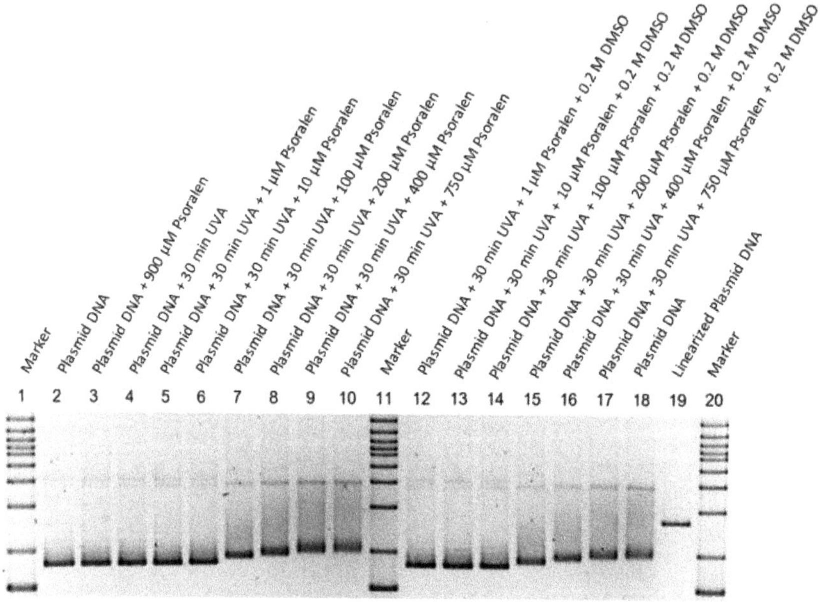

Figure 4. Fluorescent measurement of an agarose gel for combining psoralen with UV-366 nm with an additional 0.2 M DMSO for several lanes. The figure is inverted, thus showing a dark color for high intensities. Markers are placed in lanes 1, 11, and 20. Lane 2 and 18 show the untreated plasmid, whilst lane 19 contains the linear plasmid. In lanes 4–10 and 12–17, different psoralen concentrations (0, 1, 10, 100, 200, 400, 750 µM) were irradiated with UV-366 nm for 30 min, where an additional 0.2 M DMSO was present in lanes 12–18. Each lane was normalized and shows the running behavior of the different setups. In comparison to Figure 1, the additional DMSO did not produce a different behavior. Lane 3 demonstrates non-UV-activated psoralen (900 µM).

3.2. *Plasmid DNA and Psoralen with Re-188*

The CL yield of Re-188 is dose-dependent. Figure 5 shows this effect.

Additionally, Figure 6 shows the effect of varying volumes of water in combination with Re-188 visible in the measured light intensity. This is important since a certain volume of medium is needed so that the Re-188 can produce CL. For larger water volumes, an increase in light intensity was observed. A maximum volume (in the form of a 1 cm water column depending on the used setup) showed a saturating effect regarding the light intensity. This is due to the depth of the electrons in water producing the measured CL intensity. It has also been proven that DMSO does not have an influence on the light intensity, since it is not photo-activated by the resulting CL (Figure 6b).

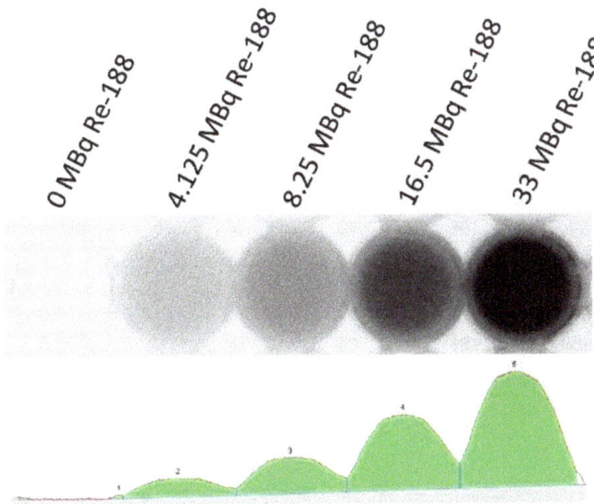

Figure 5. Light intensity of different amounts of Re-188 in a 24-well plate. The picture is inverted, thus, showing increasing intensities of light from left to right (f.l.t.r.: 0, 4.125, 8.25, 16.5, and 33 MBq). The spectrum (**bottom**) was detected by the camera and is shown visually at the top of the figure. If more Re-188 radioactivity is present, the larger the light intensity will be. All intensities were not quantifiable.

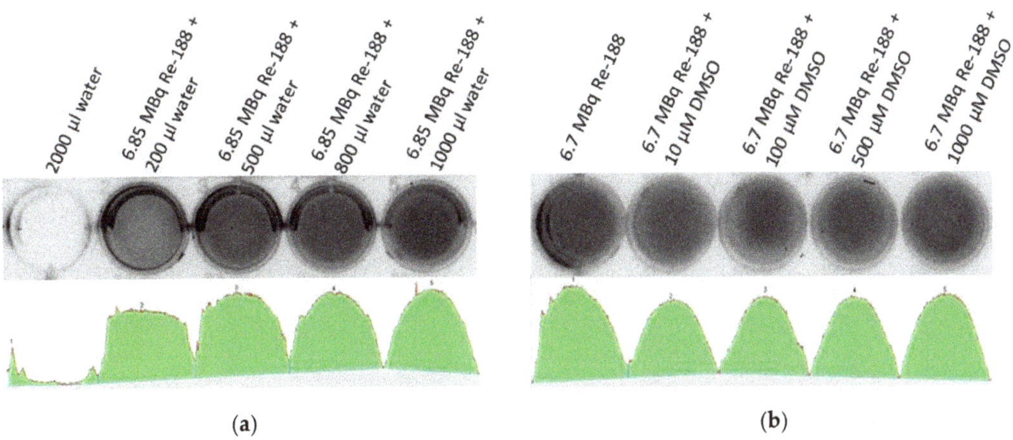

Figure 6. Light intensity of different setups of Re-188 and DMSO as preliminary investigations. The spectra show the detected light intensity in arbitrary units, thus, they are not quantifiable. (**a**) The first well did not contain radioactivity (2000 µL distilled water) and did not consequently show light intensity. The other wells contained a nearly constant amount of Re-188 (approximately 6.85 MBq) and varying volumes of water (f.l.t.r. 200, 500, 800, and 1000 µL). More light intensity was seen for larger water volumes. The spectrum showed that there was no linear behavior between the volume of the medium and the light intensity. However, a saturation was reached for a volume large enough to produce a 1 cm water column (depends on size of well). Measured light intensity: 7 Mio (background, no activity), 22 Mio, 27 Mio, 28 Mio, 29 Mio integral light volume intensity in arbitrary units for each well. (**b**) All wells contained a constant amount of Re-188 (approximately 6.7 MBq), varying amounts of DMSO, and additional water to reach a constant volume of 2000 µL. From left to right, an increasing

amount of DMSO was added to the wells (0, 10, 100, 500, and 1000 µM). The same light intensity was seen for every DMSO concentration, since it is not photo-activated by Re-188. Measured light intensity f.l.t.r.: 28 Mio, 26 Mio, 26 Mio, 27 Mio, 27 Mio integral light volume intensity in arbitrary units for each well.

Mere Re-188-induced radiotoxicity showed a decrease in SC conformations, and thus an increase in the OC fraction. The highest dose showed a transformation of OC to the L plasmid (Figure 7).

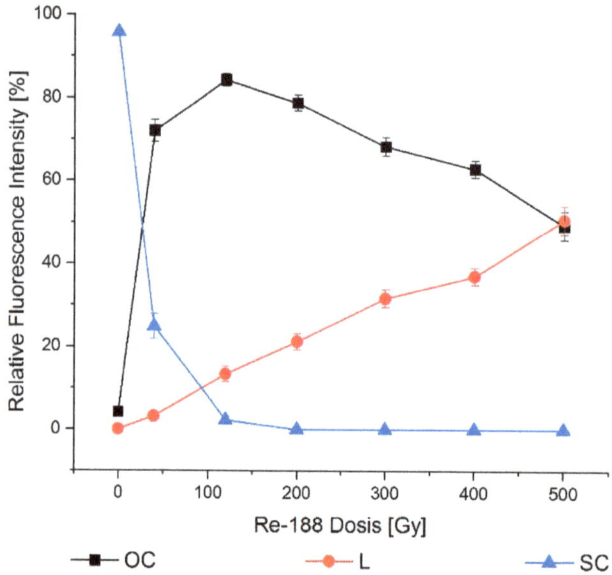

Figure 7. Dose–effect relation for Re-188 after 24 h incubation. The mere radiotoxicity of Re-188 had an influence on all conformations. Supercoiled (SC) transformed into open circular (OC) and eventually into linear (L) conformations for higher doses. The error bars show the SEM.

Regarding the effects of psoralen, it has to be considered that psoralen is dissolved in DMSO, which is a radical scavenger and thus shows a protective effect on the radical-induced DNA damage. In Figure 8, equivalent amounts of DMSO with and without psoralen are compared.

Small differences between the effects could be demonstrated with a slightly larger effect for psoralen compared to DMSO. Dose–wise comparisons were partially significant (e.g., at 200 Gy, 200 µM: $p = 0.01$).

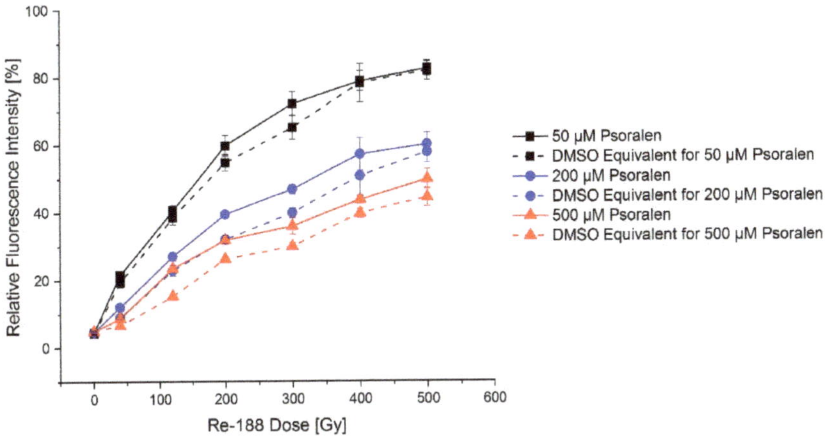

Figure 8. Exemplary dose–effect relation for Re-188 in combination with psoralen after 24 h incubation demonstrated an increasing fraction of open circular plasmid DNA. The influence of DMSO was tested. Psoralen produced a small additional effect, as visible in the difference between the continuous and dashed line. The error bars show the SEM.

3.3. Plasmid DNA and Psoralen with a Combination of UVA and Re-188

For a combination of UVA and Re-188, a slightly more visible formation of single and double strand breaks was observed when psoralen and DMSO, respectively, were not present. When psoralen was present, the effects were weakened, but still stronger than in the case where only DMSO was added (Figure 9).

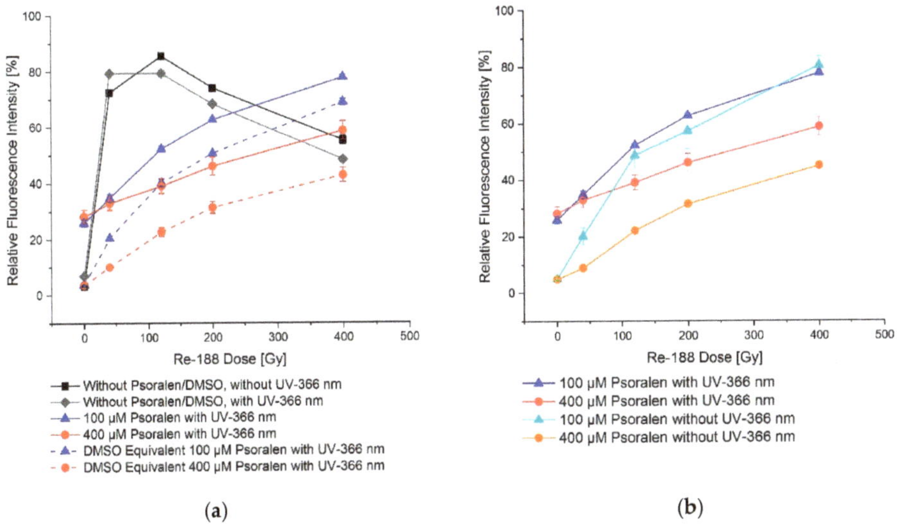

Figure 9. Dose–effect relation on plasmid DNA for Re-188 in combination with psoralen and with (**a**)/without (**b**) a 60-min UV-366 nm irradiation. (**a**) shows the additional fluorescence intensity of open circular conformation for mere radiotoxicity with/without UV-366 nm irradiation. There was a difference between the psoralen and the DMSO equivalent effect. In (**b**) the difference of the fluorescence intensity is shown by comparing UV-366 nm irradiation and no irradiation. The effect was larger with UV-366 nm irradiation. The error bars show the SEM.

3.4. FaDu Cells with Psoralen and UVA

In the course of the preliminary experiments, the impact of the time prior to incubation, meaning the time between the addition of psoralen and the UV irradiation on the clonogene survival was excluded.

Psoralen by itself is not toxic in the used concentration and does not affect the cell survival. The same holds for the light exposure. For a combination of UVA irradiation and the photosensitizer, a clear increase in cell damage was visible, depending on the psoralen concentration and the irradiation time (Figure 10).

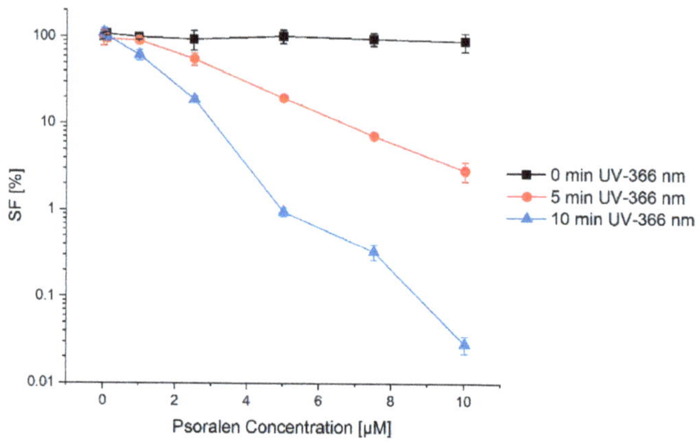

Figure 10. Concentration–effect relation for a combination of psoralen and UV-366 nm (after 24 h preincubation of psoralen in FaDu cells). The longer the irradiation with UV-366 nm, the lower the cell survival fraction (SF). This effect was enhanced by adding psoralen. Without irradiation, there was no difference in survival for any psoralen concentration. The error bars show the SD_{pooled}.

3.5. FaDu Cells with Psoralen and Re-188

Doses of 0 Gy to 12 Gy were applied, and consequently the cell survival was reduced to 0.5%. A clear increase in the effect was not observed for higher concentrations of psoralen, even though there were small differences between some psoralen and DMSO concentrations and applied doses. This can be seen in Figure 11.

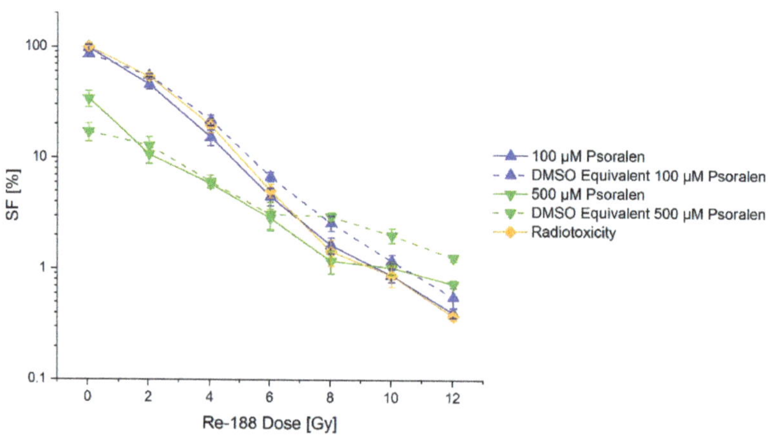

Figure 11. Concentration–effect relation for a combination of psoralen and Re-188 on the survival (SF) of FaDu cells (mere radiotoxicity is additionally shown). No significant difference between mere radiotoxicity and adding of psoralen was visible. The error bars show the SD_{pooled}.

4. Discussion

For some time, it has been issued that CL is not only usable for visual imaging [3,8,9] but can also be used for the photoactivation of photosensitizers [20]. Nonetheless, the experimental data have controversial opinions [5,21–25] and the essential principles are not clear.

Psoralen is clinically well-established for photodynamic therapy (PDT) [2] and can be used on isolated plasmid DNA as well as intact cells. Plasmid DNA has several crucial advantages. Here, it works as a basic setup, which is needed to investigate the principle methods and procedures. There is no barrier in plasmid DNA, as there would be in cells. In addition, the control over scavengers is possible, which allows for an influence on the whole system. Additionally, plasmid DNA does not have repair mechanisms, thus the caused damage is not interfered with. Using these benefits, a more precise and controllable environment was created for the experiments. We expected to find the same effects in the cell model as described for the plasmid DNA. These can become clearer due to further cell structures (cell organelles, proteins) and the impairment of DNA repair due to the formation of crosslinks.

The photochemical reaction of psoralen with the pyrimidine bases and the formation of psoralen monoadducts and crosslinks showed the well-examined and essential effects of the combination of psoralen with UVA (PUVA therapy) [19,26–30]. This again confirms the applicability of this concept.

UVA can be induced by photon irradiation in the megavolt range [10,12,31–33]. In a simple in vitro model, Yoon et al. showed that tumor cells incubated with psoralen showed a 10–20% lower survival fraction (SF) when exposed to light coming from megavolt irradiation, in addition to ionizing irradiation. This approach was adopted for the radionuclides. In plasmid DNA, it could be shown that psoralen is not toxic and does not damage the DNA. Only when exposed to light in the scope of the absorption maximum of psoralen (366 nm) did the interconnectivity of the DNA lead to a change in electrophoretic properties. However, only a small number of single strand breaks (SSBs) and no double strand breaks (DSBs) were observed. UVC (254 nm) showed a direct damage to plasmid DNA that could not be enhanced by prior incubation with psoralen nor prevented by scavenger DMSO (data not shown).

The typical effect of ionizing radiation on DNA in the form of SSBs and DSBs on plasmid DNA is known [34] and was confirmed using Re-188 (Figure 7). The emission of high energetic electrons also generates CL. Thus, radioisotopes provide the opportunity to produce CL and are usable for PDT. Since psoralen is dissolved in 1% DMSO solution, which serves as a scavenger, the combined experiments showed a much lower damage of DNA than the control without psoralen. Yoon et al. did not go into detail on this aspect, since they did not have a control without psoralen, but only as an additional effect visible for exposed cells compared to the light shielded cells. In principle, a negative control was also possible for exposed radioisotopes when using isotopes with none or only very low emissions of CL (e.g., Lu-177 with 0.141 yield per decay) [4].

Even though there was no toxic effect by only psoralen or light (λ = 366 nm), the combination of both showed an energy- and concentration-dependent decrease in cell survival. Since these effects were observed in the presence of the DMSO (as the dissolvent for psoralen), scavenger-induced processes cannot be the source of this effect.

The treatment of FaDu cells with Re-188 showed a dose-dependent reduction in the cell survival. There was a small additional cytotoxic effect visible for psoralen concentrations larger than10 µM compared to the DMSO equivalent combined with Re-188. This effect could be proven frequently. There was no increased cytotoxicity with increasing psoralen concentration and constant dose observable. Deviations for the highest psoralen concentration can be ascribable to the fact that only half of the solution was still in the culture medium, whilst the other half was the psoralen solution. The small fraction of culture medium implies a deficit of nutrition for the cells. Runge et al. [34] investigated the scavenging effect of DMSO on the PCCl3 cell line. The effect was the highest for a standard dose of 7.5 Gy and 0.2 M DMSO (SF_{+DMSO}/SF_{-DMSO} factor = 2.83). For DMSO concentrations of 0.01 mM and 0.05 mM, the SFs were 1.14 and 1.69, respectively. For a dose of 8 Gy and either 100 µM or 500 µM (corresponding to 0.013 mM and 0.065 mM DMSO, respectively) psoralen solution, the SFs resulted in values in the same magnitude: 1.79 and 2.06, respectively. Most likely, the deviation was due to the differences between the FaDu and the PCCl3 cell lines.

In conclusion, an activation of psoralen by Re-188 is rather unlikely—the crosslinks are more complex and much harder to repair compared to the scavenger-induced DNA damage, so they have an extended effect on the cell survival. The SF stayed constant for psoralen concentrations >10 µM for an incubation with Re-188, while there was a clearly visible concentration-dependent decrease for the UV exposed psoralen incubated cells. Since there already are a variety of scavengers and enzymes for reactive oxygen species in the culture medium and the cells themselves, only a small influence of the DMSO solution on the biological end point is expected. This is confirmed by the fact that the SF only merely changed for the increasing DMSO concentrations, while a large amount of SSBs and DSBs could be depleted in the plasmid model.

For plasmid DNA, the CL yield was calculated to be $8.14 * 10^{12}$ photons for an irradiation with 4.22 MBq Re-188 over 24 h in 20 µL. In comparison, a 10-min UVA irradiation (E = 20.5 W/m^2, λ = 366 nm, 5 cm distance, 0.09 cm^2 cross-sectional area) induced 0.11 J, which corresponds to 1.04×10^{17} photons [5]. Due to the 10^4-times lower light yield for an irradiation with Re-188, a measurable effect was not expected. The cell irradiation was similarly disadvantageous: a 5-min irradiation (E = 20.5 W/m^2, λ = 366 nm, 5 cm distance, 9.6 cm^2 cross-sectional area) had a 1.09×10^{19} photon yield, while 24 h of 8.22 MBq of Re-188 in 2 mL resulted in a 1.59×10^{13} photon yield. However, luminescence possibly takes up a large fraction of the light yield, which was not taken into account in this calculation [35]. Yoon et al. did not make an assumption on the photon yield in their experiments. The information on the CL irradiation with 4–8 µJ/m^2 per Gy in the spectrum of 250–850 nm [12] concluded in a photon yield in the magnitude of 10^{14} for a 6 MeV irradiation, resulting in a dose of 10 Gy. Additionally, Yoon et al. used a 3 cm block of solid water to create CL, whereas in the case of the irradiation in a 20 µL Re-188 solution, there was not enough volume to achieve the maximum Cherenkov yield of 34.9 photons per decay [4].

Hartl et al. [21] used Yttrium-90 (Y-90, a beta emitter) for PDT in combination with clinically established aminolevulinic acid (Prodrug by Protoporphyrin IX) and a modified tetraphenylporphyrin and observed small (first sample) and significant (second sample) differences. Even though Y-90 had a photon yield of 47 photons per decay (compared to 35 photons/decay for Re-188), it did not correspond to the high magnitudes. Another recent examination by Krebs et al. used the loss of the protective group on the kinase inhibitor AZD5438 as a model to prove the photoactivation [6]. A clear effect was observed for 419 MBq Y-90. A 14-h irradiation with a photon yield of 47.3 per decay corresponded to 6.6×10^{15} Cherenkov photons. Comparable 2-min irradiations with UVA corresponded to a photon number of 8.3×10^{18}.

Similar conclusions can be drawn for Kotagiri et al. [23], Quintos-Meneses et al. [36] and Chen et al. [37], who used F-18-FDG with 1.32 photons per decay, so that the photon yield was only at about 4% for Re-188 [5]. Due to their simulation calculations, Glaser et al. also concluded "that it is unlikely that the emission of Cherenkov light by radionuclides is usable source for phototherapy" [38]. A similar setting with higher activity of Re-188 could be investigated. On the other hand, a contrary conclusion was drawn by Pratt et al., where studies showed that CL was detectable in the human body and imaging was different for varying light and radioactivity. Since light is generated even in the depth of tissue this approach is suitable for PDT [39]. Additionally, a new approach compared CL imaging in the visible light range with short-waved infrared CL imaging, showing several advantages regarding the penetration depth and scattering effects [40]. Therefore, different explanations need to be found for the effects described in the literature. Only high activity concentrations of several hundred MBq/mL like those attained in selective internal radiation therapy showed a promising generation of Cherenkov photons by the exposed radionuclides.

5. Conclusions and Outlook

The phototoxicity of psoralen with light of different wavelengths (UVC: λ = 254 nm and UVA: λ = 366 nm) as well as the effect of Re-188, whose emitted beta radiation generates Cherenkov light with a yield of 35 photons per decay, was examined on plasmid DNA and a tumor cell line. Psoralen itself is not chemotoxic. Nonetheless, in combination with UVA (however, not with UVC), a concentration- and dose-dependent effect could be observed on plasmid DNA and intact cells. Thus, the models were suitable to prove phototoxicity. Both models showed a significant radiotoxicity for Re-188. However, there was no additional significant Cherenkov-induced phototoxicity observed when combined with psoralen.

Psoralen derivatives with a higher efficiency [41] or other photosensitizers better fitting the emission spectrum of the Cherenkov light [42] could be used. Furthermore, nanoparticles that supply a light amplification or a wavelength shift toward the red spectrum could be examined [25,33,43,44]. Thereby, enhancement of the effect by a factor of 3000 has been described [45]. Magnetic nanoparticles could be shifted to the tumor region by external magnetic fields to increase the efficiency [46]. In addition, it would be of advantage if simple models are used for external irradiation, from which the systematic effects can be modulated in order to compare these to the application of exposed radionuclides [38]. The concept of photodynamic therapy via Cherenkov light is promising. However, it must be better examined, explained, and understood before it could possibly be introduced into daily clinical routine.

Author Contributions: Conceptualization, C.B. and J.K.; Methodology, R.R.; Validation, R.R. and J.K.; Formal analysis, T.R. and R.F.; Investigation, T.R.; Resources, R.R.; Data curation, L.H.; Writing—original draft preparation, J.K. and L.H.; Writing—review and editing, L.H.; Visualization, L.H.; Project administration, C.B. and J.K. All authors have read and agreed to the published version of the manuscript.

Funding: This research received no external funding.

Institutional Review Board Statement: Not applicable.

Informed Consent Statement: Not applicable.

Data Availability Statement: The data presented in this study are available on request from the corresponding author.

Conflicts of Interest: The authors declare no conflict of interest.

Abbreviations

Re-188	Rhenium-188
SSB	Single strand break
PDT	Photodynamic therapy
CL	Cherenkov light
Ga-68	Gallium-68
Y-90	Yttrium-90
DMSO	Dimethyl sulfoxide
LET	Linear energy transfer
SC	Supercoiled (conformation)
OC	Open circular (conformation)
L	Linear (conformation)
SEM	Standard error of the mean
SD_{pooled}	Pooled standard deviation
SF	Survival fraction
PUVA therapy	UVA therapy with psoralen
DSB	Double strand break
ALA	Aminoevulinic acid
F-18-FDG	Fluor-18-fluorodeoxyglucose

References

1. Patrizi, A.; Raone, B.; Ravaioli, G.M. Management of atopic dermatitis: Safety and efficacy of phototherapy. *Clin. Cosmet. Investig. Dermatol.* **2015**, *8*, 511–520. [CrossRef] [PubMed]
2. Stern, R.S. Psoralen and ultraviolet a light therapy for psoriasis. *N. Engl. J. Med.* **2007**, *357*, 682–690. [CrossRef] [PubMed]
3. Zhang, P.; Wu, M.X. A clinical review of phototherapy for psoriasis. *Lasers Med. Sci.* **2018**, *33*, 173–180. [CrossRef] [PubMed]
4. Gill, R.K.; Mitchell, G.S.; Cherry, S.R. Computed Cerenkov luminescence yields for radionuclides used in biology and medicine. *Phys. Med. Biol.* **2015**, *60*, 4263–4280. [CrossRef]
5. Kotzerke, J.; Runge, R.; Gotze, P.; Wunderlich, G.; Enghardt, W.; Freudenberg, R. [Radio- and photosensitization of plasmid DNA by DNA binding ligand propidium iodide: Investigation of Auger electron induction and detection of Cherenkov-emission]. *Nuklearmedizin. Nucl. Med.* **2019**, *58*, 319–327.
6. Krebs, M.; Dobber, A.; Rodat, T.; Lutzen, U.; Zhao, Y.; Zuhayra, M.; Peifer, C. Photopharmacological Applications for Cherenkov Radiation Generated by Clinically Used Radionuclides. *Int. J. Mol. Sci.* **2021**, *22*, 9010. [CrossRef]
7. Bianfei, S.; Fang, L.; Zhongzheng, X.; Yuanyuan, Z.; Tian, Y.; Tao, H.; Jiachun, M.; Xiran, W.; Siting, Y.; Lei, L. Application of Cherenkov radiation in tumor imaging and treatment. *Future Oncol.* **2022**, *18*, 3101–3118. [CrossRef]
8. Olde Heuvel, J.; de Wit-van der Veen, B.J.; van der Poel, H.G.; Bekers, E.M.; Grootendorst, M.R.; Vyas, K.N.; Slump, C.H.; Stokkel, M.P.M. (68)Ga-PSMA Cerenkov luminescence imaging in primary prostate cancer: First-in-man series. *Eur. J. Nucl. Med. Mol. Imaging* **2020**, *47*, 2624–2632. [CrossRef]
9. Tamura, R.; Pratt, E.C.; Grimm, J. Innovations in Nuclear Imaging Instrumentation: Cerenkov Imaging. *Semin. Nucl. Med.* **2018**, *48*, 359–366. [CrossRef]
10. Jain, S.; Yoon, S.W.; Zhang, X.; Adamson, J.; Floyd, S.; Oldham, M. Evaluation of UVA emission from x-ray megavoltage-irradiated tissues and phantoms. *Phys. Med. Biol.* **2019**, *64*, 225017. [CrossRef]
11. Oldham, M.; Yoon, P.; Fathi, Z.; Beyer, W.F.; Adamson, J.; Liu, L.; Alcorta, D.; Xia, W.; Osada, T.; Liu, C.; et al. X-Ray Psoralen Activated Cancer Therapy (X-PACT). *PLoS ONE* **2016**, *11*, e0162078. [CrossRef] [PubMed]
12. Yoon, S.W.; Tsvankin, V.; Shrock, Z.; Meng, B.; Zhang, X.; Dewhirst, M.; Fecci, P.; Adamson, J.; Oldham, M. Enhancing Radiation Therapy Through Cherenkov Light-Activated Phototherapy. *Int. J. Radiat. Oncol. Biol. Phys.* **2018**, *100*, 794–801. [CrossRef] [PubMed]
13. Rangan, S.R. A new human cell line (FaDu) from a hypopharyngeal carcinoma. *Cancer* **1972**, *29*, 117–121. [CrossRef]
14. Eicheler, W.; Zips, D.; Dorfler, A.; Grenman, R.; Baumann, M. Splicing mutations in TP53 in human squamous cell carcinoma lines influence immunohistochemical detection. *J. Histochem. Cytochem.* **2002**, *50*, 197–204. [CrossRef]
15. Scott, B.R.; Pathak, M.A.; Mohn, G.R. Molecular and genetic basis of furocoumarin reactions. *Mutat. Res.* **1976**, *39*, 29–74. [CrossRef]
16. Melough, M.M.; Cho, E.; Chun, O.K. Furocoumarins: A review of biochemical activities, dietary sources and intake, and potential health risks. *Food Chem. Toxicol.* **2018**, *113*, 99–107. [CrossRef] [PubMed]
17. Cai, Z.; Chattopadhyay, N.; Liu, W.J.; Chan, C.; Pignol, J.P.; Reilly, R.M. Optimized digital counting colonies of clonogenic assays using ImageJ software and customized macros: Comparison with manual counting. *Int. J. Radiat. Biol.* **2011**, *87*, 1135–1146. [CrossRef]
18. Franken, N.A.; Rodermond, H.M.; Stap, J.; Haveman, J.; van Bree, C. Clonogenic assay of cells in vitro. *Nat. Protoc.* **2006**, *1*, 2315–2319. [CrossRef]
19. Yoakum, G.H.; Cole, R.S. Cross-linking and relaxation of supercoiled DNA by psoralen and light. *Biochim. Biophys. Acta* **1978**, *521*, 529–546. [CrossRef]
20. Kharroubi Lakouas, D.; Huglo, D.; Mordon, S.; Vermandel, M. Nuclear medicine for photodynamic therapy in cancer: Planning, monitoring and nuclear PDT. *Photodiagnosis. Photodyn. Ther.* **2017**, *18*, 236–243. [CrossRef]
21. Hartl, B.A.; Ma, H.S.W.; Hansen, K.S.; Perks, J.; Kent, M.S.; Fragoso, R.C.; Marcu, L. The effect of radiation dose on the onset and progression of radiation-induced brain necrosis in the rat model. *Int. J. Radiat. Biol.* **2017**, *93*, 676–682. [CrossRef] [PubMed]
22. Kotagiri, N.; Cooper, M.L.; Rettig, M.; Egbulefu, C.; Prior, J.; Cui, G.; Karmakar, P.; Zhou, M.; Yang, X.; Sudlow, G.; et al. Radionuclides transform chemotherapeutics into phototherapeutics for precise treatment of disseminated cancer. *Nat. Commun.* **2018**, *9*, 275. [CrossRef] [PubMed]
23. Kotagiri, N.; Sudlow, G.P.; Akers, W.J.; Achilefu, S. Breaking the depth dependency of phototherapy with Cerenkov radiation and low-radiance-responsive nanophotosensitizers. *Nat. Nanotechnol.* **2015**, *10*, 370–379. [CrossRef] [PubMed]
24. Pratx, G.; Kapp, D.S. In Regard to Yoon et al: Cherenkov-Activated Phototherapy. *Int. J. Radiat. Oncol. Biol. Phys.* **2018**, *101*, 494–495. [CrossRef]
25. Pratx, G.; Kapp, D.S. Is Cherenkov luminescence bright enough for photodynamic therapy? *Nat. Nanotechnol.* **2018**, *13*, 354. [CrossRef]
26. Ashwood-Smith, M.J.; Grant, E. Conversion of psoralen DNA monoadducts in *E. coli* to interstrand DNA cross links by near UV light (320–360 nm): Inability of angelicin to form cross links, in vivo. *Experientia* **1977**, *33*, 384–386. [CrossRef]
27. Bethea, D.; Fullmer, B.; Syed, S.; Seltzer, G.; Tiano, J.; Rischko, C.; Gillespie, L.; Brown, D.; Gasparro, F.P. Psoralen photobiology and photochemotherapy: 50 years of science and medicine. *J. Dermatol. Sci.* **1999**, *19*, 78–88. [CrossRef]
28. Chen, J.X.; Kagan, J. Sites of preferred interaction between double-stranded pBR322 DNA and 7-methylpyrido[3,4-c]psoralen. *J. Photochem. Photobiol. B* **1997**, *39*, 56–62. [CrossRef]

29. Cimino, G.D.; Gamper, H.B.; Isaacs, S.T.; Hearst, J.E. Psoralens as photoactive probes of nucleic acid structure and function: Organic chemistry, photochemistry, and biochemistry. *Ann. Rev. Biochem.* **1985**, *54*, 1151–1193. [CrossRef]
30. Xia, W.; Gooden, D.; Liu, L.; Zhao, S.; Soderblom, E.J.; Toone, E.J.; Beyer, W.F., Jr.; Walder, H.; Spector, N.L. Photo-activated psoralen binds the ErbB2 catalytic kinase domain, blocking ErbB2 signaling and triggering tumor cell apoptosis. *PLoS ONE* **2014**, *9*, e88983. [CrossRef]
31. Axelsson, J.; Davis, S.C.; Gladstone, D.J.; Pogue, B.W. Cerenkov emission induced by external beam radiation stimulates molecular fluorescence. *Med. Phys.* **2011**, *38*, 4127–4132. [CrossRef] [PubMed]
32. Clement, S.; Chen, W.; Deng, W.; Goldys, E.M. X-ray radiation-induced and targeted photodynamic therapy with folic acid-conjugated biodegradable nanoconstructs. *Int. J. Nanomed.* **2018**, *13*, 3553–3570. [CrossRef] [PubMed]
33. Scaffidi, J.P.; Gregas, M.K.; Lauly, B.; Zhang, Y.; Vo-Dinh, T. Activity of psoralen-functionalized nanoscintillators against cancer cells upon X-ray excitation. *ACS Nano* **2011**, *5*, 4679–4687. [CrossRef] [PubMed]
34. Runge, R.; Oehme, L.; Kotzerke, J.; Freudenberg, R. The effect of dimethyl sulfoxide on the induction of DNA strand breaks in plasmid DNA and colony formation of PC Cl3 mammalian cells by alpha-, beta-, and Auger electron emitters (223)Ra, (188)Re, and (99m)Tc. *EJNMMI Res.* **2016**, *6*, 48. [CrossRef] [PubMed]
35. Yamamoto, S. Discovery of the luminescence of water during irradiation of radiation at a lower energy than the Cherenkov light threshold. *Radiol. Phys. Technol.* **2021**, *14*, 16–24. [CrossRef]
36. Quintos-Meneses, H.A.; Aranda-Lara, L.; Morales-Avila, E.; Torres-Garcia, E.; Camacho-Lopez, M.A.; Sanchez-Holguin, M.; Luna-Gutierrez, M.A.; Ramirez-Duran, N.; Isaac-Olive, K. In vitro irradiation of doxorubicin with (18)F-FDG Cerenkov radiation and its potential application as a theragnostic system. *J. Photochem. Photobiol. B* **2020**, *210*, 111961. [CrossRef]
37. Chen, Y.A.; Li, J.J.; Lin, S.L.; Lu, C.H.; Chiu, S.J.; Jeng, F.S.; Chang, C.W.; Yang, B.H.; Chang, M.C.; Ke, C.C.; et al. Effect of Cerenkov Radiation-Induced Photodynamic Therapy with (18)F-FDG in an Intraperitoneal Xenograft Mouse Model of Ovarian Cancer. *Int. J. Mol. Sci.* **2021**, *22*, 4934. [CrossRef]
38. Glaser, A.K.; Zhang, R.X.; Andreozzi, J.M.; Gladstone, D.J.; Pogue, B.W. Cherenkov radiation fluence estimates in tissue for molecular imaging and therapy applications. *Phys. Med. Biol.* **2015**, *60*, 6701–6718. [CrossRef]
39. Pratt, E.C.; Skubal, M.; Mc Larney, B.; Causa-Andrieu, P.; Das, S.; Sawan, P.; Araji, A.; Riedl, C.; Vyas, K.; Tuch, D.; et al. Prospective testing of clinical Cerenkov luminescence imaging against standard-of-care nuclear imaging for tumour location. *Nat. Biomed. Eng.* **2022**, *6*, 559–568. [CrossRef]
40. Mc Larney, B.; Zhang, Q.; Pratt, E.C.; Skubal, M.; Isaac, E.; Hsu, H.T.; Ogirala, A.; Grimm, J. Shortwave infrared detection of medical radioisotope Cerenkov luminescence. *J. Nucl. Med.* **2022**, *63*. [CrossRef]
41. Buhimschi, A.D.; Gooden, D.M.; Jing, H.; Fels, D.R.; Hansen, K.S.; Beyer, W.F., Jr.; Dewhirst, M.W.; Walder, H.; Gasparro, F.P. Psoralen Derivatives with Enhanced Potency. *Photochem. Photobiol.* **2020**, *96*, 1014–1031. [CrossRef] [PubMed]
42. Clement, S.; Anwer, A.G.; Pires, L.; Campbell, J.; Wilson, B.C.; Goldys, E.M. Radiodynamic Therapy Using TAT Peptide-Targeted Verteporfin-Encapsulated PLGA Nanoparticles. *Int. J. Mol. Sci.* **2021**, *22*, 6425. [CrossRef] [PubMed]
43. Boschi, F.; Spinelli, A.E. Nanoparticles for Cerenkov and Radioluminescent Light Enhancement for Imaging and Radiotherapy. *Nanomaterials* **2020**, *10*, 1771. [CrossRef] [PubMed]
44. Squillante, M.R.; Justel, T.; Anderson, R.R.; Brecher, C.; Chartier, D.; Christian, J.F.; Cicchetti, N.; Espinoza, S.; McAdams, D.R.; Muller, M.; et al. Fabrication and characterization of UV-emitting nanoparticles as novel radiation sensitizers targeting hypoxic tumor cells. *Opt. Mater.* **2018**, *80*, 197–202. [CrossRef]
45. Niu, G.; Chen, X. When radionuclides meet nanoparticles. *Nat. Nanotechnol.* **2018**, *13*, 359–360. [CrossRef]
46. Ni, D.; Ferreira, C.A.; Barnhart, T.E.; Quach, V.; Yu, B.; Jiang, D.; Wei, W.; Liu, H.; Engle, J.W.; Hu, P.; et al. Magnetic Targeting of Nanotheranostics Enhances Cerenkov Radiation-Induced Photodynamic Therapy. *J. Am. Chem. Soc.* **2018**, *140*, 14971–14979. [CrossRef]

International Journal of
Molecular Sciences

MDPI

Article

Towards an In Vitro 3D Model for Photosynthetic Cancer Treatment: A Study of Microalgae and Tumor Cell Interactions

Christopher Holmes [1], Juan Varas [2], Sebastián San Martín [2] and José Tomás Egaña [1,*]

[1] Institute for Biological and Medical Engineering, Schools of Engineering, Medicine and Biological Sciences, Pontificia Universidad Católica de Chile, Vicuña Mackenna 4860, Santiago 7821093, Chile
[2] Biomedical Research Center, School of Medicine, Universidad de Valparaiso, Viña del Mar 2520000, Chile
* Correspondence: jte@uc.cl

Abstract: As hypoxic tumors show resistance to several clinical treatments, photosynthetic microorganisms have been recently suggested as a promising safe alternative for oxygenating the tumor microenvironment. The relationship between organisms and the effect microalgae have on tumors is still largely unknown, evidencing the need for a simple yet representative model for studying photosynthetic tumor oxygenation in a reproducible manner. Here, we present a 3D photosynthetic tumor model composed of human melanoma cells and the microalgae *Chlamydomonas reinhardtii*, both seeded into a collagen scaffold, which allows for the simultaneous study of both cell types. This work focuses on the biocompatibility and cellular interactions of the two cell types, as well as the study of photosynthetic oxygenation of the tumor cells. It is shown that both cell types are biocompatible with one another at cell culture conditions and that a 10:1 ratio of microalgae to cells meets the metabolic requirement of the tumor cells, producing over twice the required amount of oxygen. This 3D tumor model provides an easy-to-use in vitro resource for analyzing the effects of photosynthetically produced oxygen on a tumor microenvironment, thus opening various potential research avenues.

Keywords: tumor; microalgae; photosynthesis; hypoxia; 3D model

Citation: Holmes, C.; Varas, J.; San Martín, S.; Egaña, J.T. Towards an In Vitro 3D Model for Photosynthetic Cancer Treatment: A Study of Microalgae and Tumor Cell Interactions. *Int. J. Mol. Sci.* **2022**, *23*, 13550. https://doi.org/10.3390/ijms232113550

Academic Editors: Antonino Mazzaglia, Angela Scala and Enrico Caruso

Received: 25 August 2022
Accepted: 20 October 2022
Published: 4 November 2022

1. Introduction

Tumor hypoxia is one of the main factors leading to increased malignancies, such as tumor recurrence, metastasis, resistance to treatments, and overall poor patient prognosis [1]. Hypoxia in tumors does not only lead to a more aggressive tumor cell phenotype, but also dampens immune cell recognition and blocks effective immune response from taking place [2,3]. A lack of proper oxygenation also leads to an intrinsic resistance to treatments that use reactive oxygen species (ROS) as their main vehicle for killing tumor cells, such as radio or photodynamic therapies [4]. It is therefore unsurprising that recent efforts have focused on developing ways to oxygenate the tumor microenvironment by using oxygen carriers, hyperbaric oxygenation, or implantable electronic devices [5–8].

In vitro 3D models are a staple for studying tumor growth, drug delivery, and treatment efficacy [9]. At present, there are numerous methods for recreating a 3D structure to mimic the tumor microenvironment. Among these methods, those that thoroughly recreate the extracellular matrix (ECM) are the ones that most closely resemble a real tumor, due to the intrinsic relationship between tumor cells and the ECM [10]. To implement these approaches, tumor cells are embedded into a 3D matrix composed of a biocompatible polymer, such as collagen or alginate, where they grow and develop similar to real tumors. The 3D matrix, commonly made of a natural or synthetic hydrogel, imitates tissue ECM and allows cell attachment, proliferation, migration, and 3D stratification [11]. Of these hydrogels, Matrigel is the most commonly employed for 3D tumor development, mainly due to its ease of use and widespread availability. Due to the nature of its production process,

Matrigel is limited by its batch-to-batch variability and contaminants, which negatively affects the reproducibility of results [12,13]. Furthermore, Matrigel is mainly composed of proteins from the basal membrane of Engelbert–Holm–Swarm mouse sarcomas, presenting mostly type IV collagen, contrary to human tissues which are predominantly composed of type I collagen [14].

Natural collagen hydrogels are another commonly used alternative for cancer organoid production, as their composition resembles the human ECM, thus leading to the development of cell–matrix interactions which help tumor development and differentiation [15–17]. Collagen matrix cultures are highly consistent and reproducible if matrix fabrication and cell seeding is controlled and are easy to analyze by microscopy due to the autofluorescence of collagen fibers [11,18]. Collagen matrices usually also contain proteoglycans (PGs) and glycosaminoglycans (GAGs), which aid in the integration of cells with the ECM and in turn support various cellular processes, such as proliferation, migration, and angiogenesis [19]. Alterations to the ratio and presence of specific PGs/GAGs has been shown to promote carcinogenic phenotypes in cells and has been described as relevant for the development of some specific tumor types [20,21].

Over the last decade, photosynthesis has been proposed as an alternative source for oxygen to tissues in vitro [22] and in vivo [23]. By using photosynthetic microorganisms, such as microalgae and cyanobacteria, researchers have explored the induction of photosynthesis as a simple and controllable local source of oxygen for tissues. This approach has been evaluated in various in vitro and in vivo hypoxic models, such as wound healing [24], myocardial ischemia [25], 3D bioprinted constructs [26,27], and ex situ kidney preservation [28]. Moreover, a phase I clinical trial using photosynthetic collagen scaffolds with the microalgae *Chlamydomonas reinhardtii* was recently published [29], showing the safety of such photosynthetic approaches in humans.

Over the last few years, within the context of photosynthetic therapies, the use of photosynthetic microorganisms has been evaluated as an adjuvant treatment for hypoxic tumor oxygenation [30–34]. For instance, increased ROS production during radio-photodynamic therapy was achieved by coating the microalgae *Chlorella vulgaris* with erythrocyte membranes [30] or a calcium phosphate shell [31], leading to higher tumor cell death in mice. Similarly, co-encapsulation of *C. vulgaris* with perfluorocarbons in calcium-alginate hydrogels increased tumor cell death in vitro and in vivo [33] during photodynamic therapy. Moreover, it has also been shown that the cyanobacteria *Synechoccocus elongatus*, together with black phosphorus nanosheets, increases ROS production in photodynamic therapy, leading to increased tumor cell death and tumor shrinkage in mice [32]. Finally, this same cyanobacteria was also used in conjunction with sonodynamic therapy, increasing tumor cell death in vitro and in vivo [34].

Due to the promising results obtained by using photosynthesis for tumor oxygenation and treatment, the study of the interactions and functional relationships between tumor cells and photosynthetic microorganisms is a novel and highly relevant field of research with growing scientific interest. Therefore, in this work, the microalgae *C. reinhardtii* was co-cultured with a melanoma tumor cell line inside a collagen-GAG scaffold to form the 3D model, and their interaction was characterized by several means. The functional in vitro 3D model used in this study is presented as a platform to explore the relationship between these two cell types for photosynthetic cancer treatment.

2. Results

2.1. *C. reinhardtii* Survival under Cell Culture Conditions

As optimal culture conditions for *C. reinhardtii* significantly differ from human cells, the potential deleterious effect of human physiological conditions in microalgae was analyzed first (Figure 1). Osmolarity of microalgae culture media (TAP) and human cell culture media (RPMI) was measured, averaging 71.0 and 330.7 mOsm/L, respectively. Microalgae were then incubated in each media at 28 or 37 °C for 24 h. After incubation, microalgae morphology was analyzed and results showed that cell integrity was maintained in all conditions, with no obvious membrane rupturing. Interestingly, in TAP, the overall morphology of the cells was generally the same at different temperatures; in RPMI, it was slightly altered, with cells showing a reduction in cytoplasm content inside the cell wall, regardless of temperature (Figure 1A). Next, single cell area was quantified, with no significant observable differences among groups (Figure 1B). Moreover, cell death was measured and quantified by flow cytometry (Figure 1C,D), where results showed no significant effect of the culture media at 28 °C (mortality values of 5.8 ± 1.7% for TAP and 5.9 ± 1.9% for RPMI). In contrast, at 37 °C, the culture media had a significant effect as death ratio increased from 10.0 ± 2.5% in TAP to 18.0 ± 2.8% in RPMI. Next, the total chlorophyll content of the microalgae was measured in all culture conditions (Figure 1E). No effect of the culture media was observed at either temperature. In samples of 1×10^7 microalgae, 28 °C values were 5.9 ± 0.5 µg for TAP and 6.1 ± 0.5 µg for RPMI, while 37 °C values were 4.7 ± 0.1 for TAP and 4.6 ± 0.5 µg for RPMI. Finally, the proliferation capacity of the microalgae was confirmed in all groups by seeding in TAP-Agar plates (Figure 1F). As such, although these microalgae were negatively affected by RPMI media and physiological temperatures, most microalgae are still viable after 24 h of incubation.

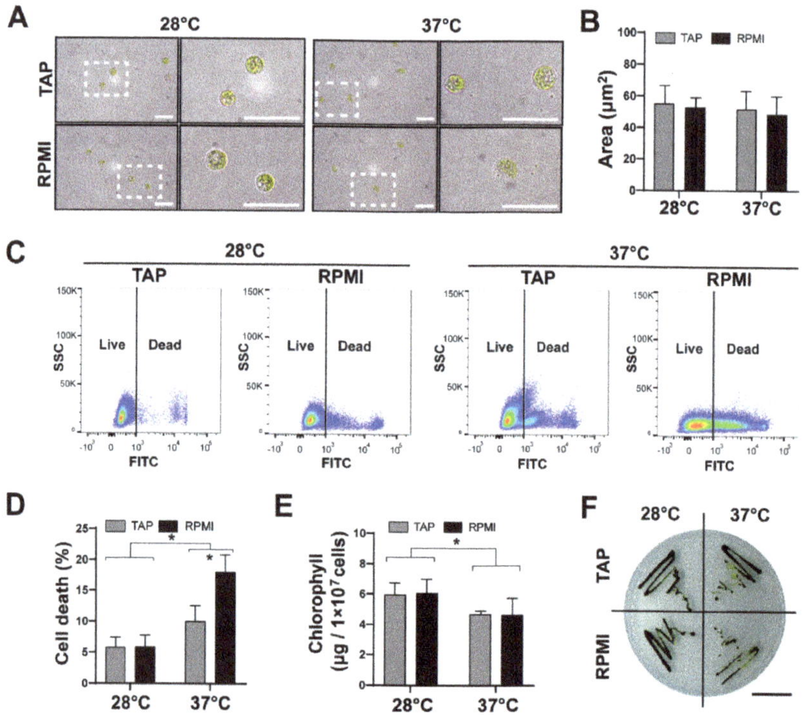

Figure 1. Effect of mammalian cell culture conditions on *C. reinhardtii*. Microalgae were cultured at 28 or 37 °C in TAP or RPMI for 24 h and key features were analyzed. Images of the microalgae show no major morphological changes (**A**). Cell area quantification shows no significant changes (**B**). Microalgae cell death was analyzed by FACS (**C**). Quantification shows that in both TAP and RPMI cell mortality was lower at 28 °C than 37 °C, being significantly higher for the cells cultured at 37 °C in RPMI (**D**). Chlorophyll content significantly decreased at 37 °C, independent of medium (**E**). When recovered, microalgae were able to proliferate in agar plates in all conditions (**F**). Scale bars represent 20 µm in (**A**) and 1 mm in (**F**). All assays were performed in at least three independent experiments and data are presented as average + standard error (SE). $p \leq 0.05$ was considered as significant using a two-way ANOVA, with different letters implying statistical differences. * = $p \leq 0.05$.

After showing that microalgae viability and morphology were not severely affected by RPMI or 37 °C, the next goal was to determine whether such conditions might have an impact on their metabolism by means of oxygen production and consumption rates. Here, microalgae were incubated in TAP or RPMI media at 28 or 37 °C for zero or 24 h in darkness, and their oxygen metabolism was measured (Figure 2). At time zero, results showed that oxygen consumption did not vary at the analyzed temperatures, with values of 0.35 ± 0.06 and 0.24 ± 0.01 nmol/(mL·s) for TAP at 28 °C and 37 °C, respectively. Similarly, no significant variations were determined for RPMI, with values of 0.3 ± 0.05 and 0.23 ± 0.03 nmol/(mL·s) for 28 °C and 37 °C, respectively. Similarly, oxygen production did not significantly vary among all conditions; values of 0.57 ± 0.03 and 0.58 ± 0.04 nmol/(mL·s) at 28 °C for TAP and RPMI, respectively, and 0.58 ± 0.03 and 0.55 ± 0.04 nmol/(mL·s) at 37 °C were recorded (Figure 2, upper panels). In contrast, after 24 h of incubation, significantly lower oxygen consumption was detected for both media at 37 °C, with values of 0.22 ± 0.01 and 0.27 ± 0.04 nmol/(mL·s) for TAP and RPMI, respectively, and 0.36 ± 0.02 and 0.31 ± 0.03 nmol/(mL·s) at 28 °C. Oxygen production was significantly reduced in all conditions, with values of 0.42 ± 0.02 and 0.43 ± 0.05 nmol/(mL·s) for TAP at 28 °C and 37 °C, respectively, and RPMI presented significantly lower values of 0.29 ± 0.01 and 0.31 ± 0.02 nmol/(mL·s) at 28 °C and 37 °C, respectively. In conclusion, at time zero, neither consumption not production was affected by any condition. After 24 h, lower consumption was reported at 37 °C independent of media, while production was overall lower than at time zero and was further reduced by RPMI and 37 °C (Figure 2, lower panels). These results illustrate that although oxygen metabolism was affected by RPMI and physiological temperatures, microalgae can still produce large quantities of oxygen even 24 h after incubation.

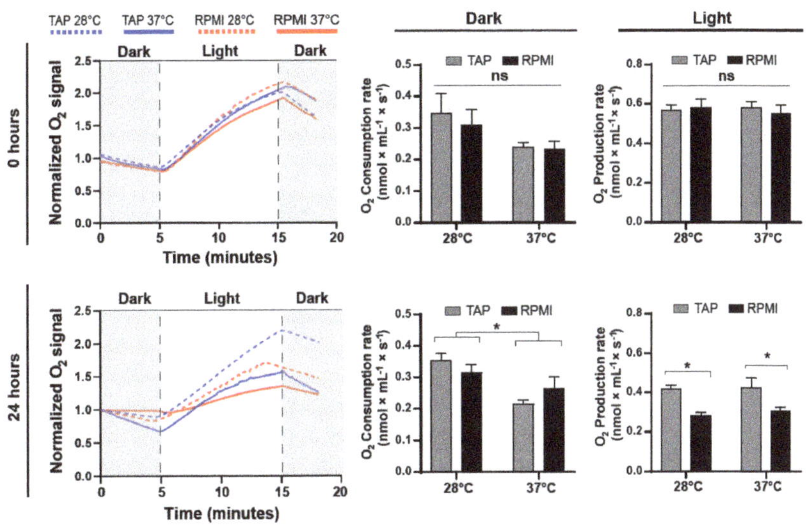

Figure 2. Effect of the mammalian culture conditions in the metabolic profile of *C. reinhardtii*. Microalgae were cultured at 28 or 37 °C in TAP or RPMI for 24 h, and their oxygen production and consumption were evaluated. At time zero, no significant differences were observed among all groups (upper panels). After 24 h in culture, oxygen consumption rate decreased at 37 °C, while production decreased in RPMI at both temperatures. All assays were performed in at least three independent experiments and data are presented as average + SE. $p \leq 0.05$ was considered as significant using a two-way ANOVA, with different letters implying statistical differences. * = $p \leq 0.05$, ns = not significant.

2.2. Biocompatibility of Microalgae and Tumor Cells in Co-Cultures

After characterizing the effect of human-like culture conditions over the microalgae, human melanoma cell line C32 and microalgae were co-cultured in 2D culture plates. First, a microscopic analysis with transmitted light was performed, showing that microalgae were often localized between the cells, leading to apparent morphological changes and detachment of C32 cells (Figure 3A, left). The microalgae were then washed out, and the total area of the plate covered by the cells was quantified (Figure 3A, right); there was a significant reduction in covered area in the co-culture (64.1 ± 1.9%) compared to the control (84.6 ± 2.3%). To further evaluate morphological changes in C32 cells, actin filaments were stained and visualized by fluorescence microscopy, and fiber distribution along the cells was quantified (Figure 3B, upper). Significant differences were detected between conditions, showing higher average actin fiber intensity along the periphery for co-cultures (0.89 ± 0.1 and 0.89 ± 0.1 AU for left and right, respectively) than control cells (0.66 ± 0.09 and 0.58 ± 0.09 AU left and right, respectively). To further analyze cell polarization, a fiber directionality map was generated (Figure 3B, lower). Quantification of fiber coherency showed a significant increase in co-cultures (coherency value of 0.18 ± 0.2 AU for the control and 0.27 ± 0.03 AU for co-cultures). These results indicate that co-cultured C32 cells became polarized, suggesting a migratory phenotype.

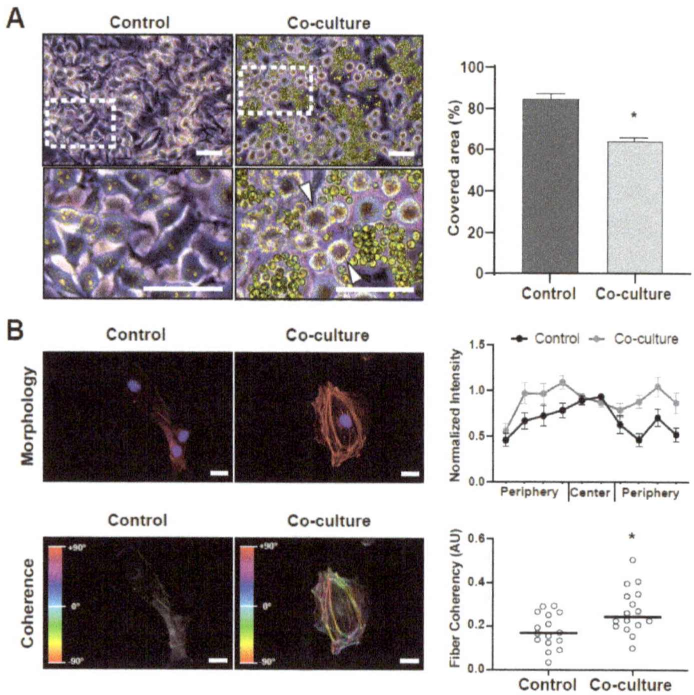

Figure 3. Co-cultures of *C. reinhardtii* and C32 tumor cells. Microalgae were added to sub-confluent C32 seeded plates and left for 24 h in standard culture conditions. A general view of the co-culture is shown in (**A**, **left**), with arrowheads showing cells that are detaching from the plate. A significantly lower surface area was found to be covered by C32 cells in co-culture conditions (**A**, **right**). A cytoskeleton analysis of the cells shows that actin fibers increased in the periphery of the cells when co-cultured (**B**, **upper right**). The actin fiber directionality analysis is color coded, and results show significantly increased fiber coherency in co-culture conditions (**B**, **lower right**). All assays were performed in at least three independent experiments and data are presented as average + SE. Scale bars represent 50 μm in (**A**) and 20 μm in (**B**). $p \leq 0.05$ was considered as significant using a Mann–Whitney test. * = $p \leq 0.05$.

Afterwards, the effect of light in co-cultures was analyzed by several means (Figure 4). First, cell viability was quantified by an MTT metabolic assay (Figure 4A, left). As expected, after 24 h, control C32 cultures showed increased metabolic activity, both in dark (175.3 ± 24.0%) and light (149.7 ± 28.9%) conditions, while in co-cultures, metabolic activity did not significantly differ from time zero control, with values of 79.3 ± 7.8% and 71.2 ± 1.8% for dark and light, respectively. In order to evaluate if the steady state condition was due to increased mortality in the co-culture, cell death was measured by LDH release into the medium (Figure 4A, right). Co-culturing with microalgae induced a small but not significant increase in tumor cell death with respect to control cultures, with values of 1.8 ± 0.7% and 3.4 ± 0.7% for dark and light conditions, respectively. Within the same experimental setting, the viability of the microalgae in co-cultures was also analyzed and further quantified by flow cytometry (Figure 4B). Interestingly, in the absence of C32 cells, the percentage of dead microalgae was significantly higher (18.0 ± 5.6%) compared to dark (3.7 ± 1.7%) or illuminated co-cultures (2.8 ± 3.5%). Finally, the capacity of microalgae to proliferate in a co-culture setting was not affected in either illumination setting (Figure 4B). These results indicate that both cell types are biocompatible with each other under long-term incubation.

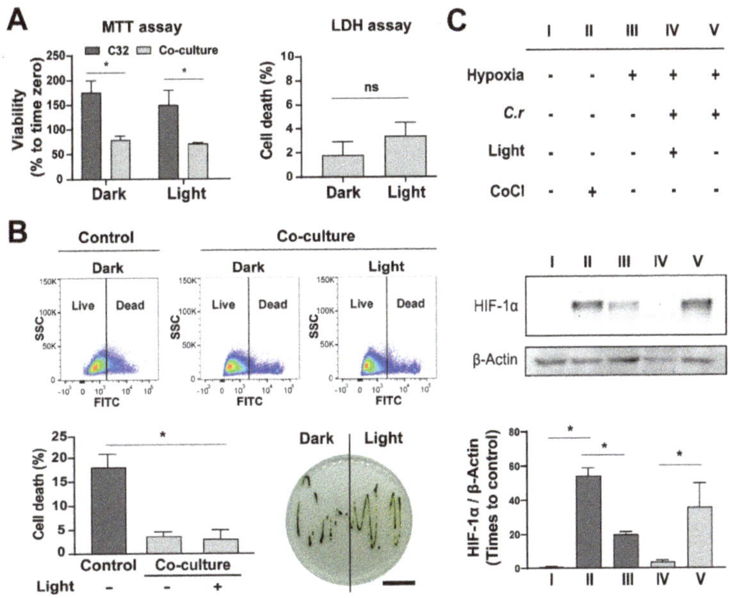

Figure 4. Effect of illumination in co-cultures of *C. reinhardtii* and C32 tumor cells. Cells were co-cultured for 24 h in the absence or presence of light. As shown by an MTT assay, regardless of the presence of light, the mitochondrial activity of C32 cells decreased significantly in co-cultures (**A, left**), while cell death did not vary amongst groups but was slightly increased from control (**A, right**). Similarly, light did not affect viability nor proliferation capacity of the microalgae in co-cultures; however, its mortality was significantly lower when compared to control monocultures in cell media (**B**). The effect of light on hypoxia was evaluated (**C**) by HIF-1α expression. Cells were grown in normoxia (I), incubated with CoCl (II) or grown in hypoxia (III). HIF-1α decreased in illuminated hypoxic co-cultures (IV) when compared to hypoxic co-cultures in the dark (V). All assays were performed in at least three independent experiments and data are presented as average + SE. The scale bar represents 1 mm in (**B**). $p \leq 0.05$ was considered as significant by using a two-way ANOVA test in an MTT viability assay, a Mann–Whitney test in an LDH release assay, and one-way ANOVA in B and C. * = $p \leq 0.05$, ns = not significant.

As microalgae and C32 cells proved to be biocompatible in co-cultures, the capability of microalgae to inhibit hypoxia in C32 cells was then analyzed (Figure 4C). For this, co-cultures were placed in a hypoxic chamber for 24 h and the levels of HIF-1α, the master protein and hallmark for the hypoxic-response, were evaluated by Western blot. In the absence of microalgae, low levels of HIF-1α were detected in normoxic C32 cells (0.7 $\pm$ 0.3), which significantly increased in hypoxia (19.9 $\pm$ 1.7) or in the presence of CoCl (54.1 $\pm$ 4.8), a chemical inductor of HIF-1α. In hypoxic co-cultures, HIF-1α levels increased in darkness (35.6 $\pm$ 14.1) but were inhibited when cultures were illuminated (3.6 $\pm$ 1.0). These results show that, in the presence of light, microalgae are able to oxygenate cell cultures under hypoxia.

2.3. Establishment and Characterization of the Photosynthetic 3D Tumor Model

Once biocompatibility was proven in a standard in vitro culture system, the next step was to engineer a scaffold-based 3D model that better resembles the tumor microenvironment. Here, C32 cells were injected into a collagen-GAG scaffold and a seeding efficiency of over 75% was obtained. In order to map and quantify the metabolic activity of the seeded scaffolds, MTT assays were performed (Figure 5A). At time zero, formazan crystals were concentrated near the injection site at the center of the scaffold, showing only some isolated metabolic clusters along the edge. However, after 24 and 48 h of incubation, metabolic activity spread within the scaffold, showing a more intense and homogeneous distribution of crystals. Compared to time zero, MTT quantification showed an increase of 180 $\pm$ 36% and 210 $\pm$ 30% at 24 and 48 h, respectively (Figure 5A, right). As no significant differences were observed between both times, 48 h was chosen for further characterization. Next, seeded scaffolds were cryosectioned and cell nuclei were stained (Figure 5B) for visualization under fluorescence microscopy. Cell distribution along the matrix was analyzed, and images were divided into 40 sections in order to count the total number of cells per section (Figure 5C). Overall, heat map analysis shows that cells were concentrated in the inner sections of the matrix, with a lower amount towards the periphery, regardless of the axis. Intra-axis quantifications show that although cells were present throughout the whole scaffold, significantly larger numbers of cells were present along the lower 50% of the Q axis (Q1 and Q2), closer to the silicone layer, and along the central sections of the D axis, between D4 to D7, both of which correlate with the approximate site of injection at the center of the scaffold. In order to get more details about the interactions between the seeded cells and the scaffold, laser confocal scanning microscopy (LCSM) studies were performed, showing that cells seemed to aggregate into clusters and form micro tumor-like structures within the scaffold (Figure 5D). White arrowheads show strong cell to cell interactions and cell proliferation (Figure 5D, upper right). Arrowheads present at the maximum intensity projection show that the cells were tightly clustered along the collagen sheets, showing direct attachment to the scaffold (Figure 5D, lower right).

After establishing the 3D tumor model, the next step was to incorporate photosynthetic microalgae into the scaffolds (co-cultured scaffolds). As shown by the overall green color, *C. reinhardtii* showed a homogeneous distribution after seeding, with a larger density at the seeding zone in the center of the scaffold (Figure 6A). Co-cultured scaffolds were incubated for 24 h and the viability of C32 cells was quantified by an MTT assay. Interestingly, just as previously observed before in 2D cultures, co-cultured scaffolds did not show differences in metabolic activity (84.2 $\pm$ 9.5%) from time zero control, while scaffolds seeded only with C32 cells (control) showed a significant increase (157.6 $\pm$ 6.0%) (Figure 6B). Aiming to characterize the interaction between both cell types and the scaffold, Scanning Electron Microscopy (SEM) images were taken, analyzed, and colored (Figure 6C). As shown by the arrowheads in the top and side views of the seeded scaffold, microalgae (green) seem to cluster among themselves or on top and in between C32 cells (red) within the collagen sheets of the scaffold. However, as delimitations between collagen sheets and C32 cells were difficult to correctly assess in SEM images, LCSM images were taken to complement this characterization, as fluorescence allows for better identification of each cell type.

Here, results obtained by SEM were confirmed, as the chlorophyll autofluorescence of the microalgae was observed in a similar pattern, showing direct interaction with the microtumors (Figure 6D).

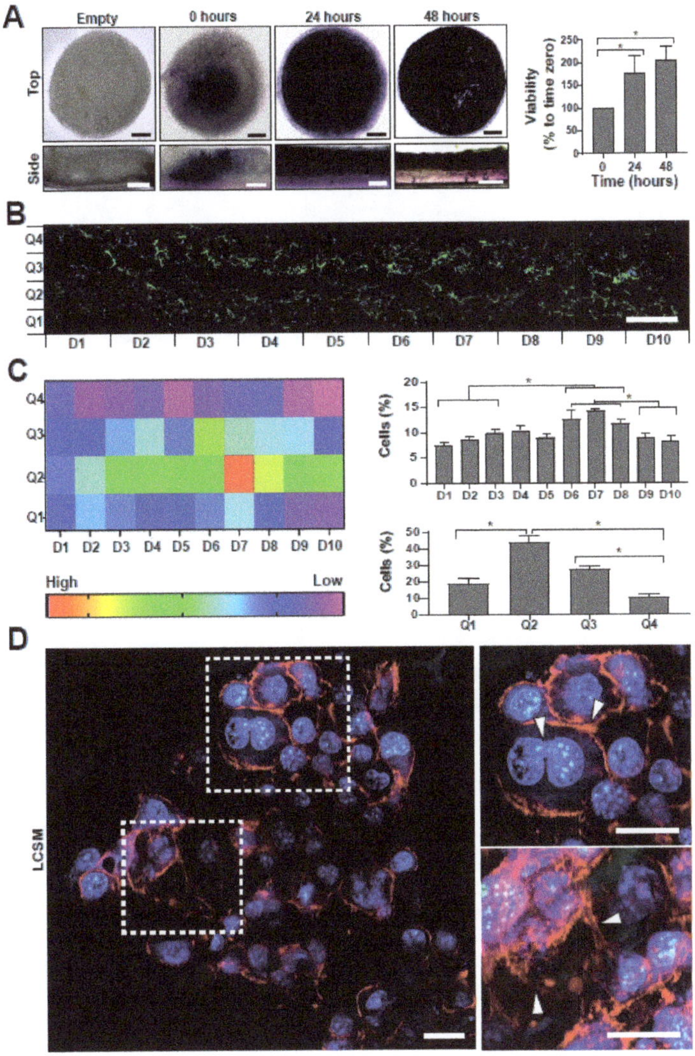

Figure 5. Establishment of a 3D tumor model. C32 cells were seeded in a collagen-GAG scaffold; 48 h post seeding, a significant increase in metabolic activity was quantified by MTT assays (**A**). 3D tumor model was cut into 5 μm slides and cell distribution was visualized by fluorescence microscopy (**B**). A heatmap (**C**, **upper left**) shows the cell distribution in the scaffold, showing a non-homogeneous internal distribution. Detailed analysis of the sections (**C**, **right**) shows that cells concentrate in the central sections of the scaffold. LCSM analysis of the seeded scaffolds shows tumor-like structures (**D**, **left**). Arrowheads in the magnified areas show cell–cell interactions and cell proliferation (**D**, **upper right**) and cell–matrix interactions (**D**, **lower right**). All assays were performed in at least three independent experiments and data are presented as average + SE. Scale bars represent 1 mm in (**A**,**B**) and 25 μm in (**D**). $p \leq 0.05$ was considered as significant using a one-way ANOVA test. * = $p \leq 0.05$.

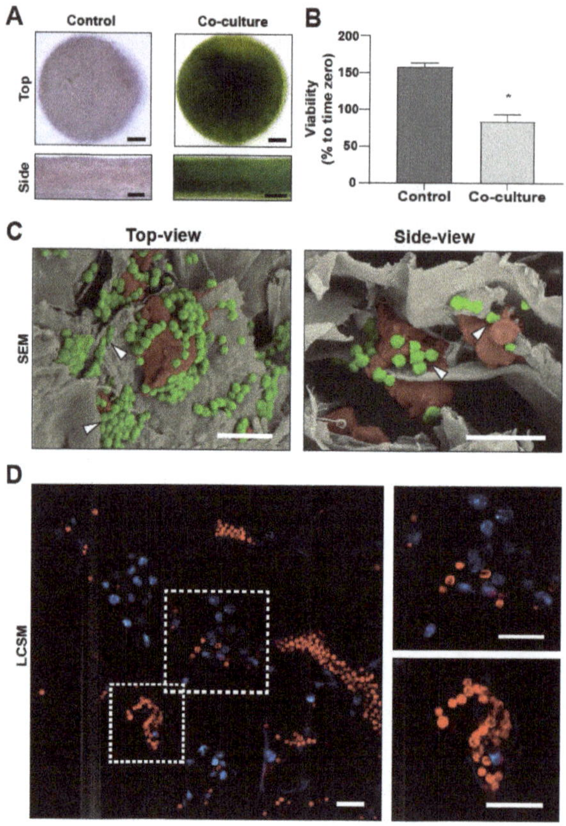

Figure 6. Morphological analysis of a photosynthetic 3D tumor model. C32 cells were seeded in a collagen-GAG scaffold and cultured for 48 h, with microalgae subsequently added. Overview images of the scaffolds show a generally homogeneous distribution of the microalgae (**A**). MTT analysis shows significantly lower metabolic activity in co-cultured scaffolds compared to control scaffolds (**B**). SEM images of 2-h co-cultured scaffolds show tumor cells in direct contact with microalgae (**C**). The top-view arrowheads (**C, left**) show microalgae in between collagen sheets, while side-view arrowheads (**C, right**) indicate microalgae interacting with cancer cells. Detailed LCSM images of 2-h co-cultured matrices (**D**) show direct physical contact between C32 cells (DAPI/blue) and microalgae (chlorophyll/red) (**D, left**). Right panels show magnifications of the indicated sections on the left panel. All assays were performed in at least three independent experiments and data are presented as average + SE. Scale bars represent 1 mm for (**A**), 20 µm for (**C**), and 50 µm for (**D**). $p \leq 0.05$ was considered as significant using a Mann–Whitney test. * = $p \leq 0.05$.

Further, co-cultured scaffolds were histologically analyzed (Figure 7A). Hematoxylin/Eosin (H&E) staining supported SEM and LCSM results, with C32 cells forming micro-tumors. As indicated by the white arrowheads in the H&E panels, microalgae were visible around and in between the tumor cell clusters. Cell proliferation and apoptosis of the C32 cells in the scaffold was evaluated by Ki-67 and Caspase-3 staining, respectively. White arrowheads show positive cells for the corresponding staining. Unexpectedly, a significant increase in Ki-67 positive nuclei was observed in the presence of the microalgae (34.3 ± 3.3%) when compared to the control (16.1 ± 1.4%), while no significant differences were observed between groups for Caspase-3, with an average of 13.56 ± 1.3% positive cells in control scaffolds and 16.3 ± 2.7% positive cells in co-cultured scaffolds. Finally, control and co-cultured scaffolds were subjected to dark and light cycles to determine if the oxygen released by the presence of microalgae could supply the metabolic requirements of the tumor model (Figure 7B). Control scaffolds showed an equal consumption of oxygen independent of light exposure, with an average of 45.0 ± 7.0 pmol/(mL·s) when in darkness and an average of 39.0 ± 8.0 pmol/(mL·s) when exposed to light. On the other hand, co-cultured scaffolds in darkness showed more than double the oxygen consumption of control scaffolds, with an average of 96.0 ± 11.0 nmol/(mL·s). When exposed to light, co-cultured scaffolds showed high oxygen production, enough to supply the oxygen consumption needed by the tumor cells and produce an average oxygen excess of 127.0 ± 17.0 nmol/(mL·s). Therefore, microalgae inside the co-cultured scaffold effectively produce enough oxygen to supply the cellular environment, greatly exceeding C32 oxygen requirements.

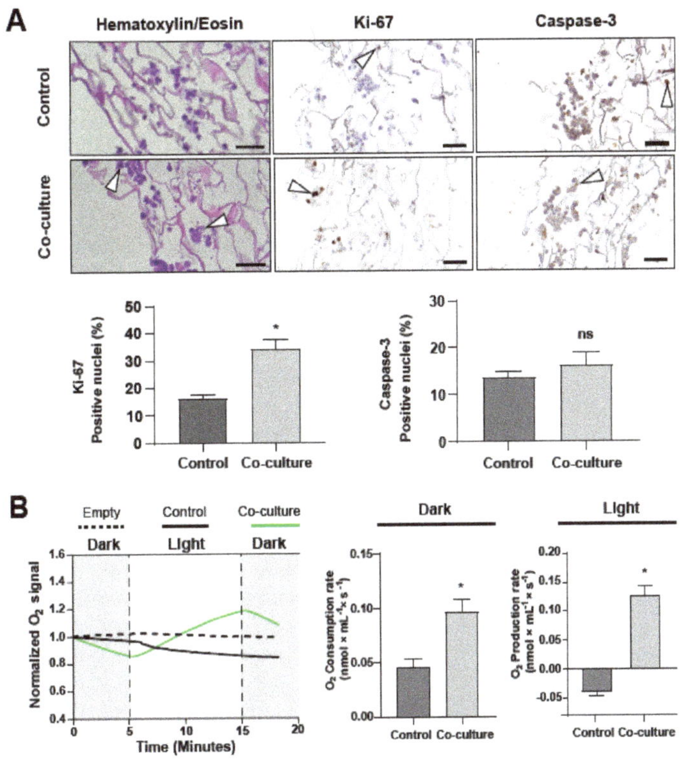

Figure 7. Characterization of a photosynthetic tumor model. Microalgae were seeded into scaffolds with grown tumors and incubated for two hours under physiological conditions. A histologic analysis of the scaffolds was conducted by staining tumor cells with Hematoxylin/Eosin against Ki-67 and Caspase-3. Arrowheads indicate the presence of microalgae clusters (H/E) and positive nuclei for the corresponding stain (Ki-67 and Caspase-3). Ki-67 quantification (**A, lower left**) shows a significant increase in tumor cell proliferation when microalgae are present, while Caspase-3 quantification (**A, lower right**) shows no significant differences were detected. Oxygraphic studies of co-cultured scaffolds show that, in darkness (**B, middle**), co-cultured scaffolds consume twice as much oxygen as control scaffolds. Under light (**B, right**), control (C32) scaffolds keep consuming oxygen, while co-cultured scaffolds produce over twice the consumed amount of oxygen. All assays were performed in at least three independent experiments and data are presented as average + SE. Scale bars in (**A**) represent 100 μm. In all experiments, $p \leq 0.05$ was considered as significant using a Mann–Whitney test. * = $p \leq 0.05$, ns = not significant.

3. Discussion

As there has been an increased interest in the treatment of hypoxic tumors with photosynthetic microorganisms [30,31,34], this work aims to generate a better understanding of the potential interactions between photosynthetic microalgae and tumor cells. As these cell types do not share a common environment in nature, their interaction is far from physiological. Nevertheless, several evolutionary examples support the notion that functional photosymbiotic relationships can be established between hetero and autotrophic systems [35]. Within this context, several groups have suggested that the induction of photosynthetic relationships could be established as a novel approach for the treatment of hypoxia-related pathological conditions [23].

Here, the interaction between the microalga *C. reinhardtii* and C32 melanoma cells was studied at different levels. Among others, *C. reinhardtii* was chosen because it is one of the pioneering microorganisms used for photosynthetic therapies [22]. It is recognized as the go-to microorganism model for studying photosynthesis as it is easy to culture and has an extensive array of molecular tools available [36]. Moreover, it does not produce exotoxins or infectious agents [37], which has propelled its use as a nutritional supplement and its application in several biomedical approaches [28,29,35]. On the other hand, human melanomas were chosen as the cellular model because they tend to develop acidic and hypoxic areas [38] that lead to increased tumor malignancy [39] and high resistance to conventional ROS induction and drug treatments [40]. Among all melanoma cells available for research, the C32 human cell line was chosen due to being amelanotic, thus not interfering with the illumination settings of the study. Moreover, they have been previously combined with photosensitizers [41,42] and used for toxicological screenings of plant compounds [43–46]. Finally, due the feasibility of further illumination and administration of microalgae in prospective patients, melanomas represent a good candidate for the translation of this concept into clinics.

Previous works have described the interactions between photosynthetic microorganisms and different human cell types, such as fibroblasts [22], cardiomyocytes [25], umbilical endothelial cells [47], and various others [30]. However, to our knowledge, this is the first study that describes in detail the behavior of photosynthetic microorganisms under 3D tumor-like conditions in vitro. Compared to optimal microalgae culture settings, such conditions significantly differ in key aspects such as temperature, medium composition, and the presence of tumor cells. Surprisingly, these conditions do not seem to inhibit the intrinsic ability of the microalgae to provide significant amounts of oxygen in the presence of light. In any case, it is interesting to note that RPMI by itself did not trigger the death of microalgae; however, human physiological temperature did, as 37 °C increased cell death by two and three times in TAP and RPMI media, respectively. Unexpectedly, the presence of C32 cells significantly lowered microalgae death and increased viability, which could be due to the removal of metabolic waste or toxic byproducts from the media or the release of certain compounds that might be beneficial for microalgae metabolism, such as CO_2. Moreover, as the protective effect was independent of light, it suggests that it is not tied to photosynthesis. This potential metabolic coupling between both cell types is interesting and deserves to be evaluated in more detail but was out of the scope of this study. Another interesting finding of this work is related to the observation that, in contrast to control conditions, after 24 h in co-culture, the metabolic activity of C32 cells did not increase, which cannot be attributed to cell death (as shown by LDH assays). Though there are no direct parallels in the literature, it has been described that colorectal cancer cell co-cultures with bacteria (*S. aureus* and *S. equisimilis*) induce a G1 arrest of the tumor cells, halting proliferation [48]. Whether these microalgae have a similar effect on tumor cell proliferation is a matter for further studies.

In terms of oxygen metabolism, even at time zero, microalgae consumption was reduced to about a third in TAP or RPMI at 37 °C. This lower consumption is contrary to what has been previously described in the literature, where oxygen consumption has been shown to increase proportionally from 28 °C to 38 °C [49]. However, in this work, the illumination setting used was different, so it cannot be directly compared to the previous studies. Concerning oxygen production, it decreases after 24-h incubation in all conditions. As this phenomenon was also observed even under control culture conditions, this decrease could be attributed to the dark pre-incubation setting and the slow kinetics of the RuBisCO enzyme [50,51]. Contrary to consumption, after 24 h of incubation, production was significantly affected by the culture media but not by temperature, which did not correlate to what was initially expected from the chlorophyll loss at 37 °C. Nevertheless, from a functional perspective, this work shows that in the presence of light microalgae were able to inhibit HIF-1α, therefore fulfilling the metabolic oxygen requirements of hypoxic melanoma cells.

The presence of *C. reinhardtii* cells affects the morphology of C32 cells, indicating possible mechanical changes and a weakening of interactions with the culture plate surface. Melanoma cells are known to easily transition into a migratory phenotype, having an overall lower expression of cadherins and integrins, as well as high expression of migration-promoting proteins when compared to non-cancerous tissue [52]. Cancer cells have also been shown to have a reduced amount of stress fibers when compared to non-cancerous tissue, leading to reduced stiffness and increased response to morphological changes. This results in a mostly amoeboid migration of the cells, though mesenchymal migration can also occur [53]. Our results suggest that, in co-culture, C32 cells show an increased presence of stress fibers in their periphery and increased fiber coherency. An increase in stress fibers suggests higher actin fiber contractility, leading to stiffer membranes. This also indicates the application of forces parallel to the adhesion surface, most likely resulting in increased mesenchymal-type migration [54]. Although the results presented here show that cells are more polarized, further experiments are needed to confirm an effective increase in migration.

With all the information described above, an in vitro 3D tumor model was engineered and characterized. Here, a collagen-GAG scaffold was chosen for cell seeding because its composition and structure resemble the tissue extracellular matrix. Moreover, this commercially available matrix has been previously shown to be biocompatible in clinical settings, as well as with several cell types in vitro [55–58], including microalgae [22] and cyanobacteria [59]. These collagen scaffolds also present unique advantages for 3D tumor cultures as they have been shown to produce hypoxic cores of cells in their center [18], a process which could be standardized for the present model. The scaffold is completely composed of type I collagen, which is found in almost all human tissues, particularly in skin [60]. On the other hand, GAGs are important for correct cell attachment and also help with scaffold water retention and compressive resistance, helping cell viability [61]. As such, this scaffold serves as a general template for the human ECM, avoiding the inherent variability and reproducibility problems of tissue-derived matrices [11]. Since this is a commercial product, it is manufactured in a controlled setting, which leads to a standardized pore size range [62].

After seeding, cells were distributed along both axes of the matrix, with larger cell density in the center of the matrix near the injection point. Melanoma cells seemed to form microtumors, which were about 100 μm in diameter and consisted of various layers of cells growing around and in between the collagen sheets of the matrix, thus closely resembling real tumor tissue. In this context, it is well known that collagen fiber directionality affects cell growth in both 2D and 3D cultures [63]. In this case, the ordered β-sheet directionality present in the scaffold results in cells aligning alongside these, which somewhat imitates a metastasis prone stroma [64]. Moreover, scaffold pores in this model serve as protected niches for microtumors and help define their shape and size, as well as affect cancer cell migration to the outside [65].

As microalgae are significantly smaller (~10 μm) than mammalian cells and the pores of the scaffold (30–120 μm [62]), they distribute homogeneously in the 3D tumor model. The LCSM and SEM images reveal that microalgae cluster inside the pores of the matrix and around the microtumors while either being in direct contact with them or in adjacent pores, confirming that this proposed model allows for the study of the interactions of both cell types in a more physiological context than standard 2D cell culture assays. Interestingly, the histological analysis showed there was significantly more cell proliferation when in the presence of the microalgae. This result was consistent in all the histological analyzed samples, but it is difficult to explain because of the short exposition time of the cells to the microalgae and the decreased metabolism observed in the co-cultures in MTT assays. Ki-67 has been described as having multiple functions during cell cycle progression [66] and is considered to be a graded marker rather than a binary one [67], which we are not able to correctly quantify with histological analysis. Thus, a more detailed cell-cycle analysis is needed to obtain more concrete results on this matter.

From a metabolic perspective, when in the presence of light, microalgae were able to produce enough oxygen to exceed the metabolic demand of C32 cells by over three times. It has been reported that mammalian tumor cells in culture have an approximate oxygen consumption range of 5 to 200 amol/(cell $\times$ s) [68], with human melanoma cells consuming an average of 37.0 $\pm$ 3.75 amol/(cell $\times$ s) when in a monolayer setting [69]. Control scaffolds consume an average of 45 pmol/s of oxygen, which can be used to roughly calculate the cell number in the scaffolds, these having roughly around 1.2 $\times$ 10^6 cells. Similarly, each co-cultured scaffold produced an average amount of 0.22 nmol/s of oxygen, and each scaffold had around 2 $\times$ 10^7 microalgae, meaning that the average oxygen produced per microalgae was around 11.15 amol/s. Thus, according to these rough calculations, to meet the metabolic requirements of C32 cells, a minimum ratio of 3.3 microalgae per melanoma cell should be used to avoid tumor hypoxia.

Undoubtedly, the 3D model presented here better resembles a real tumor than standard co-culture systems and, in its current shape, allows for the analysis of several specific responses to photosynthetic therapies in tumors. However, a drawback of this model is that this still represents an oversimplified system, as it lacks key accessory cell types present in primary tumors that are critical in shaping the tumor microenvironment, such as fibroblasts or endothelial cells. Nevertheless, this model could also be considered as a novel platform where, in further studies, other cell types could be incorporated. Similarly, this model does bring the extracellular matrix into context, which has been described to be of key importance for correctly modelling in vitro tumors [10]. Further, additional ECM components could also be incorporated.

Altogether, in this work, the behavior of the microalgae *C. reinhardtii* in tumor-like co-culture conditions has been studied and described. A reproducible and easy-to-use model is established for studying the potential therapeutic effects of photosynthetic therapies in tumors, particularly in relation to ROS-dependent therapies, such as photodynamic, chemo-, immune, and radiotherapy. Next, this model is also proposed to be used as a platform for studying the role of other key cell types in the tumor microenvironment, which can be included in the engineered microenvironment. Finally, the seeded 3D scaffolds can be implanted in animal models to study other key aspects of tumor behavior and photosynthesis in in vivo conditions. The findings and their implications should be discussed in the broadest context possible. Future research directions may also be highlighted.

4. Materials and Methods

4.1. Microalgae Culture

A cell-wall deficient UVM4-GFP *C. reinhardtii* strain was cultured as previously described [70]. Microalgae were grown photo-mixotrophically at room temperature (20–25 °C) on either solid Tris Acetate Phosphate (TAP) medium with 1.5% (w/v) agar or in liquid TAP medium placed in an orbital shaker (180 rpm). A lamp with white light was used to provide continuous light exposure of 30 $\mu E/(m^2{\cdot}s)$ [36]. Cell density was determined using a Neubauer chamber. For all illumination settings in 2D and 3D experiments, custom LED illumination equipment (Sky-Walkers Spa., Chile) was used to illuminate the co-cultures. Light intensity and distance were set so that a constant flux of 21.47 $\mu E/(m^2{\cdot}s)$

of blue light (455 nm) and 23.04 μE/(m^2·s) of red light (630 nm) reached the cultures during all experiments.

4.2. Mammalian Cell Culture

The C32 human melanoma cell line was obtained from ATCC (CRL-1585). Cells were cultured in RPMI 1640 media with glutamate (Biological Industries, Cromwell, CT, USA), 10% Fetal Bovine Serum (FBS; PAN-Biotech, Aidenbach, Germany), and 1% Penicillin/Streptomycin (Biological Industries, Cromwell, CT, USA). Cells were kept in T75 flasks under standard culture conditions (37 °C and 5% CO2). Unless specified, cells were seeded and used at 70% confluence for all experiments.

4.3. Microalgae Morphology

Microalgae were centrifuged and resuspended in control TAP media or RPMI 1640 with 10% FBS to a final density of 1×10^7 microalgae/mL. Samples were incubated for 24 h at 28 or 37 °C in darkness. For morphology analysis, microalgae samples were diluted in TAP media to a final density of 1×10^6 microalgae/mL, and 10 μL was loaded onto a glass slide and taken to the light microscope. Morphology was evaluated by optical microscopy (Dmi1, Leica, Germany) with 10×/0.4 and 40×/0.65 objectives and imaged with a standard digital camera (MS60, Mshot, Guangzhou, China). Cell area quantification was performed with ImageJ software [71], where a threshold was applied to the images and the total area of each separate microalgae was calculated.

4.4. Microalga Viability

For flow cytometry experiments, microalgae were recovered from incubations and resuspended in 500 μL PBS to a final density of 1×10^6 microalgae/mL. Samples were incubated with Sytox®-Green (Invitrogen, Waltham, MA, USA) cell death probe for 1 h at room temperature (RT) in darkness then centrifuged, washed twice with PBS, and resuspended in 500 μL PBS. Samples were then analyzed in a flow cytometer (FACS Canto II, BD Biosciences, Franklin Lakes, NJ, USA). All microalgae were first checked for chlorophyll fluorescence in the PE-Cy7 channel, where a gate was set up so that only chlorophyll-positive cells were analyzed. A dead microalgae control was prepared by exposing microalgae to repeated freezing and boiling cycles and then used to set the maximum fluorescence intensity and cutoff point for dead cells. Sytox®-Green fluorescence was quantified in the FITC channel, using the FlowJo analysis software.

4.5. Chlorophyll Quantification

C. reinhardtii microalgae were recovered from incubations, divided into aliquots of 1×10^7 cells, and centrifuged at 600× *g* for 5 min. The supernatant was discarded, and the remaining pellet was frozen at −20 °C. For quantification, pellets were thawed, washed twice with 1 mL of TAP media, resuspended in 500 μL DMSO (Merck, Branchburg, NJ, USA), and left under agitation for 45 min at RT. Absorbance was measured at 665 nm and 648 nm, and chlorophyll quantification was calculated as previously described in the literature [72].

4.6. Microalgae Proliferation Assays

C. reinhardtii microalgae were recovered from incubations, washed with TAP media, and centrifuged at 600× *g* for 5 min. Pellets were plated onto sterile TAP-Agar (1.5% *w*/*v*) and left at RT for 7 days under continuous illumination in sterile conditions. Photos were taken with a standard digital camera (ILCE-6000L, Sony, Tokyo, Japan).

4.7. Metabolic Oxygen Profile of the Microalgae

All oxygen measurements were performed with an Oxygraph+ oxygen monitoring system (Hansatech Instruments, Pentney, UK). *C. reinhardtii* were centrifuged at 1000 rpm for 5 min and resuspended in TAP media to a final density of 1×10^7 microalgae/mL.

Time zero controls were created by taking microalgae directly from liquid cultures with no prior incubation and resuspending them in the appropriate media. For each sample, 1 mL was transferred to the oxygraph chamber, previously calibrated to 28 or 37 °C, and left in darkness until the oxygen concentration reached 200 nmol/mL, after which the experiment started. Samples were subjected twice to cycles of 5 min of darkness, 10 min of light, and 5 min of darkness. Slope values were calculated for each section of the curve by a linear regression.

4.8. Cell Co-Cultures

C32 cells were seeded onto 60 mm plates (1×10^6 cells), 35 mm plates (5×10^5), 12-well multiwell plates (1×10^5 cells), or 96-well multiwell plates (1×10^4 cells) and cultured for 24 h. Microalgae cultures were centrifuged at 1000 rpm for 5 min, resuspended in RPMI 1640 with 10% FBS, and then added to grown C32 cell cultures in a 10:1 ratio of microalgae to cells. Unless specified, all co-cultures were incubated for 24 h in standard culture conditions in darkness. When illuminated, the procedure was performed as described above (see Section 4.1).

4.9. Tumor Cells, Co-Culture Imaging, and Cytoskeleton Analysis

C32 cells were grown on 60 mm plates and microalgae were added to the culture, as described in Section 4.8 above. Cell morphology was analyzed by optical microscopy (Dmi1, Leica, Wetzlar, Germany) with 10×/0.4 and 40×/0.65 objectives and a standard digital camera with 200 ms of exposure and 10 gain (MS60, Mshot, Guangzhou, China). Samples were illuminated with a backlight only, and the same light intensity was kept for all samples. The total area of the plate covered by cells was calculated in ImageJ [71], where a threshold was established, and the covered area was calculated as a percentage of the total area available in each image. For fluorescence microscopy, C32 cells were grown over 12 mm glass covers until 50% confluence was reached, and microalgae were added as described above in Section 4.8. After 24 h in the co-culture, covers were washed once with PBS Ca^{2+}/Mg^{+} and fixed with 4% paraformaldehyde (PFA) for 15 min at 37 °C. Fixed samples were washed three times and incubated with PBS-Triton X-100 0.1% for 10 min. Finally, samples were incubated with 1 μg/mL of Hoescht 3342 (Thermo, Waltham, MA, USA) and 0.17 μM of Phalloidin-AF546 (Thermo, Waltham, MA, USA) according to the manufacturer's instructions. Covers were mounted on glass slides and microscopically imaged (DM500, Leica, Wetzlar, Germany) with 20×/0.4 and 40×/0.65 objectives and a standard digital camera (MS60, Mshot, Guangzhou, China). The camera setting for Hoescht detection was 200 ms of exposure and 15 gain, while exposure was set at 220 ms and 18 gain for Phalloidin-AF546. For quantifying the actin fiber intensity of the cells, images were analyzed in ImageJ [71] by using a script previously described [73]. For calculating actin fiber coherency and making the fiber directionality map, images were analyzed in ImageJ using the plugin OrientationJ according to the protocols described in the literature [74].

4.10. Tumor Cell Viability

For viability and cell death assays, C32 cells were grown on 96-well multiwell plates and microalgae were added as described above in Section 4.8. Here, co-cultures were incubated for 24 h in the presence or absence of light. For viability analysis, an MTT assay was performed. Cells were washed once with PBS Ca^{2+}/Mg^{+} to remove the microalgae, and 10 μL of 500 μM MTT (Invitrogen, Waltham, MA, USA) and 90 μL of complete cell media were then added to each experimental point. Samples were incubated for 2 h in standard culture conditions, after which 100 μL of DMSO was added to each well. Absorbance was measured at 570/630 nm. For cell death analysis, an LDH cytotoxicity assay kit (Thermo, Waltham, MA, USA) was used. Supernatant LDH release of co-cultures and monocultures of each individual cell type was measured and calculated according to the manufacturer's instructions.

4.11. Tumor Cell Immunoblotting

Cells were cultured on 35 mm culture plates and subjected to several experimental conditions. Normoxic cultures were kept under standard culture conditions, while hypoxia was induced by incubation in a sealed chamber (STEMCELL Technologies, Vancouver, BC, Canada), where a continuous nitrogen gas flow (40 mmHg/minute) was passed through for 8 min to remove oxygen. Normoxic and hypoxic cultures were incubated at 37 °C for 24 h. The cultures and the whole hypoxia chamber were illuminated as described above in Section 4.1, and samples that required darkness were protected with aluminum foil. As positive control for HIF-1α induction, 100 μM of CoCl (Merck, Branchburg, NJ, USA) was added to the culture media and left during incubation. After incubation, plates were taken to ice and homogenized in 400 μL of RIPA buffer (NaCl 150 mM, Triton X-100 1%, Sodium Deoxycholate 0.5%, SDS 0.1%, Tris-HCl pH 8.0 50 mM) containing protease and phosphatase inhibitors, before then being incubated for 40 min under constant agitation. Supernatant protein concentration was determined using a commercially available kit (Pierce BCA Protein Assay; Thermo, Waltham, MA, USA). Afterwards, samples were loaded into 7.5% SDS-PAGE gels and electrophoresis was performed. Proteins were transferred onto a PDVF membrane then blocked with TBS-Tween-20 0.2% (BSA 1%). Membranes were incubated in TBS-Tween 0.2% overnight at 4 °C with primary anti-HIF-1α (D1S7W, Cell Signaling, Danvers, MA, USA) diluted 1:1000 or anti-β-Actin (A5060, Sigma-Aldrich, St. Louis, MO, USA) diluted 1:2000, both in TBS-Tween-20 0.2%. Afterwards, membranes were incubated with a secondary Goat anti-Rabbit-HRP (Thermo, Waltham, MA, USA) diluted 1:3000 for 1 h at RT. HRP substrate (Pierce ECL WB substrate, Thermo, Waltham, MA, USA) was added to the membranes and the signal was revealed in a chemiluminescent transilluminator (myECL Imager 62236X, Thermo, Waltham, MA, USA).

4.12. Scaffold Seeding and Co-Cultures

Integra® matrix wound dressing with a silicone layer was kindly donated by Integra Life Science and used for scaffolds (Integra LifeSciences, Princeton, NJ, USA). Six-millimeter diameter disks were cut with a punch biopsy sampler and dried with sterile gauze. Subsequently, 50 μL of cell media containing 6×10^5 cells was loaded into a 1 mL disposable syringe and injected into the center of the scaffold with the silicone layer facing down. Afterwards, seeded scaffolds were placed in 12 well plates previously covered with sterile PBS-Agar 2% and left for 30 min at RT. A total of 2 mL of complete cell culture media was then carefully added to each well and replaced daily. Unless specified, tumor cells were grown for 48 h in the scaffold before being co-cultured or being used for experiments.

For seeding the microalgae into the scaffolds, 2×10^7 microalgae were resuspended in 50 μL of RPMI 1640 (10% FBS). Cell-laden scaffolds were carefully dried with sterile gauze and the resuspended microalgae were added. Co-cultured scaffolds were transferred to PBS-Agar-covered multiwell plates and 2 mL of RPMI-10% FBS was carefully added to each sample. Co-cultured scaffolds were incubated for 2 or 24 h at 37 °C, according to the experimental setting.

4.13. Tumor Cell Viability in the Scaffold

Culture media were removed from cell-laden scaffolds and replaced with 100 μL of 500 μM MTT and 900 μL of complete cell media before then being incubated for 2 h in standard culture conditions. Optical images of scaffolds were taken on a stereomicroscope (S6D, Leica, Aidenbach, Germany) at 1.25× magnification with a standard digital camera (MS60, Mshot, Guangzhou, China). Scaffolds were kept hydrated with PBS Ca^{2+}/Mg^{+} while photos were taken. Samples were then dissolved in 2 mL of DMSO with the help of a plastic tissue homogenizer. When dissolved, the remaining silicone layer was discarded, and 200 μL of each sample was transferred into a 96-multiwell plate. Absorbance was measured at 570/630 nm.

4.14. Tumor Cell Distribution in the Scaffold

C32 cell-laden scaffolds were grown as described in Section 4.12 above and fixed with PFA 4% for 24 h at 4 °C. For cryosection analysis, scaffolds were cut into 5 µm sections (CM1520 cryostat, Leica, Germany) and deposited onto silanized glass slides. Slides were incubated for 1 h at 50 °C and then washed in dH2O. Samples were incubated with 1 µg/mL of Hoescht 3342 (Thermo, MA, USA) in PBS for 30 min and then washed twice with PBS. Slides were analyzed with an epifluorescence microscope (Axio Observer, Zeiss, Oberkochen, Germany) using the Zeiss image software and a 20×/0.35 air objective. Tile by tile mosaic images were taken and digitally stitched. For Hoescht detection, exposure of 200 ms and 15 gain were used, while for collagen autofluorescence, exposure was set at 1.2 s and gain at 22. The nuclei fluorescence channel was isolated in ImageJ [71], a threshold was established, and total nuclei per cut were counted. From the silicone layer, the D and Q axes were defined for each cut, and each axis was subdivided into ten deciles (D) and four quartiles (Q), respectively, in Adobe Illustrator. Cells were counted within each subdivision, and a heatmap was created from the individual values of each subdivision in GraphPad Prism 8. Sections belonging to the same axis were added accordingly and averages for each separate axis were calculated and plotted. All sections in the same cut were added to obtain the total cell number, which was then compared to the original count. Shown values are expressed as a percentage of the total number of cells in a cut. Results correspond to three independent scaffolds, where 10 cuts of each were selected and counted.

4.15. Laser Scanning Confocal Microscopy

Cell-laden scaffolds were fixed with PFA 4% for 24 h, permeabilized with 0.2% PBS-Triton X-100 for 20 min, and washed. Scaffolds were then incubated with 1 µg/mL of Hoescht 3342 (Invitrogen, Waltham, MA, USA) and Phalloidin-AF546 (Thermo, Waltham, MA, USA) for 1 h at RT in darkness and washed twice. Samples were taken to the microscope (Airyscan, Zeiss, Oberkochen, Germany), the silicone layer was removed, and scaffolds were imaged from the top or bottom side. Images were taken with a 40×/0.65 water objective and were analyzed with Zeiss software, where they were post-processed with the Airyscan correction.

4.16. Scanning Electron Microscopy

Cell-laden scaffolds were fixed with glutaraldehyde 2% overnight. Samples were then dehydrated in an ethanol gradient, transferred to acetone for 1 h, and air-dried for 24 h at RT. Samples were mounted, sputtered with gold, and analyzed using 15 kV of acceleration voltage (TM3000, Hitachi, Ibaraki Japan). Images were post processed in Adobe Photoshop, where contrast was increased and cell structures were colored.

4.17. Histology

Cell-laden scaffolds were fixed in PFA 4%, dehydrated in ethanol, and embedded in Paraplast (Leica, Aidenbach, Germany) at 60 °C. Sections of 5 µm in thickness were cut and adhered to glass slides using 0.1% poly-L-Lysine (Sigma-Aldrich, St. Louis, MO, USA) and further dried at RT. Prior to the immunoreaction, some samples were stained with Hematoxylin/Eosin for morphological studies. Sections were deparaffinized, rehydrated, and incubated with rabbit monoclonal primary anti-ki67 (MA5-14520 Invitrogen, Waltham, MA, USA) diluted 1:50 or rabbit polyclonal anti-cleaved caspase-3 (ab2302 Abcam, Cambridge, UK) diluted 1:250, both in PBS Tween-20 0.3%, overnight at 4 °C. Non-specific staining was blocked by 30 min of immersion in Cas-Block solution (Invitrogen, Waltham, MA, USA) and goat serum (Gibco, Carlsbad, MA, USA). After extensive rinsing in PBS, all sections were incubated for 1 h at RT with HRP-conjugated goat anti-rabbit IgG (Jackson ImmunoResearch, West Grove, PA, USA) diluted 1:500 or 1:250 in PBS. The peroxidase reaction was visualized using 3,3-Diamenobenzidine chromogen from the Envision FLEX kit (Agilent, Santa Clara, CA, USA). After immunostaining, sections were slightly stained with Harris hematoxylin (Merck, Branchburg, NJ, USA). For each immunohistochemical reaction, controls were performed either by incubating the sections with PBS or by omitting the primary antibody. Subsequently, the sections were scanned at 40× magnification equivalent resolution on a whole slide scanner (Aperio Versa, Leica, Aidenbach, Germany) and images were captured with Aperio ImageScope 12.4.6 software. Images were then analyzed in ImageJ, where total nuclei per image were counted and the positive nuclei for each stain were quantified. Results are expressed as the percentage of positive nuclei per cut.

4.18. Scaffold Metabolic Analysis

Cell-laden scaffolds were grown and co-cultures were established as previously described (see Section 4.12). The oxygraph (see Section 4.7) was calibrated at 37 °C and 1 mL of PBS 10% FBS was introduced into the chamber. Scaffolds were transferred into the chamber and subjected to the same light–dark cycles as previously discussed (see Section 4.7). Slope values were calculated for each section of the curve by linear regression.

4.19. Statistical Analysis

All assays were performed in at least three independent experiments. All statistics were obtained with GraphPad Prism 8. The statistical tests used are described in each result section. The significance threshold was set at $p \leq 0.05$.

Author Contributions: C.H. and J.T.E. conceived the general concept; C.H. planned and performed the experiments; C.H. and J.T.E. analyzed the data and wrote the paper; J.V. and S.S.M. performed the imuno-histologic analysis. All authors have read and agreed to the published version of the manuscript.

Funding: This research was supported by the National Agency of Research and Development from Chile (ANID/CONICYT), through the projects: FONDECYT (Nr. 1200280 to J.T.E.), Beca Doctorado Nacional (Nr. 21180863 to C.H.), Programa de Atracción e Inserción de Capital Humano Avanzado, Tesis de Doctorado en el Sector Productivo (Nr. T7819120008 to C.H.). This work was supported by CORFO Portafolio I+D grant 18PIDE98887.

Institutional Review Board Statement: Not applicable.

Informed Consent Statement: Not applicable.

Data Availability Statement: All data associated with this study are presented in this paper and can be shared with approved outside collaborators under a materials transfer agreement; requests should be sent to J.T.E., jte@uc.cl.

Acknowledgments: We acknowledge the Fondequip EQM12-0003 and Fondequip EQM13-116 projects for providing the in vitro analyses. We acknowledge the Advanced Microscopy Unit (UMA) of the Pontificia Universidad Católica de Chile for letting us use the laser scanning confocal and scanning electron microscopes. The authors would like to thank Rocío Corrales-Orovio, Valentina Castillo, Felipe Carvajal, Pablo Rosas, and Hilde Buzza for the critical review of this manuscript. The authors would like to thank Alexander Holmes for proofreading this manuscript.

Conflicts of Interest: All the authors declare no conflict of interest, except for J.T.E. who owns IP in the field of this application. The content of this article was expressly written by all authors listed. No ghostwriters were used to write this article.

References

1. Hanahan, D.; Weinberg, R.A. Hallmarks of Cancer: The next Generation. *Cell* **2011**, *144*, 646–674. [CrossRef] [PubMed]
2. Semenza, G.L. The Hypoxic Tumor Microenvironment: A Driving Force for Breast Cancer Progression. *Biochim. Biophys. Acta Mol. Cell Res.* **2016**, *1863*, 382–391. [CrossRef] [PubMed]
3. Vito, A.; El-Sayes, N.; Mossman, K. Hypoxia-Driven Immune Escape in the Tumor Microenvironment. *Cells* **2020**, *9*, 992. [CrossRef] [PubMed]
4. Brown, J.M. Tumor Hypoxia in Cancer Therapy. *Methods Enzymol.* **2007**, *435*, 23. [CrossRef]
5. Xiang, Y.; Bernards, N.; Hoang, B.; Zheng, J.; Matsuura, N. Perfluorocarbon Nanodroplets Can Reoxygenate Hypoxic Tumors in Vivo without Carbogen Breathing. *Nanotheranostics* **2019**, *3*, 135–144. [CrossRef]
6. Song, X.; Feng, L.; Liang, C.; Yang, K.; Liu, Z. Ultrasound Triggered Tumor Oxygenation with Oxygen-Shuttle Nanoperfluorocarbon to Overcome Hypoxia-Associated Resistance in Cancer Therapies. *Nano Lett.* **2016**, *16*, 6145–6153. [CrossRef]

7. Bennett, M.H.; Feldmeier, J.; Smee, R.; Milross, C. Hyperbaric Oxygenation for Tumour Sensitisation to Radiotherapy. *Cochrane Database Syst. Rev.* **2018**. [CrossRef]
8. Maleki, T.; Ning, C.; Song, S.H.; Kao, C.; Ko, S.-C.; Ziaie, B. An Ultrasonically Powered Implantable Micro-Oxygen Generator (IMOG). *IEEE Trans. Biomed. Eng.* **2011**, *58*, 3104–3111. [CrossRef]
9. Atat, O.E.; Farzaneh, Z.; Pourhamzeh, M.; Taki, F.; Abi-Habib, R.; Vosough, M.; El-Sibai, M. 3D Modeling in Cancer Studies. *Hum. Cell* **2021**, *35*, 23–36. [CrossRef]
10. Pape, J.; Emberton, M.; Cheema, U. 3D Cancer Models: The Need for a Complex Stroma, Compartmentalization and Stiffness. *Front. Bioeng. Biotechnol.* **2021**, *9*, 276. [CrossRef]
11. Habanjar, O.; Diab-Assaf, M.; Caldefie-Chezet, F.; Delort, L. 3D Cell Culture Systems: Tumor Application, Advantages, and Disadvantages. *Int. J. Mol. Sci.* **2021**, *22*, 12200. [CrossRef]
12. Vukicevic, S.; Kleinman, H.K.; Luyten, F.P.; Roberts, A.B.; Roche, N.S.; Reddi, A.H. Identification of Multiple Active Growth Factors in Basement Membrane Matrigel Suggests Caution in Interpretation of Cellular Activity Related to Extracellular Matrix Components. *Exp. Cell Res.* **1992**, *202*, 1–8. [CrossRef]
13. Talbot, N.C.; Caperna, T.J. Proteome Array Identification of Bioactive Soluble Proteins/Peptides in Matrigel: Relevance to Stem Cell Responses. *Cytotechnology* **2015**, *67*, 873–883. [CrossRef]
14. Seibel, M.J.; Robins, S.P.; Bilezikian, J.P. (Eds.) *Dynamics of Bone and Cartilage Metabolism: Principles and Clinical Applications*; Academic Press: San Diego, CA, USA, 2006; ISBN 9780120885626.
15. Fang, M.; Yuan, J.; Peng, C.; Li, Y. Collagen as a Double-Edged Sword in Tumor Progression. *Tumour Biol.* **2014**, *35*, 2871–2882. [CrossRef]
16. Xu, S.; Xu, H.; Wang, W.; Li, S.; Li, H.; Li, T.; Zhang, W.; Yu, X.; Liu, L. The Role of Collagen in Cancer: From Bench to Bedside. *J. Transl. Med.* **2019**, *17*, 309. [CrossRef]
17. Giussani, M.; Merlino, G.; Cappelletti, V.; Tagliabue, E.; Daidone, M.G. Tumor-Extracellular Matrix Interactions: Identification of Tools Associated with Breast Cancer Progression. *Semin. Cancer Biol.* **2015**, *35*, 3–10. [CrossRef]
18. Liverani, C.; De Vita, A.; Minardi, S.; Kang, Y.; Mercatali, L.; Amadori, D.; Bongiovanni, A.; La Manna, F.; Ibrahim, T.; Tasciotti, E. A Biomimetic 3D Model of Hypoxia-Driven Cancer Progression. *Sci. Rep.* **2019**, *9*, 12263. [CrossRef]
19. Anderegg, U.; Halfter, N.; Schnabelrauch, M.; Hintze, V. Collagen/Glycosaminoglycan-Based Matrices for Controlling Skin Cell Responses. *Biol. Chem.* **2021**, *402*, 1325–1335. [CrossRef]
20. Wei, J.; Hu, M.; Huang, K.; Lin, S.; Du, H. Roles of Proteoglycans and Glycosaminoglycans in Cancer Development and Progression. *Int. J. Mol. Sci.* **2020**, *21*, 5983. [CrossRef]
21. Theocharis, A.D.; Vynios, D.H.; Papageorgakopoulou, N.; Skandalis, S.S.; Theocharis, D.A. Altered Content Composition and Structure of Glycosaminoglycans and Proteoglycans in Gastric Carcinoma. *Int. J. Biochem. Cell Biol.* **2003**, *35*, 376–390. [CrossRef]
22. Hopfner, U.; Schenck, T.L.; Chávez, M.N.; Machens, H.G.; Bohne, A.V.; Nickelsen, J.; Giunta, R.E.; Egaña, J.T. Development of Photosynthetic Biomaterials for in Vitro Tissue Engineering. *Acta Biomater.* **2014**, *10*, 2712–2717. [CrossRef] [PubMed]
23. Chávez, M.N.; Moellhoff, N.; Schenck, T.L.; Egaña, J.T.; Nickelsen, J. Photosymbiosis for Biomedical Applications. *Front. Bioeng. Biotechnol.* **2020**, *8*, 13. [CrossRef] [PubMed]
24. Schenck, T.L.; Hopfner, U.; Chávez, M.N.; Machens, H.G.; Somlai-Schweiger, I.; Giunta, R.E.; Bohne, A.V.; Nickelsen, J.; Allende, M.L.; Egaña, J.T. Photosynthetic Biomaterials: A Pathway towards Autotrophic Tissue Engineering. *Acta Biomater.* **2015**, *15*, 39–47. [CrossRef] [PubMed]
25. Cohen, J.E.; Goldstone, A.B.; Paulsen, M.J.; Shudo, Y.; Steele, A.N.; Edwards, B.B.; Patel, J.B.; MacArthur, J.W.; Hopkins, M.S.; Burnett, C.E.; et al. An Innovative Biologic System for Photon-Powered Myocardium in the Ischemic Heart. *Sci. Adv.* **2017**, *3*, 1–11. [CrossRef] [PubMed]
26. Lode, A.; Krujatz, F.; Brüggemeier, S.; Quade, M.; Schütz, K.; Knaack, S.; Weber, J.; Bley, T.; Gelinsky, M. Green Bioprinting: Fabrication of Photosynthetic Algae-Laden Hydrogel Scaffolds for Biotechnological and Medical Applications. *Eng. Life Sci.* **2015**, *15*, 177–183. [CrossRef]
27. Maharjan, S.; Alva, J.; Cámara, C.; Rubio, A.G.; Hernández, D.; Delavaux, C.; Correa, E.; Romo, M.D.; Bonilla, D.; Santiago, M.L.; et al. Symbiotic Photosynthetic Oxygenation within 3D-Bioprinted Vascularized Tissues. *Matter* **2021**, *4*, 217–240. [CrossRef]
28. Veloso-Giménez, V.; Escamilla, R.; Necuñir, D.; Corrales-Orovio, R.; Riveros, S.; Marino, C.; Ehrenfeld, C.; Guzmán, C.D.; Boric, M.P.; Rebolledo, R.; et al. Development of a Novel Perfusable Solution for Ex Vivo Preservation: Towards Photosynthetic Oxygenation for Organ Transplantation. *Front. Bioeng. Biotechnol.* **2021**, *9*, 12. [CrossRef]
29. Obaíd, M.L.; Camacho, J.P.; Brenet, M.; Corrales-Orovio, R.; Carvajal, F.; Martorell, X.; Werner, C.; Simón, V.; Varas, J.; Calderón, W.; et al. A First in Human Trial Implanting Microalgae Shows Safety of Photosynthetic Therapy for the Effective Treatment of Full Thickness Skin Wounds. *Front. Med.* **2021**, *8*, 14. [CrossRef]
30. Qiao, Y.; Qiao, Y.; Yang, F.; Xie, T.; Du, Z.; Zhong, D.; Qi, Y.; Li, Y.; Li, W.; Li, W.; et al. Engineered Algae: A Novel Oxygen-Generating System for Effective Treatment of Hypoxic Cancer. *Sci. Adv.* **2020**, *6*, 12. [CrossRef]
31. Zhong, D.; Li, W.; Hua, S.; Qi, Y.; Xie, T.; Qiao, Y.; Zhou, M. Calcium Phosphate Engineered Photosynthetic Microalgae to Combat Hypoxic-Tumor by in-Situ Modulating Hypoxia and Cascade Radio-Phototherapy. *Theranostics* **2021**, *11*, 3580–3594. [CrossRef]
32. Qi, F.; Ji, P.; Chen, Z.; Wang, L.; Yao, H.; Huo, M.; Shi, J. Photosynthetic Cyanobacteria-Hybridized Black Phosphorus Nanosheets for Enhanced Tumor Photodynamic Therapy. *Small* **2021**, *17*, 9. [CrossRef]

33. Wang, H.; Guo, Y.; Wang, C.; Jiang, X.; Liu, H.; Yuan, A.; Yan, J.; Hu, Y.; Wu, J. Light-Controlled Oxygen Production and Collection for Sustainable Photodynamic Therapy in Tumor Hypoxia. *Biomaterials* **2021**, *269*, 120621. [CrossRef]
34. Lu, S.; Feng, W.; Dong, C.; Song, X.; Gao, X.; Guo, J.; Chen, Y.; Hu, Z. Photosynthetic Oxygenation-Augmented Sonodynamic Nanotherapy of Hypoxic Tumors. *Adv. Healthc. Mater.* **2022**, *11*, 2102135. [CrossRef]
35. Ortega, J.S.; Corrales-Orovio, R.; Ralph, P.; Egaña, J.T.; Gentile, C. Photosynthetic Microorganisms for the Oxygenation of Advanced 3D Bioprinted Tissues. *Acta Biomater.* **2022**, *in press*. [CrossRef]
36. Harris, E.H. *The Chlamydomonas Sourcebook Volume 1: Introduction to Chlamydomonas and Its Laboratory Use*; Academic Press: San Diego, CA, USA, 2013; Volume 53, ISBN 9788578110796.
37. Dauvillée, D.; Delhaye, S.; Gruyer, S.; Slomianny, C.; Moretz, S.E.; d'Hulst, C.; Long, C.A.; Ball, S.G.; Tomavo, S. Engineering the Chloroplast Targeted Malarial Vaccine Antigens in Chlamydomonas Starch Granules. *PLoS ONE* **2010**, *5*, e15424. [CrossRef]
38. Vaupel, P.; Mayer, A. Hypoxia in Cancer: Significance and Impact on Clinical Outcome. *Cancer Metastasis Rev.* **2007**, *26*, 225–239. [CrossRef]
39. Bedogni, B.; Powell, M.B. Hypoxia, Melanocytes and Melanoma—Survival and Tumor Development in the Permissive Microenvironment of the Skin. *Pigment Cell Melanoma Res.* **2009**, *22*, 166–174. [CrossRef]
40. Kozar, I.; Margue, C.; Rothengatter, S.; Haan, C.; Kreis, S. Many Ways to Resistance: How Melanoma Cells Evade Targeted Therapies. *Biochim. Biophys. Acta Rev. Cancer* **2019**, *1871*, 313–322. [CrossRef]
41. Szlasa, W.; Supplitt, S.; Drąg-Zalesińska, M.; Przystupski, D.; Kotowski, K.; Szewczyk, A.; Kasperkiewicz, P.; Saczko, J.; Kulbacka, J. Effects of Curcumin Based PDT on the Viability and the Organization of Actin in Melanotic (A375) and Amelanotic Melanoma (C32)—In Vitro Studies. *Biomed. Pharmacother.* **2020**, *132*, 110883. [CrossRef]
42. Marrelli, M.; Perri, M.R.; Amodeo, V.; Giordano, F.; Statti, G.A.; Panno, M.L.; Conforti, F. Assessment of Photo-Induced Cytotoxic Activity of Cachrys Sicula and Cachrys Libanotis Enriched-Coumarin Extracts against Human Melanoma Cells. *Plants* **2021**, *10*, 123. [CrossRef]
43. Rocha, F.D.; Soares, A.R.; Houghton, P.J.; Pereira, R.C.; Kaplan, M.A.C.; Teixeira, V.L. Potential Cytotoxic Activity of Some Brazilian Seaweeds on Human Melanoma Cells. *Phyther. Res.* **2007**, *21*, 170–175. [CrossRef] [PubMed]
44. Atjanasuppat, K.; Wongkham, W.; Meepowpan, P.; Kittakoop, P.; Sobhon, P.; Bartlett, A.; Whitfield, P.J. In Vitro Screening for Anthelmintic and Antitumour Activity of Ethnomedicinal Plants from Thailand. *J. Ethnopharmacol.* **2009**, *123*, 475–482. [CrossRef] [PubMed]
45. Conforti, F.; Ioele, G.; Statti, G.A.; Marrelli, M.; Ragno, G.; Menichini, F. Antiproliferative Activity against Human Tumor Cell Lines and Toxicity Test on Mediterranean Dietary Plants. *Food Chem. Toxicol.* **2008**, *46*, 3325–3332. [CrossRef] [PubMed]
46. Carvajal, F.; Duran, C.; Aquea, F. Effect of Alerce (Fitzroya Cupressoides) Cell Culture Extract on Wound Healing Repair in a Human Keratinocyte Cell Line. *J. Cosmet. Dermatol.* **2020**, *19*, 1254–1259. [CrossRef] [PubMed]
47. Chávez, M.N.; Schenck, T.L.; Hopfner, U.; Centeno-Cerdas, C.; Somlai-Schweiger, I.; Schwarz, C.; Machens, H.G.; Heikenwalder, M.; Bono, M.R.; Allende, M.L.; et al. Towards Autotrophic Tissue Engineering: Photosynthetic Gene Therapy for Regeneration. *Biomaterials* **2016**, *75*, 25–36. [CrossRef]
48. Klier, U.; Maletzki, C.; Göttmann, N.; Kreikemeyer, B.; Linnebacher, M. Avitalized Bacteria Mediate Tumor Growth Control via Activation of Innate Immunity. *Cell. Immunol.* **2011**, *269*, 120–127. [CrossRef]
49. Vítová, M.; Bišová, K.; Hlavová, M.; Kawano, S.; Zachleder, V.; Čížková, M. Chlamydomonas Reinhardtii: Duration of Its Cell Cycle and Phases at Growth Rates Affected by Temperature. *Planta* **2011**, *234*, 599–608. [CrossRef]
50. Graham, P.J.; Nguyen, B.; Burdyny, T.; Sinton, D. A Penalty on Photosynthetic Growth in Fluctuating Light. *Sci. Rep.* **2017**, *7*, 1–11. [CrossRef]
51. Mott, K.A.; Woodrow, I.E. Modelling the Role of Rubisco Activase in Limiting Non-steady-state Photosynthesis. *J. Exp. Bot.* **2000**, *51*, 399–406. [CrossRef]
52. D'arcy, C.; Kiel, C. Cell Adhesion Molecules in Normal Skin and Melanoma. *Biomolecules* **2021**, *11*, 1213. [CrossRef]
53. Li, Q.S.; Lee, G.Y.H.; Ong, C.N.; Lim, C.T. AFM Indentation Study of Breast Cancer Cells. *Biochem. Biophys. Res. Commun.* **2008**, *374*, 609–613. [CrossRef]
54. Friedl, P. Prespecification and Plasticity: Shifting Mechanisms of Cell Migration. *Curr. Opin. Cell Biol.* **2004**, *16*, 14–23. [CrossRef]
55. Egaña, J.T.; Fierro, F.A.; Krüger, S.; Bornhäuser, M.; Huss, R.; Lavandero, S.; Machens, H.-G. Use of Human Mesenchymal Cells to Improve Vascularization in a Mouse Model for Scaffold-Based Dermal Regeneration. *Tissue Eng. Part A* **2009**, *15*, 1191–1200. [CrossRef]
56. Danner, S.; Kremer, M.; Petschnik, A.E.; Nagel, S.; Zhang, Z.; Hopfner, U.; Reckhenrich, A.K.; Weber, C.; Schenck, T.L.; Becker, T.; et al. The Use of Human Sweat Gland–Derived Stem Cells for Enhancing Vascularization during Dermal Regeneration. *J. Investig. Dermatol.* **2012**, *132*, 1707–1716. [CrossRef]
57. Fierro, F.A.; ONeal, A.J.; Beegle, J.R.; Chávez, M.N.; Thomas, R.; Isseroff, R.R.; Egana, T. Hypoxic Pre-Conditioning Increases the Infiltration of Endothelial Cells into Scaffolds for Dermal Regeneration Pre-Seeded with Mesenchymal Stem Cells. *Front. Cell Dev. Biol.* **2015**, *3*, 1–9. [CrossRef]
58. Zhang, Z.; Ito, W.D.; Hopfner, U.; Böhmert, B.; Kremer, M.; Reckhenrich, A.K.; Harder, Y.; Lund, N.; Kruse, C.; Machens, H.G.; et al. The Role of Single Cell Derived Vascular Resident Endothelial Progenitor Cells in the Enhancement of Vascularization in Scaffold-Based Skin Regeneration. *Biomaterials* **2011**, *32*, 4109–4117. [CrossRef]

59. Chávez, M.N.; Fuchs, B.; Moellhoff, N.; Hofmann, D.; Zhang, L.; Selão, T.T.; Giunta, R.E.; Egaña, J.T.; Nickelsen, J.; Schenck, T.L. Use of Photosynthetic Transgenic Cyanobacteria to Promote Lymphangiogenesis in Scaffolds for Dermal Regeneration. *Acta Biomater.* **2021**, *126*, 132–143. [CrossRef]
60. Bosman, F.T.; Stamenkovic, I. Functional Structure and Composition of the Extracellular Matrix. *J. Pathol.* **2003**, *200*, 423–428. [CrossRef]
61. Nicolas, J.; Magli, S.; Rabbachin, L.; Sampaolesi, S.; Nicotra, F.; Russo, L. 3D Extracellular Matrix Mimics: Fundamental Concepts and Role of Materials Chemistry to Influence Stem Cell Fate. *Biomacromolecules* **2020**, *21*, 1968–1994. [CrossRef]
62. Chang, D.K.; Louis, M.R.; Gimenez, A.; Reece, E.M. The Basics of Integra Dermal Regeneration Template and Its Expanding Clinical Applications. *Semin. Plast. Surg.* **2019**, *33*, 185–189. [CrossRef]
63. Liu, G.Y.; Agarwal, R.; Ko, K.R.; Ruthven, M.; Sarhan, H.T.; Frampton, J.P. Templated Assembly of Collagen Fibers Directs Cell Growth in 2D and 3D. *Sci. Rep.* **2017**, *7*, 1–9. [CrossRef] [PubMed]
64. Fischer, R.S.; Sun, X.; Baird, M.A.; Hourwitz, M.J.; Seo, B.R.; Pasapera, A.M.; Mehta, S.B.; Losert, W.; Fischbach, C.; Fourkas, J.T.; et al. Contractility, Focal Adhesion Orientation, and Stress Fiber Orientation Drive Cancer Cell Polarity and Migration along Wavy ECM Substrates. *Proc. Natl. Acad. Sci. USA* **2021**, *118*, 12. [CrossRef] [PubMed]
65. Tien, J.; Ghani, U.; Dance, Y.W.; Seibel, A.J.; Karakan, M.Ç.; Ekinci, K.L.; Nelson, C.M. Matrix Pore Size Governs Escape of Human Breast Cancer Cells from a Microtumor to an Empty Cavity. *iScience* **2020**, *23*, 12. [CrossRef] [PubMed]
66. Sun, X.; Kaufman, P.D. Ki-67: More than a Proliferation Marker. *Chromosoma* **2018**, *127*, 175–196. [CrossRef] [PubMed]
67. Miller, I.; Min, M.; Yang, C.; Tian, C.; Gookin, S.; Carter, D.; Spencer, S.L. Ki67 Is a Graded Rather than a Binary Marker of Proliferation versus Quiescence. *Cell Rep.* **2018**, *24*, 1105–1112. [CrossRef]
68. Wagner, B.A.; Venkataraman, S.; Buettner, G.R. The Rate of Oxygen Utilization by Cells. *Free Radic. Biol. Med.* **2011**, *51*, 700–712. [CrossRef]
69. Hystad, M.E.; Rofstad, E.K. Oxygen Consumption Rate and Mitochondrial Density in Human Melanoma Monolayer Cultures and Multicellular Spheroids. *Int. J. Cancer* **1994**, *57*, 532–537. [CrossRef]
70. Neupert, J.; Karcher, D.; Bock, R. Generation of Chlamydomonas Strains That Efficiently Express Nuclear Transgenes. *Plant J.* **2009**, *57*, 1140–1150. [CrossRef]
71. Schindelin, J.; Arganda-Carreras, I.; Frise, E.; Kaynig, V.; Longair, M.; Pietzsch, T.; Preibisch, S.; Rueden, C.; Saalfeld, S.; Schmid, B.; et al. Fiji: An Open-Source Platform for Biological-Image Analysis. *Nat. Methods* **2012**, *9*, 676–682. [CrossRef]
72. Arnon, D.I. Copper Enzymes in Isolated Chloroplasts. Polyphenoloxidase in Beta Vulgaris. *Plant Physiol.* **1949**, *24*, 8. [CrossRef]
73. Zonderland, J.; Wieringa, P.; Moroni, L. A Quantitative Method to Analyse F-Actin Distribution in Cells. *MethodsX* **2019**, *6*, 2562–2569. [CrossRef]
74. Rezakhaniha, R.; Agianniotis, A.; Schrauwen, J.T.C.; Griffa, A.; Sage, D.; Bouten, C.V.C.; Van De Vosse, F.N.; Unser, M.; Stergiopulos, N. Experimental Investigation of Collagen Waviness and Orientation in the Arterial Adventitia Using Confocal Laser Scanning Microscopy. *Biomech. Model. Mechanobiol.* **2011**, *11*, 461–473. [CrossRef]

International Journal of
Molecular Sciences

MDPI

Review

BODIPYs in PDT: A Journey through the Most Interesting Molecules Produced in the Last 10 Years

Miryam Chiara Malacarne, Marzia Bruna Gariboldi and Enrico Caruso *

Department of Biotechnology and Life Sciences (DBSV), University of Insubria, Via J.H. Dunant 3, 21100 Varese, Italy
* Correspondence: enrico.caruso@uninsubria.it; Tel.: +39-0332421541

Abstract: Over the past 30 years, photodynamic therapy (PDT) has shown great development. In the clinical setting the few approved molecules belong almost exclusively to the porphyrin family; but in the scientific field, in recent years many researchers have been interested in other families of photosensitizers, among which BODIPY has shown particular interest. BODIPY is the acronym for 4,4-difluoro-4-bora-3a, 4a-diaza-s-indacene, and is a family of molecules well-known for their properties in the field of imaging. In order for these molecules to be used in PDT, a structural modification is necessary which involves the introduction of heavy atoms, such as bromine and iodine, in the beta positions of the pyrrole ring; this change favors the intersystem crossing, and increases the $^1O_2$ yield. This mini review focused on a series of structural changes made to BODIPYs to further increase $^1O_2$ production and bioavailability by improving cell targeting or photoactivity efficiency.

Keywords: 4,4-difluoro-4-bora-3a,4a-diaza-s-indacene; BODIPY; photodynamic therapy; photosensitizer

Citation: Malacarne, M.C.; Gariboldi, M.B.; Caruso, E. BODIPYs in PDT: A Journey through the Most Interesting Molecules Produced in the Last 10 Years. *Int. J. Mol. Sci.* **2022**, *23*, 10198. https://doi.org/10.3390/ijms231710198

Academic Editor: Christian Celia

Received: 25 August 2022
Accepted: 31 August 2022
Published: 5 September 2022

1. Introduction

1.1. Photodynamic Therapy (PDT)

Photodynamic therapy (PDT) [1,2] is an innovative technique that combines three components which are individually harmless but which when combined cause damage to nearby biomolecules (Figure 1).

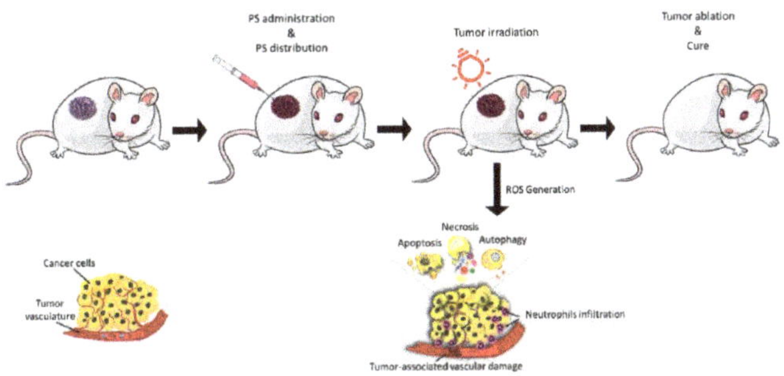

Figure 1. Schematic PDT Treatment: photodynamic strategy involves photosensitizer (PS) administration (local or systemic injection). Light irradiation is applied depending on the drug-light interval (time necessary to drug accumulation within the tumor). The activation of PS leads to the generation of $^1O_2$ and ROS that cause cancer cell death through apoptosis, necrosis or autophagy.

Nowadays, PDT is used for diverse pathologies, and it is gaining interest in cancer treatment [2–8]. PDT involves three components: a photosensitizer (PS), a molecule that has a high degree of unsaturation that is able to capture the energy provided by the second component, light at an appropriate wavelength; this energy is then transferred to the third component, molecular oxygen [1,9,10]. Specifically, once irradiated, the PS in the excited electronic state is highly unstable and will therefore tend to return to the ground state. The return to the ground state can occur through the emission of fluorescence, an excellent property in the clinical setting for imaging or photodetection, and/or heat [11]. Alternatively, and based on physico-chemical characteristics, the PS can undergo an intersystem crossing (ISC); with this step the PS will go into a more stable triplet state ($^3PS^*$) with an inverted spin of an electron. $^3PS^*$ can return to the ground state by emitting phosphorescence or it can interact with molecular oxygen (O_2) [12] thus leading to the formation of singlet oxygen ($^1O_2$) in what is called a Type II reaction [13–15]. Type I reactions can also occur; in this case the PS interacts directly with a substrate [16–20], such as the cell membrane or a molecule, thus transferring a proton or an electron to the substrate to form radicals which will then react with O_2 to form the superoxide anion ($O_2^{\bullet-}$), the hydroxyl radical ($OH^\bullet$) and hydrogen peroxide (H_2O_2) [6,21–24]. Type I and Type II reactions can occur simultaneously, even though photoreaction of type II is predominant [13,25,26]. ROS that are generated by type I and II reactions are mainly responsible for PDT-induced cell death [27,28]. These species are extremely toxic and, once formed, can strongly reduce malignant cells through necrosis, apoptosis, autophagy [4,10,29–31], and, as reported by many authors, can cause inflammatory/immune responses [32–38].

One of the three crucial elements of PDT is the PS [39]. PSs used are generally colored substances that share, as a common feature, an extensive electronic conjugation by a high number of conjugated double bonds to allow the interaction of π electrons with low energy radiations such as visible light.

The ideal PS for therapeutic application should have a selective accumulation in the tumor tissue [11]. The accumulation is favored by a certain degree of lipophilicity which favors cellular uptake but at the same time is also favored by a certain degree of hydrophilicity which facilitates its administration; the two concepts can be exemplified with the term amphiphilia [7,11,39]. PS must also have no or negligible intrinsic toxicity to preserve healthy tissues [6,39]. From a chemical point of view, PS should be easily synthesized in pure formulation, have a high molar extinction coefficient [7] and a high quantum yield of $^1O_2$ [40]. The ideal PS should be activated at wavelengths of between 680 and 800 nm for a deeper penetration of the light [41].

Over the years various generations of PSs have been obtained to meet all the requirements of the ideal PS. Historically, the PSs are divided into first, second and third generation [1,11,42].

The first generation includes naturally derived PSs belonging to the porphyrin family that were developed in the 1970s and early 1980s. The first preparations for use in PDT were based on a fairly complex mixture of Hematoporphyrin-derived (HpD) porphyrins [1,43]. HpDs showed better tissue selectivity for tumors and less photosensitizing potential of the skin [44,45], but they have limitations such as low chemical purity, poor tissue penetration and skin hypersensitivity to light [1,39,46–48].

In the 1980s, to overcome those limitations, several hundreds of substances with potential for photosensitizing had been proposed, however, only a few were used in clinical trials. The second generation includes synthetic PSs such as 5-aminolevulinic acid, benzoporphyrin derivatives, texaphyrins, chlorins, bacteriochlorins, BODIPYs, anthraquinones, xanthenes, cyanines and curcuminoids [1,5,39,49–53]. Those compounds show higher chemical purity, a higher yield of $^1O_2$ formation and better penetration [6]. The strong limitation turns out to be water solubility.

Using second generation PSs as a basis and binding them with cargos to allow targeted accumulation in the tumor site, the third generation of PSs was obtained [10,54]. In this category PSs are conjugated with organic and inorganic polymers, nanoparticles, liposomes, monoclonal antibodies, antibody fragments, protein/peptides (such as transferrin, EGF, somatostatin and insulin), carbohydrates, folic acid, and many others [1,24,55,56]. These modifications lead to enhance targeting and absorption of the PS in tumoral and drained-vasculature cells within the tumors [57]. Many PSs belonging to this generation are currently under development or undergoing clinical trial.

1.2. BODIPYs

The first member of 4,4-difluoro-4-bora-3a,4a-diaza-s-indacene dyes (hereafter abbreviated to BODIPYs) was reported by Treibs and Kreuzer in 1968 [58] although relatively little attention was given to their discovery until the end of the 1980s [59]. The potentials of this dyes for biological labeling and in many other science fields was later recognized and a series of study brought attention on them.

In addition to the high fluorescence emission (Φ), the BODIPYs are characterized by a high molar extinction coefficient (ε) in the visible region, limited changes by environmental conditions (e.g., polarity of the solvent or pH) [60], by a high lipophilicity [61] and a good resistance to photobleaching (the phenomenon of chemical destruction of a fluorophore by light [62]).

The absorption and emission characteristics of the BODIPY can be adjusted by adding suitable substituents on the main structure of the BODIPY itself. The effect is particularly evident when the substituent is bonded to one of the carbon atoms of a pyrrole ring [63]. On the contrary, the presence of an aromatic group in position 8 (meso) exerts a very weak effect, mainly explained by the weak electronic interaction between the aromatic ring and the main structure of the BODIPY [55].

Any structural difference on the *meso*-aryl group does not greatly affect the intensity and wavelength of the absorption and fluorescence bands, however the degree of polarity of this part can modulate the degree of amphiphilicity of the molecule.

The presence of substituents in positions 2 and 6 of pyrrole (β positions) involves some evident effects; for example, the presence of alkyl groups or heavy atoms has direct repercussions on the energy of the HOMO and LUMO orbitals and on the triplet state so as to influence the absorption and emission phenomena of UV-Vis radiation [64,65].

In addition to the well-known fluorescence that makes these molecules excellent biomarkers, a new important application field of BODIPYs concerns photodynamic therapy (PDT) as they can be used as photosensitizers (PS), after appropriate modifications [61].

A fundamental modification to increase the probability of ISC and the quantum yield of $^1O_2$ formation is the introduction in some specific positions of a heavy atom, such as iodine or bromine. A heavy atom can quench the fluorescence and facilitate the transition to triplet state. Heavy atoms either in solution (external Heavy Atom Quenching HAQ) or incorporated in the molecule (internal HAQ) assumed to quench fluorescence by perturbating S1 state via spin orbit interactions that modifies electron's atomic energy levels. The transition between the T_1 state and ground state are forbidden due to the spin changes, this results in a relatively long-lived and reactive excited configuration that can interact with other molecule forming complexes or radicals [40,56,61,65].

Iodine seems to severely quench fluorescence, suggesting that the intersystem crossing efficiency (Φ_{ISC}) has been enhanced by the internal heavy atom effects. Interestingly 2I-BODIPYs showed a narrow peak in NIR spectrum that is characteristic of $^1O_2$ generating molecules [40,66–71]. The incorporation of more heavy atoms in BODIPY core does not effectively enhance the $^1O_2$ quantum yield; on the contrary it appears to increase dark toxicity.

2. BODIPYs Studies

Substituents are generally attached in positions 4 (boron center); 3,5 (α-positions); 1,2,6,7 (β-positions), and 8 (*meso* position). In this context, we will show new structural derivatives of BODIPY core and their strategy to improve delivery and efficiency [56,59,66]. Principal modifications are the addition of groups that modify chemical and physical properties (hydrophilicity, absorption wavelength, ISC efficiency, etc.), or improve the interaction with biological medium (environment-response, association with targeting molecules, etc.) [56,59,66,72].

Modifications to their structures enable the tuning of their fluorescence characteristics, enhancing their capacity for $^1O_2$ generation. The stability of the core structure and its tendency to maintain photochemical properties even with complex groups as substituents in some positions makes BODIPYs an interesting class of compounds for PDT [56,59,72].

2.1. Structure-Activity

One of the important aspects in the study of the various PSs is to evaluate the relationship that exists between the chemical structure and the photodynamic activity of the PS; in fact, this could allow the design of targeted syntheses. In recent years many authors have tried to analyze this type of relationship in various series of PS such as porphyrins [73,74], the pyropheophorbides [75] and phthalocyanines [76].

Even in the case of BODIPYs this type of approach is extremely important for their application in PDT [40,71,77–85].

Recently, we characterized a panel of twenty-four new BODIPYs all united by an aromatic ring in *meso* position and by the presence of iodine atoms in β-positions [86]. The panel shows different atoms or groups as substituents of the aromatic ring. All compounds were characterized from a physico-chemical point of view (generation of $^1O_2$, fluorescence and lipophilicity) and for their activity on the human ovarian carcinoma cells SKOV3. A quantitative structure-activity relationship (QSAR) analysis was also performed to understand how the substituent could affect PS activity.

Data obtained demonstrate that the presence of an aromatic ring is essential to obtain a high production of $^1O_2$ and high phototoxicity of the compounds. Of the twenty-four compounds analyzed, two showed high activity (compounds **2** and **8**).

The first is characterized by the presence of a methoxy group in the *para* position while the other has two chlorine atoms in the *ortho* and *ortho'* position (Figure 2). The differences in regioisomerism and in the electronic effect confirm the absence of correlations between the substituents and the efficacy of the photo-induced action.

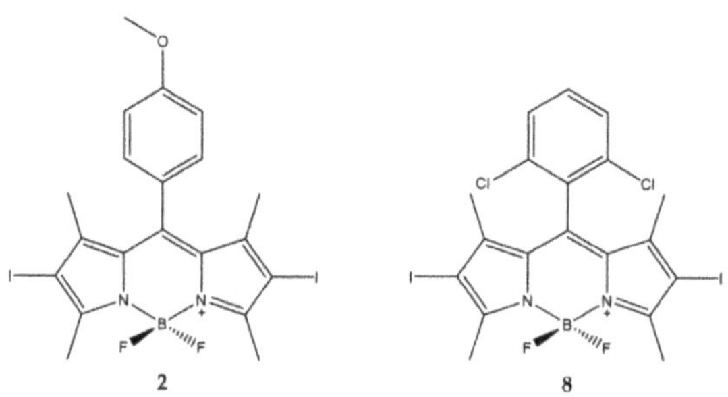

Figure 2. Chemical structures of compounds **2** and **8** [86].

The presence of iodine atoms in the β-positions of the pyrrole determine a high production of $^1O_2$. The IC_{50} were in the range of 1-2 nM for the BODIPY derivatives with the exception of compounds **2** and **8** (Table 1).

Table 1. Physico-chemical properties and biological data obtained for compounds **2** and **8** [86].

	2	8
λ_{abs} [a]	534 nm	548 nm
ε	61,500 $M^{-1}cm^{-1}$	76,400 $M^{-1}cm^{-1}$
Φ_{fluo} [b]	0.02	0.02
$^1O_2$ QY [c]	1.04	1.53
$IC_{50\ (SKOV3)}$ [d]	0.65 nM	0.77 nM

[a] 10 μM solution in Dichloromethane (DCM). [b] In DCM with fluorescein (0.1 M NaOH, 0.85) as standard. [c] In isopropanol with DPBF (1,3-diphenylisobenzofuran) as indicator and Rose Bengala as standard. [d] 2 h of irradiation with a green LED device (fluence rate 25.2 J/cm^2).

Based on these results, the effect of chains of different length on the hydroxyl groups was evaluated. Two BODIPYs with the presence of hydro-carbon chains of different lengths placed in position 4 (*para*) of the phenyl ring placed in *meso* position of the core were analyzed [87]. Specifically, compounds **3** and **4** (Figure 3) have carbon chains consisting of four and eight atoms, respectively, to obtain a different degree of lipophilicity of the two molecules. This difference should make it possible to establish a correlation between the length of the alkyl chain and the photodynamic activity. A bromine atom was presented at the end of each chain to make these molecules more versatile and eventually allow the bond with nanoparticles [88].

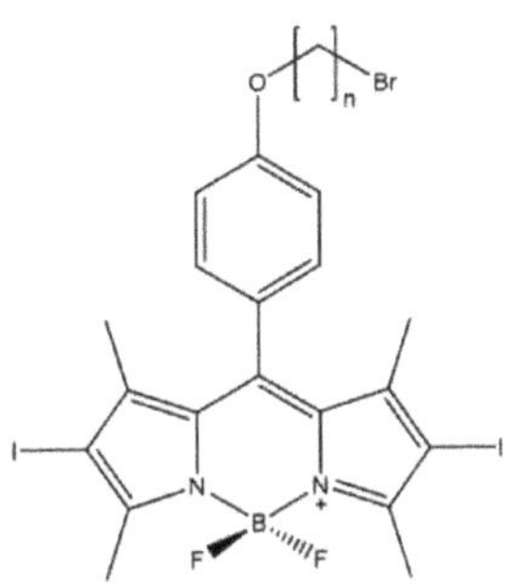

3 n=4
4 n=8

Figure 3. Chemical structures of compounds **3** and **4** [87].

The analyzes carried out confirmed that the presence of alkyl chains of different lengths affects the degree of lipophilicity and consequently also the photodynamic effect. Compound **3** has a fair degree of lipophilicity associated with an effective photodynamic action on tumor cells; on the contrary, the more lipophilic compound **4** has a lower photodynamic activity probably due to the formation of aggregates in aqueous medium (Table 2).

Table 2. Physico-chemical properties and biological data obtained for compounds **3** and **4** [87].

	3	4
λ_{abs} [a]	533 nm	534 nm
ε	66,072 $M^{-1}cm^{-1}$	41,500 $M^{-1}cm^{-1}$
λ_{em} [a]	553 nm	554 nm
Φ_{fluo} [b]	0.01	0.01
$^1O_2$ QY [c]	1.10	1.11
$IC_{50\ (HCT116)}$ [d]	2.0 nM	11.7 nM
$IC_{50\ (SKOV3)}$ [d]	3.56 nM	15.61 nM
$IC_{50\ (MCF7)}$ [d]	6.88 nM	23.11 nM

[a] 10 μM solution in DCM. [b] In DCM with fluorescein (0.1 M NaOH, 0.85) as standard. [c] In isopropanol with DPBF as indicator and Rose Bengala as standard. [d] 2 h of irradiation with a green LED device (fluence rate 25.2 J/cm^2).

As already highlighted the main problem with these PSs leads to their solubility in aqueous medium. Belfield et al. designed a BODIPY decorated in meso position with a polyethylene glycol chain to enhance solubility and to prevent aggregation in aqueous solution [89].

Compound **3** (Figure 4) was characterized as regards absorption, fluorescence and $^1O_2$ production. The ability to produce $^1O_2$ was evaluated by direct measurement of near infrared luminescence. The quantum yield of $^1O_2$ generation is higher than that of the standard used as a control (0.93 for **3**, 0.82 for acridine used as control). IC_{50} were estimated at 10 μM against Lewis lung carcinoma cells pre-treated with compound **3** and exposed to a light dose of 3.5 mW/cm² (Table 3).

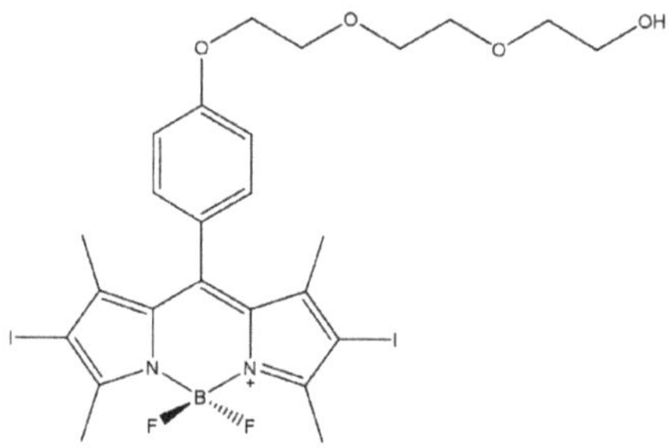

3

Figure 4. Chemical structures of compound **3** [89].

Table 3. Physico-chemical properties and biological data obtained for compound **3** [89].

	3
λ_{abs} [a]	528 nm
ε	75,500 $M^{-1}cm^{-1}$
λ_{em} [a]	546 nm
Φ_{fluo} [b]	0.02
$^1O_2$ QY [c]	0.93
$IC_{50\ (LLC)}$ [d]	10.0 μM

[a] In acetonitrile. [b] Determined by a standard relative method with Rhodamine 6 G ($\Phi_{fluo} \approx 0.94$ in ethanol) as a reference. [c] In acetonitrile using acridine as standard. [d] Irradiation with a light dose of 3.5 mW/cm² from a filtered light source (522 ± 40 nm) for different time.

In addition, the cell death mechanisms induced following photodynamic treatment were evaluated and necrosis prevails over apoptosis process. The results obtained show how it is feasible to obtain a BODIPY with a higher degree of hydrophilicity with at the same time excellent characteristics for use in PDT.

2.2. Singlet Oxygen Generation

Critical factor for the activity of BODIPY is certainly the production of $^1O_2$. The generation of $^1O_2$ in solution requires a photosensitizer (PS) which is converted to the triplet excited state (intersystem crossing, ISC) upon irradiation. The triplet state PS transfers energy to molecular oxygen in a type II process to produce $^1O_2$. The heavy atom effect has been a useful chemical approach to improve ISC in several molecules including BODIPY chromophores.

Lincoln et al., and Durantini et al., investigated the different photostability and PS efficiency for PDT of two BODIPYs bearing an acetoxymethyl substituent in the *meso* position and bromine or iodine atoms in positions 2 and 6 of the scaffold [90,91].

The presence of the substituent acetoxymethyl in meso position improves the photostability of the PS. (Figure 5). It was also observed that the presence of iodine (**I_2BOAc**) or bromine (**Br_2BOAc**) atoms in both β-positions affects the production level of $^1O_2$. The PS bearing the two bromine atoms produced a lower level of $^1O_2$ than that bearing the iodine atoms. The production of $^1O_2$ by the bromine compound conferred a slight improvement in stability mainly attributable to the lower amount of oxidant generated (Table 4).

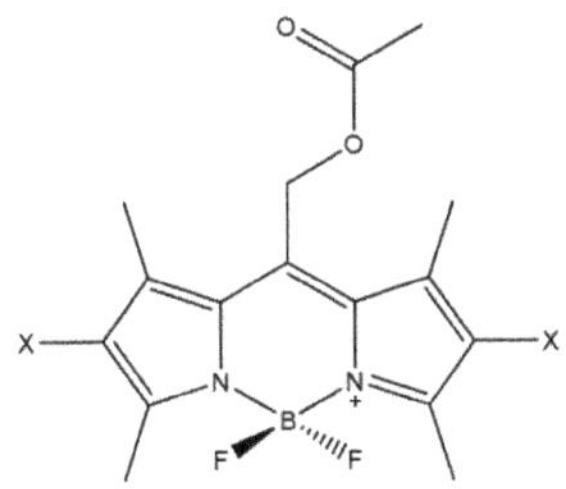

Br_2BOAc X=Br
I_2BOAc X=I

Figure 5. Chemical structures of compounds **Br_2BOAc** and **I_2BOAc** [90,91].

Table 4. Physico-chemical properties and biological data obtained for compounds **Br_2BOAc** and **I_2BOAc** [90,91].

	Br_2BOAc	**I_2BOAc**
λ_{abs} [a]	543 nm	550 nm
ε	81,000 $M^{-1}cm^{-1}$	96,000 $M^{-1}cm^{-1}$
λ_{em} [a]	562 nm	572 nm
Φ_{fluo} [b]	0.14	0.02
$^1O_2$ QY [c]	0.84	0.98
$IC_{50\ (HeLa)}$ [d]	140.0 nM	180.0 nM

[a] In acetonitrile. [b] Determined by the consumption of dimethylanthracene using Rose Bengal as a standard. [c] In acetonitrile using Rose Bengala as standard. [d] Irradiation with a 520 nm LED panel for 30 min.

Cytotoxicity studies conducted with the HeLa tumor cell line have shown that the compound bearing the bromine atoms is the most active. The localization of these compounds is mainly in the lipid membranes of the cell. Tests to evaluate the photodynamic activity of bacterial inactivation were carried out against Gram-negative *Escherichia coli*. The results obtained showed that both molecules are able to photo-kill *E. coli* already from a concentration of 5 μM, suggesting that both molecules have a PDT potential against *E. coli* and other Gram-negative strains.

A similar comparison between the two types of halogens bound in β-positions of the BODIPY was also carried out by Epelde-Elezcano et al., who studied the photo-physical properties of a series of BODIPYs bearing either two bromine atoms or two iodine in β-positions of the pyrrole of the BODIPY skeleton. Specifically, the measurement of $^1O_2$ production was carried out with an indirect assay using 9,10-dimethylanthracene (DMA) as a chemical probe and with a direct determination of the luminescence at 1276 nm of the $^1O_2$ with a NIR detector. The work demonstrates how the iodine atom favors the ISC over the bromine atom. In the same work the authors evaluate the impact of the *meso* substituent, observing how an electron donor substituent tends to decrease the production of $^1O_2$ compared to an analogous derivative bearing an electron withdrawing substituent. Furthermore, a strong decrease is observed in the case of electron donor substituents when there is free rotation of the substituent which favors the internal conversion process to the detriment of the ISC and therefore of the production of $^1O_2$ [92].

Turan et al., focused on synthesis of BODIPY derivatives that could be applied to enhanced fractional photodynamic therapy [93]. In this therapeutic strategy and to prevent photo-induced hypoxia, the light is administered intermittently (fractional PDT) to allow the replenishment of cellular oxygen with an increase in the time necessary for the effective therapy. The study reported the synthesis of a BODIPY containing 2-pyridone (Figure 6) which in the dark conditions, through a thermal cycloreversion, becomes a source of $^1O_2$ [94–98].

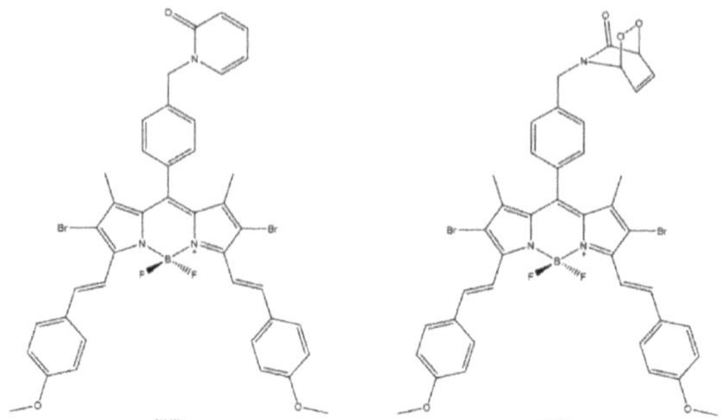

PYR
6

EPO
7

Figure 6. Chemical structures of compounds **6** (**PYR**) and **7** (**EPO**) [93].

When excited at a wavelength of 650 nm, molecule **6** (**Pyr**) generates $^1O_2$ which is partially stored in the form of 2-pyridone-endoperoxide (**7** or **EPO**). When irradiation is stopped **EPO** will undergo thermal cycloreversion to produce $^1O_2$ in the absence of light (Table 5) [99,100].

Table 5. Physico-chemical properties and biological data obtained for compounds **6** (**PYR**) and **7** (**EPO**) [93].

	6 (PYR)	7 (EPO)
λ_{abs} [a]	668 nm	667 nm
ε	22,200 $M^{-1}cm^{-1}$	74,200 $M^{-1}cm^{-1}$
λ_{em} [a]	697 nm	700 nm
Φ_{fluo} [b]	0.14	0.10
$^1O_2$ QY [c]	0.15	0.20
$IC_{50\ (HeLa)}$ [d]	49.0 nM	8.6 nM

[a] In DMSO. [b] In DMSO with Cresyl Violet ($\Phi_{fluo} \approx 0.66$ in methanol) as a reference. [c] In ethanol using Methylene blue as a reference ($^1O_2$ QY ≈ 0.52 in ethanol). [d] Irradiation cycles of light (λ = 655 nm, 10 min) and dark (50 min) with a total of 24 h of light exposure.

The molecules were tested against human cervical HeLa cells. Due to the low solubility, the compounds were prepared as micellar structures using the non-ionic surfactant cremophor EL [101]. IC_{50} was estimated at 8.6 nM and 49 nM for compound **7** and **6**, respectively. Results demonstrates how the continuous release of $^1O_2$ by the photosensitizer during the light-dark cycles has to be considered as a positive feature if we compare the cytotoxic effects with those of conventional photosensitizing agents.

Zou et al., instead focused on how the effect and configuration of the substituents of heavy atoms are important for $^1O_2$ generation [102]. A series of BODIPY derivatives was synthesized with one or two BODIPY units connected via a benzene ring; PSs thus obtained were then halogenated in β-positions with bromine or iodine atoms. The ability to produce $^1O_2$ and cytotoxic efficacy in HeLa cells was evaluated for each of the compounds.

The best producers of $^1O_2$ are compounds **4** and **6**, consisting of one and two units of iodinated BODIPY, respectively. Moreover, the insertion of two or more heavy atoms does not significantly affect the improvement of the $^1O_2$ produced (Table 6).

Table 6. Physico-chemical properties and biological data obtained for compounds **4** (**BDPI**) and **6** (**BBDPI**) [102].

	4 (BDPI)	6 (BBDPI)
λ_{abs} [a]	533 nm	540 nm
$^1O_2$ QY [b]	0.73	0.68
$IC_{50\ (HeLa)}$ [c]	1.0 μM	2.8 μM

[a] In DCM. [b] Using DPBF as the probe and methylene blue (ΦΔ = 0.57 in DCM) as standard. [c] Irradiation with a xenon lamp (40 mW/cm^2) for 8 min.

The synthesized compounds were then included in NPs to improve water solubility, required for cytotoxicity tests against HeLa cells. The MTT test showed that the most active compound is **6** (Figure 7). Given the good results obtained in vitro, compound **6** was also used for in vivo tests. In vivo, treatment with compound **6** led to a dramatic decrease in tumor growth and no tumor recurrence was observed. Biodistribution analysis showed that compound **6** accumulated in tumor cells and the liver. The selective irradiation of the tumor zone alone ensures that the liver is spared from possible photodynamic damage.

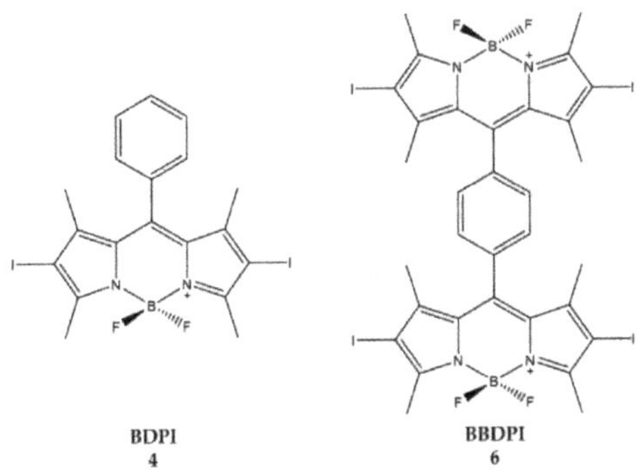

Figure 7. Chemical structures of compounds **4** (**BDPI**) and **6** (**BBDPI**) [102].

Pang et al., considered the dimerization of two BODIPYs to increase $^1O_2$ generation (Figure 8). In this case, the second BODIPY molecule is covalently linked in the beta position of the first BODIPY and can assume two positions, either orthogonal as in compounds **2b** and **2d** or angular (angle of about 30–40°) in compounds **2a** and **2c** [103].

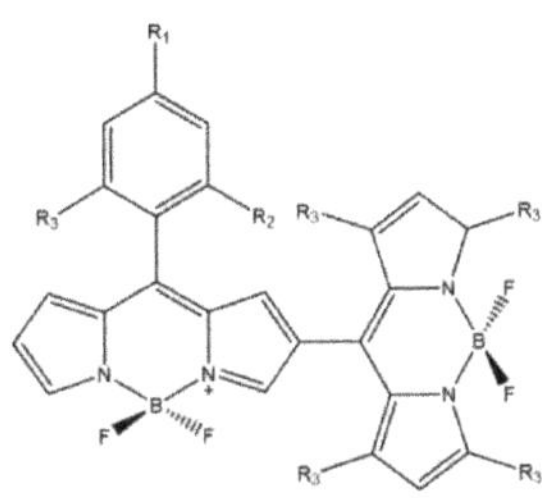

2a: R$_1$= OMe, R$_2$= H, R$_3$= H
2b: R$_1$= OMe, R$_2$= H, R$_3$= Me
2c: R$_1$= H, R$_2$= Cl, R$_3$= H
2d: R$_1$= H, R$_2$= Cl, R$_3$= Me

Figure 8. Chemical structures of compounds **2a–d** [103].

Following the synthesis, the authors took into consideration some spectroscopic properties and the $^1O_2$ production evaluated both through direct and indirect determination (Table 7).

Table 7. Physico-chemical properties and biological data obtained for compounds **2a–d** [103].

	2a	2b	2c	2d
λ_{abs} [a]	521 nm	507 nm	538 nm	507 nm
λ_{em} [a]	563 nm	540 nm	568 nm	542 nm
Φ_{fluo} [b]	0.007	0.002	0.002	0.002

[a] In DCM. [b] In DCM with fluorescein (0.1 M NaOH, 0.90) as standard.

The highest efficiency was observed for orthogonal meso–b linked dimer which confirm the data that Cakmak et al., had previously reported as well [101].

2.3. Uptake and Targeting

As reported by many authors [104–111], the aspect of cellular uptake and targeting has always been a very important aspect for all families of photosensitizers. Trying to increase cellular uptake as well as to make cellular penetration selective between healthy and tumor tissues is of primary importance, consequently cellular uptake and targeting are fundamental and interrelated factors.

2.3.1. Uptake

Shivran et al., synthesized three water soluble glucose conjugated BODIPYs and subsequently tested them on human lung cancer A549 cells. Of the three compounds, compound **4** features a glycosylated styryl appendix in position C-3 (Figure 9) [68]. The choice to insert a glucose on the PS is linked to the greater ability of the cancer cells to accumulate sugars.

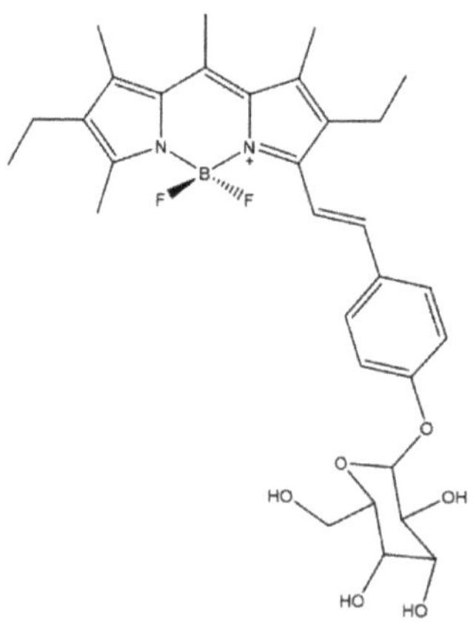

4

Figure 9. Chemical structures of compound **4** [68].

The emission spectrum of this compound is typical of a BODIPY with a high molar extinction coefficient and a high fluorescence emission. The introduction of the styryl group determines a bathochromic shift both in the absorption and in the maximum emission with respect to the BODIPY in which this group is absent. The IC_{50} value of all the compounds was evaluated and compound **4** is endowed with the greatest cytotoxicity (Table 8).

Table 8. Physico-chemical properties and biological data obtained for compound **4** [68].

	4
λ_{abs} [a]	573.8 nm
ε	8,2000 $M^{-1}cm^{-1}$
λ_{em} [a]	590 nm
Φ_{fluo} [b]	0.6
$IC_{50\ (A549)}$ [c]	2.7 μM

[a] In ethanol. [b] Relative to that of rhodamine 101 (Φ_{fluo} = 1.0 in EtOH). [c] Irradiation for 80 min (irradiance: 0.77 mW/cm^2).

Moreover, compound **4** accumulated rapidly in tumor cells. In addition, cell exposure to compound **4** led to apoptosis which is the prevailing cell death mechanism. Overall, the work in question demonstrated how it is possible to obtain a compound capable of having a selective capacity towards cancer cells in an economical and fast way.

Kuong Mai et al. [112] synthesized a water soluble BODIPY starting with a halogenated alkyl azide BODIPY to which a lactose motif is linked through an easy and straightforward CuAAC Click reaction [113]. The coupling of the lactose motif causes the compound to be soluble in water (Figure 10).

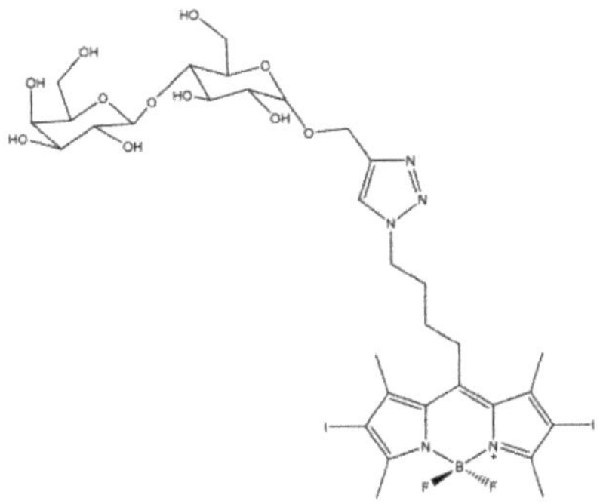

BDILa

Figure 10. Chemical structures of compound **BDILa** [112].

The emission and absorption spectra are consistent with those of the BODIPY family and the production of $^1O_2$ is high. The photodynamic activity of the PS was evaluated on three different tumor cell lines: human hepatoma Huh7, cervical cancer HeLa and breast cancer MCF7 and IC_{50} ranging from 0.6 to 0.5 μM (Table 9).

Table 9. Physico-chemical properties and biological data obtained for compound **BDILa** [112].

	BDILa
λ_{abs} [a]	526 nm
ε	41,800 $M^{-1}cm^{-1}$
λ_{em} [a]	542 nm
Φ_{fluo} [b]	0.02
$^1O_2$ QY [c]	0.47
$IC_{50\ (HeLa)}$ [d]	0.55 μM
$IC_{50\ (MCF7)}$ [d]	0.61 μM
$IC_{50\ (Huh7)}$ [d]	0.50 μM

[a] In H_2O. [b] In methanol. [c] In ethanol. [d] Irradiation for 20 min with a green LED (λ = 530 nm).

However, cell uptake was low, whatever the cell line tested, probably due to the presence of iodine atoms on the core of the BODIPY.

2.3.2. Targeting

Active targeted therapy of cancer refers to targeting the surface molecules on cancer cells using a small molecule to deliver cargo for therapeutic or diagnostic purposes. The cargo can be a cytotoxic drug, an imaging probe, or a photosensitizer [114].

Lysosomal Targeting

In 2019 Wang et al., obtained a BODIPY able to specifically targeting lysosomes [115]. The molecule obtained (**BDPI-Lyso**) has iodine atoms to generate $^1O_2$ and morpholine for targeting specifically the lysosomes (Figure 11) [116–118].

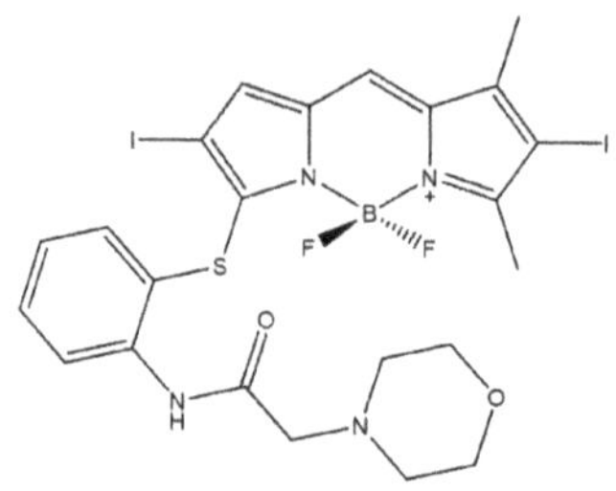

BDPI-Lyso

Figure 11. Chemical structures of compound **BDPI-Lyso** [115].

Physico-chemical analyzes showed that the absorption and fluorescence of the compound are slightly greater than that of the common BODIPYs due to the modification of the sulfur heteroatom. Moreover, under irradiation, the derivative produced higher level of $^1O_2$ in acidic versus physiological pH (Table 10).

Table 10. Physico-chemical properties and biological data obtained for compound **BDPI-Lyso** [115].

	BDPI-Lyso
λ_{abs} [a]	545 nm
ε	41,900 $M^{-1}cm^{-1}$
λ_{em} [a]	572 nm
Φ_{fluo} [a]	0.05
$^1O_2$ QY [a]	0.95
$^1O_2$ $QY_{pH=5}$ [b]	0.51
$^1O_2$ $QY_{pH=7}$ [c]	0.38
$IC_{50\,(Bel\text{-}7402)}$ [d]	0.4 μM

[a] In ethanol. [b] In ethanol/PBS = 1:1, pH = 5.10. [c] In ethanol/PBS = 1:1, pH = 7.24. [d] Irradiation for 30 min with green light (λ = 555 nm, light dose: 4 mW/cm^2).

This leads to the assumption that BODIPY exhibits greater photo-toxic effects in acid compartments such as lysosomes than in neutral organelles. In fact, the BODIPY derivative was localized in lysosomes as demonstrated by cell imaging performed with hepatoma Bel-7402 cells. Moreover, IC_{50} was estimated at 0.4 μM after cell pretreatment with the molecule and light exposition for 30 min duration.

Similarly, Li et al. [119] synthetized a BODIPY (**MBDP**) (Figure 12) containing morpholine for lysosomal targeting and iodine atoms on the BODIPY core to produce $^1O_2$. Triethylene glycol monomethyl ether benzaldehyde was added through the condensation of Knoevenagel to obtain the desired compound [120].

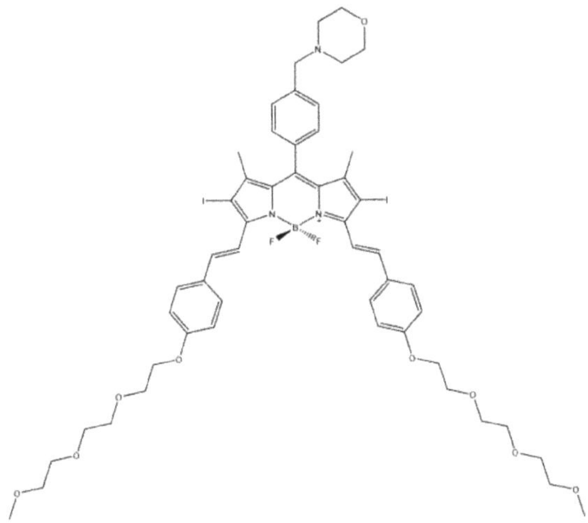

MBDP

Figure 12. Chemical structures of compound **MBDP** [119].

The analysis of the absorption spectrum of the compound showed that there is an intense cui bands at 660 nm. At 660 nm, the compound under LED light produces high levels of $^1O_2$ (Table 11).

Table 11. Physico-chemical properties and biological data obtained for compound **MBDP** [119].

	MBDP
λ_{abs} [a]	660 nm
ε	83,226 $M^{-1}cm^{-1}$
λ_{em} [a]	694 nm
Φ_{fluo} [b]	0.11
$^1O_2$ QY [c]	0.64
$IC_{50\ (MCF7)}$ [d]	0.2 μM

[a] In DCM. [b] In DCM with Cresyl Violet (Φfluo ≈ 0.66 in methanol) as a reference. [c] Using DPBF as the $^1O_2$ capture agent and methylene blue as standard (Φ = 0.57 in DCM). [d] Irradiation with 660 nm LED red light (20 mW/cm^2, 48 J/cm^2).

Cytotoxicity tests were performed on the MCF7 cell line with an estimated IC_{50} at 0.2 μM. The colocalization experiments confirmed the effective lysosomal localization of the compound. It was also possible to determine how, following irradiation, the lysosomes containing photosensitizer break down following $^1O_2$ and ROS generation. In addition, the BODIPY derivative could also be used for NIR-PDT.

Reticulum Targeting

Another target is the plasma reticulum. The BODIPY **10** (Figure 13) bearing an analogous glibenclamide motif obtained by Zhou et al., has precisely the goal of targeting towards the plasma reticulum [121].

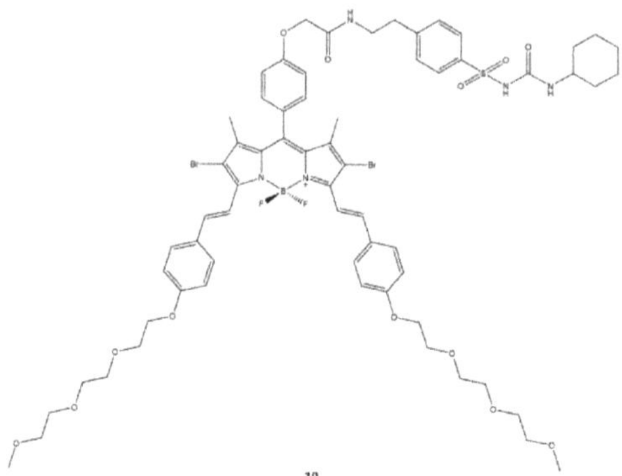

10

Figure 13. Chemical structures of compound **10** [121].

The absorption spectrum of the compound exhibits the band at 669 nm. The quantity of $^1O_2$ produced by the compound is very low when results were compared to those obtained with the reference molecule, zinc(II) phthalocyanine. The photodynamic activity has been studied on HeLa and HepG2 cells. The light source used is a 300 W halogen lamp on which a colored glass filter has been placed plus a cut-on at λ = 610 nm. The IC_{50} values obtained in both cell lines are lower than 0.2 μM (0.09 μM in HeLa and 0.16 μM in HepG2). Such low concentrations are linked to the high uptake of the compound by the cell lines being analyzed (Table 12).

Table 12. Physico-chemical properties and biological data obtained for compound **10** [121].

	10
λ_{abs} [a]	669 nm
ε	114,815 $M^{-1}cm^{-1}$
λ_{em} [a]	692 nm
Φ_{fluo} [b]	0.32
$^1O_2$ QY [c]	0.11
$IC_{50\ (Hela)}$ [d]	0.09 μM
$IC_{50\ (HepG2)}$ [d]	0.16 μM

[a] In PBS with 0.3% v/v Tween 80 and 1% v/v DMF. [b] With reference to zinc(II) phthalocyanine (Φ_{fluo} = 0.28 in DMF). [c] With reference to zinc(II) phthalocyanine ($^1O_2$ QY ≈ 0.56 in DMF). [d] Irradiation with light (λ = 610 nm, 10 min) and dark (50 min) with a total of 24 h of light exposure.

The compound was found to be associated with the endoplasmic reticulum using confocal microscopy when the results were compared to those obtained with cells treated with an analog molecule without glibenclamide. The analog molecule did not accumulate in reticulum.

Mitochondrial Targeting

In 2021 Bhattacheryya et al., thought of binding a iodine-BODIPY to a metallic ternary system to increase its solubility in water, an essential factor for in vivo applications of highly lipophilic molecules [122,123] (Figure 14).

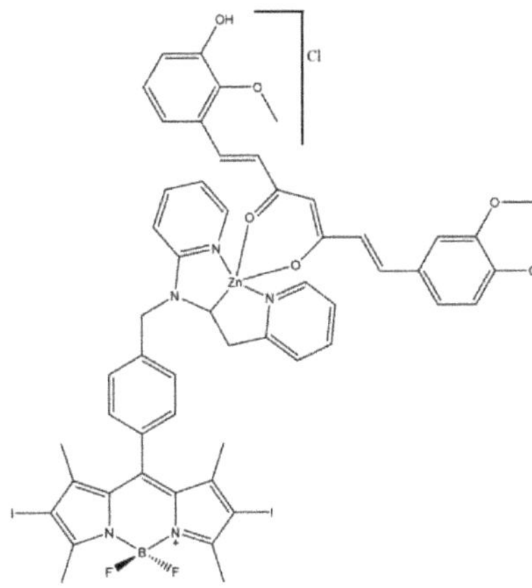

3

Figure 14. Chemical structures of compound **3** [122,123].

Curcumin was further coordinated within the system. Curcumin is well known for its tumor specific activity [124–126] but also as a photosensitizer. Curcumin in the native form has poor bioavailability and is susceptible to hydrolysis under cellular pH conditions thus reducing its therapeutic efficacy. However, curcumin bound to transition metal ions became stable and the metal complexes show significant mitochondrial localization [127]. In this work the authors show that the complex **3** having green light harvesting di-iodinated BODIPY and curcumin for mitochondrial targeting have an interesting photodynamic activity in human breast cancer (MCF7) via disruption of mitochondrial membrane (Table 13).

Table 13. Physico-chemical properties and biological data obtained for compound **3** [122,123].

	3
λ_{abs} [a]	543 nm
ε	29,400 $M^{-1}cm^{-1}$
λ_{em} [a]	506 nm
Φ_{fluo} [b]	0.02
$^1O_2$ QY [c]	0.73
$IC_{50\ (HeLa)}$ [d]	0.025 μM
$IC_{50\ (MCF7)}$ [d]	0.055 μM
$IC_{50\ (HPL1D)}$ [d]	0.230 μM

[a] In 1% DMSO-DPBS buffer medium (pH = 7.4). [b] With fluoresceinf (0.1 M NaOH, Φ_{fluo} = 0.79) as standard. [c] Using Rose Bengal as standard in DMSO ($^1O_2$ QY = 0.76). [d] Irradiation with a visible light source for 1 h (fluence rate = 2.4 mW/cm^2, light dose = 10 J/cm^2).

Cell Membrane Targeting

The targeting can be implemented not only towards cytoplasmic organelles but also towards receptors present on cell plasmic membrane.

The asialoglycoprotein receptor (ASGP) is expressed exclusively in the liver and was considered by Li et al., to design a BODIPY that specifically targeted it [111]. For this purpose, the author has synthesized a new macro-molecular PS equipped with hydrophilic appendix capable of selectively targeting liver cancer cells. The compound obtained is a galactose-functionalized BODIPY-based macromolecular photosensitizer **p(GEMA-co-BODIPYMA)-2I** with good solubility in aqueous environment (Figure 15).

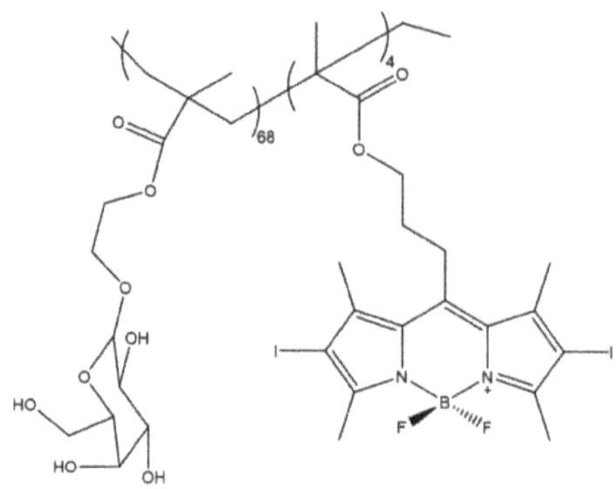

p(GEMA-co-BODIPYMA)-2I

Figure 15. Chemical structures of compound **p(GEMA-co-BODIPYMA)-2I** [111].

The compound proved to be a good $^1O_2$ producer. Biological tests were carried out on HepG2 and NIH3T3 cell lines and first concerned the specific binding towards liver cells and subsequently the cytotoxic activity on them was evaluated (Table 14).

Table 14. Physico-chemical properties and biological data obtained for compound **p(GEMA-co-BODIPYMA)-2I** [111].

	p(GEMA-co-BODIPYMA)-2I
λ_{abs} [a]	535 nm
$^1O_2$ QY [b]	0.79

[a] In H_2O. [b] Rose Bengal ($^1O_2$ = 0.76) was used as reference.

The compound was observed in the cytoplasm of HepG2 cells as demonstrated by confocal microscopy using fluorescence emission of iodine atoms. Previous studies carried out by the same group [128] have shown that the removal of galactose does not allow targeting towards HepG2 cells. The results therefore suggest that the internalization of **p(GEMA-co-BODIPYMA)-2I** depends mainly on the ASGP receptor, over expressed in HepG2 cells. The photodynamic efficacy of the compound was then evaluated. Cell survival was less than 20% following treatment with a concentration lower than 20 μM. In addition, cell exposure to the BODIPY derivative triggered cell to apoptosis. In contrast, the compound did not accumulate in NIH3T3 lacking ASGP receptor expression and the cells did not undergo cell death after light exposure.

Tyrosinase is a key regulatory enzyme in the biosynthesis of melanin through melanogenesis and is localized in the membrane of melanosome. Tyrosinase level is closely correlated with the malignancy level, so it can be considered as a biomarker of melanoma cell. Phenylthiourea (PTU) is one of the most important and well-known tyrosinase inhibitors. Kim et al., designed a BODIPY derivative with PTU pendant to enhance the PDT efficacy against melanoma cell line (Figure 16).

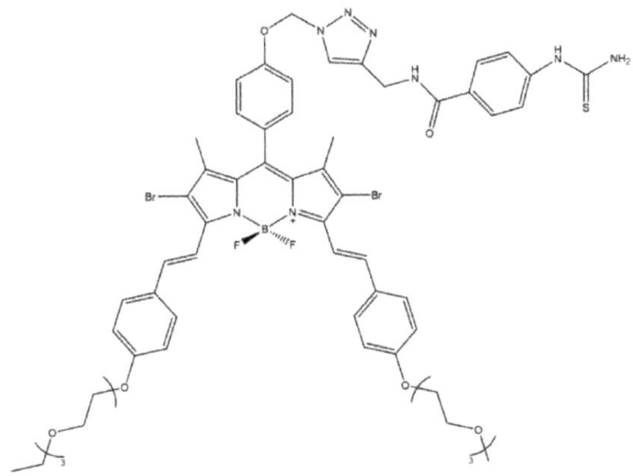

PTUBDP

Figure 16. Chemical structures of compound **PTUBDP** [129].

Cellular uptake studies demonstrated the specificity of the binding of the compound to tyrosinase overexpressed in murine B16F10 cells [129] (Table 15).

Table 15. Physico-chemical properties and biological data obtained for compound **PTUBDP** [129].

	PTUBDP
λ_{abs} [a]	667 nm
ε	54,400 $M^{-1}cm^{-1}$
λ_{em} [a]	712 nm
$^1O_2$ QY [b]	0.093

[a] In DMSO. [b] In DMSO with Rose Bengala as a reference.

Epidermal growth factor (EGFR) is another target since it is expressed in diverse tumor cells. In recent years, a series of sequences able to bind specifically to this receptor have been identified [130]; Zhao et al., selected the sequence namely D4 which was bound to polyethylene glycol and subsequently placed on a BODIPY (Figure 17) [131].

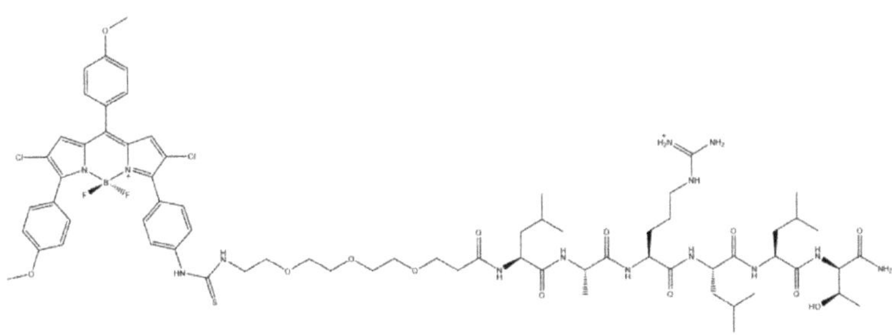

8

Figure 17. Chemical structures of compound **8** [131].

Conjugation with the peptide (**8**) shows a slight shift of the emission and absorption bands with respect to the precursor without the peptide. The quantum yield of fluorescence is five times lower than that of the compound without the peptide (Table 16).

Table 16. Physico-chemical properties and biological data obtained for compound **8** [131].

	8
λ_{abs} [a]	588 nm
ε	43,000 $M^{-1}cm^{-1}$
λ_{em} [a]	634 nm
Φ_{fluo} [b]	0.003
$IC_{50\ (HepG2)}$ [c]	74 μM

[a] In DMSO. [b] Cresyl Violet (Φfluo ≈ 0.50 in ethanol) as a reference. [c] Irradiation with halogen lamp for 20 min with a light dose of 1.5 J/cm^2.

Photodynamic activity data on HepG2 human carcinoma cell, known to express EGFR [132,133] studies show that the conjugate is more active than the precursor without the peptide this is probably due to the presence of a positive charge on the peptide sequence which should favor a faster and more efficient cell internalization [134]. Uptake kinetic demonstrated a fast absorption of the molecule over 24 h, suggesting that the presence of tri-ethylene glycol and the positive charge of the peptide sequence increase the internalization of the compound.

Burgess et al. worked on targeting the receptor tyrosine kinase [135–137]. Starting from a series of previous experiments in which specific conjugates for the receptor tyrosine kinase with cytotoxicity were identified, Burgess decided to hook these conjugates to a BODIPY to obtain a molecule (**IY-IY-PDT**) (Figure 18) usable in PDT [136].

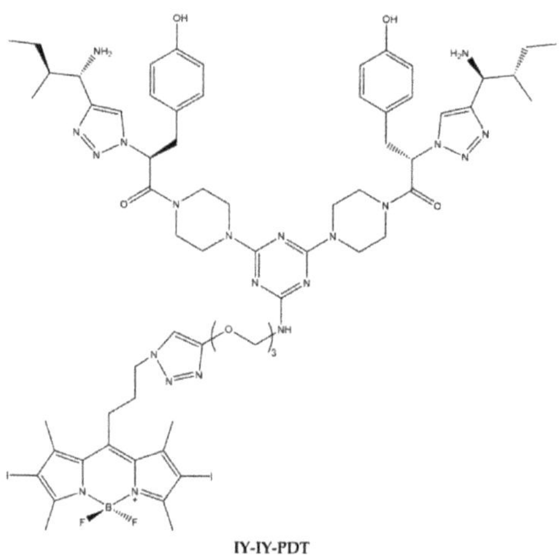

IY-IY-PDT
1

Figure 18. Chemical structures of compound **1** (**IY-IY-PDT**) [135].

Following the satisfactory results of the physico-chemical characterization, the molecule obtained was tested on NIHT3T wild-type (NIHT3T-WT) and over cell lines expressing the receptor tyrosine kinase (NIHT3T-TrkC). It was possible to detect a significant photo-induced cytotoxicity effect in over-expressing cells following treatment with the compound **IY-IY-PDT** (IC_{50} = 0.35 μM) (Table 17).

Table 17. Physico-chemical properties and biological data obtained for compound **1** (**IY-IY-PDT**) [135].

	IY-IY-PDT
λ_{abs} [a]	532 nm
λ_{em} [a]	550 nm
$IC_{50\ (NIH3T3)}$ [b]	0.35 μM

[a] In DMSO. [b] Irradiation with a halogen lamp for 10 min with a light dose of 7.3 J/cm^2.

In cellular localization it states that the compound under analysis is mainly located at the level of the lysosomes while it does not accumulate at the mitochondrial or endoplasmic reticulum level by exploiting the tyrosine kinase receptor.

The following year, the same author tested the same compound on breast cancer cell lines expressing receptor tyrosine kinase [136]. The data obtained also in this case affirm that the compound is more assimilated by the cells that express the receptor and also in this case the localization of the compound is at the lysosomal level. The first in vivo tests on mice are also carried out in this article which showed that there is a high decrease in tumor size in mice starting at four days of illumination.

In the next article published in 2016, the author reports an increased antitumor immune response after treatment with the conjugate and illumination of the area of interest [137]. The overall analysis of the data reported in the three articles published by the author leads to the hypothesis that the compound **IY-IY-PDT** acts as a therapeutic agent capable of stimulating the immune system, which makes it an excellent candidate for use in clinical practice of cancer treatment.

2.4. Sensitive Activity

Another area in which the research was concentrated was that of obtaining BODIPYs capable of being activated in specific conditions such as acid pH, typical of cancer cells, or in the presence of particular molecules such as glutathione (GSH).

Radunz et al. evaluated the activity of a pH-activable fluorescent $^1O_2$ generating BODIPY dyes (**3b**), which can be used as PS and for bioimaging (Figure 19) [138].

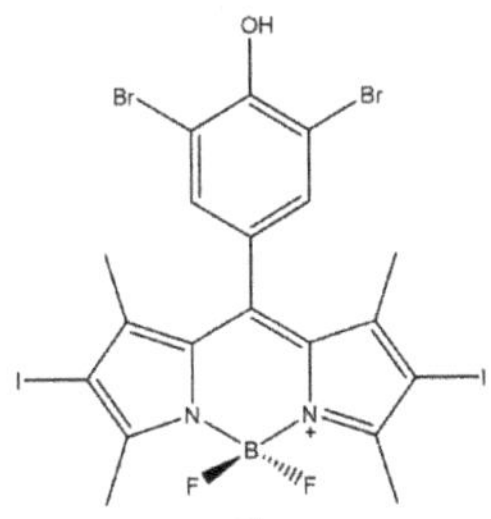

3b

Figure 19. Chemical structures of compound **3b** [138].

To ensure that the compound was able to produce a high level of $^1O_2$, the core of the BODIPY has iodine atoms in β-positions. The responsive part of the pH is instead represented by a phenolic substituent placed in the meso position of the core.

From the titrations carried out in the pH range between 4.5 and 8.2 it is observed that the absorption is minimally influenced by the pH while the intensity of the fluorescence decreases drastically at basic pH. The production of $^1O_2$ is shown in Table 18.

Table 18. Physico-chemical properties and biological data obtained for compound **3b** [138].

	3b
λ_{abs} [a]	534 nm
ε	99,000 $M^{-1}cm^{-1}$
λ_{em} [a]	550 nm
Φ_{fluo}	0.02
$^1O_2$ QY [b]	0.59
$IC_{50\,(HeLa)}$ [c]	70 nM
$IC_{50\,(HeLa,\,pH=5.5)}$ [d]	30 nM
$IC_{50\,(HeLa,\,pH=7.5)}$ [d]	150 nM

[a] In acetonitrile. [b] With DPBF as indicator and Rose Bengala as standard. [c] Irradiation with laser diode (λ = 532 nm) for 30 min. [d] Irradiation with laser diode (λ = 532 nm) for 5 min with an irradiance of 1 mW/cm^2.

As a result of the irradiation the compound showed an activity with an IC_{50} on the HeLa tumor cell line equal to 70 nM. The authors repeated the photodynamic experiment by adjusting the intracellular pH to a value of 5.5 or 7.5 before illumination [138]. The results obtained showed that cell survival was lower for cells with an acidic environment compared to those in a neutral environment (cell survival was 10% vs. 70% at a PS concentration of 0.1 μM).

Cao et al. reported the synthesis of a GSH-responsive BODIPY (**8a**) consisting of three modules: a BODIPY-based photosensitizing chromophore, a BODIPY-based quencher and a bio-reducible disulfide linker (Figure 20) [139].

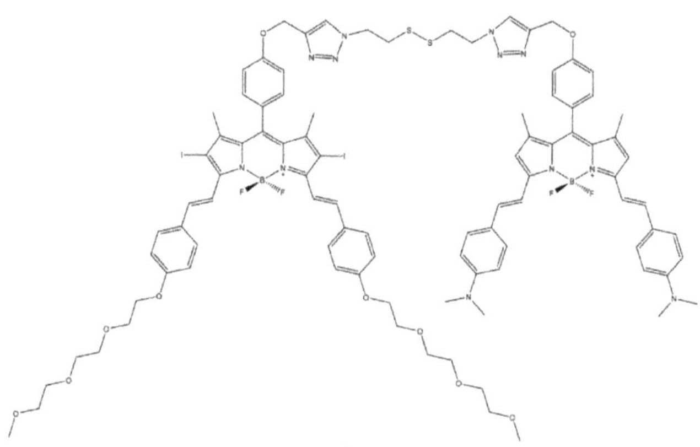

8a

Figure 20. Chemical structures of compound **8a** [139].

The compound thus obtained has shown to have excellent photophysical and photochemical characteristics such as a strong absorption in the visible region and in the NIR as well as high photostability. Overall, the compound does not show fluorescence, but after the addition of 10 μM of GSH which breaks the disulfide bond, the fluorescence is recovered.

In vitro studies have been conducted on three cell lines (A549, H22 and HeLa) known to have high intracellular concentrations of GSH. The compound is rapidly internalized by cells and effectively activated by intracellular biothiols. In the presence of light, the activity expressed as IC_{50} is of the order of μM. The compound is mainly localized in mitochondria where it is believed that GSH-mediated cleavage may also occur (Table 19).

Table 19. Physico-chemical properties and biological data obtained for compound **8a** [139].

	8a
λ_{abs} [a]	664 nm
λ_{em} [a]	768 nm
Φ_{fluo} [b]	0.052
$^1O_2$ QY [c]	0.018
$IC_{50\ (HeLa)}$ [d]	0.67 μM
$IC_{50\ (A549)}$ [d]	0.44 μM
$IC_{50\ (H22)}$ [d]	0.48 μM

[a] In DMF. [b] Relative to unsubstituted zinc(II) phthalocyanine in DMF as the reference (Φfluo ≈ 0.28). [c] Relative to unsubstituted zinc(II) phthalocyanine ($^1O_2$ QY ≈ 0.56). [d] Irradiation with a 670 nm LED lamp for 2 min with an irradiance of 20 mW/cm^2.

In vivo experiments on H22 tumor-bearing mice showed high selectivity in tumor tissue and efficacy in inhibiting tumor growth.

Jiang et al., synthetized a BODIPY responsive to both pH and thiol groups [140]. Specifically, compound (**7**) has two ferrocenyl moieties attached to a combination of ketal- and disulfide-linkers (Figure 21). The former is cleaved in acid conditions, while the latter with cellular GSH.

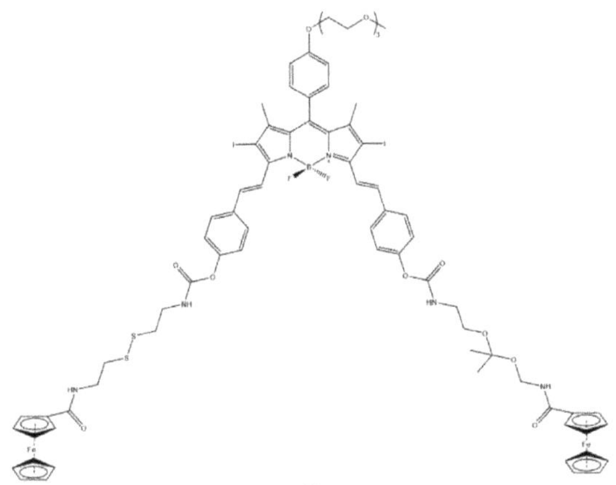

7

Figure 21. Chemical structures of compound **7** [140].

Fluorescence and $^1O_2$ production of the molecule depends on the two ferrocenyl groups, in fact at acid pH and in the presence of GSH, which break the ketal and the disulfide bond, an increase in fluorescence and $^1O_2$ is observed (Table 20).

Table 20. Physico-chemical properties and biological data obtained for compound **7** [140].

	7
λ_{abs} [a]	376, 441, 662 nm
λ_{em} [a]	686 nm
Φ_{fluo} [a]	0.03
$IC_{50\ (MCF7,\ no\ DTT}$ [b]	146.0 nM
$IC_{50\ ((MCF7,\ 2\ \mu M\ DTT)}$ [b]	140.0 nM
$IC_{50\ (MCF7,\ 4\ mM\ DTT)}$ [b]	81 nM

[a] In PBS with 0.25% Cremophor EL. [b] Irradiation with a halogen lamp for 20 min with a total fluence of 48 J/cm^2.

Photodynamic activity studies on the MCF7 cell line showed that compound **7** has dithiothreitol (DTT) dependent behavior (IC_{50} halves when DTT goes from 2 μM to 4 mM).

In vivo experiments were conducted using nude mice carrying HT29 human colorectal carcinoma. From the fluorescence images it is observed that compound **7** is localized only and exclusively in the area of the tumor, suggesting that the compound is activated only and exclusively at the tumor level where there is the cleavage of the two linkers, by means of pH and DTT, and consequent detachment of the two ferrocenyl motifs.

Funding: This research received no external funding.

Conflicts of Interest: The authors declare no conflict of interest.

References

1. Dougherty, T.J.; Gomer, C.J.; Henderson, B.W.; Jori, G.; Kessel, D.; Korbelik, M.; Moan, J.; Peng, Q. Photodynamic therapy. *J. Natl. Cancer Inst.* **1998**, *90*, 889–905. [CrossRef] [PubMed]
2. Chilakamarthi, U.; Giribabu, L. Photodynamic Therapy: Past, Present and Future. *Chem. Rec.* **2017**, *17*, 775–802. [CrossRef] [PubMed]
3. Blum, H.F. Photodynamic action and diseases caused by light. *Am. Chem. Soc. Monogr.* **1941**, *85*, 283–295. [CrossRef]
4. Zhou, Z.; Song, J.; Nie, L.; Chen, X. Reactive oxygen species generating systems meeting challenges of photodynamic cancer therapy. *Chem. Soc. Rev.* **2016**, *45*, 6597–6626. [CrossRef] [PubMed]
5. Allison, R.R.; Downie, G.H.; Cuenca, R.; Hu, X.H.; Childs, C.J.; Sibata, C.H. Photosensitizers in clinical PDT. *Photodiagnosis Photodyn. Ther.* **2004**, *1*, 27–42. [CrossRef]
6. Juzeniene, A.; Peng, Q.; Moan, J. Milestones in the development of photodynamic therapy and fluorescence diagnosis. *Photochem. Photobiol. Sci.* **2007**, *6*, 1234–1245. [CrossRef]
7. Daniell, M.D.; Hill, J.S. A history of photodynamic therapy. *Aust. N. Z. J. Surg.* **1991**, *61*, 340–348. [CrossRef] [PubMed]
8. Castellani, A.; Pace, G.P.; Concioli, M. Photodynamic effect of haematoporphyrin on blood microcirculation. *J. Pathol. Bacteriol.* **1963**, *86*, 99–102. [CrossRef] [PubMed]
9. de Gruijl, F.R.; Van der Leun, J.C. Estimate of the wavelength dependency of ultraviolet carcinogenesis in humans and its relevance to the risk assessment of a stratospheric ozone depletion. *Health Phys.* **1994**, *67*, 319–325. [CrossRef]
10. Dolmans, D.E.; Fukumura, D.; Jain, R.K. Photodynamic therapy for cancer. *Nat. Rev. Cancer* **2003**, *3*, 380–387. [CrossRef]
11. Agostinis, P.; Berg, K.; Cengel, K.A.; Foster, T.H.; Girotti, A.W.; Gollnick, S.O.; Hahn, S.M.; Hamblin, M.R.; Juzeniene, A.; Kessel, D.; et al. Photodynamic therapy of cancer: An update. *CA Cancer J. Clin.* **2011**, *61*, 250–281. [CrossRef] [PubMed]
12. Ledoux-Lebards, C. Action de la lumiere sur la toxicite de l'éosine et de quelques autres substances. *Ann. Inst. Pasteur* **1902**, *16*, 587–593.
13. Foote, C.S. Mechanisms of photosensitized oxidation. There are several different types of photosensitized oxidation which may be important in biological systems. *Science* **1968**, *162*, 963–970. [CrossRef] [PubMed]
14. Moan, J.; Wold, E. Detection of singlet oxygen production by ESR. *Nature* **1979**, *279*, 450–451. [CrossRef] [PubMed]
15. Moan, J.; Sommer, S. Oxygen dependence of the photosensitizing effect of hematoporphyrin derivative in NHIK 3025 cells. *Cancer Res.* **1985**, *45*, 1608–1610. [PubMed]
16. Devasagayam, T.P.A.; Kamat, J.P. Biological significance of singlet oxygen. *Indian J. Exp. Biol.* **2002**, *40*, 680–692.
17. Davies, M.J. Singlet oxygen-mediated damage to proteins and its consequences. *Biochem. Biophys. Res. Commun.* **2003**, *305*, 761–770. [CrossRef]
18. Juzeniene, A.; Nielsen, K.P.; Moan, J. Biophysical aspects of photodynamic therapy. *J. Environ. Pathol. Toxicol. Oncol.* **2006**, *25*, 7–28. [CrossRef] [PubMed]
19. Schmidt, R. Photosensitized Generation of Singlet Oxygen. *Photochem. Photobiol.* **2007**, *82*, 1161–1177. [CrossRef]
20. Sharman, W.M.; Allen, C.M.; van Lier, J.E. Role of activated oxygen species in photodynamic therapy. *Methods Enzymol.* **2000**, *319*, 376–400. [CrossRef]
21. Li, C. *Breast Cancer Epidemiology*, 1st ed.; Springer: New York, NY, USA, 2010; pp. 1–417.
22. Castano, A.P.; Demidova, T.N.; Hamblin, M.R. Mechanisms in photodynamic therapy: Part one-photosensitizers, photochemistry and cellular localization. *Photodiagnosis Photodyn. Ther.* **2004**, *1*, 279–293. [CrossRef]
23. Robertson, C.A.; Evans, D.H.; Abrahamse, H. Photodynamic therapy (PDT): A short review on cellular mechanisms and cancer research applications for PDT. *J. Photochem. Photobiol. B* **2009**, *96*, 1–8. [CrossRef] [PubMed]
24. Kwiatkowski, S.; Knap, B.; Przystupski, D.; Saczko, J.; Kedzierska, E.; Knap-Czop, K.; Kotlinska, J.; Michel, O.; Kotowski, K.; Kulbacka, J. Photodynamic therapy-mechanisms, photosensitizers and combinations. *Biomed. Pharmacother.* **2018**, *106*, 1098–1107. [CrossRef] [PubMed]
25. Schenck, G.O. Photosensitized reactions with molecular oxygen. *Naturwissenschaften* **1984**, *71*, 28–29.
26. Foote, C.S.; Wexler, S. Singlet Oxygen. A Probable Intermediate in Photosensitized Autoxidations. *J. Am. Chem. Soc.* **2002**, *86*, 3880–3881. [CrossRef]
27. Niedre, M.J.; Secord, A.J.; Patterson, M.S.; Wilson, B.C. In vitro tests of the validity of singlet oxygen luminescence measurements as a dose metric in photodynamic therapy. *Cancer Res.* **2003**, *63*, 7986–7994. [PubMed]

28. Weishaupt, K.R.; Gomer, C.J.; Dougherty, T.J. Identification of singlet oxygen as the cytotoxic agent in photoinactivation of a murine tumor. *Cancer Res.* **1976**, *36*, 2326–2329.
29. Fisher, A.M.; Murphree, A.L.; Gomer, C.J. Clinical and preclinical photodynamic therapy. *Lasers Surg. Med.* **1995**, *17*, 2–31. [CrossRef]
30. Dos Santos, A.F.; Terra, L.F.; Wailemann, R.A.M.; Oliveira, T.C.; de Morais Gomes, V.; Mineiro, M.F.; Meotti, F.C.; Bruni-Cardoso, A.; Baptista, M.S.; Labriola, L. Methylene blue photodynamic therapy induces selective and massive cell death in human breast cancer cells. *BMC Cancer* **2017**, *17*, 194–209. [CrossRef]
31. Auten, R.L.; Davis, J.M. Oxygen toxicity and reactive oxygen species: The devil is in the details. *Pediatr. Res.* **2009**, *66*, 121–127. [CrossRef]
32. van Duijnhoven, F.H.; Aalbers, R.I.; Rovers, J.P.; Terpstra, O.T.; Kuppen, P.J. The immunological consequences of photodynamic treatment of cancer, a literature review. *Immunobiology* **2003**, *207*, 105–113. [CrossRef]
33. Mroz, P.; Hashmi, J.T.; Huang, Y.Y.; Lange, N.; Hamblin, M.R. Stimulation of anti-tumor immunity by photodynamic therapy. *Expert Rev. Clin. Immunol.* **2011**, *7*, 75–91. [CrossRef] [PubMed]
34. Mroz, P.; Hamblin, M.R. The immunosuppressive side of PDT. *Photochem. Photobiol. Sci.* **2011**, *10*, 751–758. [CrossRef] [PubMed]
35. Reginato, E.; Wolf, P.; Hamblin, M.R. Immune response after photodynamic therapy increases anti-cancer and anti-bacterial effects. *World J. Immunol.* **2014**, *4*, 1–11. [CrossRef] [PubMed]
36. Lobo, A.C.S.; Gomes-da-Silva, L.C.; Rodrigues-Santos, P.; Cabrita, A.; Santos-Rosa, M.; Arnaut, L.G. Immune Responses after Vascular Photodynamic Therapy with Redaporfin. *J. Clin. Med.* **2019**, *9*, 104. [CrossRef] [PubMed]
37. Falk-Mahapatra, R.; Gollnick, S.O. Photodynamic Therapy and Immunity: An Update. *Photochem. Photobiol.* **2020**, *96*, 550–559. [CrossRef]
38. Nkune, N.W.; Simelane, N.W.N.; Montaseri, H.; Abrahamse, H. Photodynamic Therapy-Mediated Immune Responses in Three-Dimensional Tumor Models. *Int. J. Mol. Sci.* **2021**, *22*, 12618. [CrossRef]
39. Gilchrest, B.A. Photodynamic therapy and selected off-label uses. In *Winter Clinical Dermatology Conference-Hawaii*; Tuleya, S., Ed.; HMP Communications, LLC: Kohala Coast, HI, USA, 2010; pp. 10–12.
40. Kamkaew, A.; Lim, S.H.; Lee, H.B.; Kiew, L.V.; Chung, L.Y.; Burgess, K. BODIPY dyes in photodynamic therapy. *Chem. Soc. Rev.* **2013**, *42*, 77–88. [CrossRef]
41. Liu, J.; Zheng, L.; Li, Y.; Zhang, Z.; Zhang, L.; Shen, L.; Zhang, X.; Qiao, H. Effect of DTPP-mediated photodynamic therapy on cell morphology, viability, cell cycle, and cytotoxicity in a murine lung adenocarcinoma cell line. *Lasers Med. Sci.* **2015**, *30*, 181–191. [CrossRef]
42. Meyer-Betz, F. Untersuchungen uber die biologische (photodynamische) Wirkung des Hamatoporphyrings und anderer Derivate des Blut- und Gallenfarbstoffs. *Dtsch. Arch. Klin. Med.* **1913**, *112*, 476–503.
43. Ackroyd, R.; Kelty, C.; Brown, N.; Reed, M. The history of photodetection and photodynamic therapy. *Photochem. Photobiol.* **2001**, *74*, 656–669. [CrossRef]
44. Ormond, A.B.; Freeman, H.S. Dye Sensitizers for Photodynamic Therapy. *Materials* **2013**, *6*, 817–840. [CrossRef] [PubMed]
45. Hage, R.; Ferreira, J.; Bagnato, V.S.; Vollet, E.D.; Plapler, H. Pharmacokinetics of Photogem Using Fluorescence Spectroscopy in Dimethylhydrazine-Induced Murine Colorectal Carcinoma. *Int. J. Photoenergy* **2012**, *2012*, 1–8. [CrossRef]
46. Allison, R.R.; Sibata, C.H. Oncologic photodynamic therapy photosensitizers: A clinical review. *Photodiagnosis Photodyn. Ther.* **2010**, *7*, 61–75. [CrossRef] [PubMed]
47. Hammerer, F.; Poyer, F.; Fourmois, L.; Chen, S.; Garcia, G.; Teulade-Fichou, M.P.; Maillard, P.; Mahuteau-Betzer, F. Mitochondria-targeted cationic porphyrin-triphenylamine hybrids for enhanced two-photon photodynamic therapy. *Bioorganic Med. Chem.* **2018**, *26*, 107–118. [CrossRef] [PubMed]
48. Gilson, D.; Ash, D.; Driver, I.; Feather, J.W.; Brown, S. Therapeutic ratio of photodynamic therapy in the treatment of superficial tumours of skin and subcutaneous tissues in man. *Br. J. Cancer* **1988**, *58*, 665–667. [CrossRef] [PubMed]
49. Hausser, K.W.; Vahle, W. Die abhangigkeit des lichterythems und der pigmentbildung von der schwingungszahl (wellenlange der erregenden strahlung). *Strahlentherapie* **1931**, *13*, 41–71.
50. Chevalier, S.; Anidjar, M.; Scarlata, E.; Hamel, L.; Scherz, A.; Ficheux, H.; Borenstein, N.; Fiette, L.; Elhilali, M. Preclinical study of the novel vascular occluding agent, WST11, for photodynamic therapy of the canine prostate. *J. Urol.* **2011**, *186*, 302–309. [CrossRef]
51. Triesscheijn, M.; Ruevekamp, M.; Aalders, M.; Baas, P.; Stewart, F.A. Outcome of mTHPC mediated photodynamic therapy is primarily determined by the vascular response. *Photochem. Photobiol.* **2005**, *81*, 1161–1167. [CrossRef]
52. Detty, M.R.; Gibson, S.L.; Wagner, S.J. Current clinical and preclinical photosensitizers for use in photodynamic therapy. *J. Med. Chem.* **2004**, *47*, 3897–3915. [CrossRef]
53. Dovigo, L.N.; Pavarina, A.C.; Ribeiro, A.P.; Brunetti, I.L.; Costa, C.A.; Jacomassi, D.P.; Bagnato, V.S.; Kurachi, C. Investigation of the photodynamic effects of curcumin against Candida albicans. *Photochem. Photobiol.* **2011**, *87*, 895–903. [CrossRef]
54. Wohrle, D.; Hirth, A.; Bogdahn-Rai, T.; Schnurpfeil, G.; Shopova, M. Photodynamic therapy of cancer: Second and third generations of photosensitizers. *Russ. Chem. Bull.* **1998**, *47*, 807–816. [CrossRef]
55. Li, L.; Nguyen, B.; Burgess, K. Functionalization of the 4,4-difluoro-4-bora-3a,4a-diaza-s-indacene (BODIPY) core. *Bioorganic Med. Chem. Lett.* **2008**, *18*, 3112–3116. [CrossRef] [PubMed]

56. Awuah, S.G.; You, Y. Boron dipyrromethene (BODIPY)-based photosensitizers for photodynamic therapy. *RSC Adv.* **2012**, *2*, 11169–11183. [CrossRef]
57. Yuster, P.; Weissman, S.I. Effects of Perturbations on Phosphorescence: Luminescence of Metal Organic Complexes. *J. Chem. Phys.* **1949**, *17*, 1182–1188. [CrossRef]
58. Treibs, A.; Kreuzer, F.-H. Difluorboryl-Komplexe von Di- und Tripyrrylmethenen. *Justus Liebigs Ann. Chem.* **1968**, *718*, 208–223. [CrossRef]
59. Ulrich, G.; Ziessel, R.; Harriman, A. The chemistry of fluorescent bodipy dyes: Versatility unsurpassed. *Angew. Chem. Int. Ed. Engl.* **2008**, *47*, 1184–1201. [CrossRef]
60. Karolin, J.; Johansson, L.B.A.; Strandberg, L.; Ny, T. Fluorescence and Absorption Spectroscopic Properties of Dipyrrometheneboron Difluoride (BODIPY) Derivatives in Liquids, Lipid Membranes, and Proteins. *J. Am. Chem. Soc.* **2002**, *116*, 7801–7806. [CrossRef]
61. Yogo, T.; Urano, Y.; Ishitsuka, Y.; Maniwa, F.; Nagano, T. Highly efficient and photostable photosensitizer based on BODIPY chromophore. *J. Am. Chem. Soc.* **2005**, *127*, 12162–12163. [CrossRef]
62. Wittmershaus, B.P.; Skibicki, J.J.; McLafferty, J.B.; Zhang, Y.-Z.; Swan, S. Spectral properties of single BODIPY dyes in polystyrene microspheres and in solutions. *J. Fluoresc.* **2001**, *11*, 119–128. [CrossRef]
63. Ventura, B.; Marconi, G.; Bröring, M.; Krüger, R.; Flamigni, L. Bis(BF2)-2,2′-bidipyrrins, a class of BODIPY dyes with new spectroscopic and photophysical properties. *New J. Chem.* **2009**, *33*, 428–438. [CrossRef]
64. Kolemen, S.; Bozdemir, O.A.; Cakmak, Y.; Barin, G.; Erten-Ela, S.; Marszalek, M.; Yum, J.-H.; Zakeeruddin, S.M.; Nazeeruddin, M.K.; Grätzel, M.; et al. Optimization of distyryl-Bodipy chromophores for efficient panchromatic sensitization in dye sensitized solar cells. *Chem. Sci.* **2011**, *2*, 949–954. [CrossRef]
65. Koziar, J.C.; Cowan, D.O. Photochemical heavy-atom effects. *Acc. Chem. Res.* **2002**, *11*, 334–341. [CrossRef]
66. Loudet, A.; Burgess, K. BODIPY dyes and their derivatives: Syntheses and spectroscopic properties. *Chem. Rev.* **2007**, *107*, 4891–4932. [CrossRef] [PubMed]
67. Banfi, S.; Nasini, G.; Zaza, S.; Caruso, E. Synthesis and photo-physical properties of a series of BODIPY dyes. *Tetrahedron* **2013**, *69*, 4845–4856. [CrossRef]
68. Shivran, N.; Tyagi, M.; Mula, S.; Gupta, P.; Saha, B.; Patro, B.S.; Chattopadhyay, S. Syntheses and photodynamic activity of some glucose-conjugated BODIPY dyes. *Eur. J. Med. Chem.* **2016**, *122*, 352–365. [CrossRef] [PubMed]
69. DeRosa, M. Photosensitized singlet oxygen and its applications. *Coord. Chem. Rev.* **2002**, *233–234*, 351–371. [CrossRef]
70. Fagnoni, M. Modern Molecular Photochemistry of Organic Molecules. By Nicholas J. Turro, V. Ramamurthy and Juan C. Scaiano. *Angew. Chem. Int. Ed.* **2010**, *49*, 6709–6710. [CrossRef]
71. Pham, T.C.; Nguyen, V.N.; Choi, Y.; Lee, S.; Yoon, J. Recent Strategies to Develop Innovative Photosensitizers for Enhanced Photodynamic Therapy. *Chem. Rev.* **2021**, *121*, 13454–13619. [CrossRef]
72. Banfi, S.; Caruso, E.; Zaza, S.; Mancini, M.; Gariboldi, M.B.; Monti, E. Synthesis and photodynamic activity of a panel of BODIPY dyes. *J. Photochem. Photobiol. B* **2012**, *114*, 52–60. [CrossRef]
73. Banfi, S.; Caruso, E.; Buccafurni, L.; Murano, R.; Monti, E.; Gariboldi, M.; Papa, E.; Gramatica, P. Comparison between 5,10,15,20-tetraaryl- and 5,15-diarylporphyrins as photosensitizers: Synthesis, photodynamic activity, and quantitative structure-activity relationship modeling. *J. Med. Chem.* **2006**, *49*, 3293–3304. [CrossRef]
74. Frimayanti, N.; Yam, M.L.; Lee, H.B.; Othman, R.; Zain, S.M.; Rahman, N.A. Validation of quantitative structure-activity relationship (QSAR) model for photosensitizer activity prediction. *Int. J. Mol. Sci.* **2011**, *12*, 8626–8644. [CrossRef] [PubMed]
75. Henderson, B.W.; Bellnier, D.A.; Greco, W.R.; Sharma, A.; Pandey, R.K.; Vaughan, L.A.; Weishaupt, K.R.; Dougherty, T.J. An in vivo quantitative structure-activity relationship for a congeneric series of pyropheophorbide derivatives as photosensitizers for photodynamic therapy. *Cancer Res.* **1997**, *57*, 4000–4007.
76. Margaron, P.; Gregoire, M.J.; Scasnar, V.; Ali, H.; van Lier, J.E. Structure-photodynamic activity relationships of a series of 4-substituted zinc phthalocyanines. *Photochem. Photobiol.* **1996**, *63*, 217–223. [CrossRef] [PubMed]
77. Zagami, R.; Sortino, G.; Caruso, E.; Malacarne, M.C.; Banfi, S.; Patane, S.; Monsu Scolaro, L.; Mazzaglia, A. Tailored-BODIPY/Amphiphilic Cyclodextrin Nanoassemblies with PDT Effectiveness. *Langmuir* **2018**, *34*, 8639–8651. [CrossRef] [PubMed]
78. Barut, B.; Yalcin, C.O.; Sari, S.; Coban, O.; Keles, T.; Biyiklioglu, Z.; Abudayyak, M.; Demirbas, U.; Ozel, A. Novel water soluble BODIPY compounds: Synthesis, photochemical, DNA interaction, topoisomerases inhibition and photodynamic activity properties. *Eur. J. Med. Chem.* **2019**, *183*, 111685. [CrossRef] [PubMed]
79. Wang, C.; Qian, Y. A water soluble carbazolyl-BODIPY photosensitizer with an orthogonal D-A structure for photodynamic therapy in living cells and zebrafish. *Biomater. Sci.* **2020**, *8*, 830–836. [CrossRef]
80. Caruso, E.; Orlandi, V.T.; Malacarne, M.C.; Martegani, E.; Scanferla, C.; Pappalardo, D.; Vigliotta, G.; Izzo, L. Bodipy-Loaded Micelles Based on Polylactide as Surface Coating for Photodynamic Control of Staphylococcus aureus. *Coatings* **2021**, *11*, 223. [CrossRef]
81. Turan, I.S.; Cakmak, F.P.; Yildirim, D.C.; Cetin-Atalay, R.; Akkaya, E.U. Near-IR absorbing BODIPY derivatives as glutathione-activated photosensitizers for selective photodynamic action. *Chemistry* **2014**, *20*, 16088–16092. [CrossRef]
82. Malacarne, M.C.; Banfi, S.; Caruso, E. In vitro photodynamic treatment of cancer cells induced by aza-BODIPYs. *Photochem. Photobiol. Sci.* **2020**, *19*, 790–799. [CrossRef]

83. Lincoln, R.; Van Kessel, A.T.M.; Zhang, W.; Cosa, G. A dormant BODIPY-acrolein singlet oxygen photosensitizer intracellularly activated upon adduct formation with cysteine residues. *Photochem. Photobiol. Sci.* **2019**, *18*, 2003–2011. [CrossRef]
84. Prieto-Montero, R.; Prieto-Castaneda, A.; Sola-Llano, R.; Agarrabeitia, A.R.; Garcia-Fresnadillo, D.; Lopez-Arbeloa, I.; Villanueva, A.; Ortiz, M.J.; de la Moya, S.; Martinez-Martinez, V. Exploring BODIPY Derivatives as Singlet Oxygen Photosensitizers for PDT. *Photochem. Photobiol.* **2020**, *96*, 458–477. [CrossRef] [PubMed]
85. Ziessel, R.; Ulrich, G.; Harriman, A. The chemistry of Bodipy: A new El Dorado for fluorescence tools. *New J. Chem.* **2007**, *31*, 496–501. [CrossRef]
86. Caruso, E.; Gariboldi, M.; Sangion, A.; Gramatica, P.; Banfi, S. Synthesis, photodynamic activity, and quantitative structure-activity relationship modelling of a series of BODIPYs. *J. Photochem. Photobiol. B* **2017**, *167*, 269–281. [CrossRef] [PubMed]
87. Caruso, E.; Malacarne, M.C.; Marras, E.; Papa, E.; Bertato, L.; Banfi, S.; Gariboldi, M.B. New BODIPYs for photodynamic therapy (PDT): Synthesis and activity on human cancer cell lines. *Bioorganic Med. Chem.* **2020**, *28*, 115737. [CrossRef] [PubMed]
88. Ballestri, M.; Caruso, E.; Guerrini, A.; Ferroni, C.; Banfi, S.; Gariboldi, M.; Monti, E.; Sotgiu, G.; Varchi, G. Core-shell poly-methyl methacrylate nanoparticles covalently functionalized with a non-symmetric porphyrin for anticancer photodynamic therapy. *J. Photochem. Photobiol. B* **2018**, *186*, 169–177. [CrossRef] [PubMed]
89. Kim, B.; Sui, B.; Yue, X.; Tang, S.; Tichy, M.G.; Belfield, K.D. In Vitro Photodynamic Studies of a BODIPY-Based Photosensitizer. *Eur. J. Org. Chem.* **2016**, *2017*, 25–28. [CrossRef]
90. Lincoln, R.; Durantini, A.M.; Greene, L.E.; Martinez, S.R.; Knox, R.; Becerra, M.C.; Cosa, G. meso-Acetoxymethyl BODIPY dyes for photodynamic therapy: Improved photostability of singlet oxygen photosensitizers. *Photochem. Photobiol. Sci.* **2017**, *16*, 178–184. [CrossRef]
91. Durantini, A.M.; Greene, L.E.; Lincoln, R.; Martinez, S.R.; Cosa, G. Reactive Oxygen Species Mediated Activation of a Dormant Singlet Oxygen Photosensitizer: From Autocatalytic Singlet Oxygen Amplification to Chemicontrolled Photodynamic Therapy. *J. Am. Chem. Soc.* **2016**, *138*, 1215–1225. [CrossRef]
92. Epelde-Elezcano, N.; Martínez-Martínez, V.; Peña-Cabrera, E.; Gómez-Durán, C.F.A.; Arbeloa, I.L.; Lacombe, S. Modulation of singlet oxygen generation in halogenated BODIPY dyes by substitution at their meso position: Towards a solvent-independent standard in the vis region. *RSC Adv.* **2016**, *6*, 41991–41998. [CrossRef]
93. Turan, I.S.; Yildiz, D.; Turksoy, A.; Gunaydin, G.; Akkaya, E.U. A Bifunctional Photosensitizer for Enhanced Fractional Photodynamic Therapy: Singlet Oxygen Generation in the Presence and Absence of Light. *Angew. Chem. Int. Ed. Engl.* **2016**, *55*, 2875–2878. [CrossRef]
94. Poully, J.C.; Schermann, J.P.; Nieuwjaer, N.; Lecomte, F.; Gregoire, G.; Desfrancois, C.; Garcia, G.A.; Nahon, L.; Nandi, D.; Poisson, L.; et al. Photoionization of 2-pyridone and 2-hydroxypyridine. *Phys. Chem. Chem. Phys.* **2010**, *12*, 3566–3572. [CrossRef] [PubMed]
95. Matsumoto, M.; Yamada, M.; Watanabe, N. Reversible 1,4-cycloaddition of singlet oxygen to N-substituted 2-pyridones: 1,4-endoperoxide as a versatile chemical source of singlet oxygen. *Chem. Commun.* **2005**, *28*, 483–485. [CrossRef] [PubMed]
96. Wiegand, C.; Herdtweck, E.; Bach, T. Enantioselectivity in visible light-induced, singlet oxygen [2+4] cycloaddition reactions (type II photooxygenations) of 2-pyridones. *Chem. Commun.* **2012**, *48*, 10195–10197. [CrossRef] [PubMed]
97. Aubry, J.M.; Pierlot, C.; Rigaudy, J.; Schmidt, R. Reversible binding of oxygen to aromatic compounds. *Acc. Chem. Res.* **2003**, *36*, 668–675. [CrossRef]
98. Ayan, S.; Gunaydin, G.; Yesilgul-Mehmetcik, N.; Gedik, M.E.; Akkaya, E.U. Proof-of-principle for two-stage photodynamic therapy: Hypoxia triggered release of singlet oxygen. *Chem. Comm.* **2020**, *56*, 14793–14796. [CrossRef] [PubMed]
99. Zhao, X.; Liu, J.; Fan, J.; Chao, H.; Peng, X. Recent progress in photosensitizers for overcoming the challenges of photodynamic therapy: From molecular design to application. *Chem. Soc. Rev.* **2021**, *50*, 4185–4219. [CrossRef] [PubMed]
100. Huang, L.; Zhao, S.; Wu, J.; Yu, L.; Singh, N.; Yang, K.; Lan, M.; Wang, P.; Kim, J.S. Photodynamic therapy for hypoxic tumors: Advances and perspectives. *Coord. Chem. Rev.* **2021**, *438*, 213888. [CrossRef]
101. Cakmak, Y.; Kolemen, S.; Duman, S.; Dede, Y.; Dolen, Y.; Kilic, B.; Kostereli, Z.; Yildirim, L.T.; Dogan, A.L.; Guc, D.; et al. Designing excited states: Theory-guided access to efficient photosensitizers for photodynamic action. *Angew. Chem. Int. Ed. Engl.* **2011**, *50*, 11937–11941. [CrossRef]
102. Zou, J.; Yin, Z.; Ding, K.; Tang, Q.; Li, J.; Si, W.; Shao, J.; Zhang, Q.; Huang, W.; Dong, X. BODIPY Derivatives for Photodynamic Therapy: Influence of Configuration versus Heavy Atom Effect. *ACS Appl. Mater. Interfaces* **2017**, *9*, 32475–32481. [CrossRef]
103. Pang, W.; Zhang, X.F.; Zhou, J.; Yu, C.; Hao, E.; Jiao, L. Modulating the singlet oxygen generation property of meso-beta directly linked BODIPY dimers. *Chem. Commun.* **2012**, *48*, 5437–5439. [CrossRef]
104. Iyer, A.K.; Greish, K.; Seki, T.; Okazaki, S.; Fang, J.; Takeshita, K.; Maeda, H. Polymeric micelles of zinc protoporphyrin for tumor targeted delivery based on EPR effect and singlet oxygen generation. *J. Drug Target.* **2007**, *15*, 496–506. [CrossRef] [PubMed]
105. Huang, Z.; Xu, H.; Meyers, A.D.; Musani, A.I.; Wang, L.; Tagg, R.; Barqawi, A.B.; Chen, Y.K. Photodynamic therapy for treatment of solid tumors–potential and technical challenges. *Technol. Cancer Res. Treat.* **2008**, *7*, 309–320. [CrossRef] [PubMed]
106. Abels, C. Targeting of the vascular system of solid tumours by photodynamic therapy (PDT). *Photochem. Photobiol. Sci.* **2004**, *3*, 765–771. [CrossRef] [PubMed]
107. Boyle, R.W.; Dolphin, D. Structure and biodistribution relationships of photodynamic sensitizers. *Photochem. Photobiol.* **1996**, *64*, 469–485. [CrossRef]

108. Ricchelli, F. Photophysical properties of porphyrins in biological membranes. *J. Photochem. Photobiol. B* **1995**, *29*, 109–118. [CrossRef]
109. Konan, Y.N.; Gurny, R.; Allemann, E. State of the art in the delivery of photosensitizers for photodynamic therapy. *J. Photochem. Photobiol. B* **2002**, *66*, 89–106. [CrossRef]
110. Taillefer, J.; Jones, M.C.; Brasseur, N.; van Lier, J.E.; Leroux, J.-C. Preparation and characterization of pH-responsive polymeric micelles for the delivery of photosensitizing anticancer drugs. *J. Pharm. Sci.* **2000**, *89*, 52–62. [CrossRef]
111. Lu, Z.T.; Zhang, X.G.; Wu, Z.M.; Zhai, T.T.; Xue, Y.A.; Mei, L.; Li, C.X. BODIPY-based macromolecular photosensitizer with selective recognition and enhanced anticancer efficiency. *RSC Adv.* **2014**, *4*, 19495–19501. [CrossRef]
112. Khuong Mai, D.; Kang, B.; Pegarro Vales, T.; Badon, I.W.; Cho, S.; Lee, J.; Kim, E.; Kim, H.J. Synthesis and Photophysical Properties of Tumor-Targeted Water-Soluble BODIPY Photosensitizers for Photodynamic Therapy. *Molecules* **2020**, *25*, 3340. [CrossRef]
113. Liang, L.Y.; Astruc, D. The copper(I)-catalyzed alkyne-azide cycloaddition (CuAAC) "click" reaction and its applications. An overview. *Coord. Chem. Rev.* **2011**, *255*, 2933–2945. [CrossRef]
114. Kue, C.S.; Kamkaew, A.; Burgess, K.; Kiew, L.V.; Chung, L.Y.; Lee, H.B. Small Molecules for Active Targeting in Cancer. *Med. Res. Rev.* **2016**, *36*, 494–575. [CrossRef] [PubMed]
115. Wang, C.; Qian, Y. A novel BODIPY-based photosensitizer with pH-active singlet oxygen generation for photodynamic therapy in lysosomes. *Org. Biomol. Chem.* **2019**, *17*, 8001–8007. [CrossRef] [PubMed]
116. Kong, X.; Di, L.; Fan, Y.; Zhou, Z.; Feng, X.; Gai, L.; Tian, J.; Lu, H. Lysosome-targeting turn-on red/NIR BODIPY probes for imaging hypoxic cells. *Chem. Commun.* **2019**, *55*, 11567–11570. [CrossRef] [PubMed]
117. Wu, L.; Li, X.; Ling, Y.; Huang, C.; Jia, N. Morpholine Derivative-Functionalized Carbon Dots-Based Fluorescent Probe for Highly Selective Lysosomal Imaging in Living Cells. *ACS Appl. Mater. Interfaces* **2017**, *9*, 28222–28232. [CrossRef] [PubMed]
118. Zhang, Z.Q.; Yao, W.J.; Qiao, L.L.; Yang, X.; Shi, J.; Zhao, M.X. A Lysosome-Targetable Fluorescence Probe Based on L-Cysteine-Polyamine-Morpholine-Modified Quantum Dots for Imaging in Living Cells. *Int. J. Nanomed.* **2020**, *15*, 1611–1622. [CrossRef] [PubMed]
119. Li, M.L.; Tian, R.S.; Fan, J.L.; Du, J.J.; Long, S.; Peng, X.J. A lysosome-targeted BODIPY as potential NIR photosensitizer for photodynamic therapy. *Dye. Pigment.* **2017**, *147*, 99–105. [CrossRef]
120. Dost, Z.; Atilgan, S.; Akkaya, E.U. Distyryl-boradiazaindacenes: Facile synthesis of novel near IR emitting fluorophores. *Tetrahedron* **2006**, *62*, 8484–8488. [CrossRef]
121. Zhou, Y.; Cheung, Y.K.; Ma, C.; Zhao, S.; Gao, D.; Lo, P.C.; Fong, W.P.; Wong, K.S.; Ng, D.K.P. Endoplasmic Reticulum-Localized Two-Photon-Absorbing Boron Dipyrromethenes as Advanced Photosensitizers for Photodynamic Therapy. *J. Med. Chem.* **2018**, *61*, 3952–3961. [CrossRef] [PubMed]
122. Kue, C.S.; Ng, S.Y.; Voon, S.H.; Kamkaew, A.; Chung, L.Y.; Kiew, L.V.; Lee, H.B. Recent strategies to improve boron dipyrromethene (BODIPY) for photodynamic cancer therapy: An updated review. *Photochem. Photobiol. Sci.* **2018**, *17*, 1691–1708. [CrossRef]
123. Bhattacharyya, A.; Jameei, A.; Karande, A.A.; Chakravarty, A.R. BODIPY-attached zinc(II) complexes of curcumin drug for visible light assisted photo-sensitization, cellular imaging and targeted PDT. *Eur. J. Med. Chem.* **2021**, *220*, 113438. [CrossRef]
124. Singh, S.; Aggarwal, B.B. Activation of transcription factor NF-kappa B is suppressed by curcumin (diferuloylmethane) (*). *J. Biol. Chem.* **1995**, *270*, 24995–25000. [CrossRef] [PubMed]
125. Jobin, C.; Bradham, C.A.; Russo, M.P.; Juma, B.; Narula, A.S.; Brenner, D.A.; Sartor, R.B. Curcumin blocks cytokine-mediated NF-kappa B activation and proinflammatory gene expression by inhibiting inhibitory factor I-kappa B kinase activity. *J. Immunol.* **1999**, *163*, 3474–3483. [PubMed]
126. Plummer, S.M.; Holloway, K.A.; Manson, M.M.; Munks, R.J.; Kaptein, A.; Farrow, S.; Howells, L. Inhibition of cyclo-oxygenase 2 expression in colon cells by the chemopreventive agent curcumin involves inhibition of NF-kappaB activation via the NIK/IKK signalling complex. *Oncogene* **1999**, *18*, 6013–6020. [CrossRef] [PubMed]
127. Banerjee, S.; Chakravarty, A.R. Metal complexes of curcumin for cellular imaging, targeting, and photoinduced anticancer activity. *Acc. Chem. Res.* **2015**, *48*, 2075–2083. [CrossRef] [PubMed]
128. Lu, Z.T.; Mei, L.; Zhang, X.G.; Wang, Y.N.; Zhao, Y.; Li, C.X. Water-soluble BODIPY-conjugated glycopolymers as fluorescent probes for live cell imaging. *Polym. Chem.* **2013**, *4*, 5743–5750. [CrossRef]
129. Jung, E.; Shim, I.; An, J.; Ji, M.S.; Jangili, P.; Chi, S.G.; Kim, J.S. Phenylthiourea-Conjugated BODIPY as an Efficient Photosensitizer for Tyrosinase-Positive Melanoma-Targeted Photodynamic Therapy. *ACS Appl. Bio Mater.* **2021**, *4*, 2120–2127. [CrossRef] [PubMed]
130. Hossein-Nejad-Ariani, H.; Althagafi, E.; Kaur, K. Small Peptide Ligands for Targeting EGFR in Triple Negative Breast Cancer Cells. *Sci. Rep.* **2019**, *9*, 2723. [CrossRef]
131. Zhao, N.; Williams, T.M.; Zhou, Z.; Fronczek, F.R.; Sibrian-Vazquez, M.; Jois, S.D.; Vicente, M.G.H. Synthesis of BODIPY-Peptide Conjugates for Fluorescence Labeling of EGFR Overexpressing Cells. *Bioconjug. Chem.* **2017**, *28*, 1566–1579. [CrossRef]
132. Kiesslich, R.; Goetz, M.; Vieth, M.; Galle, P.R.; Neurath, M.F. Technology insight: Confocal laser endoscopy for in vivo diagnosis of colorectal cancer. *Nat. Clin. Pract. Oncol.* **2007**, *4*, 480–490. [CrossRef]
133. Ongarora, B.G.; Fontenot, K.R.; Hu, X.; Sehgal, I.; Satyanarayana-Jois, S.D.; Vicente, M.G. Phthalocyanine-peptide conjugates for epidermal growth factor receptor targeting. *J. Med. Chem.* **2012**, *55*, 3725–3738. [CrossRef]

134. Zhang, S.; Li, Y.; He, X.; Dong, S.; Huang, Y.; Li, X.; Li, Y.; Jin, C.; Zhang, Y.; Wang, Y. Photothermolysis mediated by gold nanorods modified with EGFR monoclonal antibody induces Hep-2 cells apoptosis in vitro and in vivo. *Int. J. Nanomed.* **2014**, *9*, 1931–1946. [CrossRef]
135. Kamkaew, A.; Burgess, K. Double-targeting using a TrkC ligand conjugated to dipyrrometheneboron difluoride (BODIPY) based photodynamic therapy (PDT) agent. *J. Med. Chem.* **2013**, *56*, 7608–7614. [CrossRef] [PubMed]
136. Kue, C.S.; Kamkaew, A.; Lee, H.B.; Chung, L.Y.; Kiew, L.V.; Burgess, K. Targeted PDT agent eradicates TrkC expressing tumors via photodynamic therapy (PDT). *Mol. Pharm.* **2015**, *12*, 212–222. [CrossRef] [PubMed]
137. Kue, C.S.; Kamkaew, A.; Voon, S.H.; Kiew, L.V.; Chung, L.Y.; Burgess, K.; Lee, H.B. Tropomyosin Receptor Kinase C Targeted Delivery of a Peptidomimetic Ligand-Photosensitizer Conjugate Induces Antitumor Immune Responses Following Photodynamic Therapy. *Sci. Rep.* **2016**, *6*, 37209. [CrossRef] [PubMed]
138. Radunz, S.; Wedepohl, S.; Rohr, M.; Calderon, M.; Tschiche, H.R.; Resch-Genger, U. pH-Activatable Singlet Oxygen-Generating Boron-dipyrromethenes (BODIPYs) for Photodynamic Therapy and Bioimaging. *J. Med. Chem.* **2020**, *63*, 1699–1708. [CrossRef]
139. Cao, J.J.; Zhang, M.S.; Li, X.Q.; Yang, D.C.; Xu, G.; Liu, J.Y. A glutathione-responsive photosensitizer with fluorescence resonance energy transfer characteristics for imaging-guided targeting photodynamic therapy. *Eur. J. Med. Chem.* **2020**, *193*, 112203. [CrossRef]
140. Jiang, X.J.; Lau, J.T.; Wang, Q.; Ng, D.K.; Lo, P.C. pH- and Thiol-Responsive BODIPY-Based Photosensitizers for Targeted Photodynamic Therapy. *Chemistry* **2016**, *22*, 8273–8281. [CrossRef]

International Journal of
Molecular Sciences

MDPI

Article

A Curcumin-BODIPY Dyad and Its Silica Hybrid as NIR Bioimaging Probes

Chiara Maria Antonietta Gangemi [1], Tania Maria Grazia Salerno [1], Anna Barattucci [1], Fabio Cucinotta [2,*], Paola Bonaccorsi [1,*], Giovanna Calabrese [1], Paola Poma [3], Maria Giovanna Rizzo [1], Sebastiano Campagna [1] and Fausto Puntoriero [1]

[1] Dipartimento di Scienze Chimiche Biologiche Farmaceutiche ed Ambientali (CHIBIOFARAM), Università degli Studi di Messina, 98166 Messina, Italy
[2] School of Natural and Environmental Sciences, Newcastle University, Newcastle upon Tyne NE1 7RU, UK
[3] Dipartimento di Scienze e Tecnologie Biologiche, Chimiche e Farmaceutiche (STEBICEF), Università degli Studi di Palermo, 90128 Palermo, Italy
* Correspondence: fabio.cucinotta@newcastle.ac.uk (F.C.); pbonaccorsi@unime.it (P.B.)

Abstract: In this paper we describe the synthesis of a novel bichromophoric system in which an efficient photoinduced intercomponent energy transfer process is active. The dyad consists of one subunit of curcumin and one of BODIPY and is able to emit in the far-red region, offering a large Stokes shift, capable of limiting light scattering processes for applications in microscopy. The system has been encapsulated in MCM-41 nanoparticles with dimensions between 50 and 80 nm. Both the molecular dyad and individual subunits were tested with different cell lines to study their effective applicability in bioimaging. MCM-41 nanoparticles showed no reduction in cell viability, indicating their biocompatibility and bio-inertness and making them capable of delivering organic molecules even in aqueous-based formulations, avoiding the toxicity of organic solvents. Encapsulation in the porous silica structure directed the location of the bichromophoric system within cytoplasm, while the dyad alone stains the nucleus of the hFOB cell line.

Keywords: luminescence; curcumin; BODIPY; NIR probes; bioimaging; bichromophoric dyad

Citation: Gangemi, C.M.A.; Salerno, T.M.G.; Barattucci, A.; Cucinotta, F.; Bonaccorsi, P.; Calabrese, G.; Poma, P.; Rizzo, M.G.; Campagna, S.; Puntoriero, F. A Curcumin-BODIPY Dyad and Its Silica Hybrid as NIR Bioimaging Probes. *Int. J. Mol. Sci.* **2022**, *23*, 9542. https://doi.org/10.3390/ijms23179542

Academic Editor: Raphaël Schneider

Received: 2 August 2022
Accepted: 22 August 2022
Published: 23 August 2022

1. Introduction

In the past two decades, considerable progress has been made in the development of non-invasive and highly sensitive biomedical imaging techniques for clinical diagnosis and therapy assessment of pathological events associated with human diseases. The diversification of imaging modalities, such as X-ray, computed tomography (CT), magnetic resonance imaging (MRI), photoacoustic imaging (PI), positron emission tomography (PET) and fluorescence imaging (FI) has facilitated early diagnoses and targeted therapies. Compared to other well-developed techniques, FI with small-molecule dyes, in the near infrared region (NIR, 700–1000 nm), has been widely applied for real-time detection of biological species and identification of cancer cells or lymph nodes, as well as for the intraoperative image-guided surgical removal of pathological tissues, owing to its high sensitivity, excellent resolution, minimum photodamage to tissues, quick feedback, non-hazardous radiation and, last, but not least, its low costs and sustainability [1–3]. Most of the traditional fluorescent probes, such as rhodamine, cyanine, or coumarin derivatives, have absorption and emission in the UV-Vis range and the interference with haemoglobin, myoglobin, and other haemoproteins leads to light scattering and impaired tissue penetration. The natural emission of light by such biological structures, upon light absorption, generates a noise background that might even wash out the image of targets. For instance, curcumin (**Curc**), a yellow pigment, naturally obtained from the rhizomes of *Curcuma longa*, with a highly delocalized π-electron symmetric structure, is reported to possess good optical properties, with the advantage of being highly biocompatible [4,5]. Curcumin-based fluorescent probes can overcome the shortcomings of organic fluorescent dyes, such as poor lipophilicity and low quantum yield, as well as display high sensitivity and molecular target ability that avoid possible interference of other substances with similar structures. However, the inability to image in the NIR region, because of the short absorption wavelengths, ranging from 410 to 430 nm, as well as the poor chemical and photochemical stability, due to the fast decomposition at pH ≥ 7, and the severe losing of brightness upon photo-illumination, have limited their use as fluorescent dyes, especially regarding in vivo applications. Synthetic organic small-molecule NIR fluorophores are advantageous because they can be rationally tailored providing creativeness in structural design, low cytotoxicity and good cell permeability [6]. Furthermore, the hydrophobic and electrostatic interactions, along with hydrogen bonding between organic fluorescent dyes and biological species can lead to better sensitivity, selectivity, and bio-imaging capacity for diagnostics. For instance, 4,4-Difluoro-4-bora-3a,4a-diaza-s-indacene (BODIPY), and its derivatives, have received attention for their excellent photo and chemical stability, high molar absorption coefficients and fluorescence quantum yields. A wide variety of biocompatible BODIPY-based fluorescent dyes have been developed [7] and some of them are commercially available. The photophysical properties of BODIPY can be easily tuned by modification of its backbone. Its derivatives have been used for many biological applications [8–12]. One common strategy exploited to achieve red- or NIR-emission wavelengths is based on the extension of the π-conjugation of the backbone of traditional fluorescent probes to lower the energy of the intramolecular charge transfer, but most of the dyes obtained by this strategy are quite large, causing a significant fading of images [6]. Furthermore, the noise background remains a disadvantage that limits the use of such NIR dyes. For their practical applications in FI, some photophysical factors, such as long emission wavelengths, high fluorescence quantum yields and large Stokes shifts, are necessary to avoid autofluorescence [13]. In this regard, the use of two different dyes that constitute an energy transfer pair or cassette, with close proximity and significant spectral overlap, is a promising strategy for tuning the emission color and enhancing the quantum efficiency of NIR dyes [14]. These bichromophoric systems allow for a high difference between the excitation and the emission energy, so suppressing photon scattering and diminishing tissue autofluorescence and have found application for the dynamic and high-contrast exploration of several pathological states. Following our research in the development of new bichromophoric systems for FI [15,16], in this paper, we describe the synthesis of a new energy donor–acceptor dyad **CB-Green** (Scheme 1), constituted of **Curc** and BODIPY **4**, that emits efficiently in the NIR region offering significant performances, both in terms of the ability to penetrate inside the cells and in terms of luminescent properties. BODIPY **2** itself has been proposed several times by us in bichromophoric systems as a significant red-emitting acceptor. On the other hand, curcumin has been previously used as a natural green-emitting donor [17] and the past results predicted that the dyad would be characterized by an efficient photoinduced energy transfer, active within the bichromophoric species. To investigate the use of a biocompatible nano carrier that could limit the drawbacks of the bichromophoric system due to its high lipophilicity, such as aggregation and shortening of emission time, and act as a "container" of both the new NIR dye and a drug, for a future theragnostic application, we decided to study the encapsulation of **CB-Green** in MCM-41 nanoparticles. These are silica particles with an average size between 50 and 80 nm and meso-structured pores with a 2D hexagonal arrangement. They exhibit singular features, such as ordered pore networks, with a very narrow and tunable pore size distribution (2–10 nm), a silanol-containing surface easily editable with different organic groups, large pore volumes (0.6–1 mL/g) and high surface areas (600–1000 m^2/g) [18]. Their textural properties have inspired the idea of introducing drugs into the pore channels to be then locally released where needed, in drug delivery systems. Encapsulation into the porous silica structure also protects the guest molecules from degradation, while the silica surface offers several functionalization routes, enabling fine-tuning of the particles' affinity towards specific environments and improving stability and uptake [19]. Furthermore, MCM-41 nanoparticles did not show reduction in cell viability, indicating their biocompatibility and bio-inertness and making them able to deliver organic molecules even in aqueous-based formulation, avoiding organic solvent toxicity.

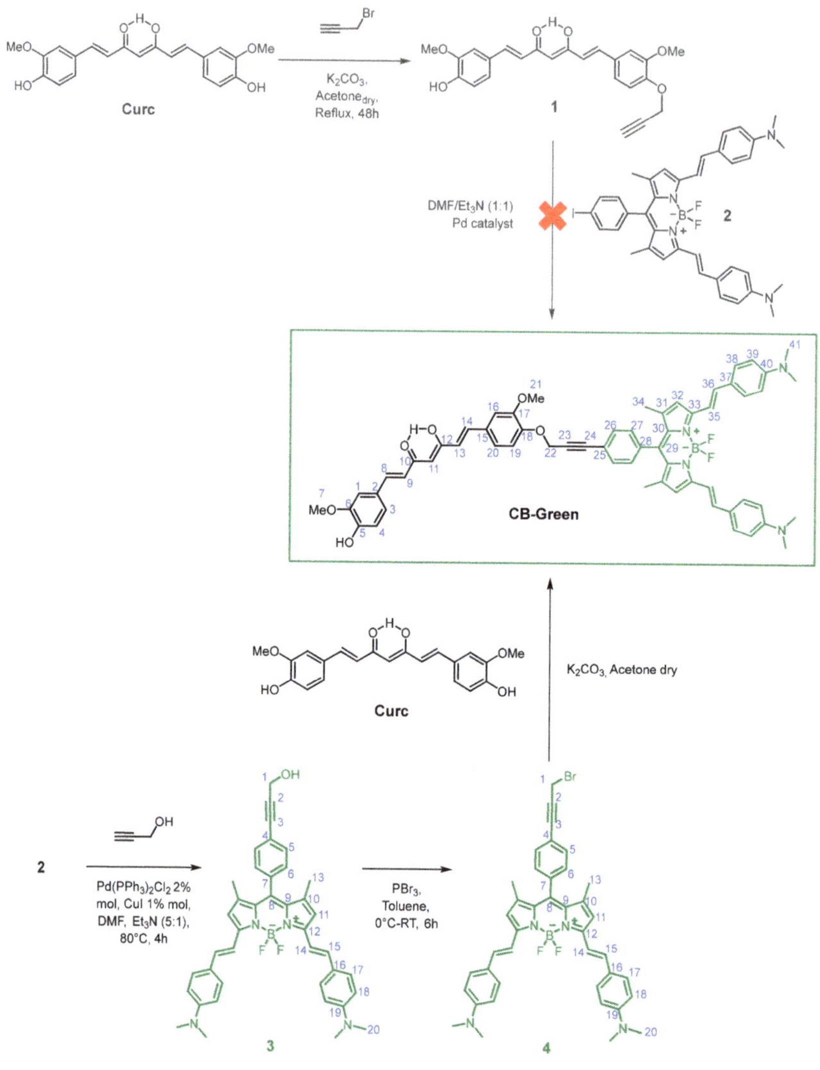

Scheme 1. Synthetic route to **CB-Green** (**Curc:** curcumin; **CB-Green:** curcumin-BODIPY dyad).

2. Results and Discussion

2.1. Synthesis of Dyad CB-Green

The synthetic approach described in the upper part of Scheme 1 to reach the bichromophoric system **CB-Green**, and already used by authors [17], did not work with BODIPY **2** [20]. The functionalized curcumin **1** [21] was involved in a Sonogashira cross-coupling with BODIPY **2** without significant results. Pd catalyst, base and conditions were changed, and the reaction was even tried in the copper-free conditions, but the dyad **CB-Green** was obtained in traces, after difficult work-up. Therefore, it was necessary to move towards an alternative synthetic strategy, described in the lower part of Scheme 1. Starting from BODIPY **2**, it was reacted with propargyl alcohol, in a classical Sonogashira reaction, to obtain

BODIPY **3** in 90% yield. BODIPY **3** was then converted into BODIPY **4** (Scheme 1) by PBr_3 in absolute toluene, with 74% yield. The bichromophoric system **CB-Green** was obtained as a unique compound, lacking even traces of the bis-substituted BODIPY **4**/curcumin which could be isolated, from the reaction of **Curc** with BODIPY **4** in the presence of a slight excess of K_2CO_3, to avoid sensible degradation of the natural polyphenol. However, yields did not exceed the 54% in compound **CB-Green**, after purification.

2.2. Photochemical Characterization of Dyad CB-Green

The absorption and emission spectra of the free compounds **Curc**, BODIPY **4** and **CB-Green** in solution are reported in Figure 1 and all the photophysical data are summarized in Table 1.

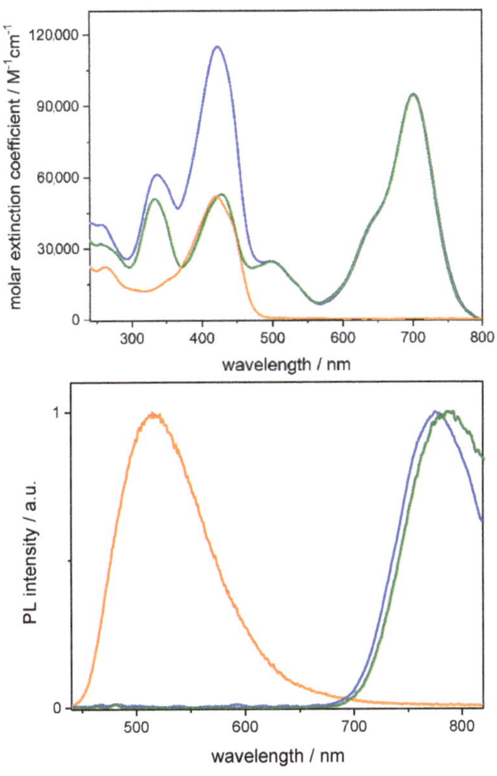

Figure 1. UV-Visible absorption (**top**) and emission (**bottom**) spectra of **Curc** (orange), **4** (green) and **CB-Green** (blue), recorded from micromolar acetonitrile solutions; emission was measured at λ_{exc} = 420 nm. Please note that the absorption maxima at lower energy lies at 418 nm for **Curc** and at 705 nm for **4** and **CB-Green**.

From the absorption spectra, it is possible to distinguish the contributions of the two chromophores. The absorption band centered at around 420 nm is assigned to the **Curc** moiety, whereas the lowest-energy absorption band around 700 nm is centered on the BODIPY subunit and it is attributed to a $\pi \rightarrow \pi^*$ transition with a partial charge transfer character, in which the lone pairs of the amine groups are involved. The emission spectrum of **CB-Green** (blue line) is dominated by a fluorescence band with a maximum at about 770 nm that is identical to the one exhibited by BODIPY **4** (green line). This, and the absence of any residual emission from the donor subunit, indicate an almost quantitative energy transfer from the higher energy absorber **Curc** to the lower-lying excited state centered on the BODIPY fragment. This is also confirmed by the comparison of the excitation spectrum of **CB-Green**, registered at 790 nm, with the absorption one (Figure 2).

Table 1. Spectroscopic data of the bichromophoric **CB-Green** species and the subunits **Curc** and **4**, recorded from acetonitrile solutions.

Compound	Absorption	Emission		
	λ_{max}, nm (ε, M^{-1} cm^{-1})	λ_{max}, nm	τ, ns	Φ
CB-Green	705 (95,000)	770	2.2	0.18
Curc	418 (52,900)	520	0.66	0.1
4	423 (116,100) 705 (94,800)	770	2.2	0.2

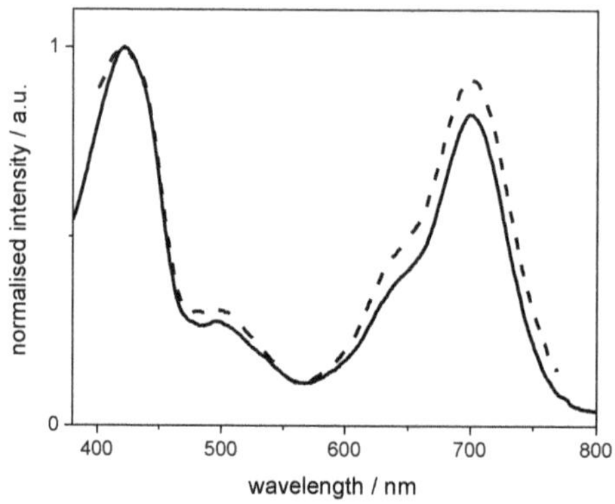

Figure 2. Comparison between the absorption spectrum (solid line) and the excitation spectrum (dashed line) of **CB-Green**; the latter was recorded at λ_{em} = 790 nm. Note that the intensity ratio between the maxima at 410 nm and 705 nm remained approximately constant between the absorption and excitation spectra.

2.3. *Synthesis of Dye-Loaded Silica Nanoparticles*

A specific type of mesoporous silica nanoparticles (MSNs) was chosen, namely the MCM-41, due to the presence of an ordered hexagonal array of pores, with mean size of 3–4 nm, where guest dyes can be encapsulated, and also due to the relatively small particle size, 70–100 nm, which is suitable for cell internalization [22]. The MSNs were doped with the bichromophoric compound **CB-Green** and the subunits, **Curc** and **4**, respectively, with the objective of studying the spectroscopic and biologic behavior of the dyes inside the nanoparticles and to determine possible variations in the energy transfer process. The experimental approach that was used for the encapsulation of the dyes took advantage of the well-established soft-template route used for all types of mesoporous silica materials. A micellar template was made by using an amphiphilic surfactant that is able to self-assemble in water and form rod-like micelles with a hexagonal arrangement. For MCM-41 particles, the surfactant of choice was cetyl trimethylammonium bromide (CTAB). After the first step, a suitable silica precursor was added, which hydrolyses and polymerizes around the template, forming the final silica network [22,23]. The precursor was tetraethyl orthosilicate (TEOS), the hydrolysis of which is facilitated in a basic aqueous solution, using 2 M NaOH. The dyes were added during the first step, in which, due to their insolubility in water, they spontaneously diffused through the CTAB micelles and entered the solvent-free, rigidified, hydrophobic core. To enable a homogeneous dispersion of the dyes, minimize intermolecular aggregation and quenching effects, prevent structural disruption of the micelles and, consequently, of the silica particles, we used a low dye loading, at a 99:1 molar ratio of CTAB:dye for each of the host–guest systems, hereafter named **Curc@MSN**, **4@MSN** and **CB-Green@MSN**. Such low content of dye was chosen on the basis of previous experiments showing that high loading of guest BODIPY species inside the nanoparticle resulted in fluorescence quenching [24,25]. Despite the relatively high temperature (80 °C) and pH (~12) used in the silica synthesis, the dyes showed excellent chemical stability and underwent no degradation.

2.4. Photochemical Characterization of the Dye-Loaded Silica Nanoparticles

The dye-loaded MSNs were characterized by UV-Visible absorption and luminescence spectroscopy, and their properties were compared to those of the free dyes, to study the effect on the energy transfer inside the nanoparticles. Optical measurements were performed from suspensions of the materials in a solvent where the particles could be finely dispersed, to limit the intense light scattering that is typically generated by the suspended particles; the best solvent for such a purpose turned out to be cyclohexane and 0.3 mg/mL suspensions were used for the measurements. However, the absorption spectra still showed pronounced light scattering; therefore, only the excitation spectra are reported. Figure 3 shows the excitation and emission spectra of **Curc@MSN**, **4@MSN** and **CB-Green@MSN** recorded from the suspensions.

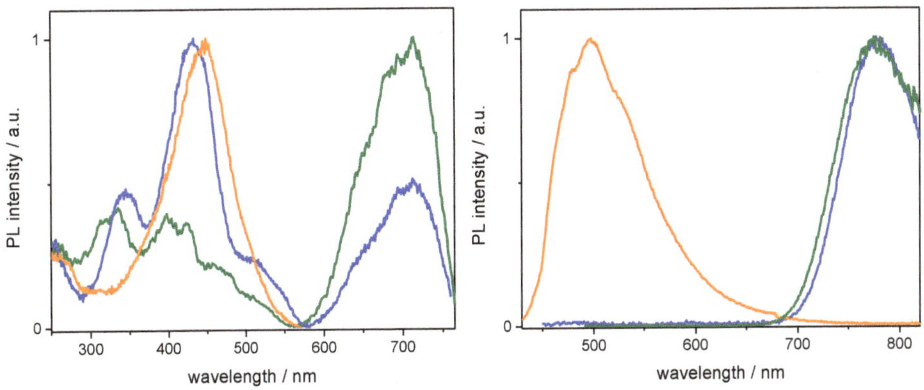

Figure 3. Normalized excitation (**left**) and emission (**right**) spectra of **Curc@MSN** (orange), **4@MSN** (green) and **CB-Green@MSN** (blue), recorded from 0.3 mg/mL cyclohexane suspensions at λ_{em} = 780 nm and λ_{exc} = 420 nm, respectively.

The spectra of the dye-loaded MSNs had an analogous trend to that observed for the dyes in solution. The excitation spectra were essentially additive, whereas the emission spectra showed only the component of the acceptor subunit BODIPY. The time-resolved fluorescence measurements also confirmed the same trend shown in solution and no residual emission nor excited state lifetime from the curcumin unit was detected.

2.5. Biocompatibility and Cellular Uptake Evaluations

We performed the biological studies on two human osteoprogenitor cell lines, the human fetal osteoblastic cells (hFOB 1.19) and human bone osteosarcoma epithelial cells (U-2 OS). Both can be considered good models of normal and cancer bone cells, respectively. We initially investigated the cytotoxicity of the free molecules **Curc**, compound **4** and **CB-Green**, and the loaded nanoparticles **Curc@MSN**, **4@MSN** and **CB-Green@MSN**. In particular, the biocompatibility of such compounds was tested by MTS cell viability assay and the results confirmed that free molecules and hybrid systems at concentrations of 50 and 5 nM did not cause any evident trace of cellular suffering even after five days from administration, thus, underlining that they are safe enough to be used for imaging purposes. (Figure 4) Data were collected for other cell lines and are shown in supplementary material (see Table S1 in Supplementary Materials) and in all cases the non-cytotoxicity of the compounds under study was confirmed.

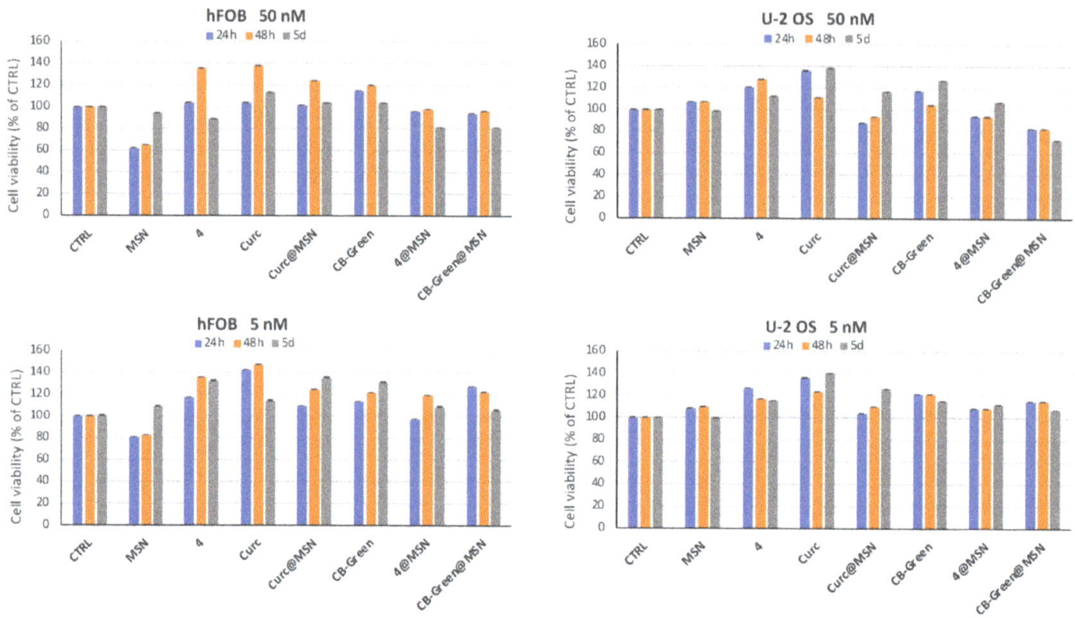

Figure 4. Cell viability test performed on hFOB and U-2 OS cells after 24 h, 48 h and 5 days of incubation with **MSN, 4, Curc, Curc@MSN, CB-Green, 4@MSN, CB-Green@MSN** molecules (50 and 5 nM). Data are represented as means ± SD of three independent experiments. Please note that the concentrations used refer to the amount of dye present in the nanostructure, as for MSN the concentration used was equal to 3.5 mg/mL for the series related to the 50 nM concentration of dyes and equal to 0.35 mg/mL for the 5 nM series (the amount of silica remained almost constant along the hybrid system series).

To investigate cellular internalization and localization of **Green** and **CB-Green@MSN** and to compare the results with those obtained for **Curc**, BODIPY **4** and their corresponding MCM-41 systems, we performed cell labeling and fluorescence microscopy. The results regarding **Curc** and **Curc@MSN** are illustrated in Figure 5. Due to its lipophilicity, **Curc** dispersed in the aqueous solvent failed to penetrate the cell membrane, remaining as luminescent green crystals on the outside of the cells, (left panel of Figure 5) while its hybrid analogous spread within the cytoplasm with a green light, (right panel), showing how encapsulation in MCM-41 led to permeabilization, and supporting the ability of the MSNs to cross the plasmatic membrane without causing any damage to the cell.

Figure 6 illustrates BODIPY **4** and the MCM-41-loaded analogous **4@MSN**, both showing enlightening of red inside the cell but with two different localizations. The comparison between DAPI stain and **4** labeling of hFOB cells (left panel) clearly suggests that BODIPY **4** penetrated the nuclear membrane of cells and focused on the nucleus, which appeared illuminated with a bright red light. The red signals of hybrid **4@MSN**, instead, were mainly localized in the cytoplasm.

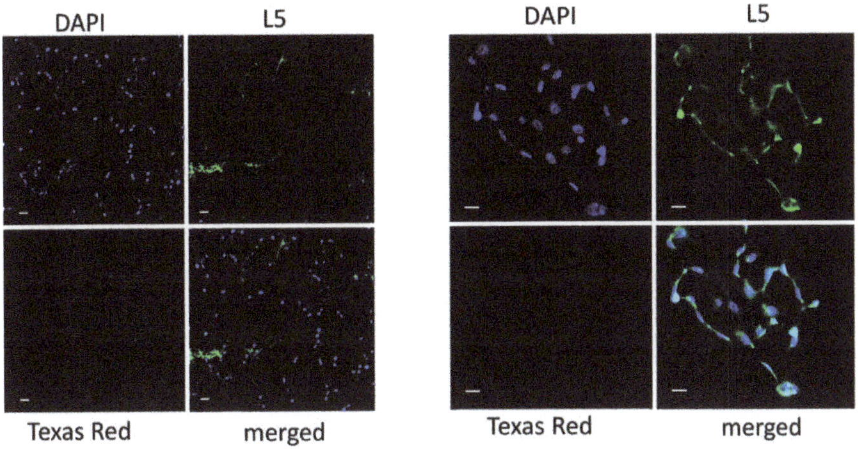

Figure 5. Fluorescence microscope images of hFOB 1.19 incubated with **Curc** (**left**) and with **Curc@MSN** (**right**). Bar = 1 µm.

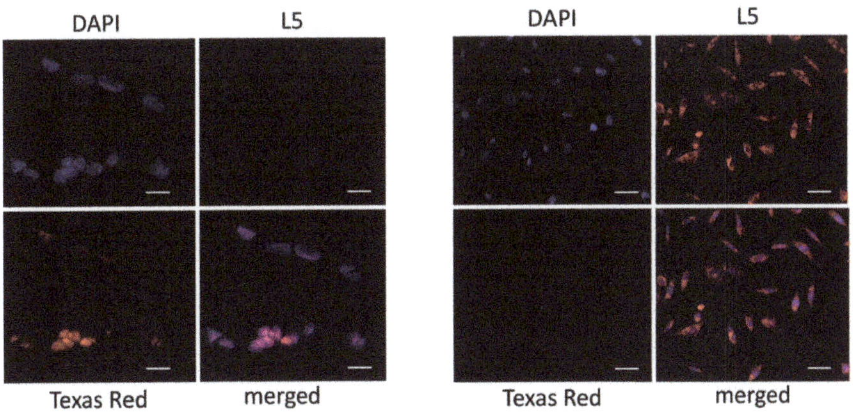

Figure 6. Fluorescence microscope images of hFOB 1.19 incubated with **4** (**left**) and with **4@MSN** (**right**). Bar = 1 µm.

Noteworthy, both **CB-Green** and **CB-Green@MSN** (Figure 7) illuminate the cellular environment with high efficiency even at concentrations of the 5 nM order, confirming that the low dosage of the dye and its hybrid system could not interfere with cell viability. Our microscope filter (Texas red filter set) did not allow reaching the wavelengths at which the two systems emitted to the maximum (790 nm), although images appeared bright and well-defined, thus, indicating that **CB-Green** and **CB-Green@MSN** performed significantly as biocompatible NIR dyes. Finally, it is worth noting that **CB-Green** illuminated with a red emission, typical of its acceptor component, BODIPY **4**, mainly in the nuclei of hFOB cells, (left Panel), while **CB-Green@MSN** was localized in the cytoplasm and, in particular, in the perinuclear region.(right Panel) The similar localization observed for **Curc@MSN** and **4@MSN** suggests that dimensions of MCM-41 nanoparticles prevented the penetration of the nuclear membrane, on one hand, offering the possibility of differentiating the localization of the light with respect to **CB-Green** within the cellular compartments, on the other hand.

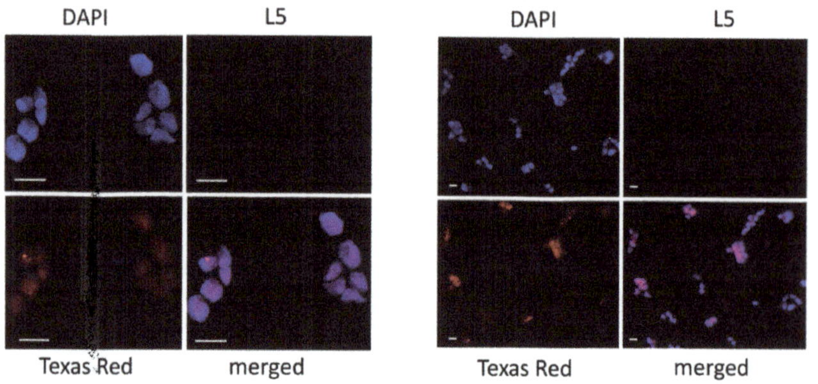

Figure 7. Fluorescence microscope images of hFOB.19 incubated with **CB-Green** (**left**) and with **CB-Green@MSN** (**right**). Bar = 1 μm.

3. Material and Methods

3.1. Chemicals

Solvents were purified according to standard procedures. All the syntheses were monitored by TLC on commercially available precoated plates (silica gel 60 F254), and the products were visualized with vanillin [1 g dissolved in MeOH (60 mL) and conc. H_2SO_4 (0.6 mL)] and UV lamp. Silica gel 60 was used for column chromatography.

3.2. Instrumentation

Proton (^{1}H) and carbon (^{13}C) NMR spectra were recorded on a Varian 500 spectrometer (at 500 MHz for ^{1}H; and 125 MHz for ^{13}C) using DMSO-d_6 as solvent. Chemical shifts were given in parts per million (ppm) (δ relative to residual solvent peak for ^{1}H and ^{13}C), coupling constants (J) were given in Hertz, and the attributions were supported by heteronuclear single-quantum coherence (HSQC) and correlation spectroscopy (COSY) experiments. Chemical shifts were reported in ppm relative to DMSO-d_6 (2.49 ppm). Numbering of Carbon atoms of compounds **3**, **4** and **CB-Green** are shown in Scheme 1. Combustion analyses were carried out with a FISONS EA1108 elemental analyzer. UV/Vis absorption spectra were taken on a Jasco V-560 spectrophotometer. For steady-state luminescence measurements, a Jobin Yvon-Spex Fluoromax 2 spectrofluorimeter was used, equipped with a Hamamatsu R3896 photomultiplier. The spectra were corrected for photomultiplier response using a program purchased with the fluorimeter. For the luminescence lifetimes, an Edinburgh OB 900 time-correlated single-photon-counting spectrometer was used. As excitation sources, a Hamamatsu PLP 2 laser diode (59 ps pulse width at 408 nm) and/or the nitrogen discharge (pulse width 2 ns at 337 nm) was employed. Emission quantum yields for de-aerated solutions were determined using the optically diluted method. As luminescence quantum yield standards, we used a trimethylammonium-phenylstyryl BODIPY species ($\phi = 0.69$ in ACN). Microscopy images were acquired using a Leica DM5000B microscope equipped with DAPI, L5 and Texas Red fluorescence filters, using Leica LAS X acquisition software.

Experimental uncertainties were as follows: elemental analysis, 0.04%; absorption maxima, ±2 nm; molar absorption coefficient, 10%.

3.3. Synthesis

Compound 3. [Pd(PPh$_3$)$_2$Cl$_2$] (0.0112 mmol) was added to a degassed solution of boron difluoro-4-(1E)-2-[5-[[5-[(1E)-2-[4-(dimethylamino) phenyl] ethenyl] -3-methyl-2*H*-pyrrol-2-ylidene-](4-iodophenyl methyl]-4-methyl -1*H*-pyrrol-2-yl-]ethenyl]-N,N-dimethyl benzenaminate (**2**) [20] (400 mg, 0.56 mmol) and propargyl alcohol (40 μL, 38 mg, 0.67 mmol) in DMF/TEA (12 mL, 5:1). In a sealed tube, the mixture was heated, at 80 °C, under argon, for 4 h, until the disappearance of **2** was observed by TLC ($CHCl_3$/Hexane 90:10). Solvents were removed under reduced pressure. The reaction crude was dissolved in DCM and then filtered over celite/silica 1:2. The solution was evaporated to dryness to obtain **3** as a deep green solid in 90% yield, without needing any purification. *Rf*: 0.60 ($CHCl_3$/Hexane 90:10^{1}H NMR (DMSO-d_6): δ 7.61–7.58 (2H, d, J = 7.8, 2× H-5), 7.48–7.40 (8H, m, 2× H-6, 2× H-14, 4× H-18), 7.31 (2H, part B of an AB system, J = 16, 2× H-15), 6.88 (2H, s, 2× H-11), 6.79 (4H, d, J = 8.8, 4× H-17), 5.37 (1H, t, J = 5.9 Hz, -OH), 4.34 (2H, d, J = 5.9 Hz, H_2-1), 3.00 (12H, 4× -NCH_3-20), 1.41 (6H, s, 2× CH_3-13). ^{13}C NMR: δ 152.7 (C- q), 151.5 (C-q), 140.7 (Cq), 137.7 and 113.7 (2× C-14, 2× C-15), 135.1 (Cq), 132.4 (Cq), 132.3 (2× C-5), 129.6 (Cq), 129.2 (2× C-6, 4× C-18,), 124.3 (Cq), 123.6 (Cq), 118.2 (2× C-11), 112.6 (4× C-17), 91.6 and 83.6 (C-2, C-3), 49.9 (C-1), 40.2 (4× C-20), 14.7 (2× C-13). Anal. Calcd. for $C_{40}H_{39}BF_2N_4O$ (640,57) C, 75.00; H, 6.14; N, 8.75. Found: C, 74.93; H, 6.12; N, 8.75.

Compound 4. A solution of **3** (50 mg, 0.08 mmol) in abs. toluene (5 mL) under argon was cooled to 0 °C in an ice-bath. PBr_3 (31 µL, 70 mg, 0.26 mmol, 3.3 eq) was added dropwise and the reaction mixture was stirred at 0 °C for 10′ and then it was allowed to warm to rt. The disappearance of **3** was followed by TLC ($CHCl_3$/Hexane 90:10). After 6 h saturated $NaCO_3$ solution was added. The phases were separated, and the organic layers were washed with water and brine, dried over $MgSO_4$, filtered and concentrated under reduced pressure. The reaction crude was purified by silica gel column chromatography (eluants: $CHCl_3$/Hexane 90:10) to obtain **4** as a deeply green solid, yield 74%. *Rf*: 0.95 ($CHCl_3$). ^{1}H NMR (DMSO-d_6): δ 7.64–7.61 (2H, d, J = 7.9, 2× H-5), 7.47–7.40 (8H, m, 2× H-6, 2× H-14, 4× H-18,), 7.29 (2H, part B of an AB system, J = 15.7, 2× H-15), 6.87 (2H, s, 2× H-11), 6.78 (4H, d, J = 8.3, 4× H-17), 4.54 (2H, s, H_2-1), 3.00 (4× -NCH_3-20), 1.39 (6H, s, 2× CH_3-13). Anal. Calcd. for $C_{40}H_{38}BBrF_2N_4$ (703,47): C, 68.29; H, 5.44; N, 7.96. Found: C, 68.49; H, 5.43; N, 7.95.

CB-Green. To a solution of **Curc** (30 mg, 0.08 mmol) in dry acetone (10 mL), **4** (75 mg, 0.106 mmol) and K_2CO_3 (22 mg, 0.16 mmol) were added. The mixture was stirred at reflux temperature for 48 h. The reaction was monitored using TLC ($CHCl_3$/MeCN 99.5:0.5) following the disappearance of **4**. The solvent was removed under reduced pressure. The crude was purified by column chromatography (eluants: $CHCl_3$ up to $CHCl_3$/MeCN 99.5:0.5) on silica gel. The column afforded compound **CB-Green** as a green solid, 54% yield. *Rf*: 0.50 ($CHCl_3$/MeCN 99.5:0.5). ^{1}H NMR (DMSO-d_6): δ 7.65–7.15 and 6.89–6.75 (28H, two m, H-1, H-3, H-4, H-8, H-9, H-13, H-14, H-16, H-19, H-20, 2× H-26, 2× H-27, 2× H-32, 2× H-35, 2× H-36, 4× H-38, 4× H-39), 6.1 (1H, s, H-11), 5.15 (2H, s, H_2-22), 3.87 and 3.83 (6H, two s, 2× [-OCH_3]), 3.00 (12H, 4× H-41), 1.4 (6H, s, 2× CH_3-34). ^{13}C NMR: δ 184.3 and 182.9 (C-10, C-12), 152.7 (Cq), 151.4 (Cq), 149.8 (Cq), 149.0 (Cq), 148.4 (C-q), 141.5 (Cq), 140.7 (Cq), 137.7 and 111.8 (2× C-35, 2× C-36), 136.1(C-q), 135.0 (Cq), 132.6 (2× C-26), 132.3 (C-8, C-14), 129.8 (Cq), 129.2 (2× C-27, 4× H-39), 129.0 (Cq), 126.7, 124.4, 123,6, 123.0, 122.8 and 122.5 (C1, C-3, C-4, C-16, C-19, C-20), 121.5 (Cq), 118.2 (Cq), 117.5 (Cq) 116.1 (2× C-32), 114.1 (Cq), 113.6 (Cq), 112.6 (C-9, C-13, 4× C-38,), 111.3(Cq), 101.4 (C-11), 86.6 and 86.4 (C-23, C-24), 57.1 (C-22), 56.1 (C-7, C-21), 40.4 (4× C-41), 14.8 (2× C-34). Anal. Calcd. for $C_{61}H_{57}BF_2N_4O_6$ (990,94): C, 73.94; H, 5.80; N, 5.65. Found: C, 74.15; H, 5.79; N, 5.67.

Dye-loaded Silica Nanoparticles. Cetyltrimethylammonium bromide (60.5 mg) was dissolved in 30 mL of water using an ultrasound bath. The dye was dissolved in a minimal amount of acetonitrile (500 µL) and transferred into the water solution of CTAB under stirring, after which a homogeneous colloidal suspension was formed. The dyes **4** (0.6 mg), **Curc** (0.31 mg) and **CB-Green** (0.83 mg) were added separately in 1:99 molar ratios in respect to CTAB. This was then followed by the addition of 2 M aqueous NaOH (0.22 mL, 0.44 mmol). The mixture was heated to 80 °C and the silica precursor tetraethylorthosilicate (283 mg, 0.3 mL, 1.37 mmol) was added, at which point silica particles started to form quickly. The suspension was kept stirring at reflux for 2 h, then stopped and allowed to cool down to room temperature. The mixture was filtered under vacuum, washed with deionized water and finally dried in a vacuum desiccator. All powders fluoresced strongly under a UV lamp.

3.4. Cell Culture and Cell Viability Assay

Human fetal osteoblastic cell line (hFOB 1.19) and human bone osteosarcoma epithelial cells (U-2 OS) were obtained from the American Type Culture Collection (ATCC, Manassas, VA, United States). HFOB 1.19 were cultured in 1:1 mixture of Ham's F12 Medium - Dulbecco's Modified Eagle's Medium (Merk Life Science S.r.l., Milan, Italy), supplemented with 2.5 mM L-glutamine (L-glu, Merk Life Science S.r.l., Milan, Italy), 0.3 mg/mL G418 (ThermoFisher, Waltham, MA USA); 10% Fetal Bovine Serum (FBS, Merk Life Science S.r.l., Milan, Italy) and 1% penicillin/streptomycin/amphotericin (PSA, Merk Life Science S.r.l., Milan, Italy). U-2 OS were grown in McCoy's 5a Medium Modified (Merk Life Science S.r.l., Milan, Italy) supplemented with 2.5 mM L-glu, 10% FBS and 1% PSA. Both cell lines were incubated in a humidified atmosphere containing 5% CO_2 at 37 °C. The medium was replaced twice a week and cells were split at about 80% of confluence. Cell viability assay was performed by MTS (3-(4,5-dimethylthiazol-2-yl)-5-(3-carboxymethoxyphenyl)-2-(4-sulfophenyl)-2H-tetrazolium) (Cell Titer96® Aqueous One Solution Proliferation Assay Kit, Promega, Madison, WI, USA) according to the manufacturing protocol. For MTS assay, 5×10^3 cells were cultured in a 96-well plate, with specific medium and incubated in a humidified atmosphere containing 5% CO_2 at 37 °C for 24 h. Then, compounds and hybrid systems under study were added separately, at two different concentrations (5 and 50 nM) and cells re-incubated for 24 h, 48 h and 5 days. After 24 h, 48 h and 5 days, MTS reagent was added to the culture medium and the plate was incubated for 1 h at 37 °C. Finally, the plate was shacked shortly and the absorbance at 490 nm measured using a synergy HT plate reader (BioTek Instruments, Inc., Winooski, VT, USA). Each biological system compound was analyzed in triplicate for each concentration and time. Data were reported as percentage of the control ± standard deviation.

The HL-60 cells were obtained from ATCC® (CCL-240, Rockville, MD, USA), while their variant, HL-60R, were derived by exposure to gradually increasing concentrations of doxorubicin. The molecular characterization of HL-60R cells was carried out previously [26] The human breast cancer cell lines MDA-MB-231 and the colorectal adenocarcinoma cell lines Caco-2 were obtained from ATCC® (respectively HTB-26™ and HTB-37™—Rockville, MD, USA). The HL-60 and HL-60R cells were routinely maintained in Roswell Park Memorial Institute (RPMI) 1640 (HyClone Europe Ltd., Cramlington UK), while MDA-MB-231 cells were cultured in Dulbecco's Modified Eagle Medium (DMEM) (HyClone Europe Ltd., Cramlington, UK) supplemented with 10% heat inactivated fetal calf serum, 2 mM L-glutamine, 100 units/mL penicillin and 100 µg/mL streptomycin (all reagents were from HyClone Europe Ltd., Cramlington, UK) in a humidified atmosphere at 37 °C in 5% CO_2. Cells with a narrow range of passage numbers were used for all experiments. The cultures were routinely tested for Mycoplasma infection.

Cells were seeded on 96-well plates at a density of 5000 cells/well and incubated overnight at 37 °C. After 24 h, at time 0 the medium was replaced with a fresh complete medium supplemented with the investigated systems. After 72 h of treatment, 15 µL of Promega Corp. commercial solution (Madison, WI, USA) containing 3-(4,5-dimethylthiazol-2-yl)-5-(3-carboxymethoxyphenyl)-2-(4-sulfophenyl)-2H-tetrazolium (MTS) and phenazine ethosulfate was added to each well and the plates were incubated at 37 °C at 5% CO_2 for 2 h. Using a microplate reader (iMark Microplate Reader; Bio-Rad Laboratories, Inc., Hercules, CA, USA), the bio-reduction of the MTS dye was evaluated by measuring the absorbance of each well at 490 nm. Cytotoxicity was expressed as a percentage of measured absorbance relative to that of control cells. Data were expressed as mean ± standard error (S.E.) of at least three different experiments performed in duplicate.

4. Conclusions

We have described the synthesis of a novel dyad in which the donor curcumin is bonded to an acceptor BODIPY to yield a bichromophoric system. **CB-Green** possesses all the desirable properties of a bioimaging probe: large Stokes shift, long emission wavelengths, high fluorescence quantum yield, and biocompatibility. As much as the conjugation of curcumin to BODIPY results in the loss of some of the inherent advantages of this subunit (cytotoxicity, antitumor activity, and so on), the new system allows the biological material to be illuminated in the near-IR by exciting it in the blue as well. This behavior is particularly interesting from an application point of view because the spectral emission range of the new species lies within the biological window.

In addition, it presents a specific intracellular staining. **CB-Green** targets the nucleus of the studied cell lines, thus supporting the idea that small-size dye molecules are desirable tools in FI since they enter the cell fast, locate in the target region and, in this specific case, do not induce any stress to cells up to 5 days. Encapsulation of **CB-Green** and the free components of the bichromophoric system, **Curc** and BODIPY **4**, inside MCM-41 silica nanoparticles allows for all three compounds to mask their lipophilic nature, overpass the cellular membrane and locate within the cytoplasm without losing brightness. The dyad maintains a very efficient energy transfer even within the hybrid silica system, which confirms the viability of the encapsulation strategy for imaging purposes.

Supplementary Materials: The following supporting information can be downloaded at: https://www.mdpi.com/article/10.3390/ijms23179542/s1.

Author Contributions: "Synthesis" T.M.G.S.; "Characterization of compound" A.B. and C.M.A.G.; "Conceptualization and writing of the draft" P.B.; "Preliminary cytotoxicity studies" P.P.; "Synthesis, characterization of Hybrid system and writing of the draft" F.C.; "Conceptualization, characterization of free compounds, Writing of the draft" F.P.; "writing and revision of the draft" S.C.; "Biological measurements" M.G.R.; "Biological data analysis" G.C. All authors have read and agreed to the published version of the manuscript.

Funding: FC gratefully acknowledges the financial support by the EPSRC (grant code EP/P015395/1).

Institutional Review Board Statement: Not applicable.

Informed Consent Statement: Not applicable.

Data Availability Statement: Not applicable.

Acknowledgments: C.M.A.G. thanks the European Union-FSE-REACT-EU, PON Research and Innovation 2014-2020 DM.1062/2021. F.P. and S.C. thank the Ministero degli Affari Esteri e della Cooperazione Internazionale, Direzione Generale per la Promozione del Sistema Paese.

Conflicts of Interest: The authors declare no conflict of interest.

References

1. Vahrmeijer, A.L.; Hutteman, M.; van der Vorst, J.R.; van de Velde, C.J.; Frangioni, J.V. Image-guided cancer surgery using near-infrared fluorescence. *Nat. Rev. Clin. Oncol.* **2013**, *10*, 507–518. [CrossRef] [PubMed]
2. Xie, D.; Gu, D.; Lei, M.; Cai, C.; Zhong, W.; Qi, D.; Wu, W.; Zeng, G.; Liu, Y. The application of indocyanine green in guiding prostate cancer treatment. *Asian J. Urol.* **2022**. [CrossRef]
3. Han, M.; Kang, R.; Zhang, C. Lymph Node Mapping for Tumor Micrometastasis. *ACS Biomater. Sci. Eng.* **2022**, *8*, 2307–2320. [CrossRef] [PubMed]
4. Li, X.; Zhu, J.; Mao, F.; Li, X.; Xia, C.; Jiang, B.; Guo, Y.; Li, J. Curcumin-based polarity fluorescent probes: Design strategy and biological applications. *Dye. Pigm.* **2020**, *177*, 108320.
5. Nocito, M.C.; De Luca, A.; Prestia, F.; Avena, P.; La Padula, D.; Zavaglia, L.; Sirianni, R.; Casaburi, I.; Puoci, F.; Chimento, A.; et al. Antitumoral Activities of Curcumin and Recent Advances to ImProve Its Oral Bioavailability. *Biomedicines* **2021**, *9*, 1476. [CrossRef]
6. Li, J.-B.; Liu, H.-W.; Fu, T.; Wang, R.; Zhang, X.-B.; Tan, W. Recent Progress in Small-Molecule Near-IR Probes for Bioimaging. *Trends Chem.* **2019**, *1*, 224–234. [CrossRef]
7. Boens, N.; Leen, V.; Dehaen, W. Fluorescent indicators based on BODIPY. *Chem. Soc. Rev.* **2012**, *41*, 1130–1172. [CrossRef]
8. Kowada, T.; Maeda, H.; Kikuchi, K. BODIPY-based probes for the fluorescence imaging of biomolecules in living cells. *Chem. Soc. Rev.* **2015**, *44*, 4953–4972. [CrossRef]

9. Bassan, E.; Gualandi, A.; Cozzi, P.G.; Ceroni, P. Design of BODIPY dyes as triplet photosensitizers: Electronic properties tailored for solar energy conversion, photoredox catalysis and photodynamic therapy. *Chem. Sci.* **2021**, *12*, 6607–6628. [CrossRef]
10. NI, Y.; Wu, J. Far-red and near infrared BODIPY dyes: Synthesis and applications for fluorescent pH probes and bio-imaging. *Org. Biomol. Chem.* **2014**, *12*, 3774–3791. [CrossRef]
11. Kamkaew, A.; Lim, S.H.; Lee, H.B.; Kiew, L.V.; Chung, L.Y.; Burgess, K. BODIPY dyes in photodynamic therapy. *Chem. Soc. Rev.* **2013**, *42*, 77. [CrossRef] [PubMed]
12. Myochin, T.; Hanaoka, K.; Komatsu, T.; Terai, T.; Nagano, T. Design Strategy for a Near-Infrared Fluorescence Probe for Matrix Metalloproteinase Utilizing Highly Cell Permeable Boron Dipyrromethene. *J. Am. Chem. Soc.* **2012**, *134*, 13730–13737. [CrossRef]
13. Sansalone, L.; Tang, S.; Garcia-Amorós, J.; Zhang, Y.; Nonell, S.; Baker, J.D.; Captain, B.; Raymo, F.M. A Photoactivatable Far-Red/Near-Infrared BODIPY To Monitor Cellular Dynamics in Vivo. *ACS Sens.* **2018**, *3*, 1347–1353. [CrossRef] [PubMed]
14. Chen, C.; Fang, C. Devising Efficient Red-Shifting Strategies for Bioimaging: A Generalizable Donor-Acceptor Fluorophore Prototype. *Chem. Asian J.* **2020**, *15*, 1514–1523. [CrossRef] [PubMed]
15. Barattucci, A.; Campagna, S.; Papalia, T.; Galletta, M.; Santoro, A.; Puntoriero, F.; Bonaccorsi, P. BODIPY on Board of Sugars: A Short Enlightened Journey up to the Cells. *ChemPhotoChem* **2020**, *4*, 647–658. [CrossRef]
16. Barattucci, A.; Gangemi, C.M.A.; Santoro, A.; Campagna, S.; Puntoriero, F.; Bonaccorsi, P. Bodipy-carbohydrate systems: Synthesis and bio- applications. *Org. Biomol. Chem.* **2022**, *20*, 2742–2763. [CrossRef]
17. Riela, S.; Barattucci, A.; Barreca, D.; Campagna, S.; Cavallaro, G.; Lazzara, G.; Massaro, M.; Pizzolanti, G.; Salerno, T.M.G.; Bonaccorsi, P.; et al. Boosting the Properties of a Fluorescent Dye by Encapsulation into Halloysite Nanotubes. *Dye. Pigm.* **2021**, *187*, 109094. [CrossRef]
18. Cheng, L.; Wang, C.; Feng, L.; Yang, K.; Liu, Z. Functional Nanomaterials for Phototherapies of Cancer. *Chem. Rev.* **2014**, *114*, 10869–10939. [CrossRef]
19. Estevao, B.M.; Miletto, I.; Hioka, N.; Marchese, L.; Gianotti, E. Mesoporous Silica Nanoparticles Functionalized with Amino Groups for Biomedical Applications. *ChemistryOpen* **2021**, *10*, 1251–1259. [CrossRef]
20. Bonaccorsi, P.; Papalia, T.; Barattucci, A.; Salerno, T.M.G.; Rosano, C.; Castagnola, P.; Viale, M.; Monticone, M.; Campagna, S.; Puntoriero, F. Localization-controlled two-color luminescence imaging via environmental modulation of energy transfer in a multichromophoric species. *Dalton Trans.* **2018**, *47*, 4733–4738. [CrossRef]
21. Massaro, M.; Poma, P.; Colletti, C.G.; Barattucci, A.; Bonaccorsi, P.M.; Lazzara, G.; Nicotra, G.; Parisi, F.; Salerno, T.M.G.; Spinella, C.; et al. Chemical and biological evaluation of cross-linked halloysite-curcumin derivatives. *Appl. Clay Sci.* **2020**, *184*, 105400. [CrossRef]
22. Kresge, C.T.; Leonowicz, M.E.; Roth, W.J.; Vartuli, J.C.; Beck, J.S. Ordered mesoporous molecular sieves synthesized by a liquid-crystal template mechanism. *Nature* **1992**, *359*, 710–712. [CrossRef]
23. Michaux, F.; Baccile, N.; Impéror-Clerc, M.; Malfatti, L.; Folliet, N.; Gervais, C.; Manet, S.; Meneau, F.; Pedersen, J.S.; Babonneau, F. In Situ Time-Resolved SAXS Study of the Formation of Mesostructured Organically Modified Silica through Modeling of Micelles Evolution during Surfactant-Templated Self-Assembly. *Langmuir* **2012**, *28*, 17477–17493. [CrossRef]
24. Manzano, H.; Gartzia-Rivero, L.; Bañuelos, J.; López-Arbeloa, I. Ultraviolet–visible dual absorption by single BODIPY dye confined in LTL zeolite nanochannels. *J. Phys. Chem. C.* **2013**, *117*, 13331–13336. [CrossRef]
25. Cucinotta, F.; Jarman, B.P.; Caplan, C.; Cooper, S.J.; Riggs, H.J.; Martinelli, J.; Djanashvili, K.; La Mazza, E.; Puntoriero, F. Light-Harvesting Antennae using the Host–Guest Chemistry of Mesoporous Organosilica. *ChemPhotoChem* **2018**, *2*, 196–206. [CrossRef]
26. Notarbartolo, M.; Cervello, M.; Poma, P.; Dusonchet, L.; Meli, M.; D'Alessandro, N. Expression of the IAPs in multidrug resistant tumor cells. *Oncol. Rep.* **2004**, *11*, 133–136. [CrossRef]

International Journal of *Molecular Sciences*

MDPI

Review

Photobiomodulation Therapy and the Glymphatic System: Promising Applications for Augmenting the Brain Lymphatic Drainage System

Farzad Salehpour [1,2], Mahsa Khademi [3], Denis E. Bragin [4] and Joseph O. DiDuro [2,*]

1 College for Light Medicine and Photobiomodulation, D-82319 Starnberg, Germany; farzadsalehpour1988@gmail.com
2 ProNeuroLIGHT LLC, Phoenix, AZ 85041, USA
3 Neurosciences Research Center (NSRC), Tabriz University of Medical Sciences, Tabriz 51666, Iran; drmahsakhademi@gmail.com
4 Department of Neurology, University of New Mexico School of Medicine, Albuquerque, NM 87131, USA; dbragin@salud.unm.edu
* Correspondence: drjoe@proneurolight.com; Tel.: +1-(845)-203-9204

Abstract: The glymphatic system is a glial-dependent waste clearance pathway in the central nervous system, devoted to drain away waste metabolic products and soluble proteins such as amyloid-beta. An impaired brain glymphatic system can increase the incidence of neurovascular, neuroinflammatory, and neurodegenerative diseases. Photobiomodulation (PBM) therapy can serve as a non-invasive neuroprotective strategy for maintaining and optimizing effective brain waste clearance. In this review, we discuss the crucial role of the glymphatic drainage system in removing toxins and waste metabolites from the brain. We review recent animal research on the neurotherapeutic benefits of PBM therapy on glymphatic drainage and clearance. We also highlight cellular mechanisms of PBM on the cerebral glymphatic system. Animal research has shed light on the beneficial effects of PBM on the cerebral drainage system through the clearance of amyloid-beta via meningeal lymphatic vessels. Finally, PBM-mediated increase in the blood–brain barrier permeability with a subsequent rise in Aβ clearance from PBM-induced relaxation of lymphatic vessels via a vasodilation process will be discussed. We conclude that PBM promotion of cranial and extracranial lymphatic system function might be a promising strategy for the treatment of brain diseases associated with cerebrospinal fluid outflow abnormality.

Keywords: photobiomodulation; near-infrared light; glymphatic system; meningeal lymphatic vessels; amyloid-beta; neurodegenerative diseases; Alzheimer's disease; Parkinson's disease

Citation: Salehpour, F.; Khademi, M.; Bragin, D.E.; DiDuro, J.O. Photobiomodulation Therapy and the Glymphatic System: Promising Applications for Augmenting the Brain Lymphatic Drainage System. *Int. J. Mol. Sci.* **2022**, *23*, 2975. https://doi.org/10.3390/ijms23062975

Academic Editors: Antonino Mazzaglia, Angela Scala and Enrico Caruso

Received: 28 December 2021
Accepted: 22 January 2022
Published: 10 March 2022

Publisher's Note: MDPI stays neutral with regard to jurisdictional claims in published maps and institutional affiliations.

1. Introduction

Photobiomodulation (PBM) therapy is the application of visible and near-infrared (NIR) light to stimulate cellular processes by changing biochemical activities of mitochondrial components at non-thermal and low-level doses [1]. Currently, PBM has obtained significant credibility and light medicine is fast becoming one of the most-accepted physical modalities. PBM has been demonstrated to be an effective approach for promoting cellular proliferation and microcirculation and for relieving pain and edema in various traumatic, acute, and chronic diseases [2]. Neuromodulation of the brain using transcranial and intranasal PBM has been shown to improve cerebral hemodynamics along with an increase in cerebral oxygenation and metabolic capacity [3,4]. Additionally, there is a rapidly increasing body of evidence to support that PBM therapy of the brain can ameliorate neuronal oxidative stress, neuroinflammation, and apoptosis, while promoting neurogenesis and synaptogenesis [5,6]. To date, no serious adverse effects have been reported in the literature for brain PBM therapy; however, caution must be considered with high-power laser sources (class 3B and 4) due to the hazard for macular lesions [7]. To deliver PBM to the brain, transcranial, intranasal, intraoral, intra-aural, and intravascular approaches have been proposed as non-invasive techniques to deliver photonic energy. Moreover, researchers have suggested that PBM therapy targeted at remote tissues (such as the abdomen or tibia) can produce indirect or abscopal effects providing neuroprotection to the brain via systemic mechanisms [8].

Recent evidence has suggested that the meningeal lymphatic vessels (MLVs) play a crucial role in maintaining brain homeostasis by draining macromolecules via both cerebral spinal fluid (CSF) and interstitial fluid (ISF) from the central nervous system (CNS) into the cervical lymph nodes (cLNs) [9–11]. An impaired cerebral lymphatic system is considered as a risk factor for neuroinflammatory diseases [12], neurovascular diseases [13], and impaired recovery from brain injuries [12]. In addition, preclinical studies have shown impaired meningeal lymphatic function in neurodegenerative diseases such as Alzheimer's disease (AD) [9] and Parkinson's disease (PD) [14]. In fact, impairment of MLVs function is a contributing factor in the development of AD and accelerates amyloid-beta (Aβ) aggregation [9]. Starting from 2018, a group of researchers from Russia has made efforts to test the modulatory actions of PBM therapy on the lymphatic drainage function of the brain. Their preliminary findings in a mouse AD model shed light on the potential effect of NIR PBM on the cerebral drainage system through the clearance of Aβ via MLVs [15]. Their follow-up studies also revealed that a PBM-mediated increase in the blood–brain barrier (BBB) permeability can result in further activation of the lymphatic clearance of Aβ from the brain. This action is most likely a result of PBM-induced relaxation of lymphatic vessels via a vasodilation process [16]. Further experiments outlined the beneficial effects of NIR PBM on the lymphatic clearance of blood products from the brain, an important strategy for the prevention of severe consequences after intracranial hemorrhage (ICH) [17].

In this review, we provide an overview of the brain lymphatic drainage system and its pathways. We also discuss the vital role of the lymphatic drainage system (cranial and extracranial lymphatics) in removing waste metabolites and toxins from the brain, maintaining CNS homeostasis and immune responses. We then review the recent animal research on the neurotherapeutic benefits of PBM on lymphatic drainage and clearance. Finally, we propose an underlying biological mechanism for the potential impacts of PBM on the cerebral lymphatic system and highlight this promising new therapeutic approach.

2. Brain Glymphatic Drainage System

2.1. The System, Its Components, and Pathways

Based on physiological findings of communication among different parts of the brain, the existence of a specific lymphatic drainage system in the brain of vertebrates has been suggested [18,19]. In 2012, Iliff et al., for the first time, identified a novel structure in the brain called the glymphatic system [20]. This system is considered as a crucial fluid-clearance system in the brain [21,22]. Studies on mouse models using different fluorescent tracers constructed this glymphatic drainage pathway in the brain [23,24].

This system consists of five main functional components, each facilitating the movement of CSF and ISF (Figure 1). The first compartment of the glymphatic system consists of the production of CSF by epithelial cells of the choroid plexus in the cerebral ventricles and circulation of CSF in the subarachnoid space, followed by the second, periarterial influx of CSF into the brain parenchyma. In fact, periarterial influx refers to the entrance of CSF into the periarterial spaces surrounding the arteries and its penetration deep into the brain tissue. Arterial pulsation caused by smooth muscle cells intensifies CSF movement inward along the periarterial space [25]. Exchange of CSF and ISF is the third component of this system, which occurs in the interstitial space of the brain parenchyma (Figure 2).

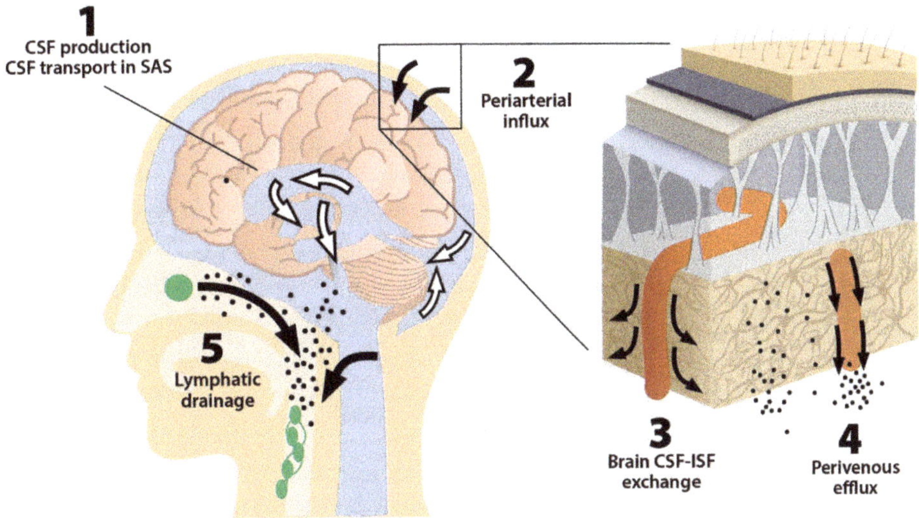

Figure 1. The five components of the glymphatic system. The fluid transport pathway is divided into five distinct segments: (1) cerebrospinal fluid (CSF) is produced by the choroid plexus and likely by extrachoroidal sources (capillary influx and metabolic water production); (2) arterial wall pulsatility drives CSF deep into brain along perivascular spaces; (3) CSF enters the brain parenchyma supported by aquaporin-4 (AQP4) water channels and disperses within the neuropil; interstitial fluid (ISF) mixes with CSF, (4) accumulates in the perivenous space, and drains out of the brain via (5) meningeal and cervical lymphatic vessels, as well as along cranial and spinal nerves Fluids from both the brain and the cribriform plate drain into the cervical lymphatic vessels, which then empty into the venous system at the level of the subclavian veins. The olfactory/cervical lymphatic drainage route is the primary bulk flow pathway.

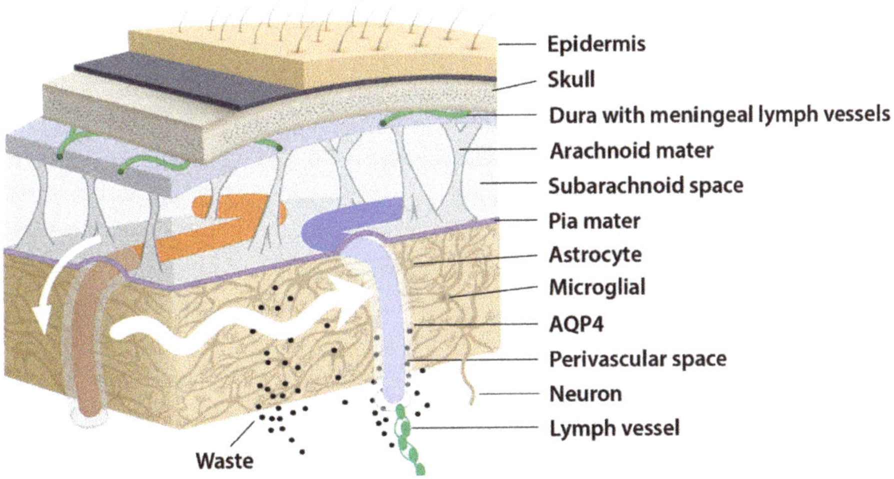

Figure 2. Periarterial influx of CSF into the brain tissue (small white arrow). CSF–ISF exchange supported by AQP4 channels in the vascular end feet plastered along the arterioles. From here, the fluid leaves the axons and moves towards the perivenous space in a path supported by astrocytes. Astrocytic AQP4 water channels facilitate this perivenous efflux of interstitial fluid, which drains to the dural lymphatic vessels.

Astrocytes are believed to facilitate the fluid movement between periarterial spaces and the interstitium through water channels such as aquaporins-4 (AQP4) [20,26]. The fourth component is the glymphatic efflux, which consists of drainage of ISF into the perivenous spaces. The meningeal lymphatic system is the fifth component and final downstream clearance of the glymphatic system. MLVs drain waste products and other solutes from the CNS [9]. This ISF then flows towards the leptomeningeal arteries located at the cortical surface (sulci) and ultimately moves into the cervical lymphatics [20].

Indeed, this system was named "glymphatic" based on the involvement of glial cells "gl" and its similar function with the "lymphatic system" [27,28]. The brain glymphatic system has several essential physiological functions such as drainage of ISF from the parenchymal section of the brain to nearby lymph nodes. It is also involved in communication with the immune system, which regulates and monitors brain responses to neuroinflammation [29]. Moreover, the glymphatic system possesses numerous physiological functions in addition to solute clearance [30]. It is hypothesized that the glymphatic system has a role in rapid lipid transportation across the blood–brain barrier (BBB) and promote glial signaling [31]. Additionally, CSF is involved in the transportation of apolipoprotein E, essential for cholesterol transport, and most notably, synaptic plasticity [32]. CSF influx is also a vehicle for glucose and other vital nutrients that are necessary for the metabolism of astrocytes and neurons [30].

Lifestyle factors, genetics, and pathological conditions can modulate brain clearance and influence the risk of developing neurodegenerative diseases [33]. Several factors such as genetic phenotypes, body posture, aging, and the sleep–wake cycle could influence these physiological functions [23] so that an impaired cerebral lymphatic system is counted as a risk factor for neurodegenerative [34], neuroinflammatory [12], and neurovascular diseases [13] and tumors, as well as impaired recovery from brain injuries [12] (Figure 3). Pathological conditions can strongly affect the brain lymphatic systems. In various vascular disorders including hypertension, atherosclerosis, and small vessel diseases [35], any alteration in the composition of the constituent proteins can result in a significant decline in vascular plasticity and decrease cerebral blood flow (CBF) into the perivascular pathways. In arterial stenosis (either cervical or intracranial), blockage of CBF and obstruction of perivascular or paravascular channels are observed [13], leading to reduced ISF flow resulting in loss of CSF clearance from the brain. Glymphatic system dysfunction has been demonstrated to be associated with many neurological diseases such as AD and PD [14,28]. The glymphatic system has been described as the "final common pathway" for neurodegenerative diseases [36].

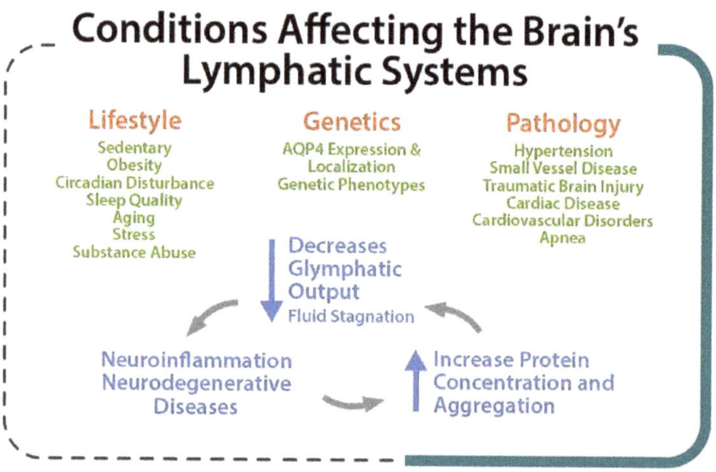

Figure 3. Lifestyle, Genetic and Pathological conditions that can strongly affect the brain lymphatic systems.

2.2. MLVs, Olfactory/Cervical Lymphatic Drainage Route, and Their Association with CSF Circulation

The absence of a conventional lymphatic vasculature in the CNS prompted a series of studies on rodents and human brains to identify MLVs as the lymphatic system of the CNS [37,38]. MLVs seem to provide a critical route for drainage of ISF and CSF. Various macromolecules and immune cells pass from CNS into the lymph nodes located in the deep cervical area [39–41]. More recently, strong evidence demonstrated that MLVs might be associated with the regulation of immune responses and also involved in the pathogenesis of neuroinflammatory diseases [42]. Animal studies have also shown impaired meningeal lymphatic function in AD [9] and PD [14]. In a neuroimaging study using a dynamic contrast-enhanced MRI, patients with idiopathic PD exhibited markedly decreased flow through the MLVs along the superior sagittal sinus and sigmoid sinus, as well as a significant delay in deep cLNs (dcLNs) perfusion [43].

Under normal physiological conditions, the olfactory/cervical lymphatic drainage route serves the primary bulk flow drainage pathway. The ethmoid bone and particularly the cribriform plate located at the anterior aspect of the brain (between the anterior cranial fossa and the nasal cavity) is considered a critical extracranial site of CSF outflow [44]. CSF in subarachnoid space passes through the cribriform plate along the olfactory nerves to the nasal lymphatics and cLNs. At the end of the route, CSF is deposited into the extracranial lymphatic system [13]. The continuous circulation and drainage of CSF are critical for removing CSF metabolic products and maintaining normal neural functions. The outflow routes of CSF are the arachnoid villi of the dural superior sagittal sinus [45], olfactory nerves, across the cribriform plate, and into the cervical lymphatic pathway [46].

The cribriform plate is a fenestrated bony plate of the ethmoid that separates the cranial and nasal cavities (Figure 4). Even though there are lymphatic vessels in the meninges [47], it has been demonstrated that CSF can drain through the cribriform plate in both humans and other mammals [48]. The main pathway by which CSF is removed from the skull is through the cribriform plate associated with the olfactory nerves [49]. The CSF is absorbed by lymphatic vessels located in the submucosa of the olfactory epithelium, in the nasal mucosa after passing the cribriform plate, and then drained into the cLNs. Any damage to the cribriform plate (by traumatic brain injuries or surgical methods) can lead to acute blockage of CSF outflow and, as a result, increase in resting intracranial pressure (ICP) and outflow resistance, emphasizing that the olfactory pathway represents the leading site for the CSF drainage [50]. There is a space between the olfactory sensory axons that provides a conduit for the outflow of CSF. Any damage to these nerves can also diminish the outflow of CSF through the cribriform plate [49] (Figure 4).

AQP are a family of small integral membrane proteins that significantly boost the permeability of cells to water and facilitate the movement of fluid down the pressure gradient in various tissues, including the brain [51]. So far, 13 AQPs have been found in mammals (AQP0–12). AQP1 maintains CSF production by the choroid plexus, and it is also expressed along the periphery of the olfactory bulb, nerve junction, and lining the foramina of the cribriform plate. Moreover, there are high levels of AQP1, 3, and 5 within the nasal cavity. These AQPs facilitate the flow of fluid out of the olfactory bulb and subarachnoid space into the nasal cavity via the extensive network of lymphatic vessels, which play an essential role in moving fluid throughout the body. AQPs are found in the meninges and at the cribriform plate and olfactory bulb junction [11]. These vessels crossing the cribriform plate play a key role in transporting fluid from the cranial cavity to the nasal cavity olfactory sensory nerves. Following CSF absorption by lymphatics, it is conveyed in larger ducts through numerous lymph nodes and eventually is deposited into the body's lymphatic system. Evidence has also shown that aging decreases the elimination rate of CSF from the nasal/cribriform plate region [52,53].

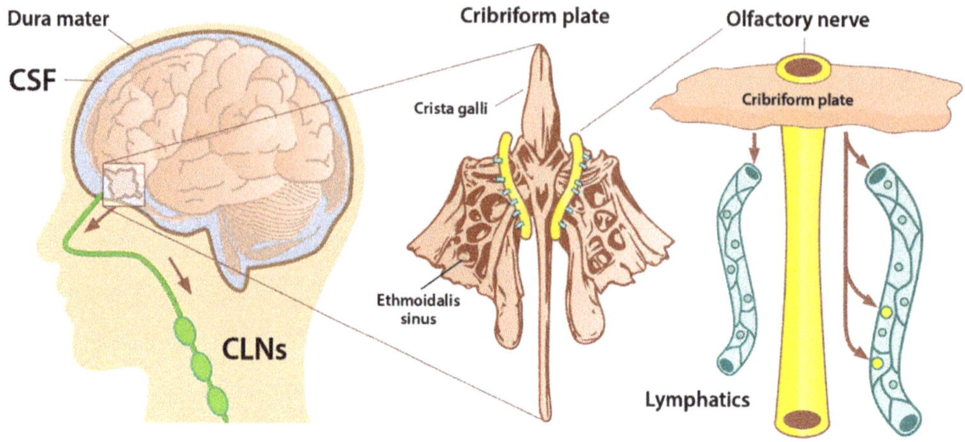

Figure 4. Perineural space surrounding olfactory nerve penetrates the nasal mucosa through the cribriform plate. The cribriform plate of the ethmoid bone is considered the key extracranial site of CSF outflow. CSF in SAS passes across the cribriform plate along olfactory nerves to nasal lymphatics and enters cervical lymph nodes (adapted from Semyachkina-Glushkovskaya et al. 2021).

2.3. Sleep and Clearance of the Brain

The glymphatic system uses convective flow between the CSF and ISF to remove toxic metabolites in/from the brain. CSF enters the brain parenchyma (functional parts) along a paraarterial route and exchanges with the ISF [54]. The ISF carries extracellular solutes from the interstitial (extracellular) space in the brain along paravenous drainage pathways (Figure 1). This activity is dramatically boosted during sleep and is related to increased interstitial volume, possibly by shrinkage of astroglial cells [55]. Emerging evidence shows that sleep is the primary driver of glymphatic clearance and is essential for the maintenance of brain function via the discharge of metabolites and neurotoxic wastes from the brain, which accumulates in the highly active brain during waking hours [36,56].

Comparing the brain ISF volume during deep sleep to wakefulness, the volume of the brain's ISF increases by 40–60% [57]. Astrocytic AQP4 water channels that encircle the brain's vasculature contribute to this increase in ISF. This increase in ISF is required for proper glymphatic function and facilitates the clearance of soluble proteins, waste products, and excess extracellular fluid. ISF increase leads to a 2-fold faster removal of neurotoxic waste products such as lactate and Aβ from the brain. This increase in the clearance of brain waste happens during non-rapid eye movement (NREM) sleep [58], and the majority of glymphatic activity occurs during deep, slow-wave sleep. Poor sleep quality and short sleep duration result in an increased amount of Aβ in the CSF as well as a risk of Aβ plaque formation [59]. In addition, tau levels have been shown to be increased in the ISF of the hippocampus following sleep deprivation [60]. Evidently, these neurobiological mechanisms can support the fact that neurodegenerative diseases such as AD, PD, Huntington disease, and frontotemporal dementias are strongly linked to sleep disturbances [61]. With the glymphatic system in mind, it is of interest to note that sleep quality decreases as a function of normal aging, and individuals over 60 years old rarely enter deep NREM (stages 3). The effectiveness of glymphatic fluid transport is directly linked to the prevalence of slow-wave activity. Therefore, the age-related impairment in sleep quality can cause a catastrophic drop in the clearance of brain waste and potentially increase the incidence risk of neurodegenerative diseases [36].

Recent evidence revealed that endocytosis occurs across the BBB during sleep, and inhibition of this process causes the need for more sleep [62]. In addition, several studies have reported that sleep deprivation can increase the activity of several pro-inflammatory mediators such as C-reactive protein, interleukin (IL)-1β, IL-6, IL-17, interferon-γ (IFN-γ), and tumor necrosis factor-alpha (TNF-α). These mediators suppress astrocytic maintenance of the BBB, causing an increase in its permeability [63,64]. Sleep deprivation has been shown to decrease influx efficiency along the perivascular space, thus impairing the function of the glymphatic system and disturbing AQP4 polarization in a mouse model [65].

3. PBM Therapy

PBM therapy or low-level light/laser therapy (LLLT) refers to the non-thermal application of visible and/or NIR light to stimulate biological processes [66]. Almost all PBM therapy procedures are applied in the wavelength range between 400 and 1300 nm from various light sources (e.g., lasers, light-emitting diodes (LEDs), or broadband light sources) [67]. One of the most recognized mechanisms for PBM has been suggested by a Russian photobiologist, Tiina Karu [68]. Her early work discovered that light-cell interaction probably should be considered as a light–mitochondria interaction [69]. Further studies with light at 600–850 nm wavelengths proved her hypothesis that the mitochondrial respiratory enzyme, cytochrome c oxidase (CCO), is the main photoacceptor responsible for the light absorption in the cells [70,71]. It is now believed that light energy is absorbed by metal centers of CCO, resulting in the excitation of electrons [72]. Along with this photoexcitation, nitric oxide (NO) is photodissociated from the CCO, leading to an increase in the mitochondrial membrane potential (MMP). This, in turn, promotes an increase in ATP production and modulates the levels of signaling molecules, including intracellular Ca^{2+} and reactive oxygen species (ROS), particularly the superoxide anion (O_2^-) and its stable product hydrogen peroxide (H_2O_2) [2]. As a secondary event, the above-mentioned primary responses change the intracellular redox potential, the intracellular pH, cyclic adenosine monophosphate (cAMP) levels, and expression of redox-sensitive factors such as nuclear factor kappa-B (NF-κB). Following this cascade of events, signal transduction processes induced by PBM will lead to activation of transcription factors and gene expression, which eventually will promote many biological functions such as cell metabolism, cell viability, proliferation, and differentiation [73–75].

Today, PBM is applied as a cutting-edge technology in numerous areas of medicine, such as wound healing, dentistry, muscle and tendon repair, dermatological conditions, and neurogenic pain [76]. In addition, recent research has focused on the application of PBM as a neuroprotective intervention for the treatment of CNS diseases [4,77], providing further breakthroughs in the field of neurorehabilitation. It has been shown that low levels of red/NIR light stimulate neuronal functions leading to neuroprotection and prevention of neuronal death, hypoxia, trauma, or neurotoxicity [4,6]. Transcranial PBM is the non-invasive delivery of light from external laser or LEDs sources (e.g., hand-held probes or wearable PBM helmets/headsets) to the head. This photonic energy is transferred onto subcranial tissues and to an extent, the cortical surface [4]. Intranasal PBM is another therapeutic approach delivering light energy through the nostrils, which has been shown to promote brain function in various CNS diseases such as depression, cerebral infarction, dementia, and Kleine–Levin syndrome [78–83].

3.1. Evidence on Potential Effects of PBM on the Brain Drainage System

Recently, a group of researchers from Russia have carried out a series of animal studies on the possible beneficial effects of PBM therapy on the lymphatic drainage function of the brain [15,16,59,84–87]. Their findings have opened up a new idea that PBM of the cranial and the extracranial lymphatics may be a promising approach for the treatment of brain disorders associated with CSF outflow abnormality [88]. Herein, we review their investigations and outline possible mechanisms of neurotherapeutic benefits of PBM on lymphatic drainage and clearance.

In 2019, they examined the idea that transcranial PBM might stimulate lymphatic drainage in an animal AD model by demonstrating improvement in the clearance of Aβ molecules from the brain following PBM therapy [15]. In the first step, they compared the effectiveness of four different skull fluencies (18, 25, 32, and 39 J/cm^2) of 1268 nm laser on the reduction in Aβ accumulation in the brain. The skull fluence of 32 J/cm^2 (cortical fluence of 4 J/cm^2) was selected as an optimal PBM fluence because it was not associated with an increase in a skull temperature or morphological alterations of the brain and was significantly effective for the reduction in Aβ depositions in the brain. Although 39 J/cm^2 effectively decreased Aβ accumulation, it resulted in a dura mater and arachnoid membrane injuries as well as a scalp temperature rise of 2 °C. In the second step, they studied the development of AD following the injection of Aβ in the hippocampus of mice and evaluated the effects of PBM (32 J/cm^2) on Aβ distribution in the brain. The confocal microscopic analysis showed that PBM actively decreased the density of small Aβ plaques, whereas the density of large Aβ plaques did not differ between the PBM and the untreated group. The accumulation of Aβ in the brain of PBM-treated mice was also accompanied by the appearance of Aβ plaques in the dcLNs, compared with the untreated group. In the third step, they explored the PBM effects on clearance of gold nanorods (GNRs) from the brain into cervical lymphatics using optical coherence tomography (OCT) in vivo to monitor the rate of GNRs accumulation into the right dcLN. OCT data showed that PBM-activated clearance of GNRs was higher for treated mice. Clearance from the cortex, lateral ventricle, cisterna magna, and hippocampus was higher in the treated mice by 3.7-, 3.9-, 6.7-, and 9.3-fold, respectively. The results of atomic absorption spectroscopy (AAS), which exhibit the level of GNRs in dcLN, were also correlated with OCT data suggesting that PBM significantly increased the clearance of GNRs from both deep (hippocampus and ventricles) and superficial (cisterna magna and cortex) regions of the brain. Results from the neurological status (tested by startle reflex, round stick balancing, and beam walk tasks) and the neurobehavioral outcomes (tested by novel object recognition task) also showed an improvement in the PBM-treated mice compared with AD mice [15].

Given the fact that disruption of MLVs is an aggravating factor in the development of AD and promotes Aβ deposition in the meninges, in 2020, they investigated the potential benefits of PBM on lymphatic pumping and contractility, which are considered the main physiological mechanisms underlying fluid transport and waste clearance from tissues [59]. They started to test the hypothesis that PBM-promoted relaxation of lymphatic vessels via vasodilation might be one of the underlying mechanisms for increasing the permeability of lymphatic endothelium, thereby allowing larger molecules to be transported through to the lymphatic vessels. Their preliminary data revealed that low PBM fluencies (5 and 10 J/cm^2) induced relaxation of the mesenteric lymphatics (extracranial and/or abdominal lymph vessels) in both systole and diastole with a decrease in contraction amplitude (with maximum response to 10 J/cm^2). These low fluencies relaxed the lymphatic vessels, while higher fluencies of 30 and 70 J/cm^2 completely blocked the contractility of the vessels. Their OCT imaging results also showed an increase in the diameter of the MLVs in systole and diastole following transcranial PBM. Data also demonstrated dilation of the MLVs and an increase in the number of macrophages inside the cavity of the vessel after transcranial PBM, most likely due to an increase in uptake of ISF (lymph). In the next phase, they studied the effects of the transcranial PBM (skull fluence of 64 J/cm^2) on the drainage function of MLVs by analyzing the clearance of GNRs from the mouse brain. They injected GNRs in different brain regions as in their previous pilot study [15] (cortex, cisterna magna, right lateral ventricle, and hippocampus) and monitored the accumulation of GNRs in dcLN before and after transcranial PBM using OCT in vivo. PBM increased clearance of GNRs from the cortex to the dcLN by 55.7-fold. From the hippocampus, cisterna magna, and left ventricle, the clearance of the GNRs to the dcLN was also higher: 14.78-fold, 4.8-fold, and 2.3-fold, respectively. Their findings provided promising evidence that transcranial and remote PBM can augment the drainage and clearance function of MLVs, providing a therapeutic target for neurological disorders such as stroke and brain trauma, as well as preventing or delaying neurodegenerative diseases [59].

In their third study [87], they explored the effects of transcranial 1268 nm laser PBM on (1) clearance of two different tracers (GNRs and Evans blue dye (EBD)) from the brain via meningeal lymphatic system into the peripheral lymphatic system and (2) on the mesenteric lymphatics permeability of the mice. Their preliminary data showed that a cortical fluence of 9 J/cm^2 (skull fluence of 32 J/cm^2) exhibited a better stimulation of clearing function of MLVs than 2 and 5 J/cm^2. Similarly, OCT data from the 9 J/cm^2 group showed a gradual increase in the speed of GNRs accumulation in dcLN after its injection into the cisterna magna. PBM-mediated dilation of mesenteric lymphatics vessels was also associated with the decrease in resistance to the lymph flow. In the next phase, to better understand the mechanisms underlying the impacts of PBM on the lymphatic vessels, they investigated the effects of 9 J/cm^2 light on the lymphatic permeability to immune cells such as macrophages. The results revealed that PBM can promote migration of macrophages from the lymphatic vessels into surrounding tissues, most likely through the decrease in transendothelial electrical resistance (TEER) integrity and an expression of tight junction (TJ) proteins (e.g., CLND, VE-cadherin, and ZO-1) [87].

In the fourth study, they aimed to test the hypothesis that the PBM-mediated opening of BBB might be one possible mechanism for the activation of Aβ clearance from the brain in AD mice [16]. First, they studied the effects of 1267 nm PBM with a cortical fluence of 9 J/cm^2 (skull fluence of 32 J/cm^2) on Aβ clearance from the mice brain using the immunohistochemical analysis of Aβ in the dcLNs. PBM-treated mice showed a pronounced Aβ level in the dcLNs, indicating the efficiency of the PBM for stimulation of Aβ clearance from the brain. The quantitative analysis also confirmed these results by representing a higher signal intensity from immunopositive Aβ plaques in the dcLNs in the PBM-treated mice. Their findings uncovered the lymphatic pathway of Aβ clearance from the brain which was also associated with the enhancement of the neurobehavioral status in AD mice. Their follow-up experiments showed that a PBM-mediated increase in the BBB permeability and subsequent increase in Aβ leakage occurs most likely as a result of PBM-induced decrease in transendothelial integrity and decrease in the expression of TJ proteins (e.g., CLND, VE-cadherin, and ZO-1) [16].

Their fifth study examined the lymphatic pathway of red blood cells (RBCs) clearance from the brain after intraventricular hemorrhage (IVH). They investigated whether transcranial PBM can improve RBCs evacuation from the ventricles to enhance the outcome after IVH [17]. First, using immunohistochemical and confocal colocalization analysis of the mouse and human brain samples (the next day after death due to IVH), they showed that RBCs moved from the ventricles into dcLNs via MLVs, confirming the lymphatic clearance of RBCs from the brain in the post-hemorrhagic period. They then studied the efficacy of transcranial 1267 nm PBM with a cortical fluence of 9 J/cm^2 (skull fluence of 60 J/cm^2) for stimulation of lymphatic clearance of tracers (GNRs and EBD) from the right lateral ventricle, mimicking the pathway of RBCs elimination from the brain of naïve mice. The transport of GNRs and EBD into dcLNs after its intraventricular injection was higher in the PBM-treated mice by 1.4- and 2.6-folds, respectively. In their follow-up experiment, they directly evaluated the effects of PBM on RBCs clearance from the mice brain after IVH and found that the number of RBCs transported into the dcLNs was significantly greater in the PBM group. They postulated that PBM facilitated RBCs drainage from the fluid-filled ventricles into the subarachnoid space where the RBCs are transported into the MLVs. This was accomplished by the PBM-mediated change in tone of the MLVs. In the final step, 3 days after surgical injection of blood into the right lateral ventricle, they treated mice with PBM 3 times in 7 days with a cortical fluence of 9 J/cm^2. They found that PBM contributed to a faster recovery of ICP after IVH, along with a 1.57-fold decrease in mortality and a significant reduction in the level of stress [17].

In the last and most recent study on AD mice [84], they tested the hypothesis that transcranial PBM stimulates Aβ clearance from the brain through the activations of cerebral lymphatic drainage and probably via an increase in cerebral energy metabolism. Application of transcranial 1268 nm laser PBM with a cortical fluence of 4 J/cm^2 (skull fluence of 32 J/cm^2) for 9 days significantly reduced Aβ plaques in the brain of AD mice along with a significant increase in clearance of Aβ via MLVs. Their further investigation shed light upon the possible involvement of PBM-induced improvement in blood oxygen saturation (SpO_2) of the brain on the stimulation of lymphatic Aβ clearance. They suggested that an increase in oxygen saturation leads to improved mitochondrial ATP production that can stimulate lymphatic contractility leading to increased drainage and clearing functions of the meningeal lymphatic system [84].

3.2. PBM and Nitric Oxide

Studies have shown increased blood flow during and after PBM both in animal models and in humans [89,90]. However, a critical question regarding this escalation of blood flow has remained open. Does this increased blood flow arise from the PBM-mediated production of NO? If so, what is the actual source? Is it NO that is photodissociated from hemoglobin in circulating RBCs, or NO that is photodissociated from other labile NO stores in the blood vessel wall, or is it derived from the dissociation of NO that has bound to CCO in the mitochondria? Nevertheless, it has been proposed that red/NIR light appears to be best for dissociating NO from CCO, thereby reversing the signaling consequences of excessive NO binding [2].

Impaired cerebral vascular perfusion has been widely known as one of the early manifestations in most of the CNS diseases [91]. Animal research has shown that PBM can improve neuronal NO levels and CBF, resulting from activation of endothelial NO synthase (eNOS) protein [90], and also can increase the blood vessel diameter [92]. It is also suggested that PBM can affect the regional CBF, most likely mediated by NO and glutamate [90]. Uozumi et al. found that the transient CBF improvement by PBM was dependent on the NOS activity and NO levels as well. They showed that transcranial 808 nm laser PBM of the naïve mice increased cortical NO levels (by 50%) immediately after starting the PBM, and gradually improved CBF in the PBM-exposed (by 30%) and the opposite hemisphere (by 19%) at 45 min after starting the irradiation [90].

PBM has also been shown to improve endothelial function through the activation of cellular pathways responsible for the modulation of inflammation and angiogenesis, as well as vasodilatation [93]. NIR light at 890 nm could significantly increase NO levels (with a peak at 5 min post-irradiation) in venous blood in healthy individuals [94]. Endothelial NOS (eNOS) is found in endothelial cells, which are the cells that line the inner surface of blood vessels as well as lymph ducts. eNOS can be activated by the pulsatile flow of blood through vessels, leading to a "shear stress" on the membrane of the endothelial cells as the column of blood in the vessel moves forward and then stops. Indeed, NO produced by eNOS maintains the diameter of the blood vessel (vasodilation) so that perfusion of various tissues (skin, bone, muscle, and nerves) is maintained at optimal levels. This eNOS-mediated NO can also activate the growth of new blood vessels (angiogenesis). NO relaxes smooth muscle cells and therefore dilates resistance vessels and lymphatics, leading to an increase in blood supply for repairing tissues and removal of the damaged cells [95]. In fact, increased lymphatic flow removes metabolic waste products and reduces edema. It has been shown that NO can be produced enzymatically following an increase in NOS activity after PBM, possibly by elevating intracellular Ca^{2+} levels [2,96].

As discussed before, it has been reported that PBM-induced relaxation of the mesenteric lymphatics endothelium is accompanied by increased permeability of lymphatic walls as well as a decreased expression of TJ proteins [87]. The TJ proteins are structural compounds of mature lymphatic vessels and play an essential role in moving ISF and immune cells through the lymphatic endothelium [88]. In fact, increased permeability of lymphatic endothelium is the main mechanism allowing for toxins to be transported by the collecting lymphatics, contacting local immune cells to activate immune responses. These effects might be related to a PBM-mediated increase in the eNOS activity. In other words, the PBM-mediated dilation of lymphatic vessels could be due to an increase in the eNOS activity [97]. Considering the fact that the improvement of endothelial NO production is a well-recognized mechanism of PBM [97] and that the lymphatic behavior is actively regulated by NO [98], it appears that PBM activates the NO synthesis in lymphatic endothelium isolated cells. Thus, the lymphatic vessel's contractility may be the possible underlying mechanisms for PBM-induced lymphatic clearance of macromolecules from the brain [88].

3.3. PBM and Neuroprotection

Researchers have explored the neuroprotective effects of PBM against toxicity induced by Aβ [99–102]. A series of studies conducted by Da Xing's research team has shown novel findings [99,100,102]. First, they proposed that activated Akt induced by laser PBM (632.8 nm, 2 J/cm^2 at culture surface) interacts with and then inactivates GSK3β upon A$\beta_{25\text{-}35}$ treatment. Following this step, due to the inhibition of GSK3β, β-catenin accumulates in the cytoplasm and then translocates into the nucleus. Subsequently, it acts as a transcriptional cofactor to improve neuronal survival [99]. Then in the follow-up research using the same laser source [100], they showed that 2 J/cm^2 protects SH-SY5Y cells against A$\beta_{25\text{-}35}$-induced toxicity only at 24 and 48 h post-PBM. After exposure to A$\beta_{25\text{-}35}$, the cell viability of both SH-SY5Y cells and hippocampal neurons was increased by PBM in a dose-dependent fashion, representing a significant increase only at the 2 and 4 J/cm^2. Their results also revealed that 2 J/cm^2 was enough to protect hippocampal neurons against A$\beta_{1\text{-}42}$ toxicity [100]. Recently, Da Xing and colleagues found that by increasing the mitochondrial CCO activity and thereby increasing the levels of cAMP and ATP, laser PBM (632.8 nm, 2 J/cm^2 at culture surface) can activate the PKA/SIRT1 signaling pathway in SH-SY5Y-APPswe cells, leading to decreased Aβ levels [102]. Duggett and Chazot have also proved the neuroprotection effects of 1068 nm wavelength, proposing that a 4.5 J/cm^2 can protect CAD neuroblastoma cells from A$\beta_{1\text{-}42}$-induced cell death [101]. A 1070 nm PBM (4.5 J/cm^2 at scalp surface with 10 Hz pulsed mode) has also been reported to decrease cerebral Aβ levels and therefore enhance cognitive performance in AD mice through microglia activation and promotion of angiogenesis. NIR PBM could trigger microglia rather than astrocyte responses with a change in morphology and increased colocalization with Aβ. The response of microglia to PBM was negatively correlated with the Aβ level, proposing that PBM decreases the Aβ deposition, probably via eliciting microglia activation and recruiting microglia to Aβ burden. The perivascular microglia were also decreased after PBM therapy, whereas an increase in cerebral vessel density was observed in PBM-treated AD mice. This increase in vessel density was positively correlated with clearance of Aβ burden, indicating that 1070 nm PBM can diminish Aβ deposition most likely through increasing cerebral vessel density [103]. In addition to these, 630 nm laser PBM therapy has been shown to decrease Aβ-disrupted flow of ISF by smashing Aβ deposition in the extracellular space and thereby reversing cognitive impairments in an APP/PS1 mouse model of AD [104].

3.4. Intranasal and Systemic PBM Therapies and Their Effects on the Brain Drainage System

Considering the fact that the blood capillaries are abundant in the nasal cavity and the blood flow is relatively slow, it is postulated that systemic effects on hematologic cells in the blood would contribute to the neuro-therapeutic benefits reported by the intranasal PBM technique [78,81]. Intranasal PBM has been shown to increase CBF [80], decrease blood viscosity [105], increase hemorheology [105], and increase blood coagulability status [106]—enhancements in blood rheology that are linked to improved cognitive functioning [107] and mood [108]. With respect to the brain drainage system, intranasal PBM might also serve as an effective treatment for obstruction of the cribriform–lymphatic route, thereby improving the CSF outflow. This is thought to occur because PBM-induced NO can modulate the lymphatic vessels contraction and subsequent increase in lymph flow [88,109]. Given this, it is speculated that the modulation of blood circulation and drainage function of the lymphatic system might be the underlying action mechanisms of intranasal PBM in the previously shown studies on dementia and other neurodegenerative diseases [78].

Recently, the possibility that PBM therapy targeted at a remote tissue (e.g., abdominal tissue) might elicit systemic mechanisms that provide neuroprotection of the brain is of great interest to researchers [83,110,111]. Today, there are a number of convincing examples of the possible systemic effects of PBM in animal models of AD [112,113] and PD [74,114,115]. Although the mechanisms underlying the phenomenon of systemic or indirect effects of PBM are not clear, we propose that the effect of PBM on the lymphatic system could be one of them. As discussed above, 1267 nm laser PBM of the mice's abdominal region results in a relaxation of the mesenteric lymphatics with a decrease in contraction amplitude leading to subsequent increases in the clearance of GNRs from the brain, partly via MLVs [59]. PBM therapy targeted to the abdominal region has also been suggested to change the gut microbiota—with the release of yet unidentified circulating mediators—promoting a neuroprotective action on the brain [113]. It is also thought that bone marrow-derived stem cells—in particular, mesenchymal stem cells—can drive remote PBM-induced neuroprotection on the brain tissue [116,117], as they can easily transmigrate across the BBB [115,118].

4. Conclusions

Emerging evidence suggests that the MLVs play a key role in maintaining brain homeostasis by draining macromolecules via both CSF and ISF from the CNS into the cLNs. An impaired cerebral lymphatic system is considered a risk factor for neuroinflammatory diseases, neurovascular diseases, and impaired recovery from brain injuries. Animal research also demonstrates impaired meningeal lymphatic function in AD and PD. In particular, MLVs dysfunction can accelerate the development of AD through Aβ aggregation. Together, PBM-mediated promotion of glymphatic and extracranial lymphatic system function might be a promising candidate for the treatment of various brain diseases associated with CSF outflow abnormality. Because of good penetration onto the subcranial, brain cortex, and even into subcortical structures, transcranial PBM can stimulate MLVs, which might be one of the mechanisms underlying the positive clinical outcomes of PBM with neurodegenerative conditions [119,120]. Transcranial PBM can affect the aqueous component of the CSF/ISF structure because of light's effect on the structure of the water molecules, creating a freer-flowing, slippery effect. Another factor is PBM-induced production of NO that can increase overall blood perfusion, increasing ISF/CSF diffusion components.

It should be noted that the therapeutic effects of transcranial PBM on subcortical regions of the human brain using biomodulatory wavelengths (e.g., 600–1300 nm) are not well understood. This is due, in part, to the poor penetration of red and NIR light through the skull/scalp into the deep brain areas. Given this, delivery of sufficient light dose to deeper structures in the human brain is still a challenge in the transcranial PBM field. Recently, nanoparticle engineering, in combination with biophotonic techniques, has been suggested as a way to overcome this problem. Considering the fact that photons in the third optical window (1550–1870 nm) have the highest penetration rate into brain tissue, it is speculated that upconverting nanoparticles (UCNPs) could help in delivering light to the deep brain by converting these photons to visible and NIR spectrum with higher energies and a greater biological effect. Of note, UCNPs exhibit good ability to cross the BBB, and also their low toxicity makes them a promising candidate for application in brain disorders.

Considering potential applications in human subjects, it is speculated that transcranial PBM can create an increase in surface temperature on the skull, thereby increasing the temperature gradient, creating a heat sink of blood to cool the sleeping brain, if and when the transcranial irradiation of the light is conducted immediately in bed before sleep. Intranasal PBM can also be an effective non-invasive approach for the treatment of cribriform plate obstruction, which is the main reason for the development of various brain pathologies due to the blocking of CSF drainage. Intranasal PBM reduces the viscosity of the blood. In turn, this would create less stacking of RBCs (rouleaux formation), which would allow a better flow of microcirculation through the areas that are smaller than capillaries. In addition, it has been suggested that brain cells might also benefit indirectly/systemically from the PBM of circulating blood or different underlying organs. A systemic PBM application using body pad LED devices across the carotid and or vertebral arteries can facilitate vasodilation and increase blood flow to the brain parenchyma in humans. Body pad-style applications of LEDs over the corpus and or cervical lymph tissue can presumably also increase the activity of the dcLNs and act as a pulling aspect of glymphatic flow. Applying PBM on the chest or abdomen area may prove effective in stimulating the mesenteric lymph vessels to augment lymphatic pull. Finally, PBM being a light energy-based therapy, has a circadian influence, so the timing of when to apply the red/NIR light is even more critical with a PBM intervention of the glymphatic and meningeal lymphatic systems. We can only speculate that future research will clarify the photonic impact of PBM on each of the five components of the glymphatic system.

Funding: This research was funded by RSF #22-45-04406.

Institutional Review Board Statement: Not applicable.

Informed Consent Statement: Not applicable.

Data Availability Statement: Not applicable.

Conflicts of Interest: The authors declare no conflict of interest.

References

1. Hamblin, M.R. Mechanisms and Mitochondrial Redox Signaling in Photobiomodulation. *Photochem. Photobiol.* **2018**, *94*, 199–212. [CrossRef] [PubMed]
2. de Freitas, L.F.; Hamblin, M.R. Proposed mechanisms of photobiomodulation or low-level light therapy. *IEEE J. Sel. Top Quan-Tum Electron* **2016**, *22*, 348–364. [CrossRef] [PubMed]
3. Gonzalez-Lima, F. Neuroprotection and Neurocognitive Augmentation by Photobiomodulation. In *Contemporary Clinical Neuroscience*; Springer: Singapore, 2021; pp. 165–207.
4. Salehpour, F.; Mahmoudi, J.; Kamari, F.; Sadigh-Eteghad, S.; Rasta, S.H.; Hamblin, M.R. Brain Photobiomodulation Therapy: A Narrative Review. *Mol. Neurobiol.* **2018**, *55*, 6601–6636. [CrossRef] [PubMed]
5. Cassano, P.; Petrie, S.R.; Hamblin, M.R.; Henderson, T.A.; Iosifescu, D.V. Review of transcranial photobiomodulation for major depressive disorder: Targeting brain metabolism, inflammation, oxidative stress, and neurogenesis. *Neurophotonics* **2016**, *3*, 031404. [CrossRef] [PubMed]
6. Hamblin, M.R. Shining light on the head: Photobiomodulation for brain disorders. *BBA Clin.* **2016**, *6*, 113–124. [CrossRef] [PubMed]
7. Caldieraro, M.A.; Cassano, P. Transcranial and systemic photobiomodulation for major depressive disorder: A systematic review of efficacy, tolerability and biological mechanisms. *J. Affect. Disord.* **2019**, *243*, 262–273. [CrossRef] [PubMed]
8. Hamblin, M.R.; Huang, Y.-Y. Photobiomodulation in the Brain: Low-Level Laser (Light) Therapy in Neurology and Neuro-science. Academic Press: Cambridge, MA, USA, 2019.
9. Da Mesquita, S.; Louveau, A.; Vaccari, A.; Smirnov, I.; Cornelison, R.C.; Kingsmore, K.M.; Contarino, C.; Onengut-Gumuscu, S.; Farber, E.; Raper, D.; et al. Functional aspects of meningeal lymphatics in ageing and Alzheimer's disease. *Nature* **2018**, *560*, 185–191. [CrossRef] [PubMed]
10. Semyachkina-Glushkovskaya, O.; Postnov, D.; Kurths, J. Blood–Brain Barrier, Lymphatic Clearance, and Recovery: Ariadne's Thread in Labyrinths of Hypotheses. *Int. J. Mol. Sci.* **2018**, *19*, 3818. [CrossRef] [PubMed]
11. Louveau, A.; Smirnov, I.; Keyes, T.J.; Eccles, J.D.; Rouhani, S.J.; Peske, J.D.; Derecki, N.C.; Castle, D.; Mandell, J.W.; Lee, K.S.; et al. Structural and functional features of central nervous system lymphatic vessels. *Nature* **2015**, *523*, 337–341. [CrossRef]
12. Dunn, G.P.; Okada, H. Principles of immunology and its nuances in the central nervous system: Figure 1. *Neuro-Oncology* **2015**, *17*, vii3–vii8. [CrossRef]
13. Sun, B.-L.; Wang, L.-H.; Yang, T.; Sun, J.-Y.; Mao, L.-L.; Yang, M.-F.; Yuan, H.; Colvin, R.A.; Yang, X.-Y. Lymphatic drainage system of the brain: A novel target for intervention of neurological diseases. *Prog. Neurobiol.* **2018**, *163–164*, 118–143. [CrossRef] [PubMed]
14. Zou, W.; Pu, T.; Feng, W.; Lu, M.; Zheng, Y.; Du, R.; Xiao, M.; Hu, G. Blocking meningeal lymphatic drainage aggravates Parkinson's disease-like pathology in mice overexpressing mutated α-synuclein. *Transl. Neurodegener.* **2019**, *8*, 1–17. [CrossRef] [PubMed]
15. Zinchenko, E.; Navolokin, N.; Shirokov, A.; Khlebtsov, B.; Dubrovsky, A.; Saranceva, E.; Abdurashitov, A.; Khorovodov, A.; Terskov, A.; Mamedova, A.; et al. Pilot study of transcranial photobiomodulation of lymphatic clearance of beta-amyloid from the mouse brain: Breakthrough strategies for non-pharmacologic therapy of Alzheimer's disease. *Biomed. Opt. Express* **2019**, *10*, 4003–4017. [CrossRef] [PubMed]
16. Zinchenko, E.; Klimova, M.; Mamedova, A.; Agranovich, I.; Blokhina, I.; Antonova, T.; Terskov, A.; Shirokov, A.; Navolokin, N.; Morgun, A.; et al. Photostimulation of Extravasation of Beta-Amyloid through the Model of Blood-Brain Barrier. *Electronics* **2020**, *9*, 1056. [CrossRef]

17. Li, D.; Liu, S.; Yu, T.; Liu, Z.; Sun, S.; Bragin, D.; Navolokin, N.; Kurths, J.; Glushkovskaya-Semyachkina, O.; Zhu, D. Photostimulation of lymphatic clearance of red blood cells from the mouse brain after intraventricular hemor-rhage. *bioRxiv*. [CrossRef]
18. Weller, R.O.; Djuanda, E.; Yow, H.-Y.; Carare, R.O. Lymphatic drainage of the brain and the pathophysiology of neurological disease. *Acta Neuropathol.* **2009**, *117*, 1–14. [CrossRef] [PubMed]
19. Matsumae, M.; Sato, O.; Hirayama, A.; Hayashi, N.; Takizawa, K.; Atsumi, H.; Sorimachi, T. Research into the physiology of cerebrospinal fluid reaches a new horizon: Intimate exchange between cerebrospinal fluid and interstitial fluid may contrib-ute to maintenance of homeostasis in the central nervous system. *Neurol. Med. Chir.* **2016**, *56*, 416–441. [CrossRef]
20. Iliff, J.J.; Wang, M.; Liao, Y.; Plogg, B.A.; Peng, W.; Gundersen, G.A.; Benveniste, H.; Vates, G.E.; Deane, R.; Goldman, S.A.; et al. A Paravascular Pathway Facilitates CSF Flow Through the Brain Parenchyma and the Clearance of Interstitial Solutes, Including Amyloid β. *Sci. Transl. Med.* **2012**, *4*, 147ra111. [CrossRef]
21. Benveniste, H.; Liu, X.; Koundal, S.; Sanggaard, S.; Lee, H.; Wardlaw, J. The Glymphatic System and Waste Clearance with Brain Aging: A Review. *Gerontology* **2018**, *65*, 106–119. [CrossRef]
22. Bakker, E.; Bacskai, B.J.; Arbel-Ornath, M.; Aldea, R.; Bedussi, B.; Morris, A.; Weller, R.O.; Carare, R.O. Lymphatic Clearance of the Brain: Perivascular, Paravascular and Significance for Neurodegenerative Diseases. *Cell. Mol. Neurobiol.* **2016**, *36*, 181–194. [CrossRef]
23. Engelhardt, B.; Carare, R.O.; Bechmann, I.; Flügel, A.; Laman, J.D.; Weller, R.O. Vascular, glial, and lymphatic immune gateways of the central nervous system. *Acta Neuropathol.* **2016**, *132*, 317–338. [CrossRef] [PubMed]
24. Carare, R.O.; Bernardes-Silva, M.; Newman, T.A.; Page, A.M.; Nicoll, J.A.R.; Perry, V.H.; Weller, R.O. Solutes, but not cells, drain from the brain parenchyma along basement membranes of capillaries and arteries: Significance for cerebral amyloid angiopathy and neuroimmunology. *Neuropathol. Appl. Neurobiol.* **2008**, *34*, 131–144. [CrossRef] [PubMed]
25. Mestre, H.; Tithof, J.; Du, T.; Song, W.; Peng, W.; Sweeney, A.M.; Olveda, G.; Thomas, J.H.; Nedergaard, M.; Kelley, D.H. Flow of cerebrospinal fluid is driven by arterial pulsations and is reduced in hypertension. *Nat. Commun.* **2018**, *9*, 1–9. [CrossRef] [PubMed]
26. Silva, I.; Silva, J.; Ferreira, R.; Trigo, D. Glymphatic system, AQP4, and their implications in Alzheimer's disease. *Neurol. Res. Pract.* **2021**, *3*, 1–9. [CrossRef] [PubMed]
27. Zhou, W.; Shen, B.; Shen, W.-Q.; Chen, H.; Zheng, Y.-F.; Fei, J.-J. Dysfunction of the Glymphatic System Might Be Related to Iron Deposition in the Normal Aging Brain. *Front. Aging Neurosci.* **2020**, *12*, 445. [CrossRef] [PubMed]
28. Rasmussen, M.K.; Mestre, H.; Nedergaard, M. The glymphatic pathway in neurological disorders. *Lancet Neurol.* **2018**, *17*, 1016–1024. [CrossRef]
29. Chen, J.; Wang, L.; Xu, H.; Wang, Y.; Liang, Q. The lymphatic drainage system of the CNS plays a role in lymphatic drain-age, immunity, and neuroinflammation in stroke. *J. Leukoc. Biol.* **2021**, *110*, 283–291. [CrossRef] [PubMed]
30. Wang, Q.; Sawyer, I.A.; Sung, M.-H.; Sturgill, D.; Shevtsov, S.P.; Pegoraro, G.; Hakim, O.; Baek, S.; Hager, G.L.; Dundr, M. Cajal bodies are linked to genome conformation. *Nat. Commun.* **2016**, *7*, 10966. [CrossRef] [PubMed]
31. Thrane, V.R.; Thrane, A.S.; Plog, B.A.; Thiyagarajan, M.; Iliff, J.J.; Deane, R.; Nagelhus, E.A.; Nedergaard, M. Paravascular microcirculation facilitates rapid lipid transport and astrocyte signaling in the brain. *Sci. Rep.* **2013**, *3*, srep02582. [CrossRef]
32. Achariyar, T.M.; Li, B.; Peng, W.; Verghese, P.B.; Shi, Y.; McConnell, E.; Benraiss, A.; Kasper, T.; Song, W.; Takano, T.; et al. Glymphatic distribution of CSF-derived apoE into brain is isoform specific and suppressed during sleep deprivation. *Mol. Neurodegener.* **2016**, *11*, 1–20. [CrossRef]
33. Kylkilahti, T.M.; Berends, E.; Ramos, M.; Shanbhag, N.C.; Töger, J.; Bloch, K.M.; Lundgaard, I. Achieving brain clearance and preventing neurodegenerative diseases—A glymphatic perspective. *Br. J. Pharmacol.* **2021**, *41*, 2137–2149. [CrossRef] [PubMed]
34. Baranello, R.J.; Bharani, K.L.; Padmaraju, V.; Chopra, N.; Lahiri, D.K.; Greig, N.H.; Pappolla, M.A.; Sambamurti, K. Amy-loid-beta protein clearance and degradation (ABCD) pathways and their role in Alzheimer's disease. *Curr. Alzheimer Res.* **2015**, *12*, 32–46. [CrossRef] [PubMed]
35. Gupta, A.; Iadecola, C. Impaired Aβ clearance: A potential link between atherosclerosis and Alzheimer's disease. *Front. Aging Neurosci.* **2015**, *7*, 115. [CrossRef] [PubMed]
36. Nedergaard, M.; Goldman, S.A. Glymphatic failure as a final common pathway to dementia. *Science* **2020**, *370*, 50–56. [CrossRef] [PubMed]
37. Lord, C.C.; Wyler, S.C.; Wan, R.; Castorena, C.M.; Ahmed, N.; Mathew, D.; Lee, S.; Liu, C.; Elmquist, J.K. The atypical antipsychotic olanzapine causes weight gain by targeting serotonin receptor 2C. *J. Clin. Investig.* **2017**, *127*, 3402–3406. [CrossRef] [PubMed]
38. Ahn, J.H.; Cho, H.; Kim, J.-H.; Kim, S.H.; Ham, J.-S.; Park, I.; Suh, S.H.; Hong, S.P.; Song, J.-H.; Hong, Y.-K.; et al. Meningeal lymphatic vessels at the skull base drain cerebrospinal fluid. *Nature* **2019**, *572*, 62–66. [CrossRef] [PubMed]
39. Louveau, A.; Herz, J.; Alme, M.N.; Salvador, A.F.; Dong, M.Q.; Viar, K.E.; Herod, S.G.; Knopp, J.; Setliff, J.C.; Lupi, A.; et al. CNS lymphatic drainage and neuroinflammation are regulated by meningeal lymphatic vasculature. *Nat. Neurosci.* **2018**, *21*, 1380–1391. [CrossRef] [PubMed]
40. King, H.H. Lymphatic Vessels Found in the Brain—Osteopathic Considerations. *J. Am. Osteopat. Assoc.* **2015**, *115*, 627. [CrossRef]
41. Dupont, G.; Iwanaga, J.; Yilmaz, E.; Tubbs, R.S. Connections between amyloid beta and the meningeal lymphatics as a possi-ble route for clearance and therapeutics. Lymphat. *Res. Biol.* **2020**, *18*, 2–6.

42. Wang, X.; Tian, H.; Liu, H.; Liang, D.; Qin, C.; Zhu, Q.; Wang, X. Impaired Meningeal Lymphatic Flow in NMOSD Patients with Acute Attack. *Front. Immunol.* **2021**, *12*, 2239. [CrossRef]
43. Ding, X.-B.; Wang, X.-X.; Xia, D.-H.; Liu, H.; Tian, H.-Y.; Fu, Y.; Chen, Y.-K.; Qin, C.; Wang, J.-Q.; Xiang, Z.; et al. Impaired meningeal lymphatic drainage in patients with idiopathic Parkinson's disease. *Nat. Med.* **2021**, *27*, 411–418. [CrossRef] [PubMed]
44. Ghandili, M.; Munakomi, S. Neuroanatomy, Putamen. StatPearls [Internet]. 2021. Available online: https://www.ncbi.nlm.nih.gov/books/NBK542170/ (accessed on 25 December 2021).
45. Tripathi, R. Tracing the bulk outflow route of cerebrospinal fluid by transmission and scanning electron microscopy. *Brain Res.* **1974**, *80*, 503–506. [CrossRef]
46. Kida, S.; Pantazis, A.; Weller, R.O. CSF drains directly from the subarachnoid space into nasal lymphatics in the rat. Anatomy, histology and immunological significance. *Neuropathol. Appl. Neurobiol.* **1993**, *19*, 480–488. [CrossRef] [PubMed]
47. De Leon, M.J.; Li, Y.; Okamura, N.; Tsui, W.H.; Saint-Louis, L.A.; Glodzik, L.; Osorio, R.; Fortea, J.; Butler, T.; Pirraglia, E.; et al. Cerebrospinal Fluid Clearance in Alzheimer Disease Measured with Dynamic PET. *J. Nucl. Med.* **2017**, *58*, 1471–1476. [CrossRef] [PubMed]
48. Ma, Q.; Ries, M.; Decker, Y.; Müller, A.; Riner, C.; Bücker, A.; Fassbender, K.; Detmar, M.; Proulx, S.T. Rapid lymphatic efflux limits cerebrospinal fluid flow to the brain. *Acta Neuropathol.* **2019**, *137*, 151–165. [CrossRef] [PubMed]
49. Norwood, J.N.; Zhang, Q.; Card, D.; Craine, A.; Ryan, T.M.; Drew, P.J. Anatomical basis and physiological role of cerebro-spinal fluid transport through the murine cribriform plate. *elife* **2019**, *8*, e44278. [CrossRef] [PubMed]
50. Wostyn, P. COVID-19 and chronic fatigue syndrome: Is the worst yet to come? *Med. Hypotheses* **2021**, *146*, 110469. [CrossRef] [PubMed]
51. Rash, J.E.; Davidson, K.G.V.; Kamasawa, N.; Yasumura, T.; Kamasawa, M.; Zhang, C.; Michaels, R.; Restrepo, D.; Ottersen, O.P.; Olson, C.O.; et al. Ultrastructural localization of connexins (Cx36, Cx43, Cx45), glutamate receptors and aquaporin-4 in rodent olfactory mucosa, olfactory nerve and olfactory bulb. *J. Neurocytol.* **2005**, *34*, 307–341. [CrossRef]
52. Brady, M.; Rahman, A.; Combs, A.; Venkatraman, C.; Kasper, R.T.; McQuaid, C.; Kwok, W.-C.E.; Wood, R.W.; Deane, R. Cerebrospinal fluid drainage kinetics across the cribriform plate are reduced with aging. *Fluids Barriers CNS* **2020**, *17*, 1–16. [CrossRef] [PubMed]
53. Ma, Q.; Ineichen, B.V.; Detmar, M.; Proulx, S.T. Outflow of cerebrospinal fluid is predominantly through lymphatic vessels and is reduced in aged mice. *Nat. Commun.* **2017**, *8*, 1–13. [CrossRef] [PubMed]
54. Hauglund, N.L.; Pavan, C.; Nedergaard, M. Cleaning the sleeping brain–the potential restorative function of the glymphatic system. *Curr. Opin. Physiol.* **2020**, *15*, 1–6. [CrossRef]
55. Benveniste, H.; Heerdt, P.M.; Fontes, M.; Rothman, D.L.; Volkow, N.D. Glymphatic System Function in Relation to Anesthesia and Sleep States. *Anesthesia Analg.* **2019**, *128*, 747–758. [CrossRef] [PubMed]
56. Fultz, N.E.; Bonmassar, G.; Setsompop, K.; Stickgold, R.A.; Rosen, B.R.; Polimeni, J.R.; Lewis, L.D. Coupled electrophysiological, hemodynamic, and cerebrospinal fluid oscillations in human sleep. *Science* **2019**, *366*, 628–631. [CrossRef] [PubMed]
57. Xie, L.; Kang, H.; Xu, Q.; Chen, M.J.; Liao, Y.; Thiyagarajan, M.; O'Donnell, J.; Christensen, D.J.; Nicholson, C.; Iliff, J.J.; et al. Sleep Drives Metabolite Clearance from the Adult Brain. *Science* **2013**, *342*, 373–377. [CrossRef] [PubMed]
58. Reddy, O.C.; Van Der Werf, Y.D. The sleeping brain: Harnessing the power of the glymphatic system through lifestyle choic-es. *Brain Sci.* **2020**, *10*, 868. [CrossRef] [PubMed]
59. Semyachkina-Glushkovskaya, O.; Abdurashitov, A.; Dubrovsky, A.; Klimova, M.; Agranovich, I.; Terskov, A.; Shirokov, A.; Vinnik, V.; Kuzmina, A.; Lezhnev, N.; et al. Photobiomodulation of lymphatic drainage and clearance: Perspective strategy for augmentation of meningeal lymphatic functions. *Biomed. Opt. Express* **2020**, *11*, 725–734. [CrossRef] [PubMed]
60. Holth, J.K.; Fritschi, S.K.; Wang, C.; Pedersen, N.P.; Cirrito, J.R.; Mahan, T.E.; Finn, M.B.; Manis, M.; Geerling, J.C.; Fuller, P.M.; et al. The sleep-wake cycle regulates brain interstitial fluid tau in mice and CSF tau in humans. *Science* **2019**, *363*, 880–884. [CrossRef] [PubMed]
61. Raggi, A.; Ferri, R. Sleep disorders in neurodegenerative diseases. *Eur. J. Neurol.* **2010**, *17*, 1326–1338. [CrossRef] [PubMed]
62. Cuddapah, V.; Zhang, S.L.; Sehgal, A. Regulation of the Blood–Brain Barrier by Circadian Rhythms and Sleep. *Trends Neurosci.* **2019**, *42*, 500–510. [CrossRef] [PubMed]
63. Yehuda, S.; Sredni, B.; Carasso, R.L.; Kenigsbuch-Sredni, D. REM sleep deprivation in rats results in inflammation and interleukin-17 elevation. *J. Interferon Cytokine Res.* **2009**, *29*, 393–398. [CrossRef] [PubMed]
64. Hurtado-Alvarado, G.; Becerril-Villanueva, E.; De Oca, A.C.-M.; Domínguez-Salazar, E.; Salinas-Jazmín, N.; Pérez-Tapia, S.; Pavon, L.; Velázquez-Moctezuma, J.; Gómez-González, B. The yin/yang of inflammatory status: Blood-brain barrier regula-tion during sleep. *Brain Behav. Immun.* **2018**, *69*, 154–166. [CrossRef] [PubMed]
65. Liu, D.-X.; He, X.; Wu, D.; Zhang, Q.; Yang, C.; Liang, F.-Y.; He, X.-F.; Dai, G.-Y.; Pei, Z.; Lan, Y.; et al. Continuous theta burst stimulation facilitates the clearance efficiency of the glymphatic pathway in a mouse model of sleep deprivation. *Neurosci. Lett.* **2017**, *653*, 189–194. [CrossRef] [PubMed]
66. Hamblin, M.R.; Carroll, J.D.; De Freitas, L.F.; Huang, Y.-Y.; Ferraresi, C. *Low-Level Light Therapy: Photobiomodulation*; SPIE: Bellingham, WA, USA, 2018.
67. Hamblin, M.R.; Agrawal, T.; de Sousa, M. *Handbook of Low-Level Laser Therapy*; Jenny Stanford Publishing: New York, NY, USA, 2016. [CrossRef]

68. Karu, T.I. Mitochondrial Signaling in Mammalian Cells Activated by Red and Near-IR Radiation. *Photochem. Photobiol.* **2008**, *84*, 1091–1099. [CrossRef] [PubMed]
69. Karu, T. Photobiology of Low-power Laser Effects. *Health Phys.* **1989**, *56*, 691–704. [CrossRef] [PubMed]
70. Karu, T.; Kolyakov, S. Exact Action Spectra for Cellular Responses Relevant to Phototherapy. *Photomed. Laser Surg.* **2005**, *23*, 355–361. [CrossRef] [PubMed]
71. Wong-Riley, M.T.; Liang, H.L.; Eells, J.T.; Chance, B.; Henry, M.M.; Buchmann, E.; Kane, M.; Whelan, H.T. Photobiomodula-tion directly benefits primary neurons functionally inactivated by toxins: Role of cytochrome c oxidase. *J. Biol. Chem.* **2005**, *280*, 4761–4771. [CrossRef] [PubMed]
72. Santana-Blank, L.; Rodríguez-Santana, E.; Santana-Rodriguez, K. Theoretic, experimental, clinical bases of the water oscilla-tor hypothesis in near-infrared photobiomodulation. *Photomed. Laser Surg.* **2010**, *28*, S41–S52. [CrossRef] [PubMed]
73. Karu, T.I. Cellular and Molecular Mechanisms of Photobiomodulation (Low-Power Laser Therapy). *IEEE J. Sel. Top. Quantum Electron.* **2013**, *20*, 143–148. [CrossRef]
74. Ganeshan, V.; Skladnev, N.V.; Kim, J.Y.; Mitrofanis, J.; Stone, J.; Johnstone, D.M. Pre-conditioning with remote photobio-modulation modulates the brain transcriptome and protects against MPTP insult in mice. *Neuroscience* **2019**, *400*, 85–97. [CrossRef]
75. George, S.; Hamblin, M.R.; Abrahamse, H. Photobiomodulation-Induced Differentiation of Immortalized Adipose Stem Cells to Neuronal Cells. *Lasers Surg. Med.* **2020**, *52*, 1032–1040. [CrossRef]
76. Chung, H.; Dai, T.; Sharma, S.K.; Huang, Y.-Y.; Carroll, J.D.; Hamblin, M.R. The nuts and bolts of low-level laser (light) ther-apy. *Ann. Biomed. Eng.* **2012**, *40*, 516–533. [CrossRef] [PubMed]
77. Ramezani, F.; Neshasteh-Riz, A.; Ghadaksaz, A.; Fazeli, S.M.; Janzadeh, A.; Hamblin, M.R. Mechanistic aspects of photobi-omodulation therapy in the nervous system. *Lasers Med. Sci.* **2021**, 1–8, *Epub ahead of print*. [CrossRef]
78. Salehpour, F.; Gholipour-Khalili, S.; Farajdokht, F.; Kamari, F.; Walski, T.; Hamblin, M.R.; DiDuro, J.O.; Cassano, P. Thera-peutic potential of intranasal photobiomodulation therapy for neurological and neuropsychiatric disorders: A narrative re-view. *Rev. Neurosci.* **2020**, *31*, 269–286. [CrossRef] [PubMed]
79. Caldieraro, M.A.; Sani, G.; Bui, E.; Cassano, P. Long-term near-infrared photobiomodulation for anxious depression compli-cated by Takotsubo cardiomyopathy. *J. Clin. Psychopharmacol.* **2018**, *38*, 268–270. [CrossRef] [PubMed]
80. Xiao, X.; Guo, Y.; Chu, X.; Jia, S.; Zheng, X.; Zhou, C. Effects of low power laser irradiation in nasal cavity on cerebral blood flow perfusion of patients with brain infarction. *Chin. J. Rehabil. Med.* **2003**, *12*, wpr-573842.
81. Hennessy, M.; Hamblin, M.R. Photobiomodulation and the brain: A new paradigm. *J. Opt.* **2017**, *19*, 013003. [CrossRef] [PubMed]
82. Hamper, M.; Cassano, P.; Lombard, J. Treatment of Kleine-Levin Syndrome with Intranasal Photobiomodulation and Meth-ylene Blue. *Cureus* **2021**, *13*, e18596. [CrossRef] [PubMed]
83. Salehpour, F.; Hamblin, M.R.; DiDuro, J.O. Rapid Reversal of Cognitive Decline, Olfactory Dysfunction, and Quality of Life Using Multi-Modality Photobiomodulation Therapy: Case Report. *Photobiomodul. Photomed. Laser Surg.* **2019**, *37*, 159–167. [CrossRef] [PubMed]
84. Semyachkina-Glushkovskaya, O.; Klimova, M.; Iskra, T.; Bragin, D.; Abdurashitov, A.; Dubrovsky, A.; Khorovodov, A.; Terskov, A.; Blokhina, I.; Lezhnev, N.; et al. Transcranial Photobiomodulation of Clearance of Beta-Amyloid from the Mouse Brain: Effects on the Meningeal Lymphatic Drainage and Blood Oxygen Saturation of the Brain. *Adv. Exp. Med. Biol.* **2021**, *1269*, 57–61. [CrossRef] [PubMed]
85. Semyachkina-Glushkovskaya, O.; Fedosov, I.; Shirokov, A.; Vodovozova, E.; Alekseeva, A.; Khorovodov, A.; Blokhina, I.; Terskov, A.; Mamedova, A.; Klimova, M.; et al. Photomodulation of lymphatic delivery of liposomes to the brain bypassing the blood-brain barrier: New perspectives for glioma therapy. *Nanophotonics* **2021**, *10*, 3215–3227. [CrossRef]
86. Li, D.; Liu, S.; Yu, T.; Liu, Z.; Sun, S.; Bragin, D.; Shirokov, A.; Navolokin, N.; Bragina, O.; Kurths, J.; et al. Noninvasive technology of photostimulation of lymphatic clearance of red blood cells from the mouse brain after intraventricular hemorrhage. *Res. Sq.* **2021**, 1–24. [CrossRef]
87. Semyachkina-Glushkovskaya, O.; Abdurashitov, A.; Klimova, M.; Dubrovsky, A.; Shirokov, A.; Fomin, A.; Terskov, A.; Agranovich, I.; Mamedova, A.; Khorovodov, A.; et al. Photostimulation of cerebral and peripheral lymphatic functions. *Transl. Biophotonics* **2020**, *2*, 201900036. [CrossRef]
88. Semyachkina-Glushkovskaya, O.; Postnov, D.; Lavrova, A.; Fedosov, I.; Borisova, E.; Nikolenko, V.; Penzel, T.; Kurths, J.; Tuchin, V. Biophotonic Strategies of Measurement and Stimulation of the Cranial and the Extracranial Lymphatic Drainage Function. *IEEE J. Sel. Top. Quantum Electron.* **2020**, *27*, 1–13. [CrossRef]
89. Saucedo, C.L.; Courtois, E.C.; Wade, Z.S.; Kelley, M.N.; Kheradbin, N.; Barrett, D.W.; Gonzalez-Lima, F. Transcranial laser stimulation: Mitochondrial and cerebrovascular effects in younger and older healthy adults. *Brain Stimul.* **2021**, *14*, 440–449. [CrossRef] [PubMed]
90. Uozumi, Y.; Nawashiro, H.; Sato, S.; Kawauchi, S.; Shima, K.; Kikuchi, M. Targeted increase in cerebral blood flow by tran-scranial near-infrared laser irradiation. *Lasers Surg. Med.* **2010**, *42*, 566–576. [CrossRef] [PubMed]
91. Borghammer, P.; Cumming, P.; Østergaard, K.; Gjedde, A.; Rodell, A.; Bailey, C.J.; Vafaee, M.S. Cerebral oxygen metabolism in patients with early Parkinson's disease. *J. Neurol. Sci.* **2012**, *313*, 123–128. [CrossRef] [PubMed]
92. Litscher, G.; Min, L.; Passegger, C.A.; Litscher, D.; Li, M.; Wang, M.; Ghaffari-Tabrizi-Wizsy, N.; Stelzer, I.; Feigl, G.; Gaischek, I.; et al. Transcranial Yellow, Red, and Infrared Laser and LED Stimulation: Changes of Vascular Parameters in a Chick Embryo Model. *Integr. Med. Int.* **2015**, *2*, 80–89. [CrossRef]

93. Colombo, E.; Signore, A.; Aicardi, S.; Zekiy, A.; Utyuzh, A.; Benedicenti, S.; Amaroli, A. Experimental and clinical applica-tions of red and near-infrared photobiomodulation on endothelial dysfunction: A Review. *Biomedicines* **2021**, *9*, 274. [CrossRef] [PubMed]
94. Mitchell, U.H.; Mack, G.L. Low-level laser treatment with near-infrared light increases venous nitric oxide levels acutely: A single-blind, randomized clinical trial of efficacy. *Am. J. Phys. Med. Rehabil.* **2013**, *92*, 151–156. [CrossRef] [PubMed]
95. Weihrauch, D.; Keszler, A.; Lindemer, B.; Krolikowski, J.; Lohr, N.L. Red light stimulates vasodilation through extracellular vesicle trafficking. *J. Photochem. Photobiol. B Biol.* **2021**, *220*, 112212. [CrossRef]
96. Lubart, R.; Eichler, M.; Lavi, R.; Friedman, H.; Shainberg, A. Low-Energy Laser Irradiation Promotes Cellular Redox Activity. *Photomed. Laser Surg.* **2005**, *23*, 3–9. [CrossRef] [PubMed]
97. Hamblin, M. The role of nitric oxide in low level light therapy. In *Mechanisms for Low-Light Therapy III*; International Society for Optics and Photonics: Bellingham, WA, USA, 2008; Volume 6846, p. 684602.
98. Scallan, J.P.; Zawieja, S.D.; Castorena-Gonzalez, J.; Davis, M.J. Lymphatic pumping: Mechanics, mechanisms and malfunction. *J. Physiol.* **2016**, *594*, 5749–5768. [CrossRef] [PubMed]
99. Liang, J.; Liu, L.; Xing, D. Photobiomodulation by low-power laser irradiation attenuates Aβ-induced cell apoptosis through the Akt/GSK3β/β-catenin pathway. *Free Radic. Biol. Med.* **2012**, *53*, 1459–1467. [CrossRef] [PubMed]
100. Meng, C.; He, Z.; Xing, D. Low-Level Laser Therapy Rescues Dendrite Atrophy via Upregulating BDNF Expression: Implications for Alzheimer's Disease. *J. Neurosci.* **2013**, *33*, 13505–13517. [CrossRef] [PubMed]
101. Duggett, N.A.; Chazot, P.L. Low-Intensity Light Therapy (1068 nm) Protects CAD Neuroblastoma Cells from β-Amyloid-Mediated Cell Death. *Biol. Med.* **2014**, *1*, 2.
102. Zhang, Z.; Shen, Q.; Wu, X.; Zhang, D.; Xing, D. Activation of PKA/SIRT1 signaling pathway by photobiomodulation thera-py reduces Aβ levels in Alzheimer's disease models. *Aging Cell* **2020**, *19*, e13054. [CrossRef] [PubMed]
103. Tao, L.; Liu, Q.; Zhang, F.; Fu, Y.; Zhu, X.; Weng, X.; Wei, X. Microglia modulation with 1070-nm light attenuates Aβ burden and cognitive impairment in Alzheimer's dis-ease mouse model. *Light Sci. Appl.* **2021**, *10*, 1–14. [CrossRef] [PubMed]
104. Yue, X.; Mei, Y.; Zhang, Y.; Tong, Z.; Cui, D.; Yang, J.; Wang, A.; Wang, R.; Fei, X.; Ai, L.; et al. New insight into Alzheimer's disease: Light reverses Aβ-obstructed interstitial fluid flow and ameliorates memory decline in APP/PS1 mice. *Alzheimer's Dement. Transl. Res. Clin. Interv.* **2019**, *5*, 671–684. [CrossRef] [PubMed]
105. Liu, T.C.-Y.; Cheng, L.; Su, W.-J.; Zhang, Y.-W.; Shi, Y.; Liu, A.-H.; Zhang, L.-L.; Qian, Z.-Y. Randomized, Double-Blind, and Placebo-Controlled Clinic Report of Intranasal Low-Intensity Laser Therapy on Vascular Diseases. *Int. J. Photoenergy* **2012**, *2012*, 1–5. [CrossRef]
106. Gao, X.; Zhi, P.-K.; Wu, X.-J. Low-energy semiconductor laser intranasal irradiation of the blood improves blood coagulation status in normal pregnancy at term. *J. South. Med Univ.* **2008**, *28*, 1400–1401.
107. Elwood, P.; Pickering, J.; Gallacher, J.E.J. Cognitive function and blood rheology: Results from the Caerphilly cohort of older men. *Age Ageing* **2001**, *30*, 135–139. [CrossRef] [PubMed]
108. Gao, Z.; Zhang, L.; Qin, C. The relationship between hemorheological changes and the anxiety and depression symptoms in schizophrenia. *Clin. J. Hemorheol.* **2004**, 1.
109. Chakraborty, S.; Davis, M.J.; Muthuchamy, M. Emerging trends in the pathophysiology of lymphatic contractile function. *Semin. Cell Dev. Biol.* **2015**, *38*, 55–66. [CrossRef] [PubMed]
110. Johnstone, D.M.; Gordon, L.C. Remote photobiomodulation: An emerging strategy for neuroprotection. *Neural Regen. Res.* **2019**, *14*, 2086–2087. [CrossRef] [PubMed]
111. Caldieraro, M.A.; Salehpour, F.; Cassano, P. *Transcranial and Systemic Photobiomodulation for the Enhancement of Mito-Chondrial Metabolism in Depression, in Clinical Bioenergetics*; Elsevier: Amsterdam, The Netherlands, 2021; pp. 635–651. [CrossRef]
112. Farfara, D.; Tuby, H.; Trudler, D.; Doron-Mandel, E.; Maltz, L.; Vassar, R.J.; Frenkel, D.; Oron, U. Low-Level Laser Therapy Ameliorates Disease Progression in a Mouse Model of Alzheimer's Disease. *J. Mol. Neurosci.* **2015**, *55*, 430–436. [CrossRef] [PubMed]
113. Blivet, G.; Meunier, J.; Roman, F.J.; Touchon, J. Neuroprotective effect of a new photobiomodulation technique against Aβ 25–35 peptide–induced toxicity in mice: Novel hypothesis for therapeutic approach of Alzheimer's disease suggested. *Alzheimer's Dement. Transl. Res. Clin. Interv.* **2018**, *4*, 54–63. [CrossRef] [PubMed]
114. Kim, B.; Mitrofanis, J.; Stone, J.; Johnstone, D.M. Remote tissue conditioning is neuroprotective against MPTP insult in mice. *IBRO Rep.* **2018**, *4*, 14–17. [CrossRef] [PubMed]
115. Johnstone, D.M.; El Massri, N.; Moro, C.; Spana, S.; Wang, X.S.; Torres, N.; Mitrofanis, J. Indirect application of near infrared light induces neuroprotection in a mouse model of parkinsonism–an abscopal neuroprotective effect. *Neuroscience* **2014**, *274*, 93–101. [CrossRef]
116. Tuby, H.; Maltz, L.; Oron, U. Implantation of Low-Level Laser Irradiated Mesenchymal Stem Cells into the Infarcted Rat Heart Is Associated with Reduction in Infarct Size and Enhanced Angiogenesis. *Photomed. Laser Surg.* **2009**, *27*, 227–233. [CrossRef]
117. Tuby, H.; Maltz, L.; Oron, U. Induction of autologous mesenchymal stem cells in the bone marrow by low-level laser therapy has profound beneficial effects on the infarcted rat heart. *Lasers Surg. Med.* **2011**, *43*, 401–409. [CrossRef]
118. Matsushita, T.; Kibayashi, T.; Katayama, T.; Yamashita, Y.; Suzuki, S.; Kawamata, J.; Honmou, O.; Minami, M.; Shimohama, S. Mesenchymal stem cells transmigrate across brain microvascular endothelial cell monolayers through transiently formed inter-endothelial gaps. *Neurosci. Lett.* **2011**, *502*, 41–45. [CrossRef] [PubMed]

119. Salehpour, F.; Khademi, M.; Hamblin, M.R. Photobiomodulation Therapy for Dementia: A Systematic Review of Pre-Clinical and Clinical Studies. *J. Alzheimer's Dis.* **2021**, *83*, 1431–1452. [CrossRef] [PubMed]
120. Berman, M.H.; Nichols, T.W. Treatment of Neurodegeneration: Integrating Photobiomodulation and Neurofeedback in Alzheimer's Dementia and Parkinson's: A Review. *Photobiomodul. Photomed. Laser Surg.* **2019**, *37*, 623–634. [CrossRef] [PubMed]

International Journal of
Molecular Sciences

MDPI

Review

Exploitation of the Antibacterial Properties of Photoactivated Curcumin as 'Green' Tool for Food Preservation

Zunaira Munir [1], Giuliana Banche [2], Lorenza Cavallo [2], Narcisa Mandras [2,*], Janira Roana [2], Raffaele Pertusio [1], Eleonora Ficiarà [1,*], Roberta Cavalli [3] and Caterina Guiot [1]

[1] Department of Neurosciences, University of Turin, 10124 Turin, Italy; zunaira.munir@unito.it (Z.M.); raffaele.pertusio@unito.it (R.P.); caterina.guiot@unito.it (C.G.)
[2] Bacteriology and Mycology Laboratory, Department of Public Health and Pediatric Science, University of Torino, Via Santena 9, 10126 Turin, Italy; giuliana.banche@unito.it (G.B.); lorenza.cavallo@unito.it (L.C.); janira.roana@unito.it (J.R.)
[3] Department of Drug Science and Technology, University of Turin, 10125 Turin, Italy; roberta.cavalli@unito.it
* Correspondence: narcisa.mandras@unito.it (N.M.); eleonora.ficiara@unito.it (E.F.)

Abstract: In the search for non-chemical and green methods to counteract the bacterial contamination of foods, the use of natural substances with antimicrobial properties and light irradiation at proper light waves has been extensively investigated. In particular, the combination of both techniques, called photodynamic inactivation (PDI), is based on the fact that some natural substances act as photosensitizers, i.e., produce bioactive effects under irradiation. Notably, curcumin is a potent natural antibacterial and effective photosensitizer that is able to induce photodynamic activation in the visible light range (specifically for blue light). Some practical applications have been investigated with particular reference to food preservation from bacterial contaminants.

Keywords: curcumin; photoactivation; food preservation; antibacterial properties

Citation: Munir, Z.; Banche, G.; Cavallo, L.; Mandras, N.; Roana, J.; Pertusio, R.; Ficiarà, E.; Cavalli, R.; Guiot, C. Exploitation of the Antibacterial Properties of Photoactivated Curcumin as 'Green' Tool for Food Preservation. *Int. J. Mol. Sci.* **2022**, *23*, 2600. https://doi.org/10.3390/ijms23052600

Academic Editors: Antonino Mazzaglia, Angela Scala and Enrico Caruso

Received: 25 January 2022
Accepted: 23 February 2022
Published: 26 February 2022

Publisher's Note: MDPI stays neutral with regard to jurisdictional claims in published maps and institutional affiliations.

1. Introduction

The concept of food preservation has its origins in ancient times when our ancestors tried to find methods to keep food fresh and edible (e.g., sun-drying, salting and pasteurization). The possibility of optimally preserving food for long periods is increasingly relevant: if food products are not subjected to treatments that ensure their proper preservation and decontamination by microorganisms, they are subject to more rapid deterioration and can be responsible for serious foodborne infections [1]. Classical food preservation techniques rely on physical and physicochemical action treatments, such as heat, freezing, radiation, and ultrasound [2]. However, some disadvantages have been discovered in these procedures, such as reductions in food volume and texture and losses of nutrients and organic properties, leading to huge overall losses in food products [3].

Therefore, scientific and technological researchers have been increasingly interested in the study of new antimicrobial materials and new food sanitization methods that can extend shelf-lives while ensuring their safety for humans. Thanks to their broad spectrum of activity, natural antimicrobials of plant origin are promising candidates to prevent food contamination by harmful microorganisms.

Light irradiation is another interesting 'green' approach. It has been extensively used as a food decontaminant via X-rays, ultrasound, UV, and visible light, among others. A detailed description can be found in the work of Cossu et al. [4].

More recently, light-emitting diodes (LED) in the range of 400–460 nm were found to very effective against bacteria. The mechanism that is assumed to play a major role in their effectiveness is the excitation of endogenous porphyrins that catalyze the production of intracellular reactive oxygen species (ROS), inducing cell oxidation and death [5]. The different responses of Gram-positive and -negative bacteria could be therefore explained by the different contents of porphyrins. To produce effective results in some food matrices, especially fruits and juices, LED exposition should last several hours unless photosensitizers (PS) are used. PS are non-toxic molecules that interact with irradiating light by generating radical species responsible for massive and aggressive oxidation, finally resulting in cell lysis and death. This effect is enhanced by the presence of molecular oxygen [6].

Both natural antimicrobial properties and PS are shared by curcumin, a natural component of *Curcuma longa*, and its antimicrobial properties against a wide variety of microorganisms, including bacteria and fungi, have been demonstrated in vitro [7]. Curcumin, in addition to exhibiting remarkable antioxidant, anti-inflammatory, and anticancer activities [8], benefits from a high availability on the market, safety even at high doses, and low cost [9]. In addition, due to its ability to emit fluorescence, its antimicrobial efficacy can be enhanced by photoactivation.

Curcumin is one of the strongest natural PS, with a light absorption peak between 400 and 500 nm suitable for excitation by blue light. Moreover, the irradiation breaks down the curcumin molecule, thus limiting its presence in the decontaminated final product [10].

However, the potential of this natural extract is limited by its poor water solubility and instability at physiological pH, leading to poor absorption and rapid degradation by hydrolysis and molecular fragmentation [7].

2. Principal Microbial Contaminant of Food

Foods responsible for illness, hospitalization, and death include meat, poultry, dairy, fruit, vegetables, seafood, grains, and nuts. Pathogenic microorganisms, particularly bacteria or fungi, can contaminate food at different stages such as during production and processing, as well as during the storage and transportation of the final product. In the course of replication, some contaminating bacteria can produce virulence factors and harmful substances such as toxins, which are involved in the pathogenesis of various infections [11].

The bacterium responsible for most foodborne infections is *Staphylococcus aureus*, a Gram-positive, commensal, and opportunistic pathogenic microorganism that can cause a wide spectrum of infections. *S. aureus* is known to be resistant to several chemical and physical factors: in fact, it grows in environments with a temperature ranging from 7 to 48.5 °C (optimal: 30–37 °C), a pH from 4.2 to 9.3 (optimal: 7–7.5), and a sodium chloride (NaCl) concentration of up to 15%.

Another etiologic agent of foodborne illness is *Escherichia coli*, specifically the enterohemorrhagic serogroup O157:H7. This Gram-negative species harmlessly colonizes the gastrointestinal tracts of humans and animals. However, there are strains that have become pathogenic because they have acquired virulence factors through plasmids, transposons, bacteriophages, and/or islands of pathogenicity. Unfortunately, there is no specific treatment for *E. coli* O157:H7 infection, and the use of antibiotics may be discouraged; consequently, the most widely used approach is supportive therapy to limit symptoms, halt disease evolution, and prevent systemic inconveniences [12].

Another category of bacteria that has become very important in recent times is represented by enterococci, which are very resistant by nature and able to survive, even for long periods of time, in a wide range of hostile environmental conditions. Enterococci are common commensals of the gastrointestinal tract, play important roles in food maturation and the development of specific flavors (such as in various cheeses), and can cause spoilage in some meats. However, these Gram-positive bacteria are the etiological agents of several nosocomial infections and can occur as food contaminants. Among them, the predominant species is *Enterococcus faecalis*, a Gram-positive bacterium often associated with several pathological conditions including urinary tract infections, bacteremia, meningitis, wound infections, dental diseases, and neonatal infections [13].

Further details on their characteristics, including how their infections suffer from the development of antibiotic resistance (which is one of the most severe and harmful modern-day health problems) are provided in Table 1.

Table 1. Some of the main bacteria contaminant of food.

Bacteria	Gram	Contaminate Food	Health Damage Mechanism	Antibiotic Resistance
E. faecalis	+	Cheeses, fermented sausages, andready-to-eat food.	Enterococcal surface proteins, hyaluronidase, gelatinase, and biofilm production [14].	The most prevalent antibiotic resistances were found to be tetracycline, minocycline, erythromycin kanamycin, and chloramphenicol [15].
S. aureus	+	Fruits, meat, egg products, milk and its derivatives, salads, and baked goods such as pastries and cream-filled desserts.	Staphylococcal enterotoxins (SEs) causing Staphylococcal foodborne disease (SFD) [16,17].	Foodborne strain methicillin-resistant *S. aureus* (MRSA) with the highest resistance rate to penicillin G, ampicillin, and erythromycin [18,19].
E. coli (specifically the enterohemor-rhagic serogroup O157:H7)	-	Dairy products, delicatessen products, salads, spices, cream cakes, and fresh fruit and vegetables.	Shiga toxins, pathogenicity island products, and F-like plasmid pO157 products.	Colistin had the lowest prevalence (0.8%) and amoxicillin had the highest (70.5%), a recent and significant increase in ciprofloxacin resistance [20,21].

In addition to bacterial contamination, fungal and viral contamination can seriously reduce food conservation and threaten human health. Fungal food spoilage plays a pivotal role in the deterioration of food and feed systems, and some fungi are also able to produce toxic compounds for humans and animals. The mycotoxins produced by fungi can cause serious health hazards, including Kashin–Beck disease and cancerogenic, immunotoxic, teratogenic, neurotoxic, nephrotoxic, and hepatotoxic effects. Additionally, fungal spoilage/pathogens can cause losses of the marketable quality and hygiene of foodstuffs, resulting in major economic problems throughout the world [22–25].

A recent review focused on mycotoxins produced mainly by *Aspergillus*, *Fusarium*, *Penicillium* and *Alternaria* fungi that affect cereals and spices, as well as on the need for safe decontaminants [26,27]. According to Liu et al., most of these mycotoxins are stable, and non-thermal approaches are required to counteract post-harvest fungi contamination [28]. Fresh food is often contaminated by viruses, such as noroviruses and hepatitis. A review of Yeargin and Gibson focused on viruses contaminating leafy greens, red fruits, and mollusks that require specific surveillance [29]. The current prevention strategies were extensively described in Shukla et al. [30].

The pathogenic bacteria described so far are not the only ones able to contaminate food. In fact, various food products can be subject to contamination by many other microorganisms responsible for infections. For this reason, in recent times, scientific and technological research has been increasingly focused on the study of new antimicrobial materials and new food sanitization methods in order to reduce contamination by microorganisms and lengthen the shelf-lives of food products while ensuring their safety to humans.

There have been few studies of foodborne pathogens in the literature, but the consequences of exposure through food can be severe and a considerable number of people suffer from them every year [31].

3. Curcumin: Chemical Characterization

Since ancient times, turmeric/curcumin has been used as a spice, especially in India and other Asian countries. In addition, it is commonly used for medicinal purposes, especially to treat inflammatory conditions [32]. Curcumin ($C_{21}H_{20}O_6$; 1,7-bis-4-hydroxy-3-methoxyphenyl-1,6-heptadiene-3,5-dione, molecular weight = 368,38 g/M) is a polyphenolic compound derived from the rhizomes of *Curcuma longa* L.

Regarding its chemical structure, curcumin contains two highly polar aromatic rings connected by means of a seven-carbon aliphatic chain and two α, β-unsaturated carbonyl groups (β-diketone). This chain is responsible for the hydrophobic nature of curcumin, and it additionally occurs with two tautomeric conformations: ketone and enol. The presence of these two forms is due to the intramolecular transfer of hydrogen atoms throughout the β-diketone molecule. The enol form is energetically more stable in solution and the solid state than the ketone form due to strong intramolecular hydrogen bonding. The tautomeric ketone form prevails in acidic/neutral aqueous solutions and cell membranes, whereas the enolic form is predominantly found in alkaline environments. Several factors determine the relative contribution of tautomeric forms in solution, including solvent polarity, temperature, and aromatic ring substitution [7]. According to some advanced nuclear magnetic resonance (NMR) studies, the dominance of the enolic form of curcumin in most organic solvents has been confirmed [33]. However, the likelihood of the re-equilibration of the two forms under specific conditions (e.g., in an acidic environment) is not excluded. Intramolecular hydrogen transfer determines the conversion of curcumin to the enolic form in nonpolar, aprotic solvents (such as deuterated chloroform), whereas the breakdown of an intramolecular hydrogen bond with an intermolecular hydrogen bond in protic solvents (such as methanol) leads to the conversion of curcumin in the enolic form to the ketonic form [7]. This step is critical because the poor solubility of curcumin in aqueous solutions could be due to the presence of these inter- and intra-hydrogen bonds [34]. Generally, the presence of diverse functional groups in a curcumin molecule (such as β-dicotyl groups, carbon–carbon double bonds, and phenyl rings with various hydroxyl and methoxyl substituents) determines its various biological activities. For example, curcumin exhibits a high antioxidant capacity, as it is able to neutralize several reactive species, such as superoxide anions, nitrogen dioxide radicals, and ROS [35]. The neutralization of these substances serves to prevent damage to bio-macromolecules. However, curcumin has both antioxidant and pro-oxidant activity—in fact, it is able to produce cytotoxic reactive oxygen species when exposed to light [7]. The oxidation mechanism and antioxidant capacity of curcumin are determined by the number of hydroxyl groups and their position in the aromatic ring; they also make the enolic form more susceptible to oxidation than the ketone form.

Though curcumin has been considered to be a promising antibacterial drug with therapeutic potential, the major drawbacks of taking curcumin alone appear to be mostly due to its poor absorption, quick metabolism, and rapid elimination [36]. Curcumin bioavailability has been a source of controversy in recent years and has been improved by combining certain compounds to generate a curcumin complex. Piperine, for example, increases the bioavailability of curcumin by 20 times [37]. Curcumin's water dispersibility, chemical stability, bioaccessibility, absorption, and overall bioavailability can all be improved with micelles and microemulsions. For example, curcumin microemulsions made from food-grade components such Tween 20, lecithin, vitamin E, and ethanol were found to improve curcumin water dispersibility by 1000–10,000 times [38]. Additionally, curcumin-loaded nanoemulsions can be successfully incorporated into commercial food products such as milk [39], as it was found that the addition of emulsified curcumin to milk reduced lipid oxidation.

The chemical structure of curcumin is also responsible for its ability to absorb light. This property is mainly due to the presence of alternating single and double bonds in the carbon chain. Curcumin possesses a broad absorption spectrum, especially in the visible region (420 nm), and it is capable of absorbing light up to a wavelength of about 430 nm. However, the transition to the ketone form of curcumin results in maximum absorption in the UV region (approximately 389 nm) [7].

4. Mechanisms of Antimicrobial Action of Curcumin

Curcumin exhibits antimicrobial activity, particularly against bacteria, and several studies have been conducted to demonstrate this antibacterial activity. For example, in one study, the antibacterial activity of turmeric oil against *Bacillus subtilis*, *B. coagulans*, *B. cereus*, *S. aureus*, *E. coli*, and *Pseudomonas aeruginosa* was tested and verified [40]. Curcumin has also shown inhibitory activity on methicillin-resistant *S. aureus* (MRSA) strains [41]. The antibacterial activity of curcumin comes from its ability to inhibit the assembly of the FtsZ protein in the Z loop, resulting in the blockage of bacterial cell division [42]. In addition, curcumin possesses anti-biofilm activity against bacteria. This was demonstrated in a study conducted on plant and animal models (*Arabidopsis thaliana* and *Caenorhabditis elegans*) infected with *P. aeruginosa*, in which the ability of curcumin to inhibit the production of genes involved in the early stages of biofilm formation was demonstrated [43].

The mechanism of action of curcumin is mainly based on the perturbation of the functions of FtsZ, a protein involved in bacterial cell division that is homologous to the cytoskeletal protein tubulin of eukaryotes. In fact, curcumin can inhibit the assembly of FtsZ and the Z loop, which are essential for the cytokinesis of bacteria, resulting in the blockage of bacterial proliferation [42]. In addition, Gram-positive bacteria show greater sensitivity to curcumin than Gram-negative bacteria. Sensitivity to curcumin varies depending on the structure of the cell envelope of the bacteria (Figure 1). In fact, Gram-negative bacteria possess different cellular components that confer this resistance to curcumin [44]; they possess an inner cytoplasmic membrane surrounded by a thin layer of peptidoglycan and an outer membrane containing lipopolysaccharides. The outer membrane serves as a permeability barrier, controls the entry and exit of various substances (such as ions, nutrients, and environmental toxins), and contributes to osmoprotection. Gram-positive bacteria possess an inner plasma membrane and a thicker layer of peptidoglycan but lack a protective outer membrane [45]. This is why Gram-negative bacteria exhibit greater resistance to antimicrobial agents than Gram-positive bacteria [44].

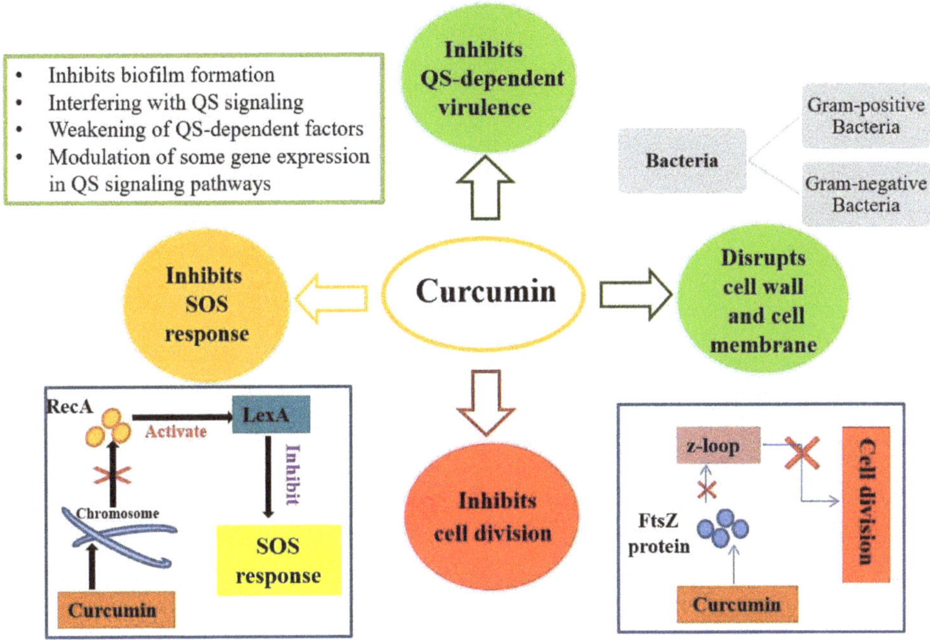

Figure 1. Mechanism of action of curcumin in bacterial growth. Curcumin can inhibit bacterial growth by targeting the bacterial cell membrane, cell wall, protein, DNA, and other cellular structures or by inhibiting bacterial growth through the quorum sensing (QS) system.

Several studies have been conducted to evaluate the antimicrobial activity of curcumin against different bacterial strains. For example, a study was conducted on MRSA to confirm the antibacterial activity of curcumin on this Gram-positive bacterium [46]. Curcumin has been shown to bind to the cell wall, thus altering bacterial integrity. In addition, it causes damage to the cell membrane: after 8 h of curcumin treatment, transmission electron microscopy (TEM) was used to observe that a cytoplasmic membrane was damaged and, as a result, cell lysis occurred [46]. Another study conducted on *B. subtilis* demonstrated similar results. Following curcumin treatment, the alteration of several proteins—mainly involved in bacterial cell division, cell wall biosynthesis, fatty acid synthesis, and central metabolism—was found. Indeed, fluorescence microscopy showed significant alterations in bacterial cell morphology [47]. A recent study was conducted on a strain of *E. coli* to evaluate the antimicrobial effect of curcumin on this Gram-negative bacterium. At high concentrations of curcumin, a baptismal response similar to apoptosis was recorded. Indeed, at these concentrations, curcumin induced an accumulation of ROS, alteration of membrane potential, depolymerization of the membrane itself, DNA fragmentation, and cell apoptosis [48]. Therefore, curcumin can inhibit bacterial growth by targeting the bacterial cell membrane, cell wall, protein, DNA, and other cellular structures or by inhibiting bacterial growth through the quorum sensing (QS) system (Figure 1).

To monitor cell density and species complexity in a population, QS (a communication process between microbial cells) uses chemical signals (self-inducers) that are generated and detected. By coordinating collective activities, QS enables bacteria to behave as a coherent community [49]. Curcumin inhibits QS activity through a variety of mechanisms (Figure 1). It can inhibit QS-dependent factors such as exopolysaccharide synthesis in pathogenic strains such as *E. coli*, *P. aeruginosa*, *Proteus mirabilis*, and *S. aureus*. Curcumin has also been shown to diminish several phenotypes associated with QS inhibition such as swimming, clustering, and motility because the swimming and clustering activities of bacteria are major harmful aspects of biofilm development in different bacterial strains. Curcumin can reduce virulence characteristics via the QS system. The signaling molecules of the QS system can modulate virulence factors [50]. Curcumin also inhibits the production of biofilms through the bacterial QS system, not by killing bacteria or destroying the mature biofilm (Figure 1) but by inhibiting the biofilm generation process. Biofilms are bacterial, extracellular, macromolecule-encased clusters of microbial tissues. Biofilm formation is a dynamic process that comprises bacterial adhesion, biofilm growth, and maturation, all of which are regulated by the QS system [51].

5. Enhancement of Curcumin Antimicrobial Effects by Photoactivation

Photoactivation is based on the local or systemic activation of PS by light with an oxygen source at an appropriate wavelength. The generation of ROS-singlet oxygen and radical species is a result of this activation. Photodynamic therapy is based on a non-thermal photochemical process that requires the presence of a photosensitizing agent (photosensitizer), oxygen, and visible light [52]. The procedure is illustrated in Figure 2.

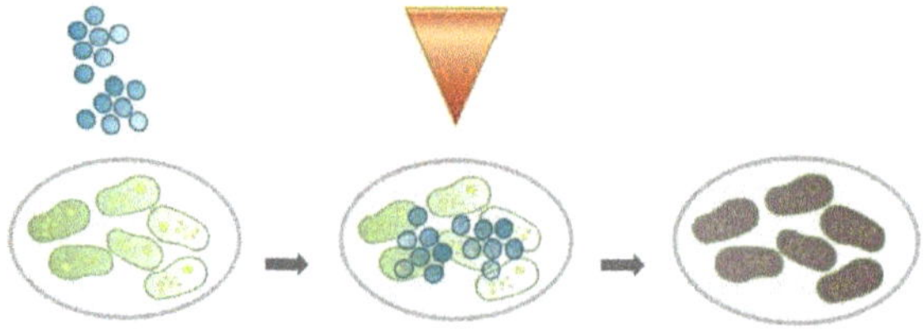

Figure 2. The experimental procedure is graphically depicted. Curcumin interferes with harmful bacterial microorganisms, and the photoactivation of curcumin by means of LEDs enhances the toxic effect on microorganisms.

Recently, researchers have reported interest in enhancing the antimicrobial activity of curcumin by photoexcitation [7,53]. In aqueous solutions, curcumin is photodegradable (light-sensitive) and self-degradable in the dark [54]. However, via illumination (400–500 nm wavelength), curcumin produces ROS, superoxide anions, and hydroxyl radicals, which have lethal effects on bacterial cells, because it is a natural photosensitizer [53].

Photodynamic inactivation (PDI), which has the benefits of safety, environmental protection, and low energy consumption, has been highlighted as a promising strategy against foodborne microorganisms [55,56]. The basic idea of PDI is that particular wavelengths of light activate PS that then form ROS such as superoxide anions, hydroxyl radicals, and hydrogen peroxide, which cause cytotoxic responses that lead to cell death [6]. There have been some attempts to elucidate the mechanism of PDI in foodborne bacteria [57]. ROS may cause DNA damage by targeting guanine nucleotides according to [58], and the authors of [59] reported that curcumin-mediated PDI damaged cytoplasmic DNA and proteins. On the other hand, oxidative bursts of ROS disrupt microorganism homeostasis and the antioxidant enzyme system generates a key cellular defensive network: particularly superoxide dismutase (SOD), catalase (CAT), and glutathione peroxidase (GPx) [60]. Several studies have demonstrated the photodynamic microbial inactivation of some pathogenic bacteria by exploiting an LED light source [61]. At a wavelength of 425 nm, LEDs are able to photoactivate curcumin, as it is a natural photosensitizer and would be very useful in the decontamination of food surfaces. In contrast to traditional light sources, LEDs have several advantages: they are very energy-efficient, cost-effective, durable, and emit little heat. In addition, upon the photoactivation of curcumin, LEDs result in the production of ROS such as superoxide anions, hydroxyl radicals and many others that lead to bacterial cell death. Therefore, the combination of PS and LED illumination would result in the reasonable inactivation of microorganisms on the surface of food. Furthermore, the use of photoactivated curcumin would be a viable option because it is safe, abundant in nature, and does not significantly affect the organoleptic characteristics of food products [53].

The food industry, at present, appears to be very interested in curcumin due to its natural chlorination, antimicrobial and flavoring properties, low cost, and availability. Moreover, because it is a natural photosensitizer, it could be used against a wide range of microorganisms to decontaminate the surface of food. This is the reason why it is considered a "green" and safe alternative compared to common antimicrobial drugs produced by the pharmaceutical industries [7].

Preliminary data obtained by our group [62] showed that curcumin already has a robust effect against some bacterial strains, with low minimal inhibition concentration (MIC) and minimal bactericidal concentration (MBC) values. Moreover, the possible synergistic effect of curcumin combined with photodynamic treatment for 3 h by blue visible light (LED) was evaluated on the same bacteria (two Gram-positive bacteria, *S. aureus* ATCC 29213 and *E. faecalis* ATCC 29212, and one Gram-negative bacterium, *E. coli* ATCC 25922); see Table 2.

Table 2. MIC and MBC of curcumin alone tested against *E. faecalis*, *S. aureus*, and *E. coli* with or without photodynamic treatment on 3 different bacterial strains.

Bacteria	MIC (mg/mL) No LED	MIC (mg/mL) 3 h LED	MBC (mg/mL) No LED	MBC (mg/mL) 3 h LED
E. faecalis	0.125 ± 0.063	0.0037 ± 0.0019	>0.25	0.0037 ± 0.0019
S. aureus	0.06 ± 0.03	0.0075 ± 0.063	>0.25	0.0075 ± 0.0038
E.coli	0.125 ± 0.063	0.125 ± 0.063	>0.25	>0.25

The values were confirmed by three independent measures. As the MIC is a discrete value and the overall sensitivity allows for the discrimination of the doubling and halving of the stated value, the error on each measure in the table corresponded to 50%.

Finally, the antimicrobial effect of curcumin was directly tested by the same research group [63] on samples of fresh fruit (blueberries), which were settled in three containers:

1. blueberries washed with a solution of cyclodextrin (1 mg/mL) (control); 2. blueberries washed with a curcumin and cyclodextrin solution without physical treatment; and 3. blueberries washed with a curcumin and cyclodextrin solution and irradiated daily for 6 h with LED. The three trays, containing a single layer of fruit, were placed in a refrigerator. Each day, at the same time, 3–4 fruits were randomly collected from each tray, transferred to tubes filled with sterile water, and shaken for a few minutes. To assess the presence of microorganisms, the washing water, after fruit extraction with sterile tweezers, was in part directly plated on Brain Heart Infusion Agar (BHA) for bacterial counts and Sabouraud dextrose agar (SAB) to assess the fungal presence and in part filtered through a vacuum filtration unit while changing the unit for each tube, i.e., the filter. Each filter was cut in half: one half was placed on the BHA, and the other half was placed on SAB. All agar plates were incubated at 37 °C for 24–48 h to permit microbial growth. The following day, the plates were compared to assess the microbial load in the washing water. To check their organoleptic characteristics (sight, smell, touch, and taste) the fruits were tasted following extraction from the water. Specifically, volunteer subjects A, B, C and D were asked to taste blueberries and evaluate the appearance of the fruit by sight, smell, touch, and taste; they then assigned a score of 0 if the appearance was altered or 1 if the appearance was normal.

In Figures 3 and 4, a visible reduction in microbial growth, both bacteria or fungi, is evidenced under the third experimental condition, i.e., blueberries washed with a curcumin and cyclodextrin solution and irradiated daily for 6 h with LED.

Figure 3. Microbial growth from filter on BHA and SAB plates: 1. Blueberries washed with a cyclodextrin solution (control); 2. blueberries washed with a curcumin and cyclodextrin solution; and 3. blueberries washed with a curcumin and cyclodextrin solution irradiated daily for 6 h with blue LEDs.

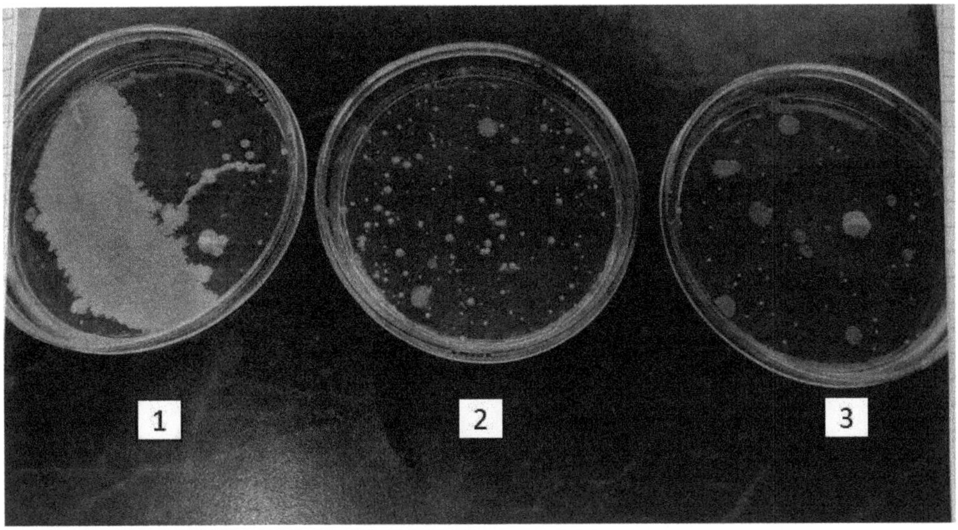

Figure 4. Direct microbial growth on BHA plates: 1. Blueberries washed with cyclodextrin (control); 2. blueberries washed with a curcumin and cyclodextrin solution; and 3. blueberries washed with a curcumin and cyclodextrin solution irradiated daily for 6 h with blue LEDs.

Table 3 shows the results of the organoleptic qualities analysis of the three samples. Sample 1 did not change because it was not rinsed with curcumin. After being washed with curcumin and not exposed to physical treatments, sample 2 exhibited no organoleptic alterations. Finally, sample 3 exhibited no changes in organoleptic properties after being rinsed with curcumin and exposed to blue LEDs for 6 h.

Table 3. Organoleptic aspects of three samples: 1. Blueberries washed with water (control); 2. blueberries subjected to washing with curcumin; and 3. blueberries subjected to washing with curcumin irradiated for 6 h with blue LEDs. V: sight; O: olfaction; T: touch; G: taste; A–D signify subjects who ate blueberries to evaluate their organoleptic characteristics. 0 = alteration of the characteristic; 1 = no alteration of the characteristic.

	1	2	3
Subject	**V/O/T/G**	**V/O/T/G**	**V/O/T/G**
A	1/1/1/1	1/1/1/1	1/1/1/1
B	1/1/1/1	1/1/1/1	1/1/1/1
C	1/1/1/1	1/1/1/1	1/1/1/1
D	1/1/1/1	1/1/1/1	1/1/1/1

6. Further Considerations on Food Treatment

The most efficient approach to reduce the risks of disease outbreaks and postharvest spoilage is to prevent contamination from primary sources with appropriate sanitation and decontamination methods. Curcumin can have antimicrobial and antibacterial activity, and, because it is a natural photosensitizer, this effect can be enhanced by photodynamic treatment. Various applications have been proposed in recent years, indicating that photodynamic activation can be a very promising sterilization method.

PDI is a technique that relies on the occurrence of non-thermal photophysical and photochemical reactions, requiring light and PS in the presence of oxygen [64]. It works on the principle that PS can be activated in certain wavelengths to produce ROS with strong oxidation to inactivate malignant cells and pathogenic microorganisms [65]. The advantage of PDI is that it does not produce toxic chemicals; the only energy required is the light source, and the possibility of causing microbial resistance is low due to its multi-target nature [66].

Typical light sources are LEDs, lasers, and halogen lamps, and the wavelength of light is a crucial factor for PDI. LEDs have the advantages of low cost, wider emission bands, ease of use, and greater flexibility in irradiation time.

It is known that curcumin may both exhibit antimicrobial properties and produce photodynamic effects to further potentiate its antimicrobial efficacy [67]. Curcumin and illumination were found to synergistically inhibit pineapple slice microorganisms [68].

Other encouraging preliminary results have been obtained for red fruits: following curcumin administration and lighting with LEDs, MIC at a very low concentration of curcumin—i.e., 0.125, <0.0075, and <0.0037 mg/mL against *E. coli*, *S. aureus*, and *E. faecalis*, respectively—were found (see Table 2).

Such data showed that curcumin combined with photodynamic treatment by blue visible light (LED) led to impressive variations in MIC (one or even two orders of magnitude) in antibacterial activities against Gram-positive bacteria.

Note that the minimal concentration of curcumin required for *S. aureus* (MIC corresponding to 20.3 μM with LED irradiation) is far lower than that reported in Corrêa et al. [10] on apples (80 μM) and beef, chicken, and pork meat (40 μM).

In addition, curcumin treatment with and without LED photoactivation was used on a fruit sample (blueberries), and microbiological and organoleptic (sight, smell, touch, and taste) investigations showed that it was effective without altering taste (see Table 3).

Microbial decontamination via curcumin photoinactivation has been successfully applied to oysters [69], showing that the food matrix was minimally oxidated [70] and could therefore maintain its organoleptic properties. The same procedure was extended to other seafood [71].

As far as fruit is concerned, experiments were performed on dates and showed that small curcumin concentrations (few nM) could greatly prolong the shelf-life period [72]. Investigations into apples have also been reported [73].

Very recently, curcumin's natural properties and their enhancement by photodynamical inactivation have been tested on various fungal strains [74,75] such as *Candida albicans*, *Aspergillum* (*niger* and *flavus*), and aflatoxins.

Regarding viruses, the same approach was tested against feline calicivirus (FCV) and murine norovirus (MNV) with satisfactory results [76].

In general, all the above-mentioned studies show that the effectiveness of PDI is strongly dependent on the bacterial strain, e.g. particular Gram-negative strains are far less affected. Unfortunately, very few researchers have investigated why the mechanism of the antibacterial activity of curcumin seems to differ depending on the bacterial strain being studied. Tyagi et al. [32] showed that curcumin attacks both Gram-positive and Gram-negative bacteria in a similar manner by causing membrane permeabilization. Aurum et al. [53] showed that LED illumination alone had negligible antibacterial effects but photoactivated curcumin exhibited strong bactericidal activities against *E. coli*. In our study, a slight minor efficacy was evidenced in *E. coli*, and this result could be attributable to the different bacteria inocula, curcumin formulations, and experimental conditions we used. A study by Shlar et al. [44] provided strong evidence that the mode of biological activity of curcumin depends on the properties of the delivery system.

These variations are possible due to the different cell wall compositions of Gram-positive and Gram-negative bacteria. Gram-positive bacteria are easily killed due to their porous structure and easy penetration of PS. On the other hand, Gram-negative bacteria are not easily killed due to their outer membrane that gradually block the activity of curcumin.

Similar results have been published by other authors since the first studies [64,77–79], and it has been speculated that different results have resulted from different membrane structures. Nevertheless, when curcumin is delivered by nanovectors (which are expected to facilitate membrane crossing [80,81]), the outer membrane of Gram-negative bacteria protects them from curcumin diffusion and photosensitization.

Curcumin-loaded nanoparticles and chitosan-modified curcumin-loaded nanoparticles were prepared by Agel et al. to quantify their antibacterial photodynamic effects against *Staphylococcus saprophyticus* subsp. *bovis* and *E. coli* DH5 alpha. The researchers demonstrated that neither irradiation alone nor curcumin in the absence of light could lead to significant growth reductions, confirming the photodynamic effect of curcumin. The increased adherence of the chitosan-modified nanoparticles to bacteria and structural damage upon photodynamic treatment were evident and confirmed the results of in vitro studies. Moreover, curcumin-loaded nanoparticles in combination with LED irradiation resulted in a *S. saprophyticus* survival rate of less than 0.0001% (>6.2 log10 reduction). In the presence of nanoparticles loaded with chitosan-modified curcumin in combination with light, the efficacy was greater, with a survival rate of 0.0000045%. In contrast, in the presence of *E. coli*, the photoactivated curcumin-loaded nanoparticles showed a survival rate of 0.13% (CFU only ~2.9 log10) and an increase in antibacterial efficacy with the curcumin-loaded nanoparticles modified with chitosan with LED, with an up to 5.9 log10 reduction in CFU (0.00013% survival) [80].

7. Photoactivated Curcumin as 'Green' Tool for Food Preservation: Future Perspective

In conclusion, this review discusses the emerging role of curcumin, alone or together with photodynamic light, in enhancing antimicrobial and antibacterial activity and finding effective strategies to preserve food. The preliminary results of the proposed research can be promising in the fight against harmful microorganisms in fruits, potentially improving food quality.

The development of innovative systems based on natural products and physical methods (such as PDI) in the food industry is considered a promising future perspective for alternative "green" techniques in food preservation.

Before proposing applications at the industrial scale, this technique still requires the optimization of the typical parameters that must be considered in experiments with LEDs, i.e., the light wavelength (nm), the emission spectra, the irradiance ($J \cdot cm^{-2}$), and LED potency (W).

Along with light sources, the temperature and acidity of food affect the efficacy of PS and antibacterial effects, so attention should be paid to all phases of food treatment and shelf life. However, due to its low cost and simple applicability, photoactivated curcumin is a promising and biofriendly technique for preserving food by bacterial contamination.

Author Contributions: Conceptualization, C.G. and R.C.; methodology, Z.M., R.C., N.M., G.B. and R.P.; validation, Z.M. and J.R.; formal analysis, L.C. and J.R..; investigation, Z.M., E.F. and R.P.; resources, C.G.; data curation, Z.M.; writing—original draft preparation, Z.M. and C.G.; writing—review and editing, C.G., E.F., R.C., N.M. and G.B.; supervision, C.G. and R.C. All authors have read and agreed to the published version of the manuscript.

Funding: This project was carried out within the TOINPROVE/2020 Call of University of Turin, issued by the MISE and managed by Invitalia.

Institutional Review Board Statement: Not applicable.

Informed Consent Statement: Not applicable.

Data Availability Statement: The presented data are available on request from the corresponding author.

Conflicts of Interest: The authors declare no conflict of interest.

References

1. Yoon, J.-H.; Lee, S.-Y. Review: Comparison of the Effectiveness of Decontaminating Strategies for Fresh Fruits and Vegetables and Related Limitations. *Crit. Rev. Food Sci. Nutr.* **2018**, *58*, 3189–3208. [CrossRef] [PubMed]
2. Sridhar, A.; Ponnuchamy, M.; Kumar, P.S.; Kapoor, A. Food Preservation Techniques and Nanotechnology for Increased Shelf Life of Fruits, Vegetables, Beverages and Spices: A Review. *Environ. Chem. Lett.* **2021**, *19*, 1715–1735. [CrossRef] [PubMed]
3. Deng, L.-Z.; Mujumdar, A.S.; Pan, Z.; Vidyarthi, S.K.; Xu, J.; Zielinska, M.; Xiao, H.-W. Emerging Chemical and Physical Disinfection Technologies of Fruits and Vegetables: A Comprehensive Review. *Crit. Rev. Food Sci. Nutr.* **2020**, *60*, 2481–2508. [CrossRef]
4. Cossu, M.; Ledda, L.; Cossu, A. Emerging Trends in the Photodynamic Inactivation (PDI) Applied to the Food Decontamination. *Food Res. Int.* **2021**, *144*, 110358. [CrossRef]
5. Wang, Y.; Wang, Y.; Wang, Y.; Murray, C.K.; Hamblin, M.R.; Hooper, D.C.; Dai, T. Antimicrobial Blue Light Inactivation of Pathogenic Microbes: State of the Art. *Drug Resist. Updates* **2017**, *33–35*, 1–22. [CrossRef] [PubMed]
6. Hamblin, M.R. Antimicrobial Photodynamic Inactivation: A Bright New Technique to Kill Resistant Microbes. *Curr. Opin. Microbiol.* **2016**, *33*, 67–73. [CrossRef]
7. Seidi Damyeh, M.; Mereddy, R.; Netzel, M.E.; Sultanbawa, Y. An Insight into Curcumin-Based Photosensitization as a Promising and Green Food Preservation Technology. *Compr. Rev. Food Sci. Food Saf.* **2020**, *19*, 1727–1759. [CrossRef]
8. Alizadeh, N.; Malakzadeh, S. Antioxidant, Antibacterial and Anti-Cancer Activities of β-and γ-CDs/Curcumin Loaded in Chitosan Nanoparticles. *Int. J. Biol. Macromol.* **2020**, *147*, 778–791. [CrossRef]
9. Zorofchian Moghadamtousi, S.; Abdul Kadir, H.; Hassandarvish, P.; Tajik, H.; Abubakar, S.; Zandi, K. A Review on Antibacterial, Antiviral, and Antifungal Activity of Curcumin. *Biomed. Res. Int.* **2014**, *2014*, e186864. [CrossRef]
10. Corrêa, T.Q.; Blanco, K.C.; Garcia, É.B.; Perez, S.M.L.; Chianfrone, D.J.; Morais, V.S.; Bagnato, V.S. Effects of Ultraviolet Light and Curcumin-Mediated Photodynamic Inactivation on Microbiological Food Safety: A Study in Meat and Fruit. *Photodiagnosis Photodyn. Ther.* **2020**, *30*, 101678. [CrossRef]
11. Martinović, T.; Andjelković, U.; Gajdošik, M.Š.; Rešetar, D.; Josić, D. Foodborne Pathogens and Their Toxins. *J. Proteom.* **2016**, *147*, 226–235. [CrossRef] [PubMed]
12. Lim, J.Y.; Yoon, J.; Hovde, C.J. A Brief Overview of *Escherichia Coli* O157:H7 and Its Plasmid O157. *J. Microbiol. Biotechnol.* **2010**, *20*, 5–14. [CrossRef] [PubMed]
13. Anderson, A.C.; Jonas, D.; Huber, I.; Karygianni, L.; Wölber, J.; Hellwig, E.; Arweiler, N.; Vach, K.; Wittmer, A.; Al-Ahmad, A. Enterococcus Faecalis from Food, Clinical Specimens, and Oral Sites: Prevalence of Virulence Factors in Association with Biofilm Formation. *Front. Microbiol.* **2016**, *6*, 1534. [CrossRef]
14. Tolker-Nielsen, T. Biofilm Development. *Microbiol. Spectr.* **2015**, *3*, 2. [CrossRef] [PubMed]
15. Jamet, E.; Akary, E.; Poisson, M.-A.; Chamba, J.-F.; Bertrand, X.; Serror, P. Prevalence and Characterization of Antibiotic Resistant Enterococcus Faecalis in French Cheeses. *Food Microbiol.* **2012**, *31*, 191–198. [CrossRef] [PubMed]
16. Kadariya, J.; Smith, T.C.; Thapaliya, D. Staphylococcus Aureus and Staphylococcal Food-Borne Disease: An Ongoing Challenge in Public Health. *Biomed. Res. Int.* **2014**, *2014*, e827965. [CrossRef] [PubMed]
17. Argudín, M.Á.; Mendoza, M.C.; Rodicio, M.R. Food Poisoning and Staphylococcus Aureus Enterotoxins. *Toxins* **2010**, *2*, 1751–1773. [CrossRef]
18. Jia, K.; Fang, T.; Wang, X.; Liu, Y.; Sun, W.; Wang, Y.; Ding, T.; Wang, J.; Li, C.; Xu, D.; et al. Antibiotic Resistance Patterns of Staphylococcus Aureus Isolates from Retail Foods in Mainland China: A Meta-Analysis. *Foodborne Pathog. Dis.* **2020**, *17*, 296–307. [CrossRef]
19. Chajęcka-Wierzchowska, W.; Zadernowska, A.; Łaniewska-Trokenheim, Ł. *Staphylococcus aureus* from ready-to-eat food as a source of multiple antibiotic resistance genes. *Cbu Int. Conf. Proc.* **2017**, *5*, 1108–1112. [CrossRef]
20. Pormohammad, A.; Nasiri, M.J.; Azimi, T. Prevalence of antibiotic resistance in Escherichia coli strains simultaneously isolated from humans, animals, food, and the environment: A systematic review and meta-analysis. *IDR* **2019**, *12*, 1181–1197. [CrossRef]
21. Barlaam, A.; Parisi, A.; Spinelli, E.; Caruso, M.; Taranto, P.D.; Normanno, G. global Emergence of Colistin-Resistant *Escherichia Coli* in Food Chains and Associated Food Safety Implications: A Review. *J. Food Prot.* **2019**, *82*, 1440–1448. [CrossRef] [PubMed]
22. Pawlowska, A.M.; Zannini, E.; Coffey, A.; Arendt, E.K. Chapter 5—"Green Preservatives": Combating Fungi in the Food and Feed Industry by Applying Antifungal Lactic Acid Bacteria. In *Advances in Food and Nutrition Research*; Henry, J., Ed.; Academic Press: Cambridge, MA, USA, 2012; Volume 66, pp. 217–238.
23. Pietsch, C.; Müller, G.; Mourabit, S.; Carnal, S.; Bandara, K. Occurrence of Fungi and Fungal Toxins in Fish Feed during Storage. *Toxins* **2020**, *12*, 171. [CrossRef]
24. Singh, B.K.; Tiwari, S.; Dubey, N.K. Essential Oils and Their Nanoformulations as Green Preservatives to Boost Food Safety against Mycotoxin Contamination of Food Commodities: A Review. *J. Sci. Food Agric.* **2021**, *101*, 4879–4890. [CrossRef] [PubMed]
25. Gil-Serna, J.; García-Díaz, M.; Vázquez, C.; González-Jaén, M.T.; Patiño, B. Significance of Aspergillus Niger Aggregate Species as Contaminants of Food Products in Spain Regarding Their Occurrence and Their Ability to Produce Mycotoxins. *Food Microbiol.* **2019**, *82*, 240–248. [CrossRef] [PubMed]
26. Luo, S.; Du, H.; Kebede, H.; Liu, Y.; Xing, F. Contamination Status of Major Mycotoxins in Agricultural Product and Food Stuff in Europe. *Food Control.* **2021**, *127*, 108120. [CrossRef]

27. Thanushree, M.P.; Sailendri, D.; Yoha, K.S.; Moses, J.A.; Anandharamakrishnan, C. Mycotoxin Contamination in Food: An Exposition on Spices. *Trends Food Sci. Technol.* **2019**, *93*, 69–80. [CrossRef]
28. Liu, Y.; Galani Yamdeu, J.H.; Gong, Y.Y.; Orfila, C. A Review of Postharvest Approaches to Reduce Fungal and Mycotoxin Contamination of Foods. *Compr. Rev. Food Sci. Food Saf.* **2020**, *19*, 1521–1560. [CrossRef] [PubMed]
29. Yeargin, T.; Gibson, K.e. Key Characteristics of Foods with an Elevated Risk for Viral Enteropathogen Contamination. *J. Appl. Microbiol.* **2019**, *126*, 996–1010. [CrossRef]
30. Shukla, S.; Cho, H.; Kwon, O.J.; Chung, S.H.; Kim, M. Prevalence and Evaluation Strategies for Viral Contamination in Food Products: Risk to Human Health—A Review. *Crit. Rev. Food Sci. Nutr.* **2018**, *58*, 405–419. [CrossRef]
31. Thakali, A.; MacRae, J.D. A Review of Chemical and Microbial Contamination in Food: What Are the Threats to a Circular Food System? *Environ. Res.* **2021**, *194*, 110635. [CrossRef]
32. Tyagi, P.; Singh, M.; Kumari, H.; Kumari, A.; Mukhopadhyay, K. Bactericidal Activity of Curcumin I Is Associated with Damaging of Bacterial Membrane. *PLoS ONE* **2015**, *10*, e0121313. [CrossRef] [PubMed]
33. Payton, F.; Sandusky, P.; Alworth, W.L. NMR Study of the Solution Structure of Curcumin. *J. Nat. Prod.* **2007**, *70*, 143–146. [CrossRef] [PubMed]
34. Heger, M.; van Golen, R.F.; Broekgaarden, M.; Michel, M.C. The Molecular Basis for the Pharmacokinetics and Pharmacodynamics of Curcumin and Its Metabolites in Relation to Cancer. *Pharm. Rev.* **2014**, *66*, 222–307. [CrossRef] [PubMed]
35. Menon, V.P.; Sudheer, A.R. Antioxidant and anti-inflammatory properties of curcumin. In *The Molecular Targets and Therapeutic Uses of Curcumin in Health and Disease*; Aggarwal, B.B., Surh, Y.-J., Shishodia, S., Eds.; Advances in experimental medicine and biology; Springer: Boston, MA, USA, 2007; pp. 105–125. ISBN 978-0-387-46401-5.
36. Anand, P.; Kunnumakkara, A.B.; Newman, R.A.; Aggarwal, B.B. Bioavailability of Curcumin: Problems and Promises. *Mol. Pharm.* **2007**, *4*, 807–818. [CrossRef]
37. Shoba, G.; Joy, D.; Joseph, T.; Majeed, M.; Rajendran, R.; Srinivas, P.S.S.R. Influence of Piperine on the Pharmacokinetics of Curcumin in Animals and Human Volunteers. *Planta Med.* **1998**, *64*, 353–356. [CrossRef]
38. Bergonzi, M.C.; Hamdouch, R.; Mazzacuva, F.; Isacchi, B.; Bilia, A.R. Optimization, Characterization and in Vitro Evaluation of Curcumin Microemulsions. *Lwt Food Sci. Technol.* **2014**, *59*, 148–155. [CrossRef]
39. Joung, H.J.; Choi, M.-J.; Kim, J.T.; Park, S.H.; Park, H.J.; Shin, G.H. Development of Food-Grade Curcumin Nanoemulsion and Its Potential Application to Food Beverage System: Antioxidant Property and In Vitro Digestion. *J. Food Sci.* **2016**, *81*, N745–N753. [CrossRef] [PubMed]
40. Negi, P.S.; Jayaprakasha, G.K.; Jagan Mohan Rao, L.; Sakariah, K.K. Antibacterial Activity of Turmeric Oil: A Byproduct from Curcumin Manufacture. *J. Agric. Food Chem.* **1999**, *47*, 4297–4300. [CrossRef]
41. Mun, S.-H.; Joung, D.-K.; Kim, Y.-S.; Kang, O.-H.; Kim, S.-B.; Seo, Y.-S.; Kim, Y.-C.; Lee, D.-S.; Shin, D.-W.; Kweon, K.-T.; et al. Synergistic Antibacterial Effect of Curcumin against Methicillin-Resistant Staphylococcus Aureus. *Phytomed. Int. J. Phytother. Phytopharm.* **2013**, *20*, 714–718. [CrossRef]
42. Rai, D.; Singh, J.K.; Roy, N.; Panda, D. Curcumin Inhibits FtsZ Assembly: An Attractive Mechanism for Its Antibacterial Activity. *Biochem. J.* **2008**, *410*, 147–155. [CrossRef]
43. Rudrappa, T.; Bais, H.P. Curcumin, a Known Phenolic from Curcuma Longa, Attenuates the Virulence of Pseudomonas Aeruginosa PAO1 in Whole Plant and Animal Pathogenicity Models. *J. Agric. Food Chem.* **2008**, *56*, 1955–1962. [CrossRef] [PubMed]
44. Shlar, I.; Droby, S.; Choudhary, R.; Rodov, V. The Mode of Antimicrobial Action of Curcumin Depends on the Delivery System: Monolithic Nanoparticles vs. Supramolecular Inclusion Complex. *Rsc Adv.* **2017**, *7*, 42559–42569. [CrossRef]
45. Rajagopal, M.; Walker, S. Envelope Structures of Gram-Positive Bacteria. In *Protein and Sugar Export and Assembly in Gram-Positive Bacteria*; Bagnoli, F., Rappuoli, R., Eds.; Current Topics in Microbiology and Immunology; Springer International Publishing: Cham, Switzerland, 2017; pp. 1–44. ISBN 978-3-319-56014-4.
46. Mun, S.-H.; Kim, S.-B.; Kong, R.; Choi, J.-G.; Kim, Y.-C.; Shin, D.-W.; Kang, O.-H.; Kwon, D.-Y. Curcumin Reverse Methicillin Resistance in Staphylococcus Aureus. *Molecules* **2014**, *19*, 18283–18295. [CrossRef] [PubMed]
47. Reddy, P.J.; Sinha, S.; Ray, S.; Sathe, G.J.; Chatterjee, A.; Prasad, T.S.K.; Dhali, S.; Srikanth, R.; Panda, D.; Srivastava, S. Comprehensive Analysis of Temporal Alterations in Cellular Proteome of Bacillus Subtilis under Curcumin Treatment. *PLoS ONE* **2015**, *10*, e0120620. [CrossRef]
48. Yun, D.G.; Lee, D.G. Antibacterial Activity of Curcumin via Apoptosis-like Response in *Escherichia coli*. *Appl. Microbiol. Biotechnol.* **2016**, *100*, 5505–5514. [CrossRef]
49. Hawver, L.A.; Jung, S.A.; Ng, W.-L. Specificity and Complexity in Bacterial Quorum-Sensing Systems. *Fems Microbiol. Rev.* **2016**, *40*, 738–752. [CrossRef]
50. Zheng, D.; Huang, C.; Huang, H.; Zhao, Y.; Khan, M.R.U.; Zhao, H.; Huang, L. Antibacterial Mechanism of Curcumin: A Review. *Chem. Biodiv.* **2020**, *17*, e2000171. [CrossRef]
51. Kievit, T.R.D.; Gillis, R.; Marx, S.; Brown, C.; Iglewski, B.H. Quorum-Sensing Genes in Pseudomonas Aeruginosa Biofilms: Their Role and Expression Patterns. *Appl. Environ. Microbiol.* **2001**, *67*, 1865–1873. [CrossRef]
52. Kübler, A.C. Photodynamic Therapy. *Med Laser Appl.* **2005**, *20*, 37–45. [CrossRef]
53. Aurum, F.S.; Nguyen, L.T. Efficacy of Photoactivated Curcumin to Decontaminate Food Surfaces under Blue Light Emitting Diode. *J. Food Process. Eng.* **2019**, *42*, e12988. [CrossRef]

54. Mondal, S.; Ghosh, S.; Moulik, S.P. Stability of Curcumin in Different Solvent and Solution Media: UV-Visible and Steady-State Fluorescence Spectral Study. *J. Photochem. Photobiol. BBiol.* **2016**, *158*, 212–218. [CrossRef] [PubMed]
55. Gonzales, J.C.; Brancini, G.T.P.; Rodrigues, G.B.; Silva-Junior, G.J.; Bachmann, L.; Wainwright, M.; Braga, G.Ú.L. Photodynamic Inactivation of Conidia of the Fungus Colletotrichum Abscissum on Citrus Sinensis Plants with Methylene Blue under Solar Radiation. *J. Photochem. Photobiol. B Biol.* **2017**, *176*, 54–61. [CrossRef] [PubMed]
56. Majiya, H.; Galstyan, A. Dye Extract of Calyces of Hibiscus Sabdariffa Has Photodynamic Antibacterial Activity: A Prospect for Sunlight-Driven Fresh Produce Sanitation. *Food Sci. Nutr.* **2020**, *8*, 3200–3211. [CrossRef] [PubMed]
57. Ghate, V.; Leong, A.L.; Kumar, A.; Bang, W.S.; Zhou, W.; Yuk, H.-G. Enhancing the Antibacterial Effect of 461 and 521 Nm Light Emitting Diodes on Selected Foodborne Pathogens in Trypticase Soy Broth by Acidic and Alkaline PH Conditions. *Food Microbiol.* **2015**, *48*, 49–57. [CrossRef] [PubMed]
58. Kim, M.-J.; Yuk, H.-G. Antibacterial Mechanism of 405-Nanometer Light-Emitting Diode against Salmonella at Refrigeration Temperature. *Appl. Environ. Microbiol.* **2016**, *83*, e02582-16. [CrossRef]
59. Huang, J.; Chen, B.; Li, H.; Zeng, Q.-H.; Wang, J.J.; Liu, H.; Pan, Y.; Zhao, Y. Enhanced Antibacterial and Antibiofilm Functions of the Curcumin-Mediated Photodynamic Inactivation against Listeria Monocytogenes. *Food Control.* **2020**, *108*, 106886. [CrossRef]
60. Montllor-Albalate, C.; Colin, A.E.; Chandrasekharan, B.; Bolaji, N.; Andersen, J.L.; Wayne Outten, F.; Reddi, A.R. Extra-Mitochondrial Cu/Zn Superoxide Dismutase (Sod1) Is Dispensable for Protection against Oxidative Stress but Mediates Peroxide Signaling in Saccharomyces Cerevisiae. *Redox Biol.* **2019**, *21*, 101064. [CrossRef]
61. Endarko, E.; MacLean, M.; Timoshkin, I.; MacGregor, S.; Anderson, J. High-Intensity 405 Nm Light Inactivation of Listeria Monocytogenes. *Photochem. Photobiol.* **2012**, *88*, 1280–1286. [CrossRef]
62. Munir, Z.; Cavallo, L.; Menotti, F.; Tullio, V.; Comini, S.; Cavalli, R.; Guiot, C.; Cuffini, A.M.; Banche, G.; Mandras, N. Evaluation of Chitosan/Curcumin loaded Nanobubbles with and without Photoactivated light for Food Preservation. Poster Communication for Società Italiana di Microbiologia 2021, Virtual Congress, 16–17 and 20–21 September 2021.
63. Munir, Z.; Cavalli, R.; Banche, G.; Mandras, N.; Guiot, C. Application of Curcumin-Loaded Nanobubbles with and without Photodynamic Light (LED) for Food Preservation. Poster Communication for Nanoinnovation 2021, Rome, Italy, 21–24 September 2021.
64. Penha, C.B.; Bonin, E.; da Silva, A.F.; Hioka, N.; Zanqueta, É.B.; Nakamura, T.U.; de Abreu Filho, B.A.; Campanerut-Sá, P.A.Z.; Mikcha, J.M.G. Photodynamic Inactivation of Foodborne and Food Spoilage Bacteria by Curcumin. *Lwt Food Sci. Technol.* **2017**, *76*, 198–202. [CrossRef]
65. Jiang, Y.; Liang, Y.; Zhang, H.; Zhang, W.; Tu, S. Preparation and Biocompatibility of Grafted Functional β-Cyclodextrin Copolymers from the Surface of PET Films. *Mater. Sci. Eng. C* **2014**, *41*, 1–7. [CrossRef] [PubMed]
66. Costa, L.; Tomé, J.P.C.; Neves, M.G.; Tomé, A.C.; Cavaleiro, J.A.S.; Faustino, M.A.F.; Cunha, Â.; Gomes, N.C.M.; Almeida, A. Evaluation of Resistance Development and Viability Recovery by a Non-Enveloped Virus after Repeated Cycles of APDT. *Antivir. Res.* **2011**, *91*, 278–282. [CrossRef] [PubMed]
67. Neelakantan, P.; Cheng, C.Q.; Ravichandran, V.; Mao, T.; Sriraman, P.; Sridharan, S.; Subbarao, C.; Sharma, S.; Kishen, A. Photoactivation of Curcumin and Sodium Hypochlorite to Enhance Antibiofilm Efficacy in Root Canal Dentin. *Photodiagnosis Photodyn. Ther.* **2015**, *12*, 108–114. [CrossRef] [PubMed]
68. Zou, Y.; Yu, Y.; Cheng, L.; Li, L.; Zou, B.; Wu, J.; Zhou, W.; Li, J.; Xu, Y. Effects of Curcumin-Based Photodynamic Treatment on Quality Attributes of Fresh-Cut Pineapple. *LWT* **2021**, *141*, 110902. [CrossRef]
69. Wu, J.; Hou, W.; Cao, B.; Zuo, T.; Xue, C.; Leung, A.W.; Xu, C.; Tang, Q.-J. Virucidal Efficacy of Treatment with Photodynamically Activated Curcumin on Murine Norovirus Bio-Accumulated in Oysters. *Photodiagnosis Photodyn. Ther.* **2015**, *12*, 385–392. [CrossRef]
70. Liu, F.; Li, Z.; Cao, B.; Wu, J.; Wang, Y.; Xue, Y.; Xu, J.; Xue, C.; Tang, Q.J. The Effect of a Novel Photodynamic Activation Method Mediated by Curcumin on Oyster Shelf Life and Quality. *Food Res. Int.* **2016**, *87*, 204–210. [CrossRef] [PubMed]
71. Chen, B.; Huang, J.; Li, H.; Zeng, Q.-H.; Wang, J.J.; Liu, H.; Pan, Y.; Zhao, Y. Eradication of Planktonic Vibrio Parahaemolyticus and Its Sessile Biofilm by Curcumin-Mediated Photodynamic Inactivation. *Food Control.* **2020**, *113*, 107181. [CrossRef]
72. Al-Asmari, F.; Mereddy, R.; Sultanbawa, Y. The Effect of Photosensitization Mediated by Curcumin on Storage Life of Fresh Date (*Phoenix Dactylifera* L.) Fruit. *Food Control.* **2018**, *93*, 305–309. [CrossRef]
73. Song, L.; Zhang, F.; Yu, J.; Wei, C.; Han, Q.; Meng, X. Antifungal Effect and Possible Mechanism of Curcumin Mediated Photodynamic Technology against Penicillium Expansum. *Postharvest Biol. Technol.* **2020**, *167*, 111234. [CrossRef]
74. Schamberger, B.; Plaetzer, K. Photofungizides Based on Curcumin and Derivates Thereof against Candida Albicans and Aspergillus Niger. *Antibiotics* **2021**, *10*, 1315. [CrossRef]
75. Nguenha, R.J.; Damyeh, M.S.; Hong, H.T.; Chaliha, M.; Sultanbawa, Y. Effect of Solvents on Curcumin as a Photosensitizer and Its Ability to Inactivate Aspergillus Flavus and Reduce Aflatoxin B1 in Maize Kernels and Flour. *J. Food Process. Preserv.* **2022**, *46*, e16169. [CrossRef]
76. Randazzo, W.; Aznar, R.; Sánchez, G. Curcumin-Mediated Photodynamic Inactivation of Norovirus Surrogates. *Food Environ. Virol.* **2016**, *8*, 244–250. [CrossRef] [PubMed]
77. Dahl, T.A.; McGowan, W.M.; Shand, M.A.; Srinivasan, V.S. Photokilling of Bacteria by the Natural Dye Curcumin. *Arch. Microbiol.* **1989**, *151*, 183–185. [CrossRef] [PubMed]

78. de Oliveira, E.F.; Tosati, J.V.; Tikekar, R.V.; Monteiro, A.R.; Nitin, N. Antimicrobial Activity of Curcumin in Combination with Light against *Escherichia Coli* O157:H7 and Listeria Innocua: Applications for Fresh Produce Sanitation. *Postharvest Biol. Technol.* **2018**, *137*, 86–94. [CrossRef]
79. Maisch, T.; Szeimies, R.-M.; Jori, G.; Abels, C. Antibacterial Photodynamic Therapy in Dermatology. *Photochem. Photobiol. Sci.* **2004**, *3*, 907–917. [CrossRef] [PubMed]
80. Agel, M.R.; Baghdan, E.; Pinnapireddy, S.R.; Lehmann, J.; Schäfer, J.; Bakowsky, U. Curcumin Loaded Nanoparticles as Efficient Photoactive Formulations against Gram-Positive and Gram-Negative Bacteria. *Colloids Surf. B Biointerfaces* **2019**, *178*, 460–468. [CrossRef]
81. Ryu, V.; Ruiz-Ramirez, S.; Chuesiang, P.; McLandsborough, L.A.; McClements, D.J.; Corradini, M.G. Use of Micellar Delivery Systems to Enhance Curcumin's Stability and Microbial Photoinactivation Capacity. *Foods* **2021**, *10*, 1777. [CrossRef]

International Journal of *Molecular Sciences*

MDPI

Article

Influence of Incubation Time on Ortho-Toluidine Blue Mediated Antimicrobial Photodynamic Therapy Directed against Selected *Candida* Strains—An In Vitro Study

Rafał Wiench [1], Joanna Nowicka [2], Magdalena Pajączkowska [2], Piotr Kuropka [3], Dariusz Skaba [1], Anna Kruczek-Kazibudzka [4], Anna Kuśka-Kiełbratowska [1] and Kinga Grzech-Leśniak [5,6,*]

1 Department of Periodontal Diseases and Oral Mucosa Diseases, Faculty of Medical Sciences in Zabrze, Medical University of Silesia, 40-055 Katowice, Poland; rwiench@sum.edu.pl (R.W.); dskaba@sum.edu.pl (D.S.); anna.kuska@hotmail.com (A.K.-K.)
2 Department of Microbiology, Faculty of Medicine, Wroclaw Medical University, 50-368 Wroclaw, Poland; joanna.nowicka@umed.wroc.pl (J.N.); magdalena.pajaczkowska@umed.wroc.pl (M.P.)
3 Department of Histology and Embryology, Wroclaw University of Environmental and Life Sciences, 50-375 Wroclaw, Poland; piotr.kuropka@upwr.edu.pl
4 Private Practice, 44-100 Gliwice, Poland; kruczeka00@gmail.com
5 Laser Laboratory Dental Surgery Department, Medical University of Wroclaw, 50-425 Wroclaw, Poland
6 Department of Periodontics, School of Dentistry, Virginia Commonwealth University, Richmond, VA 23284, USA
* Correspondence: kgl@periocare.pl

Citation: Wiench, R.; Nowicka, J.; Pajączkowska, M.; Kuropka, P.; Skaba, D.; Kruczek-Kazibudzka, A.; Kuśka-Kiełbratowska, A.; Grzech-Leśniak, K. Influence of Incubation Time on Ortho-Toluidine Blue Mediated Antimicrobial Photodynamic Therapy Directed against Selected *Candida* Strains—An In Vitro Study. *Int. J. Mol. Sci.* **2021**, 22, 10971. https://doi.org/10.3390/ijms222010971

Academic Editors: Antonino Mazzaglia, Angela Scala and Enrico Caruso

Received: 2 September 2021
Accepted: 3 October 2021
Published: 11 October 2021

Abstract: (1) Background and the aim: The appropriate incubation time in the antimicrobial photodynamic therapy protocol seems to have a huge impact on the efficacy of this process. This is particularly important in relation to *Candida* strains, due to the size of these cells and the presence of the cell wall. The aims of this study were to determine the optimal incubation time needed for the absorption of toluidine blue by cells of *C. albicans*, *C. glabrata*, *C. krusei* and *C. parapsilosis* using direct observation by optical microscopy, and to evaluate the efficacy of TBO-mediated aPDT on planktonic cells of these strains. (2) Methods: The microscopic evaluation consisted of taking a series of images at a magnification of 600× and counting the % of stained cells. The in vitro effect of TBO-mediated aPDT combined with a diode laser (635 nm, 400mW, 12 J/cm^2, CW) on the viability of yeast cells with different incubation times was evaluated. (3) Results: The presence of TBO within the cytoplasm was observed in all tested *Candida* strains and at all microscopic evaluation times. However, the highest percentages of cells were stained at 7 and 10 min. The highest % reduction of CFU/mL after TBO-mediated aPDT against *Candida* was obtained for the strain *C. albicans* ATCC 10,231 and it was 78.55%. (4) Conclusions: TBO-mediated aPDT against *Candida* was effective in reducing the number of CFU/mL at all assessed incubation times. However, the most efficient period for almost all strains was 7–10 min.

Keywords: aPDT; diode laser; TBO; yeasts; oral candidiasis; oral microbiome

1. Introduction

Photodynamic therapy (PDT) is a kind of light therapy [1] which causes irreversible damage to target cells (diseased host cells or microorganisms) and has been shown to be a potential approach for cancer treatment [2–4] and for treating microbial infection induced by Gram-positive and Gram-negative bacteria, including antibiotic-resistant strains [5–8] as well as yeasts [9,10]. If the cells being destroyed are microorganisms, this form of therapy is called antimicrobial photodynamic therapy (aPDT).

The basis of PDT is the cooperation of three elements: photosensitizer (PS), light, and oxygen. These components are harmless by themselves but combined they lead to selective destruction of pathogenic cells [11]. It is important to take into consideration that different PSs have varying absorption maximums. The PS is usually an organic, aromatic dye molecule capable of delocalizing π electrons and is non-toxic in the absence of light. It is an external chromophore that absorbs light in the red or near-infrared range (600–800 nm); wavelengths >800 nm do not have enough internal photonic energy to induce a photodynamic reaction, while wavelengths <600 nm have less tissue penetration [12]. The antimicrobial effect of a PS is strictly dependent on its physical properties (absorption peak (k_{max}), intensity of absorption (e_{max}), and quantum yield for singlet oxygen) and chemical parameters (lipophilicity/hydrophilicity balance (logP) and the presence of charged groups that determine the mechanism of cellular uptake). As a rule, hydrophilicity and the presence of charged groups are better for cellular uptake [13].

The light sources used for PDT are typically lasers, light-emitting diodes (LEDs), or lamps with a broad spectrum of wavelengths. Lasers and LEDs seem to be the most appropriate sources of light. They can provide the most exact match to the absorption maximum of a PS, which results in less heating of the tissues and an optimal course of the photodynamic reaction [14]. The absorption peak of the PS changes with the concentration of the aqueous solution, and therefore when selecting a light source for therapy or experiments, particular attention should be paid to this parameter [15].

The interest in the possibility of using aPDT for treatment of oral candidiasis results from the infectious nature of these etiopathogenic factors and the more and more frequent multiple resistance to popular antimycotics [16]. aPDT against *Candida* (C.) spp. is more difficult than against bacteria, related to the larger size of fungal cells (on average 25–50 times larger than bacterial cells) and their more complex structure [17]. Particularly important is the additional thick cell wall consisting of glucans, mannans, chitin, and lipoproteins [18–20] and the presence of a cell nucleus separated from the cytoplasm by a nuclear membrane. Yeast could also produce antioxidant enzymes such as superoxide dismutase (SOD) and catalase. SOD converts damaging superoxide radicals (O_2^-), one type of reactive oxygen species (ROS), to the less damaging hydrogen peroxide (H_2O_2) which can be converted into water by catalase [21–23]. Moreover, many species of fungi, including *Candida* spp., have specific defense mechanisms against aPDT including inhibition of PS uptake, stimulation of its excretion from the cell, or intracellular metabolism by inducing enzymatic remodeling of PS particles [23]. Therefore, aPDT against yeasts requires application of only selected PS which should also be used in higher concentrations, for longer incubation times, and with higher parameters of the laser physical settings (power density, fluence) [24,25]. One of the most used PS is toluidine blue ortho (TBO) also known as tolonium chloride, a basic metachromatic thiazide dye. Its small size, good solubility in water, cationic form, hydrophilic character, and tendency to form dimers facilitate its binding to microorganisms' cell membranes [26]. A large difference in affinity for the surface of yeast and host cells provides selectivity, and no DNA damage to keratinocytes in vitro was observed [27,28]. Its potential to kill yeast cells is to damage the cell wall where PS molecules, thanks to their cationic form, create ionic bonds between amino groups on the surface [29,30]. As a result of photoactivation, the oxidation processes change the architecture of the cell so that the TBO reaches the cell membrane and the cytoplasm [24]. These processes require longer contact times of TBO with the surface of the yeast cell wall and membrane so that the large and biochemically complex PS molecules can pass through the narrow *Candida* membrane channels. Practically, this translates into use of a sufficiently long incubation time (pre-irradiation time, PIT) to accumulate a sufficient amount of PS on the surface of the yeast cell and inside the cytoplasm so that the efficacy of TBO-mediated aPDT is as high as possible [14]. The selection of the most efficacious photosensitizer, as well as the laser dosimetry, to be used in the elimination of *Candida* have been the subject of many studies [29,30].

The least studied element of the aPDT procedure for treating candidiasis is the optimal incubation time with the PS. Therefore, the aims of this study were to determine the optimal incubation time needed for the absorption of TBO by cells of selected *Candida* strains using direct observation by optical microscopy, and to evaluate the efficacy of TBO-mediated aPDT on planktonic cells of these strains.

2. Results

In the first part of the experiment, direct microscopic evaluation of the absorption of TBO by cells of *Candida*, the presence of cells containing photosensitizer was observed in all the strains and at all sampling times. Figures 1–5 show an exemplary series of original micrographs for each strain and the same representative series after processing the original images in ImageJ to increase contrast.

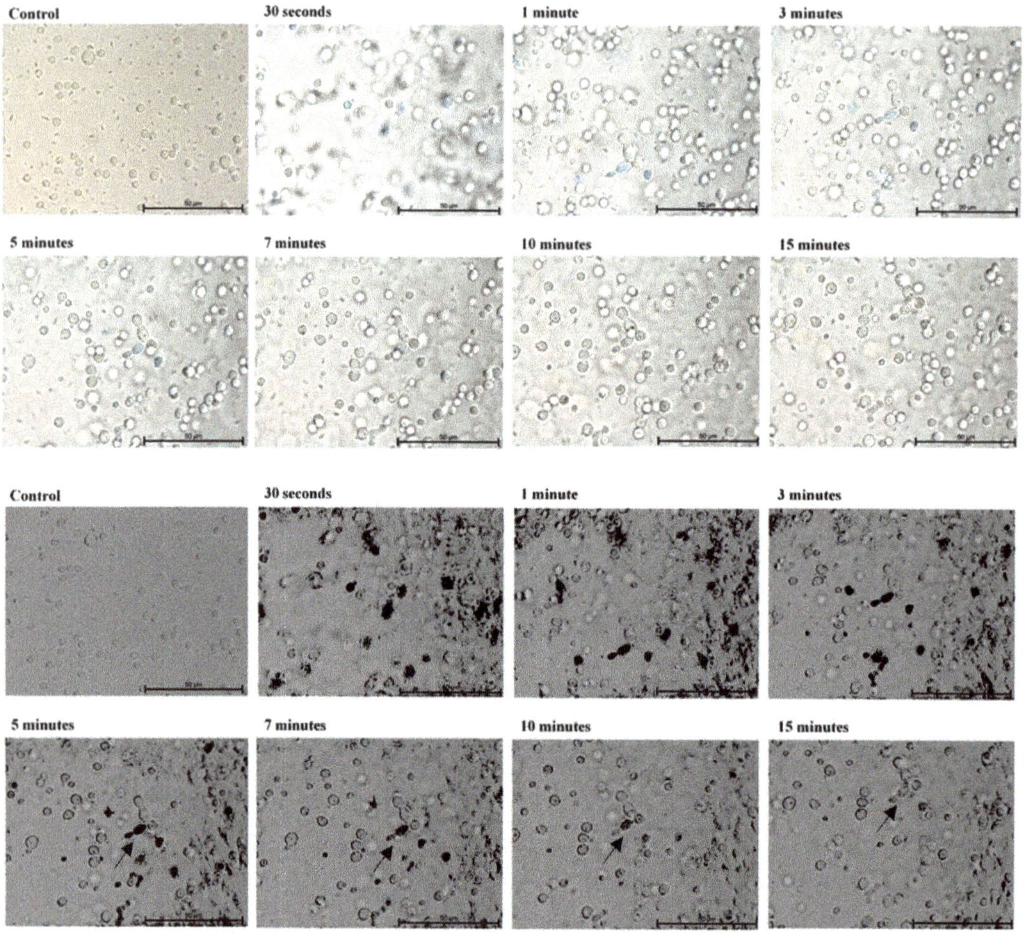

Figure 1. A representative series of original images (two upper rows) and images after isolating indigo and blue colors and converting to black in ImageJ (two lower rows) showing the number of *Candida albicans* ATCC 90,028 cells with absorbed ortho-toluidine blue (TBO) at individual times relative to the control. Magnification 600×. Scale bars = 50 µm. Black arrows indicate yeast cells with gradually disappearing dye in the cytoplasm. American Type Culture Collection (ATCC).

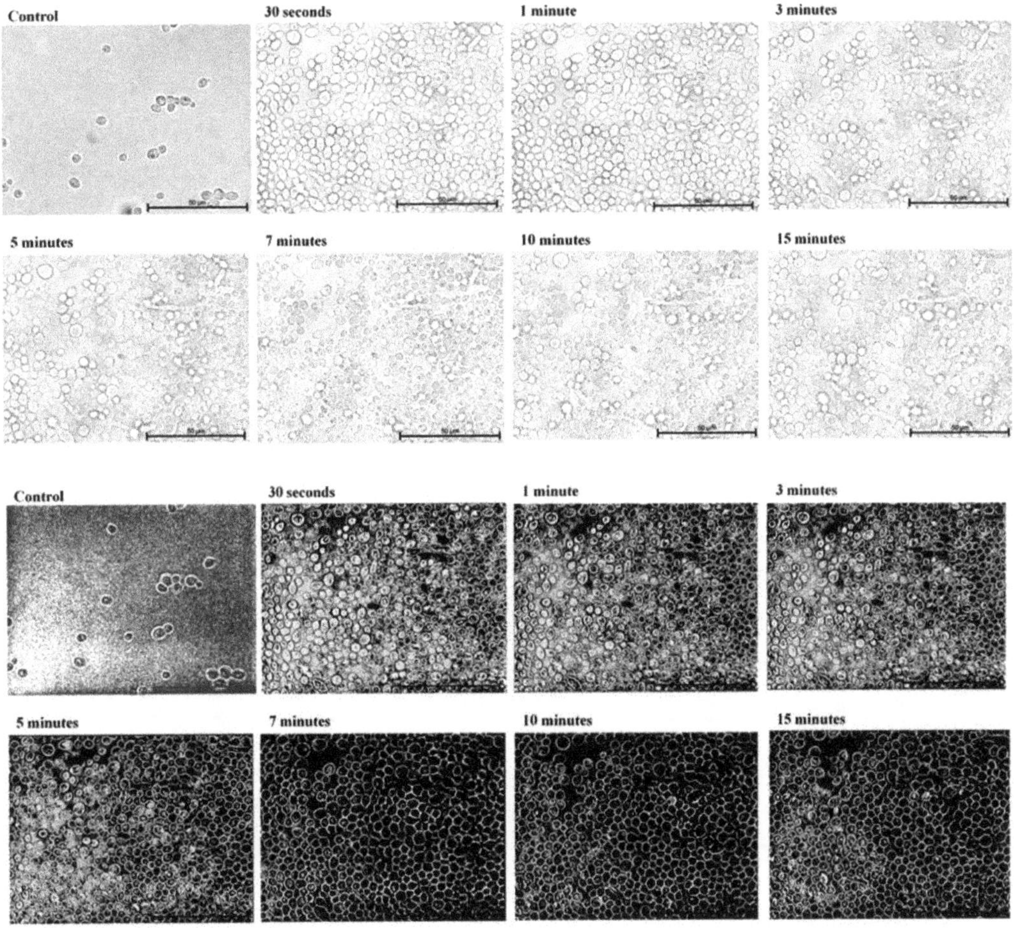

Figure 2. A representative series of original images (two upper rows) and images after isolating indigo and blue colors and converting to black in ImageJ (two lower rows) showing the number of *Candida albicans* ATCC 10,231 cells with absorbed ortho-toluidine blue (TBO) at individual times relative to the control. Magnification 600×. Scale bars = 50 µm. American Type Culture Collection (ATCC).

Descriptive statistics expressed as the percentage of stained cells visible in the field of view of the microscope at particular observation times are presented in Figure 6. Large differences in PS absorption by different strains of the yeast were found. The highest percentage of TBO-containing cells (mean) was observed in *C. albicans* ATCC 10,231 and was −97.53%. *C. parapsilosis* ATCC 90,018 −47.0%, *C. krusei* ATCC 6258 −46.29% and *C. glabrata* ATCC 90,030 −40.15% were recorded consecutively. The lowest result was achieved by *C. albicans* ATCC 90,028 −23.48%. The time required to obtain the maximum adsorption also differed for individual strains and ranged from 5 min for *C. albicans* ATCC 90,028, 7 min for *C. albicans* ATCC 10,231, *C. glabrata* ATCC 90,030, *C. parapsilosis* ATCC 90,028, to 10 min for *C. krusei* ATCC 6258.

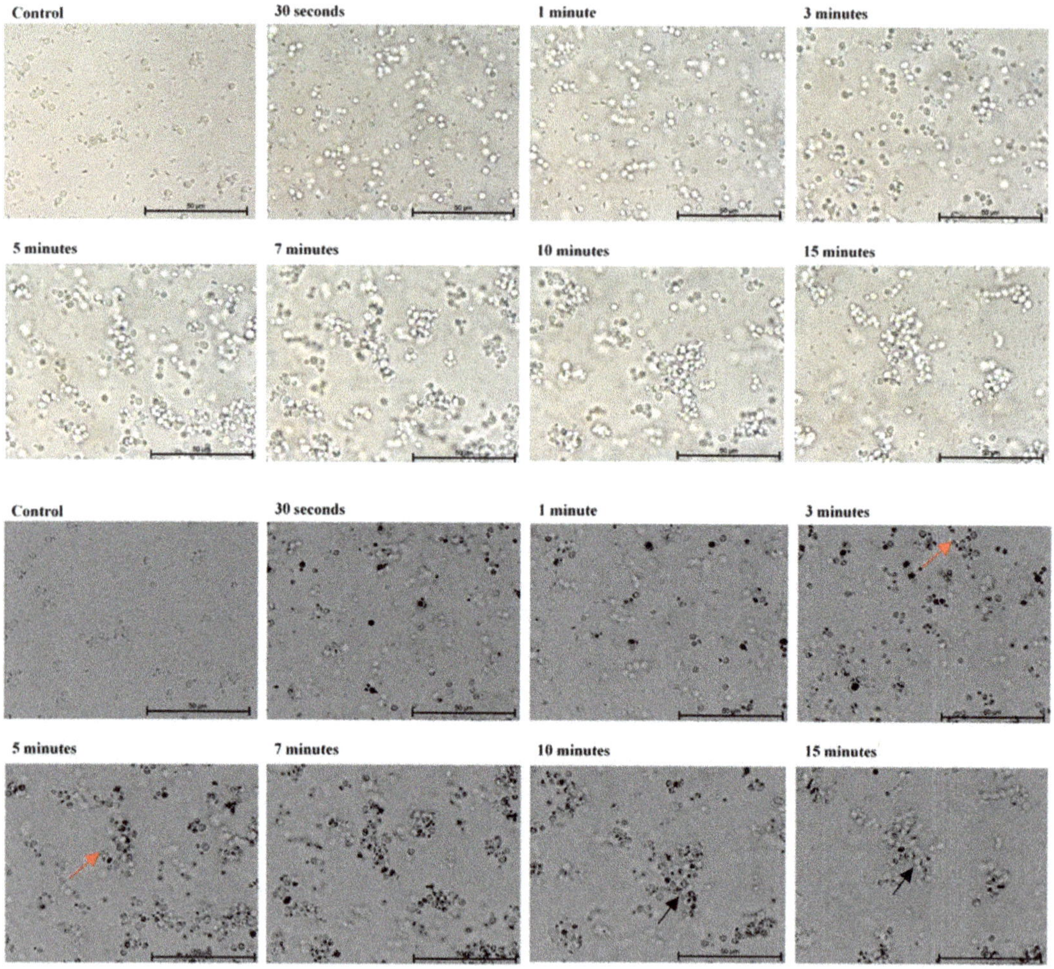

Figure 3. A representative series of original images (two upper rows) and images after isolating indigo and blue colors and converting to black in ImageJ (two lower rows) showing the number of *Candida glabrata* ATCC 90,030 cells with absorbed ortho-toluidine blue (TBO) at individual times relative to the control. Magnification 600×. Scale bars = 50 μm. Black arrows indicate yeast cells with gradually disappearing dye in the cytoplasm. Red arrows indicate representative early aggregation structures of *Candida glabrata* ATCC 90,030. American Type Culture Collection (ATCC).

In the next experiments, an evaluation of the photodynamic inhibitory effects on the growth of planktonic *Candida* spp. cultures was made after 1, 3, 5, 7, 10 and 15 min of incubation in all test and control groups. Figures 7–11 show the results as viable counts (colony-forming units per milliliter) (CFU/mL, mean ± SD). TBO-mediated aPDT significantly reduced the cell number (CFUs) of all tested *Candida* strains in comparison to the other two treatments and to the control group. The reduction depended on the time of incubation. There was no significant difference between CFU counts of groups (L-P-) and (L+P-) and (L-P+). These results indicate that laser irradiation alone or the use of the photosensitizer alone had no fungicidal effect. The highest % reduction of CFU/mL was obtained for the strain *C. albicans* ATCC 10,231 and it was 76.89%. Successively for *C. glabrata* ATCC 90,030 −61.36%, *C. krusei* ATCC 6258 −59.39%, *C. parapsilosis* ATCC 90,018- 46.81%. The lowest result was obtained for C *albicans* ATCC 90,028 and it was only

23.01% (Figure 12). Table 1 summarizes comparison of the results achieved in both parts of our experiment—the efficacy of TBO-mediated aPDT presented as the highest % reduction of CFU/mL to % of cells stained by the presence of photosensitizer in their cytoplasm visible in the microscope field for individual *Candida* strains tested.

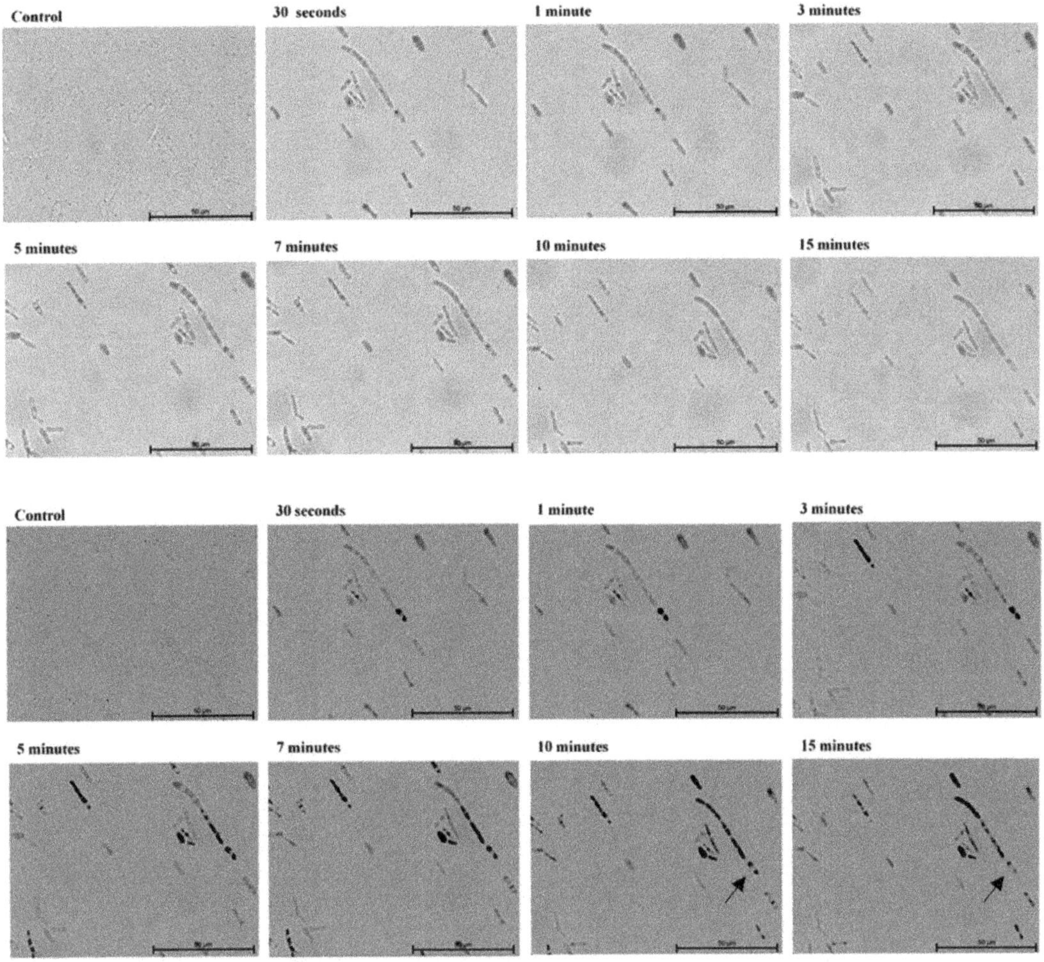

Figure 4. A representative series of original images (two upper rows) and images after isolating indigo and blue colors and converting to black in ImageJ (two lower rows) showing the number of *Candida krusei* ATCC 6258 cells with absorbed ortho-toluidine blue (TBO) at individual times relative to the control. Magnification 600×. Scale bars = 50 µm. Black arrows indicate yeast cells with gradually disappearing dye in the cytoplasm. American Type Culture Collection (ATCC).

Table 1. Comparison of the results achieved in both parts of our experiment—the efficacy of TBO-mediated aPDT presented as the highest % reduction of CFU/mL to % of cells stained by the presence of photosensitizer in their cytoplasm visible in the microscope field for individual *Candida* strains tested.

Candida Strain (The Most Efficient Incubation Time)	The Highest Reduction of CFU/mL [%]	Assessment of TBO Uptake by Cells over Time [%]
C. albicans ATCC 90,028 (7 min)	23.02	19.93
C. albicans ATCC 10,231 (10 min)	76.89	94.43
C. glabrata (7 min)	61.37	40.15
C. krusei (3 min)	59.40	23.39
C. parapsilosis (7 min)	46.81	47.00

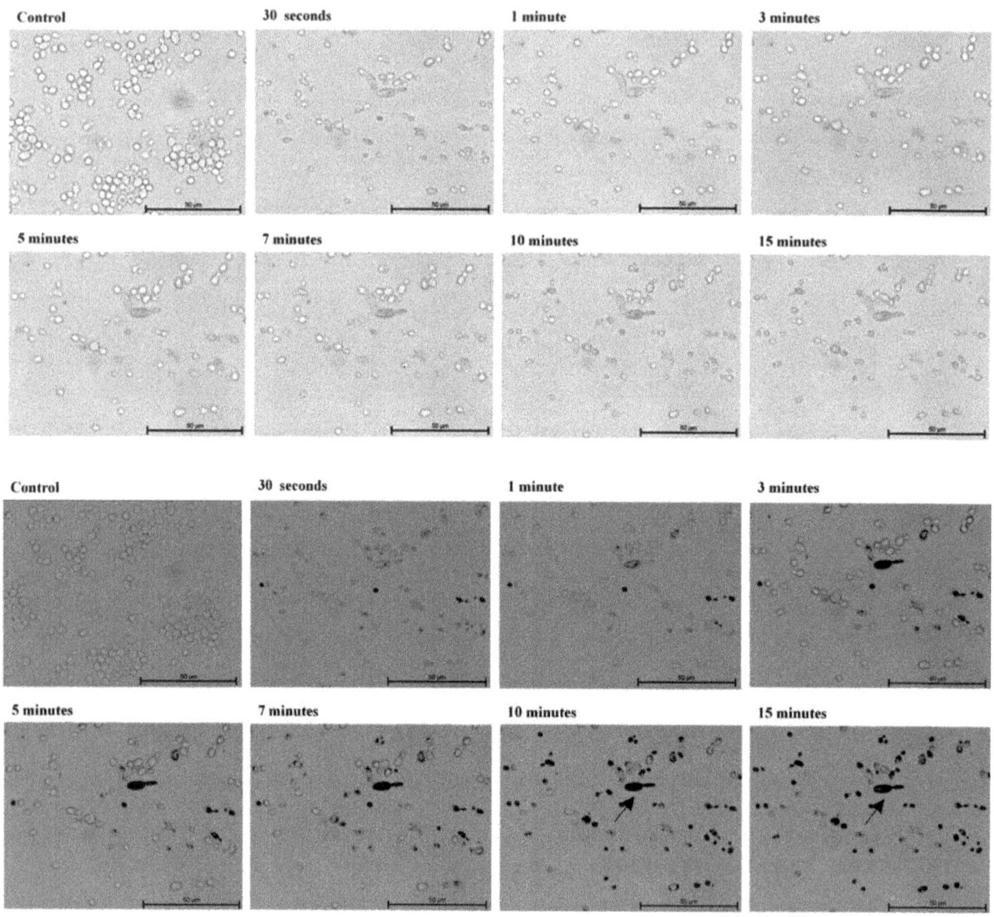

Figure 5. A representative series of original images (two upper rows) and images after isolating indigo and blue colors and converting to black in ImageJ (two lower rows) showing the number of *Candida parapsilosis* ATCC 90,018 cells with absorbed ortho-toluidine blue (TBO) at individual times relative to the control. Magnification 600×. Scale bars = 50 μm. Black arrows indicate yeast cells with gradually disappearing dye in the cytoplasm. American Type Culture Collection (ATCC).

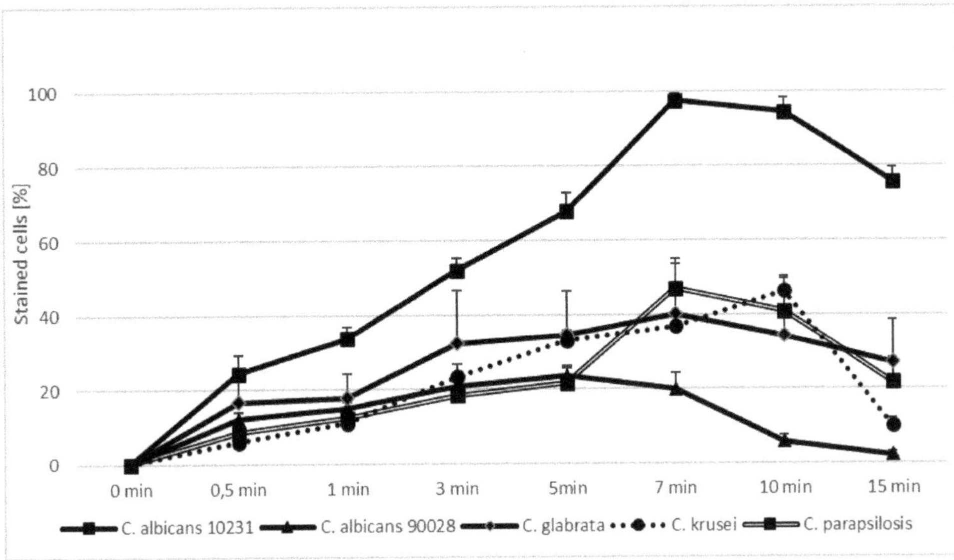

Figure 6. TBO uptake by *Candida albicans* ATCC 10,231, *Candida albicans* ATCC 90,028, *Candida glabrata* ATCC 90,030, *Candida krusei* ATCC 6258 and *Candida parapsilosis* ATCC 90,018 over time (0–15 minutes) by direct observation in an optical microscope. Data are mean values and standard deviations from six replicate experiments. American Type Culture Collection (ATCC).

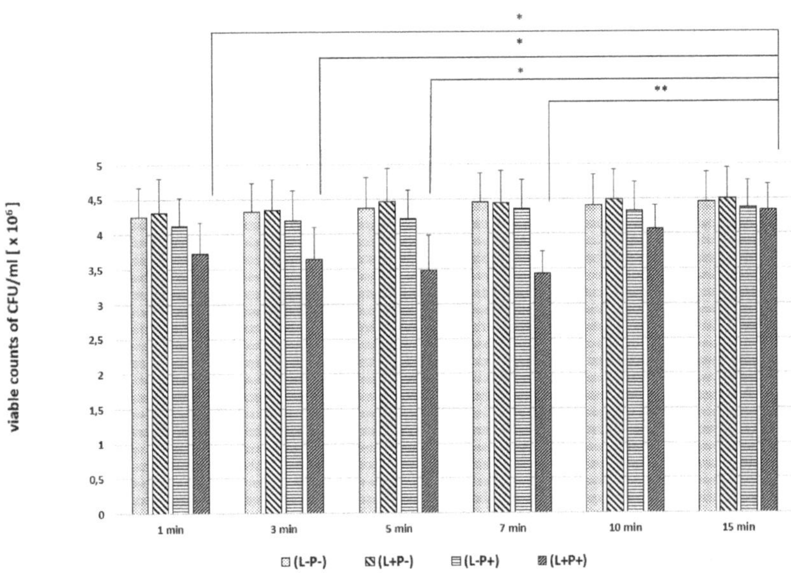

Figure 7. Influence of different incubation times (1–15 min) on the effect of photodynamic inactivation in reduction of number of viable cells (CFU/mL) of *C. albicans* ATCC 90,028 in planktonic form. CFU/mL was determined after TBO-mediated aPDT (L+P+), treatment with light alone (L+P-) or treatment with the photosensitizer alone (L-P+) and compared to negative control treatment (L-P-). Data are mean values and standard deviations from six replicate experiments. * $p < 0.05$, ** $p < 0.01$. *Candida* (C.). American Type Culture Collection (ATCC).

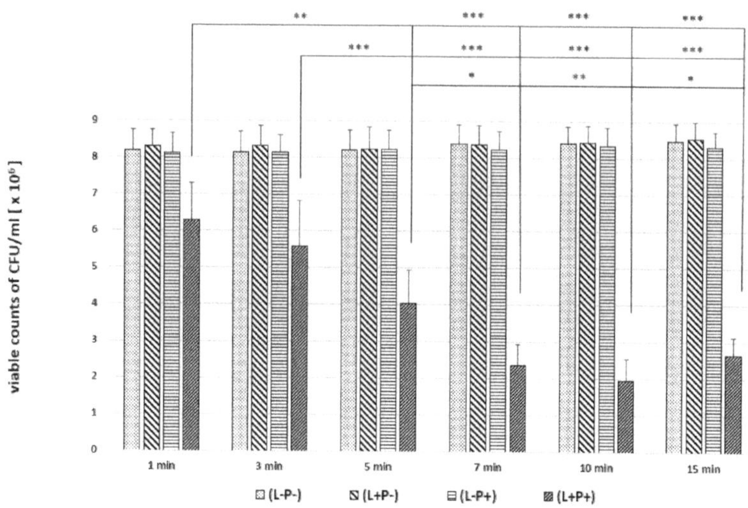

Figure 8. Influence of different incubation times (1–15 min) on the effect of photodynamic inactivation in reduction of number of viable cells (CFU/mL) of *C. albicans* ATCC 10,231 in planktonic form. CFU/mL was determined after TBO-mediated aPDT (L+P+), treatment with light alone (L+P-) or treatment with the photosensitizer alone (L-P+) and compared to negative control treatment (L-P-). Data are mean values and standard deviations from six replicate experiments. * $p < 0.05$, ** $p < 0.01$, *** $p < 0.001$. *Candida* (C.). American Type Culture Collection (ATCC).

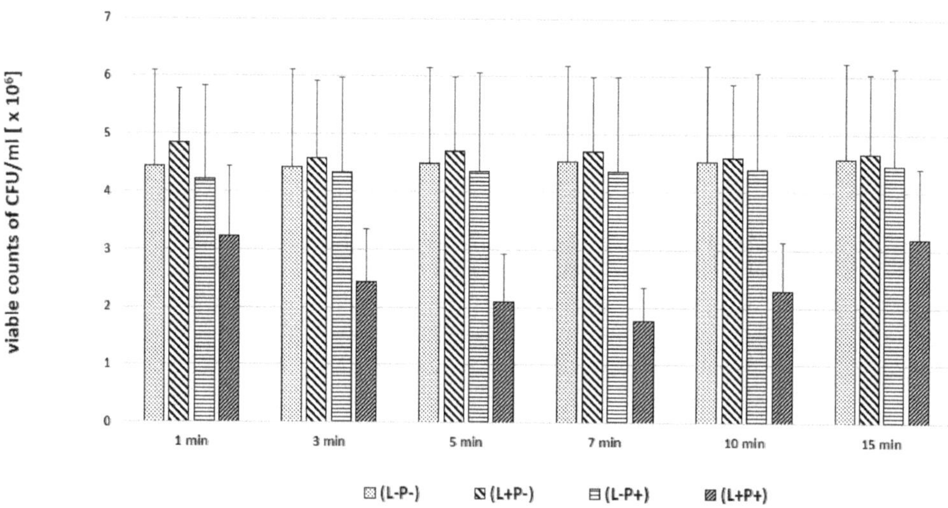

Figure 9. Influence of different incubation times (1–15 min) on the effect of photodynamic inactivation in reduction of number of viable cells (CFU/mL) of *C. glabrata* ATCC 90,030 in planktonic form. CFU/mL was determined after TBO-mediated aPDT (L+P+), treatment with light alone (L+P-) or treatment with the photosensitizer alone (L-P+) and compared to negative control treatment (L-P-). Data are mean values and standard deviations from six replicate experiments. *Candida* (C.). American Type Culture Collection (ATCC).

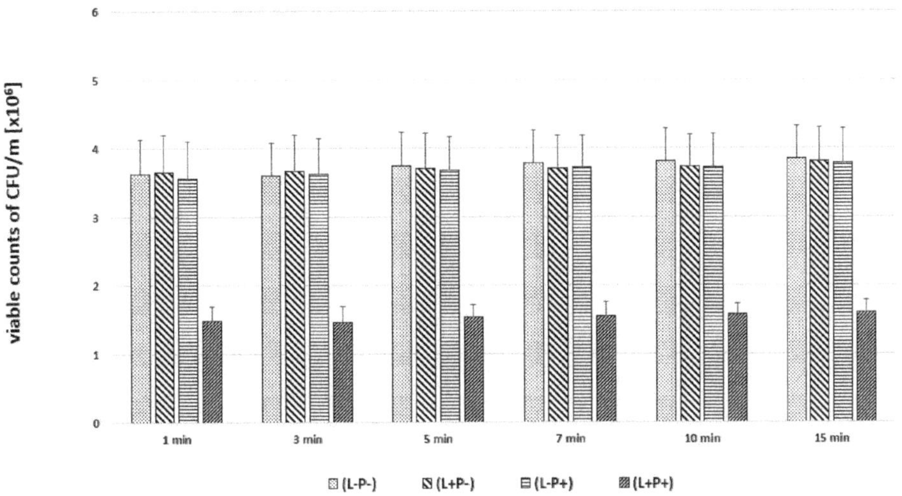

Figure 10. Influence of different incubation times (1–15 min) on the effect of photodynamic inactivation in reduction of number of viable cells (CFU/mL) of *C. krusei* ATCC 6258 in planktonic form. CFU/mL was determined after TBO-mediated aPDT (L+P+), treatment with light alone (L+P-) or treatment with the photosensitizer alone (L-P+) and compared to negative control treatment (L-P-). Data are mean values and standard deviations from six replicate experiments. *Candida* (C.). American Type Culture Collection (ATCC).

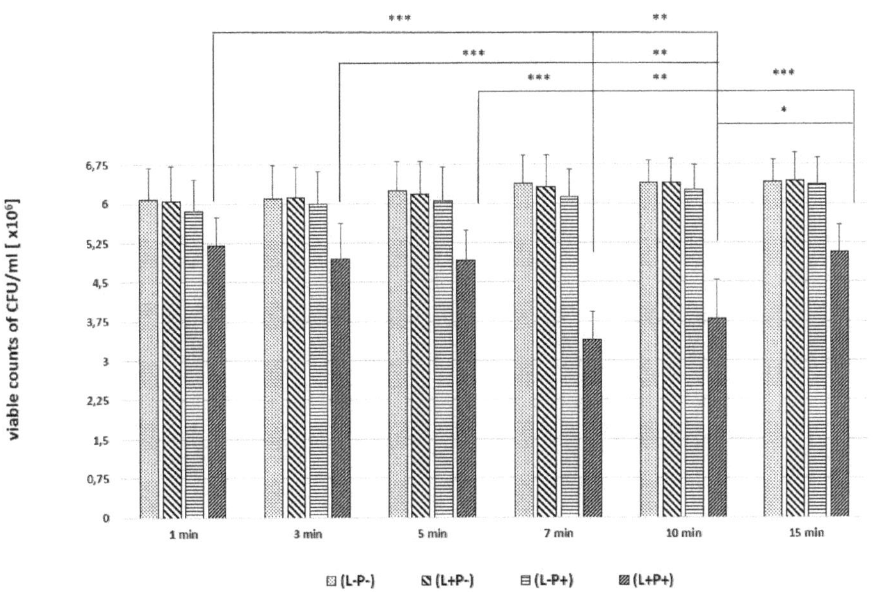

Figure 11. Influence of different incubation times (1–15 min) on the effect of photodynamic inactivation in reduction of number of viable cells (CFU/mL) of *C. parapsilosis* ATCC 90,018 in planktonic form. CFU/mL was determined after TBO-mediated aPDT (L+P+), treatment with light alone (L+P-) or treatment with the photosensitizer alone (L-P+) and compared to negative control treatment (L-P-). Data are mean values and standard deviations from six replicate experiments. * $p < 0.05$, ** $p < 0.01$, *** $p < 0.001$. *Candida* (C.). American Type Culture Collection (ATCC).

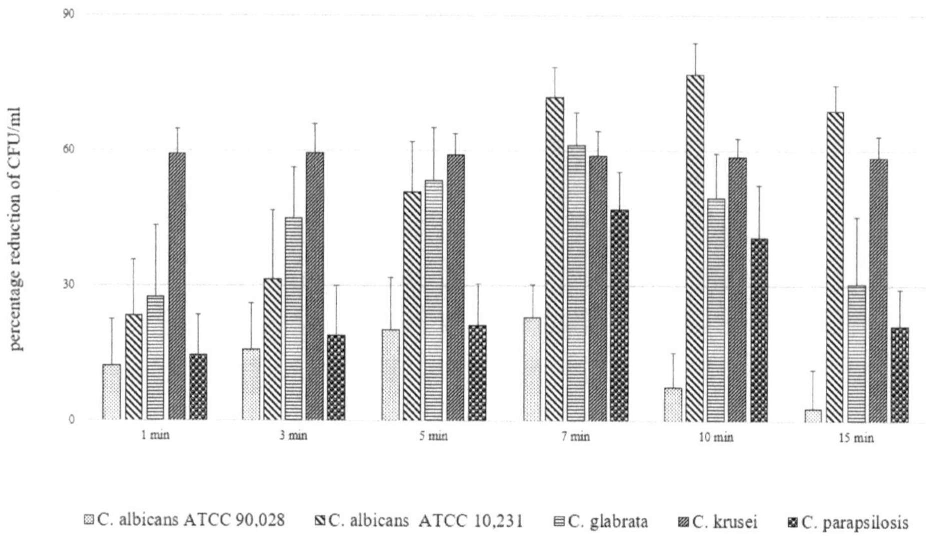

Figure 12. Influence of different incubation times (1–15 min) on the effect of photodynamic inactivation in % reduction of CFU/mL of *C. albicans* ATCC 90,028, *C. albicans* ATCC 10,231, *C. glabrata* ATCC 90,030, *C. krusei* ATCC 6258, *C. parapsilosis* ATCC 90,018 in planktonic form. CFU/mL was determined after TBO-mediated aPDT (L+P+). Data are mean values and standard deviations from six replicate experiments. *Candida* (C.). American Type Culture Collection (ATCC).

3. Discussion

The initial component of photodynamic therapy protocols for efficient TBO-mediated aPDT against yeast, the optimal incubation time, appears to have a major impact on its efficacy. The results of our study confirm this impact, both by direct microscopic observation of in vivo preparations, allowing real-time evaluation of the amount of TBO absorbed by selected *Candida* strains after time intervals, as well as by the efficacy of aPDT (considered as the reduction in the number of CFU cells/mL) carried out with the same parameters of laser settings and the same PS concentration, where the only variable was the incubation time (1–15 min).

The microscopic analysis showed that for *C. albicans* ACTT 10,231, *C. glabrata* and *C. parapsilosis* the incubation time to obtain the highest number of TBO-containing stained cells was 7 min, and the remaining strains reached the maximum after 5 min. (*C. albicans* ATCC 90,018) or 10 min (*C. krusei*). However, in both cases after 7 min the mean % of stained cells was not statistically different from that achieved at the maximum (Figure 6). In the literature, only the study by Chien et al. included direct observation of *C. albicans* by confocal microscopy after 10, 30, 60 min waiting time before irradiation from the moment of mixing the suspension with 0.2 mM TBO [28]. In the present study, the amount of TBO was assessed by spectrophotometric measurement. After 10 min a large amount of dye in the cytoplasm of the cells was noted, by Chien et al. which did not significantly increase after 30 and 60 min. This observation does not correspond to our results, where in most of the strains studied (*C. albicans* 90,028, *C. glabrata*, *C. krusei* and *C. parapsilosis*) PS gradually disappeared from the cytoplasm after 5 min (Figure 1—black arrows) and 10 min (Figures 3–5—black arrows) which would confirm the reports of enzymatic alteration of PS or the possible induction of drug efflux pumps by *Candida* [13,31]. We also noticed that *C. glabrata* uses an additional mechanism of protection against the effects of TBO, consisting of aggregation of individual cells into large, clustered structures that may prevent the PS from reaching cells in the center (Figure 3—black arrows). These structures appeared as early as 3–5 min after the addition of TBO to the medium (Figure 3—red arrows). A similar mechanism was not observed in the other strains.

We found only a few published studies evaluating the efficacy of aPDT with TBO in which one of the variables assessed was the incubation time [24,32,33]. In one of these, a study by Jackson et al. using a He-Ne laser with a wavelength of 632.8 nm (21 J energy) and TBO at a concentration of 25 μg/mL, it was found that the most appropriate PIT to inhibit the growth of planktonic forms of various *C. albicans* strains (including those resistant to azoles) is 5 min. A shorter time (2 min) resulted in a very poor efficacy of aPDT. Of the longer times tested (10, 60 and 180 min), the most favorable was 10 min, and the result was only slightly worse than that achieved in 5 min. In addition, it was noticed that during 180 min in the group where the effect of PS alone was assessed without laser irradiation, TBO had a detrimental effect on the viability of cells [32]. We obtained similar results in the second part of our experiments; the highest efficacy of aPDT (expressed as reduction of CFU/mL) was obtained after a 7 min incubation time (*C. albicans* ATCC 90,018, *C. glabrata*, *C. parapsilosis*) and after 10 min for *C. albicans* ATCC 10,231 and 3 min for *C. krusei* (Figure 12). For *C. krusei*, the efficacy with a 7 and 10 min incubation time did not differ statistically from that after 15 min waiting time (Figure 10). The results of the evaluation of the most favorable incubation time from both parts of our experiment coincide and indicate a 7–10 min PIT period (Table 1). The differences in the results between individual strains indicate the importance of the morphological structure of the cell wall and the resulting sensitivity of the strain to aPDT (like the phenomenon of sensitivity in pharmacotherapy).

A similar 10 min incubation time was recommended in the study by Chien et al., based on the highest efficiency of photodynamic therapy performed with the use of a 635 nm LED (fluence 50 J/cm^2) and TBO at a concentration of 0.2 mM; extending the PIT to 30 or 60 min did not change the percentage of surviving aPDT cells [24]. Completely different results were obtained in the study by Donnelly et al. using a Paterson lamp emitting 635 nm light and TBO at a concentration of 5 mg/mL. The highest % reduction of CFU/mL for planktonic *C. albicans* was obtained with an incubation time of 30 min while a PIT of 5 min (with the same fluency of 100 J/cm^2) resulted in a smaller but statistically significant difference [33]. The appropriate incubation time seems to be even more important in the case of experiments carried out on strains organized in a biofilm. This is connected, among other factors, with the presence of EPS (extracellular polymeric substances) which makes it difficult for a PS to reach individual cells, and with the variety of yeast strains and forms [34]. Studies show that aPDT treatment of biofilms requires longer incubation times and higher TBO concentrations, which results in improved efficacy visible as a reduction of the number of cells in both the yeast and filamentous forms [35,36] and a significant reduction in the mass of EPS (LED 635 nm, fluence 50 J/cm^2, concentration of TBO 2.5 mM, PIT 30 min) [37]. In other studies, using planktonic solutions and biofilms of *Candida* strains with TBO-mediated aPDT, the PIT was not a variable of the protocol; the authors used 30 s to 180 min with the most common of 5 min (nine studies) [25,32,33,35,36,38–41] or 30 min (seven studies) [24,33,37,42–45]. Other incubation times used in that study were 1 min [46,47], 10 min [48] and 20 min [29]. In two other studies of TBO-mediated aPDT, the applied PIT was not reported [49,50]. All these studies showed the efficacy of the therapy as a reduction in the number of cells or % CFU/mL, but only that by Nielsen et al. showed complete elimination of *C. albicans* from the planktonic solution (LED 635 nm, 37.7 J/cm^2, 400 mW, TBO concentration 0.226 mM, PIT 1 min) [47] and the others showed only a partial reduction. No statistically significant difference in efficacy versus control was reported only in the study by Merigo et al. carried out on a *C. albicans* biofilm (diode laser 650 nm, fluence 10 J/cm^2, 30 mW, concentration of TBO 0.1 mM, PIT 5 min) [41]. The reductions in growth for the tested *Candida* strains in our experiment (Figure 12) are not the most effective in comparison to other published studies [36,47]. However, it was important to us in this experiment, to evaluate the incubation time itself as the only protocol variable. The results and experience will be used for other variables of the TBO-mediated aPDT protocol directed against *Candida* strains, which are the parameters of laser settings.

The literature also shows a slightly better efficacy of aPDT with TBO against *C. albicans* than against other *Candida* strains [26,32,40,42,45,46,51], especially those showing clinical resistance to azoles [24,32,40,45] despite the use of comparable experimental conditions. In both parts of our experiment also there was a noticeable difference in the percentages achieved between individual strains (Table 1): in microscopic observation (from 98.1% for *C. albicans* ACTT 10,231 to 23.6% for *C. albicans* ATCC 90,018) and after aPDT (from 78.5% for *C. albicans* ATCC 10,231 to 21.56% for *C. albicans* ATCC 90,018). This large range is probably due to the differences in structure and metabolic activity of individual strains, and a similar phenomenon is observed in the sensitivity of *Candida* strains to various antibiotics and antifungal drugs [35]. Surprisingly, the biggest difference in our results (both parts of our experiment) is that between the two different strains of *C. albicans* used. In many studies comparing the efficacy of TBO-mediated aPDT against *Candida* and bacteria, attention is drawn to the shorter incubation time (1–3 min) required for bacteria, especially Gram-positive [45,50,52,53] which is related to the different morphological structure of bacteria, the lack of membrane enzymes, and the easy access of PS [13]. In our experiments in which a commercial TBO preparation distributed for use in periodontitis and periimplantitis was used to reproduce a clinical setting, it was noted that the photosensitizer is found inside individual cells of all tested strains after 30 s and that the number of such cells gradually increases over time. However, after 3 min the % of the maximum value reached is only (*C. albicans* ACTT 10,231 −53.15%, *C. albicans* ACTT 90,018 −21.2%, *C. glabrata* −31.23%, *C. krusei* −23.7% and *C. parapsilosis* −19.0%). Therefore, when planning experiments in animal and clinical models using aPDT to treat oral candidiasis, care should be taken not to use the short incubation times recommended for bacteria.

4. Materials and Methods

4.1. Organisms and Growth Conditions

This research was carried out on reference strains of *Candida* fungi from the American Type Culture Collection (ATCC, Manassas, VA, USA): *C. albicans* ATCC 90,028, *C. albicans* ATCC 10,231, *C. glabrata* ATCC 90,030, *C. krusei* ATCC 6258 and *C. parapsilosis* ATCC 90,018. These represent the most common strains causing oral candidiasis [52]. Cultures of each strain were placed separately onto Sabouraud dextrose agar plates with addition of 4% glucose (BTL, Łódź, Poland) and incubated in atmospheric air at 37 °C. After 24 h of incubation, a sample of colonies was removed from the surface of the plate and suspended in sterile physiological solution (0.9% NaCl). The number of viable cells in suspension was counted in a spectrophotometer Densimat (bioMerieux, Marcy I'Etoile, France) at a wavelength of 950 nm and using the optical density of McFarland standard number 0.5 equivalent to 106 viable cells/mL.

4.2. Photosensitizer and Laser

Toluidine blue ortho—a watery solution of tolonium chloride whose concentration is proprietary information viscous fluid PAD Smart Solution, (Denfotex, London, UK)—was used for the sensitization of *Candida* strains.

The light source used was a diode laser with a wavelength of 635 nm Smart M Pro (Lasotronix, Piaseczno, Poland), mode-continuous wave (CW), output power of 400 mW, spot size of approx. 0.5 cm^2, flat glass tip 8 mm diameter, energy density (fluence) of 12 J/cm^2, time 30 s. These parameters were based on our previous research [46].

4.3. Microscopic Evaluation of the Absorption of Photosensitizer Particles by Planktonic Cells of Candida Strains in Real Time

First, 50 μL aliquots of a fresh suspension of the *Candida* strains were transferred with a sterile pipette to a glass slide and then 5 μL of TBO solution was added. A coverslip was placed on top, and excess dye was removed with an absorbent paper and towel. For all the strains, 6 replicates were performed, and images were recorded on a Nikon Eclipse 80i optical microscope at 600× magnification after 30 s and 1, 3, 5, 7, 10, and 15 min. Images of the planktonic cells of each strain without the addition of PS were used as controls. To assess the absorption of TBO by counting the % of stained cells in images corresponding to the field of view of the microscope, the original images were further processed and analyzed in ImageJ [53]; the indigo and blue areas (420–500 nm) in the original image were isolated and converted to black for better contrast to detect even small amounts of TBO before counting the cells. Cells with absorbed PS were defined as those whose cytoplasm was at least 50% full of TBO, and only those whose entire circumference was visible were counted. The counting was done manually by two independent researchers and any differences in the scores were resolved through common counting. The results were used for further statistical analysis.

4.4. Experimental Groups and Photodynamic Inactivation of *Candida* spp. *In Vitro*

In total, 720 assays were prepared, 144 for each *Candida* strain tested. These were divided into the following experimental groups: (L+P+) laser irradiation with the presence of photosensitizer (TBO) ($n = 6$); (L-P+) treated only with TBO without laser irradiation ($n = 6$); (L+P-) treated only with laser irradiation without TBO ($n = 6$); (L-P-) no exposure to laser light or TBO used as a negative control group ($n = 6$). In each of the study and control groups, the influence of different incubation times on the effect of reduction of number of viable cells (CFU/mL) were assessed as follows: 200 µL of suspension of each *Candida* strain were added using a sterile pipette to 24 flat-bottom wells of sterile, black 96-well microtiter plates with lids (Nunc, Denmark), leaving 1 well empty between successive samples to avoid cross-diffusion of light. Next, 20 µL of TBO were added to groups (L+P+) and (L-P+). In groups (L+P-) and (L-P-), 20 µL of physiological solution (0.9% NaCl) was added. Then the plates were shaken for 1 min, at 350 rpm at 35 °C in a thermo-shaker PST-60HL-4 (Biosan, Riga, Latvia). In the dark, at room temperature, the cover of the plate was removed and the wells, one by one at the right times (1, 3, 5, 7, 10, and 15 min) were irradiated according to the described protocol, in groups (L+P+) and (L+P-). During the irradiation the laser tip with the flat end (gaussian profile) was mounted on a rack just above the cell suspension (1 mm from the plate surface) and the other wells were covered with a black, matte screen with a hole whose diameter matched the diameter of the laser tip to prevent the spreading of light to neighboring regions. Immediately after irradiation, serial dilutions (10^{-1} to 10^{-6}) were prepared and 100 µL of each dilution were seeded in duplicate onto Sabouraud dextrose agar with addition of 4% glucose (BTL, Łodź, Poland) and incubated for 48 h at 37 °C. After incubation, the number of colony forming units (CFU) were counted with Anacolyte automatic counter (Synbiosis, Cambridge, UK), and the colony-forming units per milliliter CFU/mL were calculated. All tests were performed 6 times and the mean values were used for further statistical analysis.

5. Statistical Analysis

Values were expressed as means ± standard deviation (SD). Statistical differences were evaluated by analysis of variance (ANOVA) and post-hoc comparison with the Newman–Keuls test. A p value of ≤ 0.05 was considered to indicate a statistically significant difference. The statistical analysis was performed using the software Statistica v. 7.1 PL (StatSoft, Krakow, Poland).

6. Conclusions

The optimal incubation time needed for the uptake of TBO by the cells of selected *Candida* strains is 7–10 min. This is confirmed both by direct observation by optical microscopy and by evaluation of the efficacy of TBO-mediated aPDT on planktonic cells of these strains.

Author Contributions: Conceptualization, R.W. and K.G.-L.; methodology R.W.; J.N.; M.P.; P.K.; and K.G.-L.; validation, R.W.; J.N.; D.S.; K.G.-L.; formal analysis R.W.; J.N.; A.K.-K. (Anna Kuśka-Kiełbratowska); A.K.-K. (Anna Kruczek-Kazibudzka) investigation R.W.; J.N.; M.P.; A.K.-K. (Anna Kruczek-Kazibudzka); A.K.-K. (Anna Kuśka-Kiełbratowska); K.G.-L.; data curation R.W.; D.S.; K.G.-L.; writing—original draft preparation, R.W.; J.N.; P.K.; K.G.-L.; writing—review and editing, R.W.; D.S.; K.G.-L.; visualization R.W.; P.K.; A.K.-K. (Anna Kuśka-Kiełbratowska); A.K.-K. (Anna Kruczek-Kazibudzka); supervision D.S.; K.G.-L.; project administration R.W.; K.G.-L. All authors have read and agreed to the published version of the manuscript.

Funding: This research received no external funding.

Institutional Review Board Statement: Not applicable.

Informed Consent Statement: Not applicable.

Data Availability Statement: Not applicable.

Conflicts of Interest: The authors declare no conflict of interest.

References

1. Kwiatkowski, S.; Knap, B.; Przystupski, D.; Saczko, J.; Kędzierska, E.; Knap-Czop, K.; Kotlińska, J.; Michel, O.; Kotowski, K.; Kulbacka, J. Photodynamic therapy–mechanisms, photosensitizers and combinations. *Biomed. Pharmacother.* **2018**, *106*, 1098–1107. [CrossRef]
2. Champeau, M.; Vignoud, S.; Mortier, L.; Mordon, S. Photodynamic therapy for skin cancer: How to enhance drug penetration? *J. Photochem. Photobiol. B Biol.* **2019**, *197*, 111544. [CrossRef]
3. Wu, H.; Minamide, T.; Yano, T. Role of photodynamic therapy in the treatment of esophageal cancer. *Dig. Endosc.* **2019**, *31*, 508–516. [CrossRef] [PubMed]
4. Horne, T.K.; Cronje, M.J. Cancer Tissue Classification, Associated Therapeutic Implications and PDT as an Alternative. *Anticancer. Res.* **2017**, *37*, 2785–2807. [CrossRef]
5. Salvi, G.E.; Stähli, A.; Schmidt, J.C.; Ramseier, C.A.; Sculean, A.; Walter, C. Adjunctive laser or antimicrobial photodynamic therapy to non-surgical mechanical instrumentation in patients with untreated periodontitis: A systematic review and meta-analysis. *J. Clin. Periodontol.* **2019**, *47*, 176–198. [CrossRef]
6. Grzech-Leśniak, K.; Gaspirc, B.; Sculean, A. Clinical and microbiological effects of multiple applications of antibacterial photodynamic therapy in periodontal maintenance patients. A randomized controlled clinical study. *Photodiagnosis Photodyn. Ther.* **2019**, *27*, 44–50. [CrossRef]
7. Grzech-Leśniak, K.; Nowicka, J.; Pajączkowska, M.; Matys, J.; Szymonowicz, M.; Kuropka, P.; Rybak, Z.; Dobrzyński, M.; Dominiak, M. Effects of Nd:YAG laser irradiation on the growth of Candida albicans and Streptococcus mutans: In Vitro Study. *Lasers Med. Sci.* **2018**, *34*, 129–137. [CrossRef] [PubMed]
8. Park, D.; Kim, M.; Choi, J.W.; Baek, J.-H.; Lee, S.H.; Baek, K. Antimicrobial photodynamic therapy efficacy against specific pathogenic periodontitis bacterial species. *Photodiagnosis Photodyn. Ther.* **2020**, *30*, 101688. [CrossRef] [PubMed]
9. Lohse, M.B.; Gulati, M.; Johnson, A.D.; Nobile, C.J. Development and regulation of single- and multi-species Candida albicans biofilms. *Nat. Rev. Microbiol.* **2018**, *16*, 19–31. [CrossRef] [PubMed]
10. Carrera, E.T.; Dias, H.B.; Corbi, S.C.T.; Marcantonio, R.A.; Bernardi, A.C.A.; Bagnato, V.S.; Hamblin, M.R.; Rastelli, A.N.S. The application of antimicrobial photodynamic therapy (aPDT) in dentistry: A critical review. *Laser Phys.* **2016**, *26*, 12300. [CrossRef]
11. Castano, A.P.; Demidova, T.N.; Hamblin, M.R. Mechanisms in photodynamic therapy: Part one—Photosensitizers, photochemistry and cellular localization. *Photodiagnosis Photodyn. Ther.* **2004**, *1*, 279–293. [CrossRef]
12. Calzavara-Pinton, P.; Rossi, M.T.; Sala, R.; Venturini, M. Photodynamic Antifungal Chemotherapy. *Photochem. Photobiol.* **2012**, *88*, 512–522. [CrossRef] [PubMed]
13. Mylona, V.; Anagnostaki, E.; Parker, S.; Cronshaw, M.; Lynch, E.; Grootveld, M. Laser-Assisted aPDT Protocols in Randomized Controlled Clinical Trials in Dentistry: A Systematic Review. *Dent. J.* **2020**, *8*, 107. [CrossRef] [PubMed]
14. D'Ilario, L.; Martinelli, A. Toluidine blue: Aggregation properties and structural aspects. *Model. Simul. Mater. Sci. Eng.* **2006**, *14*, 581–595. [CrossRef]
15. Abrahamse, H.; Hamblin, M.R. New photosensitizers for photodynamic therapy. *Biochem. J.* **2016**, *473*, 347–364. [CrossRef]
16. Hamblin, M.R.; Hasan, T. Photodynamic therapy: A new antimicrobial approach to infectious disease? *Photochem. Photobiol. Sci.* **2004**, *3*, 436–450. [CrossRef]
17. Rodrigues, C.F.; Rodrigues, M.E.; Henriques, M. Promising Alternative Therapeutics for Oral Candidiasis. *Curr. Med. Chem.* **2019**, *26*, 2515–2528. [CrossRef]
18. Garcia-Rubio, R.; De Oliveira, H.C.; Rivera, J.; Trevijano-Contador, N. The Fungal Cell Wall: Candida, Cryptococcus, and Aspergillus Species. *Front. Microbiol.* **2020**, *10*, 2993. [CrossRef]
19. Chaffin, W.L. Candida albicans Cell Wall Proteins. *Microbiol. Mol. Biol. Rev.* **2008**, *72*, 495–544. [CrossRef]
20. Free, S.J. Fungal Cell Wall Organization and Biosynthesis. *Adv. Genet.* **2013**, *81*, 33–82. [CrossRef]

21. Linares, C.E.B.; Giacomelli, S.R.; Altenhofen, D.; Alves, S.H.; Morsch, V.M.; Schetinger, M.R.C. Fluconazole and amphotericin-B resistance are associated with increased catalase and superoxide dismutase activity in Candida albicans and Candida dubliniensis. *Rev. Soc. Bras. Med. Trop.* **2013**, *46*, 752–758. [CrossRef]
22. Johnson, F.; Giulivi, C. Superoxide dismutases and their impact upon human health. *Mol. Asp. Med.* **2005**, *26*, 340–352. [CrossRef] [PubMed]
23. Martchenko, M.; Alarco, A.-M.; Harcus, D.; Whiteway, M. Superoxide Dismutases inCandida albicans: Transcriptional Regulation and Functional Characterization of the Hyphal-inducedSOD5Gene. *Mol. Biol. Cell* **2004**, *15*, 456–467. [CrossRef] [PubMed]
24. Chien, H.-F.; Chen, C.-P.; Chen, Y.-C.; Chang, P.-H.; Tsai, T.; Chen, C.-T. The Use of Chitosan to Enhance Photodynamic Inactivation against Candida albicans and Its Drug-Resistant Clinical Isolates. *Int. J. Mol. Sci.* **2013**, *14*, 7445–7456. [CrossRef]
25. Souza, R.C.; Junqueira, J.C.; Rossoni, R.D.; Pereira, C.A.; Munin, E.; Jorge, A.O.C. Comparison of the photodynamic fungicidal efficacy of methylene blue, toluidine blue, malachite green and low-power laser irradiation alone against Candida albicans. *Lasers Med. Sci.* **2009**, *25*, 385–389. [CrossRef] [PubMed]
26. Wiench, R.; Skaba, D.; Matys, J.; Grzech-Leśniak, K. Efficacy of Toluidine Blue—Mediated Antimicrobial Photodynamic Therapy on *Candida* spp. A Systematic Review. *Antibiotics* **2021**, *10*, 349. [CrossRef]
27. Zeina, B.; Greenman, J.; Purcell, W.; Das, B. Killing of cutaneous microbial species by photodynamic therapy. *Br. J. Dermatol.* **2001**, *144*, 274–278. [CrossRef]
28. Zeina, B.; Greenman, J.; Corry, D.; Purcell, W. Cytotoxic effects of antimicrobial photodynamic therapy on keratinocytes in vitro. *Br. J. Dermatol.* **2002**, *146*, 568–573. [CrossRef]
29. Demidova, T.N.; Hamblin, M.R. Effect of Cell-Photosensitizer Binding and Cell Density on Microbial Photoinactivation. *Antimicrob. Agents Chemother.* **2005**, *49*, 2329–2335. [CrossRef]
30. Donnelly, R.F.; McCarron, P.; Tunney, M. Antifungal photodynamic therapy. *Microbiol. Res.* **2008**, *163*, 1–12. [CrossRef]
31. Prates, R.A.; Kato, I.T.; Ribeiro, M.S.; Tegos, G.P.; Hamblin, M.R. Influence of multidrug efflux systems on methylene blue-mediated photodynamic inactivation of Candida albicans. *J. Antimicrob. Chemother.* **2011**, *66*, 1525–1532. [CrossRef]
32. Jackson, Z.; Meghji, S.; MacRobert, A.; Henderson, B.; Wilson, M. Killing of the Yeast and Hyphal Forms of Candida albicans Using a Light-Activated Antimicrobial Agent. *Lasers Med. Sci.* **1999**, *14*, 150–157. [CrossRef]
33. Donnelly, R.F.; McCarron, P.; Tunney, M.; Woolfson, A.D. Potential of photodynamic therapy in treatment of fungal infections of the mouth. Design and characterisation of a mucoadhesive patch containing toluidine blue O. *J. Photochem. Photobiol. B Biol.* **2007**, *86*, 59–69. [CrossRef] [PubMed]
34. Guo, D.; Yue, H.; Wei, Y.; Huang, G. Genetic regulatory mechanisms of Candida albicans biofilm formation. *Sheng Wu Gong Cheng Xue Bao = Chin. J. Biotechnol.* **2017**, *33*, 1567–1581.
35. Pinto, A.P.; Rosseti, I.B.; Carvalho, M.L.; da Silva, B.G.M.; Alberto-Silva, C.; Costa, M.S. Photodynamic Antimicrobial Chemotherapy (PACT), using Toluidine blue O inhibits the viability of biofilm produced by Candida albicans at different stages of development. *Photodiagn. Photodyn. Ther.* **2018**, *21*, 182–189. [CrossRef] [PubMed]
36. Da Silva, B.G.M.; Carvalho, M.L.; Rosseti, I.B.; Zamuner, S.; Costa, M.S. Photodynamic antimicrobial chemotherapy (PACT) using toluidine blue inhibits both growth and biofilm formation by Candida krusei. *Lasers Med. Sci.* **2018**, *33*, 983–990. [CrossRef]
37. Huang, M.-C.; Shen, M.; Huang, Y.-J.; Lin, H.-C.; Chen, C.-T. Photodynamic Inactivation Potentiates the Susceptibility of Antifungal Agents against the Planktonic and Biofilm Cells of Candida albicans. *Int. J. Mol. Sci.* **2018**, *19*, 434. [CrossRef]
38. Carvalho, G.G.; Felipe, M.P.; Costa, M.S. The photodynamic effect of methylene blue and toluidine blue on Candida albicans is dependent on medium conditions. *J. Microbiol.* **2009**, *47*, 619–623. [CrossRef]
39. Garcia, B.A.; Panariello, B.H.D.; de Freitas Pontes, K.M.; Duarte, S. Regimen and different surfaces interfere with photodynamic therapy on Candida albicans biofilms. *J. Microbiol. Methods* **2020**, *178*, 106080. [CrossRef]
40. Soares, B.M.; da Silva, D.L.; Sousa, G.R.; Amorim, J.C.F.; de Resende, M.A.; Pinotti, M.; Cisalpino, P.S. In vitro photodynamic inactivation of Candida spp. growth and adhesion to buccal epithelial cells. *J. Photochem. Photobiol. B Biol.* **2009**, *94*, 65–70. [CrossRef]
41. Merigo, E.; Chevalier, M.; Conti, S.; Ciociola, T.; Fornaini, C.; Manfredi, M.; Vescovi, P.; Doglio, A. Antimicrobial effect on Candida albicans biofilm by application of different wavelengths and dyes and the synthetic killer decapeptide KP. *Laser Ther.* **2019**, *28*, 180–186. [CrossRef]
42. Sherwani, M.A.; Tufail, S.; Khan, A.A.; Owais, M. Gold Nanoparticle-Photosensitizer Conjugate Based Photodynamic Inactivation of Biofilm Producing Cells: Potential for Treatment of C. albicans Infection in BALB/c Mice. *PLoS ONE* **2015**, *10*, e0131684. [CrossRef]
43. Dai, T.; De Arce, V.J.B.; Tegos, G.P.; Hamblin, M.R. Blue Dye and Red Light, a Dynamic Combination for Prophylaxis and Treatment of Cutaneous Candida albicans Infections in Mice. *Antimicrob. Agents Chemother.* **2011**, *55*, 5710–5717. [CrossRef]
44. Lin, C.-H.; Chien, H.-F.; Lin, M.-H.; Chen, C.-P.; Shen, M.; Chen, C.-T. Chitosan Inhibits the Rehabilitation of Damaged Microbes Induced by Photodynamic Inactivation. *Int. J. Mol. Sci.* **2018**, *19*. [CrossRef]
45. Rodrigues, G.B.; Baruffi, M.D.; Holman, N.; Wainwright, M.; Braga, G. In vitro photodynamic inactivation of Candida species and mouse fibroblasts with phenothiazinium photosensitisers and red light. *Photodiagnosis Photodyn. Ther.* **2013**, *10*, 141–149. [CrossRef] [PubMed]

46. Wiench, R.; Skaba, D.; Stefanik, N.; Kępa, M.; Gilowski, Ł.; Cieślar, G.; Kawczyk-Krupka, A. Assessment of sensitivity of selected Candida strains on antimicrobial photodynamic therapy using diode laser 635 nm and toluidine blue–In vitro research. *Photodiagnosis Photodyn. Ther.* **2019**, *27*, 241–247. [CrossRef]
47. Nielsen, H.K.; Garcia, J.; Væth, M.; Schlafer, S. Comparison of Riboflavin and Toluidine Blue O as Photosensitizers for Photoactivated Disinfection on Endodontic and Periodontal Pathogens In Vitro. *PLoS ONE* **2015**, *10*, e0140720. [CrossRef] [PubMed]
48. Rosseti, I.B.; Chagas, L.R.; Costa, M.S. Photodynamic antimicrobial chemotherapy (PACT) inhibits biofilm formation by Candida albicans, increasing both ROS production and membrane permeability. *Lasers Med. Sci.* **2014**, *29*, 1059–1064. [CrossRef] [PubMed]
49. Barbério, G.S.; da Costa, S.V.; Silva, M.D.S.; de Oliveira, T.M.; Silva, T.C.; Machado, M.A.D.A.M. Photodynamic inactivation of Candida albicans mediated by a low density of light energy. *Lasers Med. Sci.* **2013**, *29*, 907–910. [CrossRef]
50. Decraene, V.; Pratten, J.; Wilson, M. Cellulose Acetate Containing Toluidine Blue and Rose Bengal Is an Effective Antimicrobial Coating when Exposed to White Light. *Appl. Environ. Microbiol.* **2006**, *72*, 4436–4439. [CrossRef]
51. Javed, F.; Samaranayake, L.P.; Romanos, G.E. Treatment of oral fungal infections using antimicrobial photodynamic therapy: A systematic review of currently available evidence. *Photochem. Photobiol. Sci.* **2014**, *13*, 726–734. [CrossRef] [PubMed]
52. Hu, L.; He, C.; Zhao, C.; Chen, X.; Hua, H.; Yan, Z. Characterization of oral candidiasis and the Candida species profile in patients with oral mucosal diseases. *Microb. Pathog.* **2019**, *134*, 103575. [CrossRef] [PubMed]
53. Rasband, W.S. ImageJ.US National Institutes of Health, Bethesda, Maryland, USA, 1997–2018. Available online: https://imagej.nih.gov/ij/ (accessed on 14 March 2021).

International Journal of *Molecular Sciences*

MDPI

Article

Nitrogen Functionalities of Amino-Functionalized Nitrogen-Doped Graphene Quantum Dots for Highly Efficient Enhancement of Antimicrobial Therapy to Eliminate Methicillin-Resistant *Staphylococcus aureus* and Utilization as a Contrast Agent

Wen-Shuo Kuo [1,2,3,4,†], Ping-Ching Wu [5,6,7,8,†], Chi-Yao Hung [9,†], Chia-Yuan Chang [10], Jiu-Yao Wang [3,4], Pei-Chi Chen [3,4], Miao-Hsi Hsieh [3,4], Sheng-Han Lin [11,*], Chan-Chi Chang [12,*] and Yen-Sung Lin [13,14,*]

1 School of Chemistry and Materials Science, Nanjing University of Information Science and Technology, Nanjing 210044, China; wskuo88@gmail.com
2 State Key Laboratory for Chemistry and Molecular Engineering of Medicinal Resources, Guangxi Normal University, Guilin 541004, China
3 Allergy & Clinical Immunology Research Center, National Cheng Kung University Hospital, College of Medicine, National Cheng Kung University, Tainan 701, Taiwan; a122@mail.ncku.edu.tw (J.-Y.W.); simple48686@gmail.com (P.-C.C.); karinadrift@gmail.com (M.-H.H.)
4 Allergy Immunology and Microbiome Center, China Medical University Children's Hospital, China Medical University, Taichung 404, Taiwan
5 Department of Biomedical Engineering, National Cheng Kung University, Tainan 701, Taiwan; wbcxyz@bme.ncku.edu.tw
6 Institute of Oral Medicine and Department of Stomatology, National Cheng Kung University Hospital, College of Medicine, National Cheng Kung University, Tainan 701, Taiwan
7 Center of Applied Nanomedicine, National Cheng Kung University, Tainan 701, Taiwan
8 Medical Device Innovation Center, Taiwan Innovation Center of Medical Devices and Technology, National Cheng Kung University Hospital, College of Medicine, National Cheng Kung University, Tainan 701, Taiwan
9 Department of Physical Medicine and Rehabilitation, An Nan Hospital, China Medical University, Tainan 709, Taiwan; chiyaohung05@gmail.com
10 Department of Mechanical Engineering, National Cheng Kung University, Tainan 701, Taiwan; cychang0829@gs.ncku.edu.tw
11 Department of Anesthesiology, E-Da Hospital, Kaohsiung 824, Taiwan
12 Department of Otolaryngology, National Cheng Kung University Hospital, College of Medicine, National Cheng Kung University, Tainan 701, Taiwan
13 Division of Pulmonary and Critical Care Medicine, An Nan Hospital, China Medical University, Tainan 709, Taiwan
14 Department of Nursing, Chung Hwa University of Medical Technology, Tainan 717, Taiwan
* Correspondence: ed111667@edah.org.tw (S.-H.L.); 109s0015@gs.ncku.edu.tw (C.-C.C.); chestlin@gmail.com (Y.-S.L.)
† The authors contributed equally to this work.

Citation: Kuo, W.-S.; Wu, P.-C.; Hung, C.-Y.; Chang, C.-Y.; Wang, J.-Y.; Chen, P.-C.; Hsieh, M.-H.; Lin, S.-H.; Chang, C.-C.; Lin, Y.-S. Nitrogen Functionalities of Amino-Functionalized Nitrogen-Doped Graphene Quantum Dots for Highly Efficient Enhancement of Antimicrobial Therapy to Eliminate Methicillin-Resistant *Staphylococcus aureus* and Utilization as a Contrast Agent. *Int. J. Mol. Sci.* **2021**, *22*, 9695. https://doi.org/10.3390/ijms22189695

Academic Editors: Antonino Mazzaglia, Angela Scala and Enrico Caruso

Received: 12 July 2021
Accepted: 2 September 2021
Published: 7 September 2021

Abstract: There is an urgent need for materials that can efficiently generate reactive oxygen species (ROS) and be used in photodynamic therapy (PDT) as two-photon imaging contrast probes. In this study, graphene quantum dots (GQDs) were subjected to amino group functionalization and nitrogen doping (amino-N-GQDs) via annealing and hydrothermal ammonia autoclave treatments. The synthesized dots could serve as a photosensitizer in PDT and generate more ROS than conventional GQDs under 60-s low-energy (fixed output power: 0.07 $W \cdot cm^{-2}$) excitation exerted by a 670-nm continuous-wave laser. The generated ROS were used to completely eliminate a multidrug-resistant strain of methicillin-resistant *Staphylococcus aureus* (MRSA), a Gram-positive bacterium. Compared with conventional GQDs, the amino-N-GQDs had superior optical properties, including stronger absorption, higher quantum yield (0.34), stronger luminescence, and high stability under exposure. The high photostability and intrinsic luminescence of amino-N-GQDs contribute to their suitability as contrast probes for use in biomedical imaging, in addition to their bacteria tracking and localization abilities. Herein, the dual-modality amino-N-GQDs in PDT easily eliminated multidrug-resistant bacteria, ultimately revealing their potential for use in future clinical applications.

Keywords: graphene quantum dot with nitrogen doping and amino group functionalization; photodynamic therapy; multidrug-resistant methicillin-resistant *Staphylococcus aureus*; reactive oxygen species; contrast probe

1. Introduction

Materials that are primarily based on graphene have exceptional chemical stability and mechanical, thermal, and electronic properties. Thus, these materials have great potential for use in nanodevices. However, currently available graphene-based materials produced by typical physical and chemical routes, including micromechanical cleavage, reduction in exfoliated graphene oxide, and solvothermal synthesis, are generally micrometer-sized graphene sheets, limiting their direct applications in nanodevices [1]. As a result, there is an urgent need to develop effective routes for cutting large graphene sheets into nanometer-sized pieces with a well-confined shape, such as graphene nanoribbons and graphene quantum dots (GQDs) [2]. GQDs are a novel class of carbon materials that are smaller than 10 nm. They were first reported in a 2004 study where single-walled carbon nanotubes were purified through preparative electrophoresis [3]. GQDs have become increasingly crucial members of the nanocarbon family owing to their benignity, abundance, and low cost. Carbon is usually black and has been recently thought to have weak fluorescence and low solubility in water [4]. As GQDs exhibit strong fluorescence, they have attracted considerable attention, attaining the moniker "fluorescent carbons" [5]. Compared with conventional organic dyes, photoluminescent GQDs possess superior properties, such as aqueous solubility, easy functionalization, high resistance to photobleaching, excellent chemical inertness, high biocompatibility, and low toxicity [6]. Consequently, their potential for use in biological labeling, bioimaging, and drug delivery has been widely explored. GQDs have been found to emit visible to near-infrared (NIR) photoluminescence (PL) under light excitation, which is of particular interest. NIR PL emissions from GQDs excited using NIR light are particularly critical for bionanotechnology applications due to body tissue transparency in the NIR window for water [7]. PL emissions from GQDs can be efficiently quenched using electron acceptors or donor molecules in solution, which reveal the excellent electron acceptance and donation abilities of photoexcited GQDs. Owing to their notable photoinduced electron transfer properties, GQDs may be suitable for use in photovoltaic devices and light energy conversion, among other applications [8]. GQDs are also promising nanoprobes for sensitive detection [9].

Doping is crucial for semiconductors as it can considerably change the carrier density of a material and, thus, completely alter its intrinsic electrical and optical properties [10,11]. Nitrogen (N) doping or functionalization of GQDs is an extremely helpful method for altering the intrinsic properties of GQDs. N atoms, with atomic sizes comparable to those of carbon atoms, have five valence electrons and are highly accepting of electrons; thus, they provide the adjacent carbon atoms in graphene with a comparatively high level of positive charge density. In molecular systems with an sp^2 bond, carbon can be replaced with nitrogen, which can enable the formation of heterocyclic aromatic compounds. The inherent features of materials that are based on carbon, including their electronic characteristics and local and surface chemical features, can be effectively modified by atom doping. N-atom doping of GQDs (size < 10 nm), which exhibit notable quantum confinement and edge effects, results in the modulation of the dots' chemical composition and bandgap. Because the properties of N-doped GQDs (N-GQDs) can be modulated, these dots should exhibit improved electrochemical, electrocatalytic, and photochemical activity and possess tunable luminescence, which is critical for optoelectronic and bioimaging applications [12]. Accordingly, chemical doping can be considered an effective approach for tailoring the optical, chemical, and electronic properties of graphene. GQDs subjected to amino group functionalization and nitrogen doping (amino-N-GQDs) have electronic properties strongly influenced by chemical modifications that enable strong electron donation in primary amine molecules (a phenomenon referred to as amino group functionalization) [13]. When amino-N-GQDs undergo singlet–triplet splitting, the corresponding intersystem crossing is sufficiently efficient to compete with internal conversion between multiplicity-identical states, which results in simultaneous PL and the creation of reactive oxygen species (ROS) involved in photodynamic therapy (PDT) [14].

In PDT, ROS are formed when molecular oxygen reacts with a photoexcited photosensitizer (PS) exposed to a suitable wavelength of light source and energy [15]. Photosensitized reactions involving oxygen are categorized as type I or II. A light-sensitized (excited) PS can directly react with a suitable substrate (unsaturated lipids, proteins, or nucleic acids) to produce unstable radicals through proton or electron transfer (type I reaction), leading to oxygenated products in the presence of oxygen, such as a superoxide anion radical (O_2^-), hydroxyl radicals (OH.), or hydrogen peroxide (H_2O_2). Then, they react with molecular oxygen to form singlet oxygen ($^1O_2$) through energy transfer (type II reaction) [16]. Cells or bacteria can be irreversibly damaged when ROS undergo oxidative reactions with adjacent biological substrates. In addition to the application of conventional and newly synthesized PSs, the use of materials for PDT is a potentially favorable approach for improving the efficacy of therapy. However, because materials are used in combination with PSs, some studies have reported the induction of PDT mechanisms [11,14]. Few studies have directly employed amino-N-GQDs as PSs for ROS generation in PDT [11,14]. Accordingly, in the present study, amino-N-GQDs were employed in PDT as PSs to eliminate a multidrug-resistant (MDR) strain of methicillin-resistant *Staphylococcus aureus* (MRSA), a Gram-positive bacterium, using a 670-nm continuous-wave laser. A low concentration (0.5 $\mu g \cdot mL^{-1}$) of amino-N-GQDs and a photoexcitation time of 60 s with an energy irradiation dose of 0.07 $W \cdot cm^{-2}$ were found to result in the elimination of all bacteria [17,18]. However, using GQDs instead of amino-N-GQDs in the same treatment resulted in 37% bacterial elimination. The properties of the PL (Em: 696 nm)—optimal irradiation penetration, low energy absorption, and slight scattering—emitted in the NIR region revealed that the derived amino-N-GQDs are promising contrast probes for tracking and localizing amino-N-GQD-treated bacteria and could provide further information regarding the status of the irradiated bacteria. The combination of dual-modality PDT, a contrast agent, and amino-N-GQDs was determined to be promising for eliminating and tracking MDR bacteria.

2. Results and Discussion

The modified Hummers method [19] was employed to prepare a graphene oxide sheet, which was subjected to an ultrasonic shearing reaction to synthesize amino-N-GQDs [20]. The mean lateral size of the amino-N-GQDs was approximately 8.3 nm (Figure S1a), as determined using low-magnification (Figure 1a) and high-resolution transmission electron microscopy (HR-TEM) images (Figure 1b). The interlayer spacing of the as-prepared amino-N-GQDs was 0.213 nm, which corresponded to the *d*-spacing of the $\{1\bar{1}00\}$ lattice planes of graphene (Figure 1b) [6]. Figure 1c displays the crystallinity analysis-derived X-ray diffraction (XRD) spectrum. The diffraction angle of the amino-N-GQD material was nearly $2\theta = 24.3^\circ$, indicating appropriate layer regularity as well as a well-ordered lamellar structure exhibiting a 0.360-nm interlayer distance. The interfringe distances were 0.213 and 0.360 nm, which correspond to the in-plane lattice spacing and basal plane distance of graphite, respectively. The values for amino-N-GQDs are consistent with those reported previously (0.340–0.403 nm) [5]. However, based on the XRD findings, the basal plane was not considerably functionalized. This finding aligns with the theory that, compared with in-plane carbons, graphene has considerably more active edges, and functional groups are present at the amino-N-GQD edges [15]. The large basal spacing of the amino-N-GQDs is due to the accommodation of several oxygen species, such as epoxy, hydroxyl, and amino groups, and alteration of the carbon hexahedron grid plane, resulting in greater graphene layer spacing. Various peaks were present in the UV-vis absorption spectrum of the amino-N-GQDs, specifically at 315 nm ($n \to \pi^*$ transition of the C=O shoulder and C–N) and 212 nm ($\pi \to \pi^*$ transition of the aromatic C=C bonds; Figure 1d). The absorption peaks in the spectrum of amino-N-GQDs containing oxygen corresponded to the π electron transition and revealed that the GQDs were successfully doped with N. The crystallinity of the amino-N-GQDs was investigated using Raman spectroscopy. The peak at 1384 cm^{-1} (D band) was identified to be associated with the vibration of the sp3-carbon atoms in disordered graphite, while the peak at 1606 cm^{-1} (G band) corresponded with the vibration of the sp2 hybrid carbon atoms. Further, the integrated D to G band intensity ratio (I_D/I_G), which indicates the degree of graphitization, was approximately 0.90, indicating that the amino-N-GQDs were highly distorted (Figure 1e) [21]. The I_D/I_G ratios were used as inputs in the Raman calculations to determine the mean sp^2 domain size of the GQD-based specimens [22]. The estimated size almost matched the one obtained from HR-TEM, although the value obtained from the Raman calculations (~8.0 nm) was slightly lower due to the Raman estimation (Equations (S2) and (S3)) [22]. The exposed functional groups of the amino-N-GQDs were directly analyzed using Fourier-transform infrared (FTIR) spectroscopy, revealing characteristic bands corresponding to C–O stretching at approximately 1023 cm^{-1} (band 1), C–N stretching at approximately 1150 cm^{-1} (band 2), N–C=O stretching at approximately 1212 cm^{-1} (band 3), tertiary alcohol C–OH bonding at approximately 1448 cm^{-1} (band 4), a C=C ring at approximately 1611 cm^{-1} (band 5), N–H bonding and amide at approximately 1786 cm^{-1} (band 6), C=O stretching at approximately 1833 cm^{-1} (band 7), N–H stretching at approximately 2971 cm^{-1} (band 8), and N–H vibration at approximately 3455 cm^{-1} (band 9), ultimately indicating the carbonyl, hydroxyl, and amino groups exposed on the material (Figure 1f). Because of these exposed functional groups, the amino-N-GQDs had a surface charge of 16.4 mV (in ddH_2O) based on the zeta potential findings. X-ray photoelectron spectroscopy (XPS) was also performed to evaluate the surface chemistry of the amino-N-GQDs. The deconvoluted C(1s) spectrum indicated a non-oxygenated ring (C–C/C=C, 285.0 eV), C–N (286.6 eV), C–O (287.1 eV), and carbonyl (C=O, 288.0 eV) bonds (Figure 1g). Moreover, the deconvoluted N(1s) spectrum indicated pyridinic N (398.3 eV), amino N (NH_2, 398.9 eV), pyrrolic N (399.6 eV), quaternary N (400.4 eV), and amide N (O=C–N, 401.2 eV; Figure 1h). The table in Figure 1 summarizes the bonding composition and atomic ratio of amino-N-GQDs. The characterization results indicated the successful preparation of amino-N-GQDs.

The continuous increase in MDR bacterial strains is a serious medical problem due to their ability to develop resistance to antibiotics through different pathways [23]. Because of the growing resistance of the Gram-positive bacteria, MRSA, to conventional antimicrobial agents, it is desirable to develop alternative approaches to eliminate MDR bacterial strains. This study employed an experimental MRSA template. The surface protein, protein A, on the cell wall of MRSA was considered. Accordingly, the material was coated with an $Ab_{\text{protein A}}$ antibody to form the material, $Ab_{\text{protein A}}$ hybrid, assuming it would possess enhanced specificity, selectivity, and efficiency. A colony-forming unit (CFU) assay was performed to examine material biocompatibility. The number of bacterial cells was calculated as $\log_{10}$ CFU per milliliter ($\log_{10}$ CFU/mL) and is expressed as a percentage (Figure 2a,b). The CFU assay revealed that the amino-N-GQDs were highly biocompatible. Additionally, the toxicity of the material was found to contribute to the elimination of bacteria in PDT.

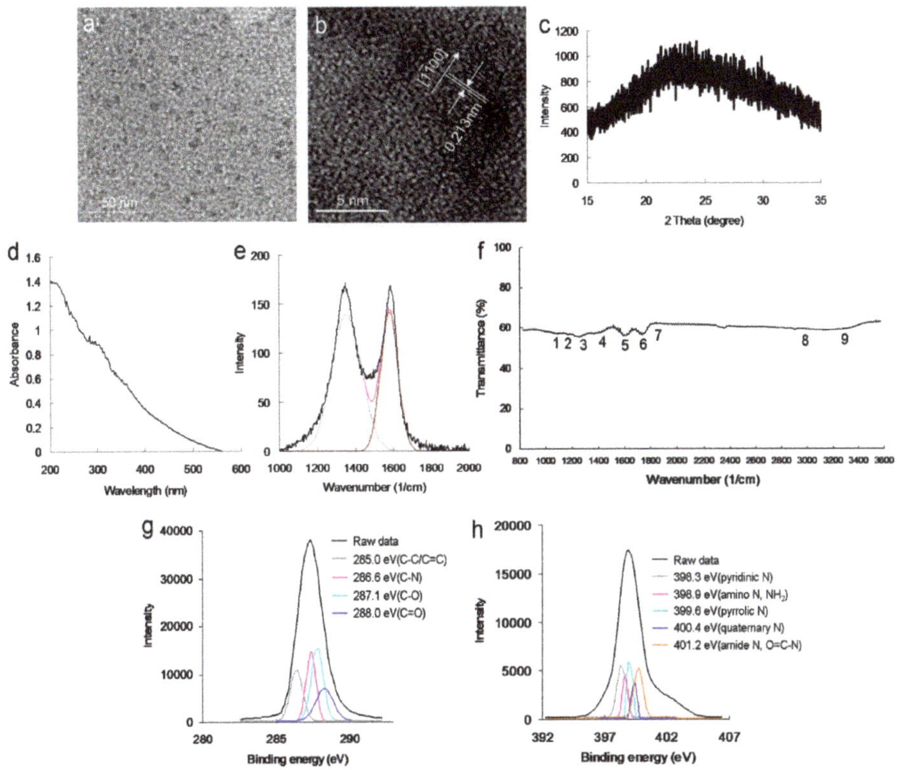

Atomic ratio O(1s)/C(1s)		Carbon bonding composition (%)			
		C-C/C=C	C-N	C-O	C=O
28.4%		22	27	32	19
N(1s)/C(1s)	Nitrogen bonding composition (%)				
	pyridinic	amino	pyrrolic	quaternary	amide
5.2%	25	15	22	11	27

Figure 1. (**a**) Low-magnified transmission electron microscopy (TEM) image; (**b**) High-resolution (HR) TEM image captured for a single amino-N-graphene quantum dot (GQD) (mean lateral size = ~8.3 nm; interlayer spacing = 0.213 nm); the size distribution of its histogram was determined using dynamic light scattering (DLS) (Figure S1a). (**c**) X-ray diffraction (XRD) pattern of amino-N-GQDs. (**d**) Raman spectrum of amino-N-GQDs, with the gray and brown lines representing the spectrum decomposed and fitted to the D- and G-band peaks (at ~1384 and 1606 cm^{-1}, respectively; black line: raw data; pink line: decomposed spectrum). (**e**) Ultraviolet-visible (UV-vis), and (**f**) Fourier-transform infrared (FTIR) spectra of amino-N-GQDs. (**g**,**h**) X-ray photoelectron spectroscopy (XPS) spectrum of amino-N-GQDs obtained to determine the changes in the chemical state. In the deconvoluted C(1s) spectrum, the peaks were fitted using the Gaussian function corresponding to C–C/C=C, C–N, C–O, and C=O. In the deconvoluted N(1s) spectrum, the peaks were fitted using the Gaussian function corresponding to pyridinic N, amino N, pyrrolic N, quaternary N, and amide N. The table summarizes the bonding composition and atomic ratio of the amino-N-GQDs. The O(1s)/C(1s) and N(1s)/C(1s) atomic ratios were 28.4% and 5.2%, respectively. Delivered dose: 0.5 μg·mL^{-1} to 5 mg·mL^{-1} material.

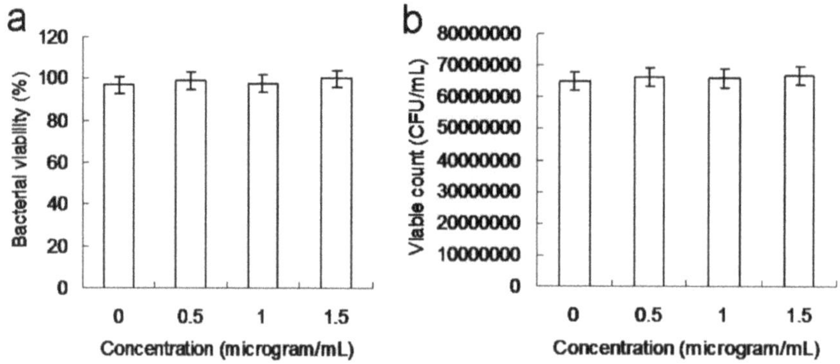

Figure 2. Number of surviving (**a**) amino-N-GQD–$Ab_{protein\ A}$-treated methicillin-resistant *Staphylococcus aureus* (MRSA) cells (expressed as a percentage) based on the colony-forming unit (CFU) counting assay, and (**b**) MRSA cells (unit: CFU/mL). Delivered dose: OD_{600} = ~0.05 μg·mL^{-1} for bacteria and 0–1.5 μg·mL^{-1} for material–$Ab_{protein\ A}$. Data are presented as mean ± SD (n = 6).

The experiment would have been compromised if the amino-N-GQD samples were exposed to white light. To prevent the possibility of confounding the PDT process by inadvertently exposing the amino-N-GQD samples to white light, experiments related to PDT were conducted in the dark. Indeed, the viabilities of amino-N-GQD (0.5–1.5 μg·mL^{-1})-treated MRSA decreased from approximately 7% to 15%, which was somewhat influenced by light illumination (not in the dark) (Figure S2a,b). A low dose (0.5 μg·mL^{-1}) of amino-N-GQDs that had been incubated for 3 h with MRSA (OD_{600}: ~0.05) at 37 °C in the dark was employed in all subsequent experiments.

To evaluate the antimicrobial potential of the material, PDT experiments were conducted against MRSA using low-dose energy irradiation (0.07 W·cm^{-2}) with an exposure time of 0–60 s. In addition, the effect of the photoexcited material, $Ab_{protein\ A}$, on the viability of MRSA was determined. No bactericidal effects were observed for MRSA alone, with or without laser exposure, or on the panel of material-treated bacteria when laser treatment was not applied (Figure 3a). After photoexcitation for 30 s, the viability of amino-N-GQD–$Ab_{protein\ A}$ was relatively low (approximately 36%); this value corresponded to a reduction of approximately 0.449 $\log_{10}$ (Figure S2a). In contrast, the bacterial viability was higher in the panel treated with amino-N-GQDs (51%, approximately 0.031 $\log_{10}$ reduction; Figure S3a) without antibody coating. This finding indicates that the $Ab_{protein\ A}$ antibody coated the material, resulting in greater selectivity and specificity. After 60 s, the amino-N-GQD–$Ab_{protein\ A}$-treated bacteria were completely eliminated (100% elimination; Figure 3b). Subsequently, we investigated whether bacterial viability was affected by N-bonding composition and amino functional groups. In identical photoexcitation experiments, the amino-N-free GQDs (characterizations, Figure 4; Figures S1b, S2a,b, S4 and S5; the characteristics of single GQDs and amino-N-GQDs were compared and are shown in Table S1) had lower bactericidal ability than the amino-N-GQDs (Figure 3). Specifically, viability levels of 75% and 37% (~0.114 and ~0.429 $\log_{10}$ reductions, respectively; Figure S3b) were observed for the GQD–$Ab_{protein\ A}$-treated MRSA after different laser exposure times. Furthermore, because lipopolysaccharide (LPS) is a major component of the outer membrane of Gram-negative *Escherichia coli* (*E. coli*), GQD-Ab_{LPS} and amino-N-GQD-Ab_{LPS} also eliminated *E. coli* (Figures S2c,d, S6 and S7), exhibiting a trend similar to that shown in Figure 3.

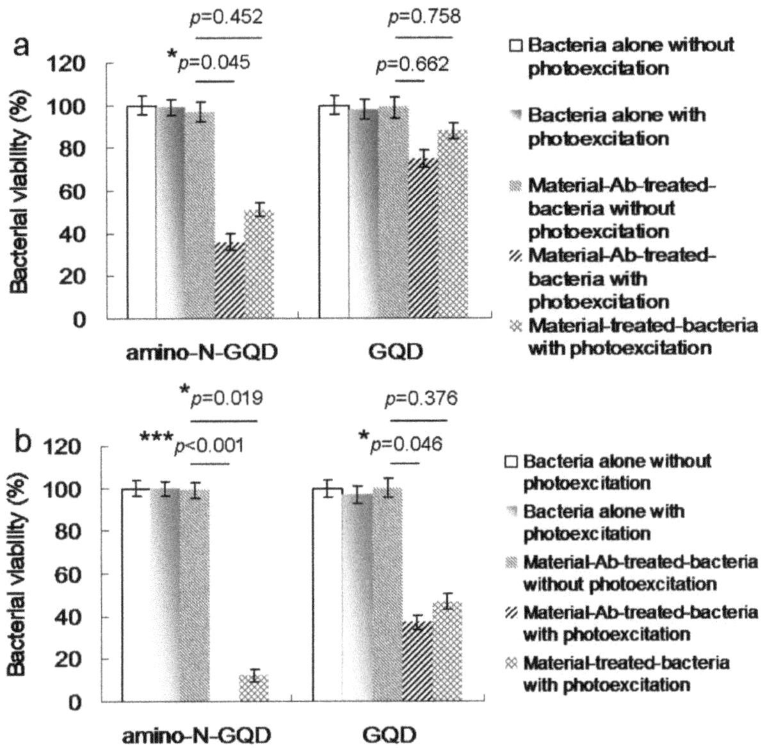

Figure 3. MRSA viability (%) based on the viable count of material–Ab$_{\text{protein A}}$-treated MRSA obtained using the CFU assay under a short 670-nm laser excitation (0.07 W·cm^{-2}) for (**a**) 30 and (**b**) 60 s. Delivered dose: OD600 = ~0.05 μg·mL^{-1} for bacteria and 0.5 μg·mL^{-1} for material–Ab$_{\text{protein A}}$. Data are presented as mean ± SD (n = 6). For amino-N-GQD- and GQD–Ab$_{\text{protein A}}$-treated MRSA with photoexcitation, (**a**) p = 0.045 and p = 0.662, and (**b**) p < 0.001 and p = 0.046, respectively. For amino-N-GQD- and GQD-treated MRSA with photoexcitation, (**a**) p = 0.425 and p = 0.758, and (**b**) p = 0.019 and p = 0.376, respectively. * p value was calculated using the Student's t test (* p < 0.05, ** p < 0.01, *** p < 0.001).

The different incubation times had no influence on the bacterial viability; however, the viability level visibly decreased with dose (Table 1). Based on the results, the amino-N-GQDs performed PDT more efficiently than amino-N-free GQDs, amino-group-free N-GQDs, and amino-GQDs, indicating their superior efficacy. With respect to the amino-N-GQD-eliminated bacteria, which were subjected to the same treatment procedure, the experiments with the nitrogen and amino functional groups demonstrated superior bactericidal capability compared with those without these compositions.

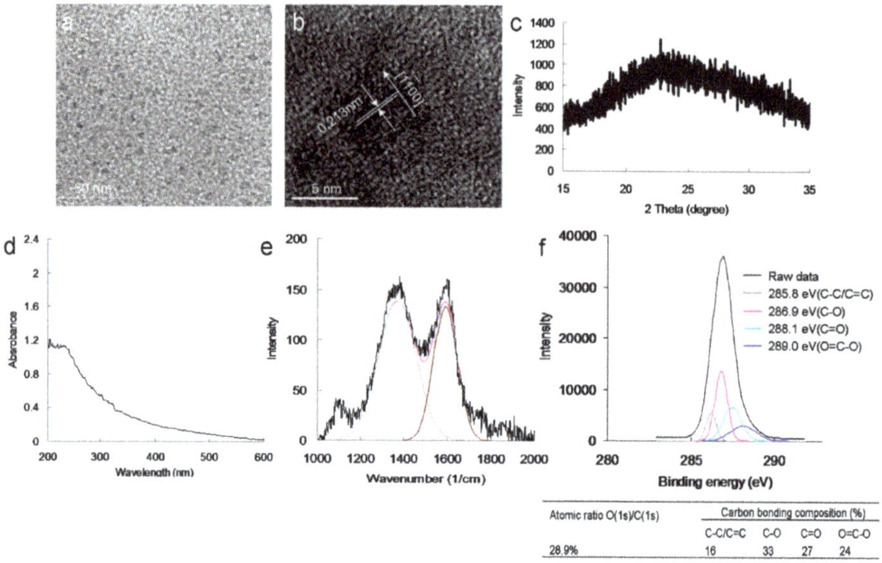

Figure 4. (**a**) Low-magnified TEM image; (**b**) HR-TEM image of a single GQD exhibiting the graphene {1$\bar{1}$00} lattice plane (mean lateral size = 8.0 nm; *d*-spacing = 0.213 nm); the size distribution of its histogram was determined using DLS (Figure S1b). (**c**) XRD spectrum of GQDs, indicating a diffraction angle 2θ of nearly 24.3° and revealing a well-ordered lamellar structure with favorable layer regularity and an interlayer distance of 0.360 nm. (**d**) UV-vis spectrum of GQDs. The peak at 211 nm corresponds to the $\pi \rightarrow \pi^*$ transition of aromatic C=C bonds, whereas that at 312 nm corresponds to the $n \rightarrow \pi^*$ transitions of the C=O shoulder. (**e**) Raman spectrum of GQDs. The gray and brown lines indicate the spectrum decomposed and fitted to D- and G-band peaks (at ~1383 and 1607 cm^{-1}, respectively; black line: raw data; pink line: decomposed spectrum). The ID/IG integrated intensity ratio was 0.89. The diameter obtained from the Raman calculations (Equations (S2) and (S3)) was approximately 7.8 nm. The estimated size almost matched that obtained from the HR-TEM calculations. (**f**) Deconvoluted C(1s) XPS spectra with Gaussian function-fitted peaks: nonoxygenated ring (C–C/C=C, 285.8 eV), hydroxyl (C–O, 286.9 eV), carbonyl (C=O, 288.1 eV), and carboxylate (O=C–O, 289.0 eV) groups. The table summarizes the bonding composition and atomic ratio of the GQDs. Delivered dose: from 0.5 μg·mL^{-1} to 5 mg·mL^{-1} material.

Table 1. MRSA viability (%), determined from the viable count of material–Ab$_{\text{protein A}}$-treated MRSA obtained using the CFU assay under a short 670-nm laser excitation. Data are presented as mean ± SD (n = 6).

0.5 μg·mL^{-1}	**GQD**		**N-GQD**		**amino-GQD**		**amino-N-GQD**	
	3 h-Incubation	**6 h-Incubation**	**3 h-Incubation**	**6 h-Incubation**	**3 h-Incubation**	**6 h-Incubation**	**3 h-Incubation**	**6 h-Incubation**
0-s photoexcitation	99 ± 1%	99 ± 2%	100 ± 1%	99 ± 1%	98 ± 2%	99 ± 2%	100 ± 1%	99 ± 1%
30-s photoexcitation	75 ± 4%	76 ± 3%	66 ± 2%	68 ± 3%	67 ± 4%	70 ± 2%	36 ± 3%	35 ± 2%
60-s photoexcitation	37 ± 3%	35 ± 3%	24 ± 2%	23 ± 2%	24 ± 3%	26 ± 2%	0%	0%
1.0 μg·mL^{-1}	**GQD**		**N-GQD**		**amino-GQD**		**amino-N-GGD**	
	3 h-Incubation	**6 h-Incubation**	**3 h-Incubation**	**6 h-Incubation**	**3 h-Incubation**	**6 h-Incubation**	**3 h-Incubation**	**6 h-Incubation**
0-s photoexcitation	99 ± 2%	100 ± 1%	98 ± 1%	99 ± 2%	99 ± 1%	98 ± 2%	99 ± 2%	100 ± 1%
30-s photoexcitation	43 ± 3%	44 ± 3%	35 ± 3%	35 ± 4%	34 ± 4%	35 ± 5%	17 ± 2%	16 ± 2%
60-s photoexcitation	19 ± 2%	20 ± 3%	10 ± 1%	11 ± 1%	10 ± 2%	12 ± 3%	0%	0%

The aforementioned results correspond to the generated ROS, which originate from the PSs of the materials and may be influential in PDT. The oxidative stress of ROS may contribute to failure in the normal redox reaction functioning of the biological substrates surrounding bacteria treated with photoexcited material and may result in damage to DNA, ultimately causing death of the bacteria. Therefore, $^1O_2$ and O_2^- involved in PDT (photoexcited using a 670-nm laser) were detected. The intensity of $^1O_2$ was measured by monitoring the fluorescence intensity of the singlet oxygen sensor green (SOSG) reagent and trans-1-(2′-methoxyvinyl)pyrene (*t*-MVP). Moreover, the intensity of O_2^- was detected using the 470-nm absorbance of 2,3-bis (2-methoxy-4-nitro-5-sulfophenyl)-2H-tetrazolium-5-carboxanilide (XTT) and the absorbance of γ-L-glutamyl-L-cysteinyl-glycine (GSH) after the bacteria were incubated with the material [15,17]. The materials generated $^1O_2$ and O_2^- in a photoexcitation time-dependent manner (photoexcitation time, 0–60 s; Table S2). However, false-positive ROS signals may have occurred due to material–ROS reagent interaction (ROS reagents included SOSG reagent, *t*-MVP, XTT, and GSH), and these false positives may have compromised the results [24]. Therefore, material–$Ab_{protein\ A}$-treated MRSA was introduced and monitored (Table 2) to detect the amount of ROS formed from the laser-irradiated material. The results revealed a trend similar to that shown in Table S2, and that of material Ab_{LPS}-treated *E. coli* (Table S3). However, the amino-N-GQDs were associated with greater ROS formation capability than the standard GQDs, regardless of the detection method. The amount of generated ROS was reduced after treatment with MRSA and *E. coli* without antibody coating (Tables S4 and S5). In addition, the antibody was successfully coated on the material, and its functions were performed favorably. The material demonstrated its potential application in PDT because of its notable antibacterial effect. To confirm that ROS were involved in the PDT effect of amino-N-GQDs, ROS neutralization was achieved using α-tocopherol [25]. The quantity of ROS decreased after the addition of α-tocopherol, and viability increased, as expected. After this experiment, amino-N-free GQDs were observed using XPS (Figure 4f). The amino-N-free GQDs were less capable of forming $^1O_2$ and O_2^- than amino-N-GQDs (Tables 2 and S2–S5). This result is consistent with the $^1O_2$ phosphorescence signal emitted by the material at 1270 nm (Figure 5a). The amino-N-GQDs generated greater amounts of $^1O_2$ and O_2^- than amino-N-free GQDs. Moreover, the amino-N-GQDs had a higher Φ_Δ (~0.55) than the GQDs (~0.29; for reference, $\Phi_\Delta = 0.64$ is the quantum yield (QY) of meso-tetra [4-sulfonatophenyl] porphine dihydrochloride dissolved in D_2O [18]).

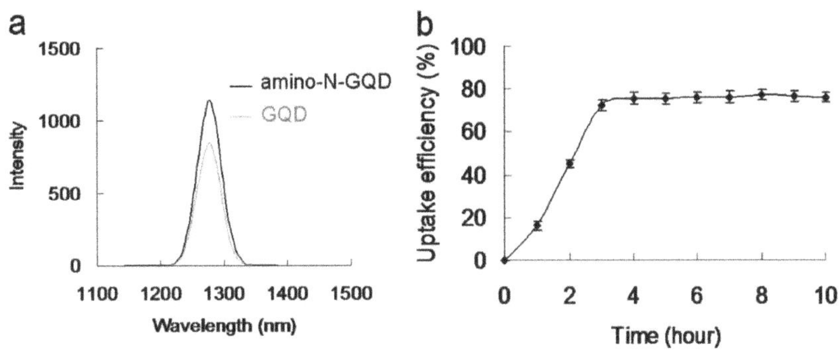

Figure 5. (**a**) Phosphorescence spectra measured at 1270 nm for amino-N-GQDs and GQDs. Delivered dose: 0.5 μg·mL^{-1} for material. (**b**) Uptake assay of MRSA and amino-N-GQD–$Ab_{protein\ A}$, conducted at 37 °C for 10 h. Delivered dose: OD_{600} = ~0.05 μg·mL^{-1} for bacteria and 0.5 μg·mL^{-1} for amino-N-GQD–$Ab_{protein\ A}$. Data are presented as mean ± SD (*n* = 6).

Table 2. Amount of ROS generated by conducting photoexcitation (0.07 W cm^{-2}; Ex: 670 nm) and by using materials–Ab-treated bacteria was maintained in the dark and monitored. Delivered dose: OD_{600} = ~0.05 for MRSA and 0.5 μg·mL^{-1} for material. Data are presented as mean ± SD (n = 6) [6,14,22].

			$^1O_2$ (by SOSG) [c]				
	Negative Control [ac]	**ROS Neutralization** [abc]	**Positive Control** [cd]	**amino-N-GQD**	**ROS Neutralization** [bc]	**GQD**	**ROS Neutralization** [bc]
30-s photoexcitation	239 ± 10	241 ± 12	1956 ± 81	1024 ± 39	241 ± 12	505 ± 23	240 ± 9
60-s photoexcitation	241 ± 11	242 ± 9	2689 ± 137	1735 ± 55	241 ± 11	878 ± 30	242 ± 10
			$^1O_2$ (by *t*-MVP) [e]				
	Negative control [ac]	**ROS neutralization** [abe]	**Positive control** [cd]	**amino-N-GQD**	**ROS neutralization** [be]	**GQD**	**ROS neutralization** [be]
30-s photoexcitation	239 ± 10	241 ± 12	1956 ± 81	1024 ± 39	241 ± 12	505 ± 23	240 ± 9
60-s photoexcitation	241 ± 11	242 ± 9	2689 ± 137	1735 ± 55	241 ± 11	878 ± 30	242 ± 10
30-s photoexcitation	323 ± 19	322 ± 19	6043 ± 141	4107 ± 102	322 ± 21	2149 ± 57	323 ± 20
60-s photoexcitation	326 ± 22	327 ± 21	8584 ± 175	5362 ± 133	328 ± 20	3006 ± 82	327 ± 21
			O_2^- (by XTT) [f]				
	Negative control [ac]	**ROS neutralization** [abf]	**Positive control** [cd]	**amino-N-GQD**	**ROS neutralization** [bf]	**GQD**	**ROS neutralization** [bf]
30-s photoexcitation	0	0	1.13 ± 0.08	0.85 ± 0.07	0.02 ± 0.01	0.55 ± 0.06	0.02 ± 0.01
60-s photoexcitation	0	0	1.99 ± 0.14	1.40 ± 0.11	0.03 ± 0.02	0.87 ± 0.07	0.03 ± 0.01
			O_2^- (by GSH) [g]				
	Negative control [ac]	**ROS neutralization** [abg]	**Positive control** [cd]	**amino-N-GQD**	**ROS neutralization** [bg]	**GQD**	**ROS neutralization** [bg]
30-s photoexcitation	0	0	1.13 ± 0.08	0.85 ± 0.07	0.02 ± 0.01	0.55 ± 0.06	0.02 ± 0.01
60-s photoexcitation	0	0	1.99 ± 0.14	1.40 ± 0.11	0.03 ± 0.02	0.87 ± 0.07	0.03 ± 0.01

[a] Negative control: only treat using reagent and laser radiation without using any material (0 μg·mL^{-1}). [b] ROS neutralization includes nanomaterial treatment, laser irradiation, and 30 ppm of antioxidant α-Tocopherol/methyl linoleate. [c] The singlet oxygen sensor green (SOSG) reagent (Ex/Em: 488/525 nm) has a specific reactivity to generate fluorescence that is recorded using a PL spectrometer. [d] Positive control: treatment of 50 μM of *tert*-butyl hydroperoxide and laser irradiation. [e] trans-1-(2′-methoxyvinyl)pyrene (*t*-MVP) (Ex/Em: 352/465 nm) can react with $^1O_2$, and form a dioxetane intermediate that generates fluorescence upon decomposition to 1-pyrenecarboxaldehyde. The process is monitored using a PL spectrometer. [f] 2,3-bis (2-methoxy-4-nitro-5-sulfophenyl)-2H-tetrazolium-5-carboxanilide (XTT) can interact with O_2^- and produce XTT-formazan that generates strong absorption (wavelength: 470 nm). [g] γ-L-glutamyl-L-cysteinyl-glycine (GSH) containing a thiol tripeptide can prevent damages to cellular or bacterial components caused by stress of oxidation. The thiol group from GSH can be oxidized to the disulfide bond, thus converting GSH to glutathione disulfide. GSH oxidation was used to determine the generated O_2^-. Loss of GSH (%) = (difference between of the absorbance of the sample and negative control/absorbance of negative control) × 100%.

The aforementioned results indicate that N atoms can replace carbon atoms (through atom doping) in sp^2-bonded molecular systems, generating heterocyclic aromatic compounds. Doping is, thus, a productive approach that can be executed to alter the inherent features of materials that are based on C, including electronic features and local and surface chemical characteristics [11]. Moreover, the NH_2 groups that exist at the edge of functionalized amino groups and N-atom-doped GQDs have been reported to have the highest occupied molecular orbital due to a strong orbital interaction with the primary amine [14]. Therefore, the resonance between the delocalized π orbital and the primary amine molecular orbital narrows the orbital bandgap. Consequently, owing to their extraordinary edge and quantum confinement effects, the amino-N-GQDs were determined to have different chemical compositions to the GQDs, enhancing their photochemical, electrochemical, and electrocatalytic activities in optoelectronic and biomedical applications [2,6]. The increased $^1O_2$ due to the amino-N-GQDs may, thus, have been the result of the triplet state yield of the amino-N-GQDs and more intersystem crossing [14]. These factors led to the superior antimicrobial activity of GQDs through the PDT mechanism. TEM was used to obtain images of MRSA treated with amino-N-GQDs exposed to a laser. After MRSA (Figure 6a) was incubated with amino-N-GQD–$Ab_{protein\ A}$ for 3 h (Figure 6b), numerous materials were adsorbed on the bacterial surface [26]. Bacteria must assimilate nutrients and filter external ions to maintain and develop cell wall physiological functions. The uptake assay results revealed that a considerable amount of material was adsorbed on the bacterial surface [26,27], with a burst rate of approximately 72% within the first 3 h of incubation (Figure 5b). This rate saturated from 3 to 10 h. Therefore, the material was adsorbed on the bacterial surface and consequently constituted an external barrier. After 3 h of incubation, the material–$Ab_{protein\ A}$-treated MRSA did not exhibit any exceptional morphology. However, the shape of the MRSA changed considerably after 4 days of incubation (Figure 6c), with a viability equal to nearly 62% (Figure 6e). This finding implies that the bacteria could not function normally and began to die. In particular, the material exhibited bacteriostatic or bactericidal abilities after 4 days of incubation. Moreover, after 3 h of additional incubation, the photoexcited material–$Ab_{protein\ A}$-treated MRSA appeared substantially damaged, which resulted in an abnormal morphology (Figure 6d) corresponding to an elimination rate of almost 100% (Figure 6e).

The surface groups of the amino-N-GQDs may contribute to intrinsic and defect state emissions that involve a PL mechanism [2,7,8,11]. Because of the amino functional groups on the amine-N-GQD surface, radiative electron–hole recombination was induced, which increased the intrinsic state emission [11,12]. The relative QY of amino-N-GQDs was estimated to be 0.34 (as reference, QY_{ref} = 0.28 for Cy5.5 in dimethyl sulfoxide [26,27]), whereas its absolute QY [28,29] was estimated to be 0.33. In contrast to the mentioned amino-N-GQDs, the GQDs were determined to have smaller relative and absolute QYs of 0.11 and 0.13, respectively. The PL spectrum of amino-N-GQDs strongly emitting NIR fluorescence at 696 nm, obtained using a fluorescence spectrometer (excitation: 670 nm), is displayed in Figure 7a. The laser-treated MRSA alone displayed no fluorescence emission (Figure 7b). When irradiation was applied, fluorescence from the materials on the bacterial surface was observed (Figure 7c). After 60 s of irradiation, the bacteria exhibited severe morphological damage. This behavior led to material desorption from the surface of the bacteria and caused an apparent decrease in fluorescence (Figure 7d). The amino-N-GQDs in this study also exhibited photostability, as indicated by their emission intensity after exposure. This photostability resulted in reduced photobleaching (Figure 7e) and high stability in physiological environments, even after 8 weeks (Table S6, which corresponds to Figures S8 and S9). MRSA viability was determined by fluorescence and quantification (Figure 8) [15]. After photoexcitation, the generated ROS required additional incubation to effectively process the PDT action. The green fluorescence indicative of living bacteria (Figure 8a) revealed that the bacteria exposed to laser treatment alone with 3 h of additional incubation were almost completely undamaged, which is consistent with the results presented in Figures 6a and 7b. The photoexcited amino-N-GQD-Ab_{LPS}-treated bacteria without additional incubation also showed nearly no damage (Figure 8b). After 3 h of additional incubation, the same panel revealed that the dead bacteria were distinguishable to an extent (represented by the red fluorescence in Figure 8c), which is consistent with the results presented in Figures 6d and 7d. Bacterial viability was quantified for further antimicrobial testing. The viability tests indicated almost complete elimination of the material-treated bacteria (Figure 8d). Viability was also quantified using a CFU assay (Figures 3b and S3b) to demonstrate the efficient antibacterial effect of the amino-N-GQDs in PDT. No other photochemical activity (i.e., photothermal effect) after photoexcitation was detected in this study.

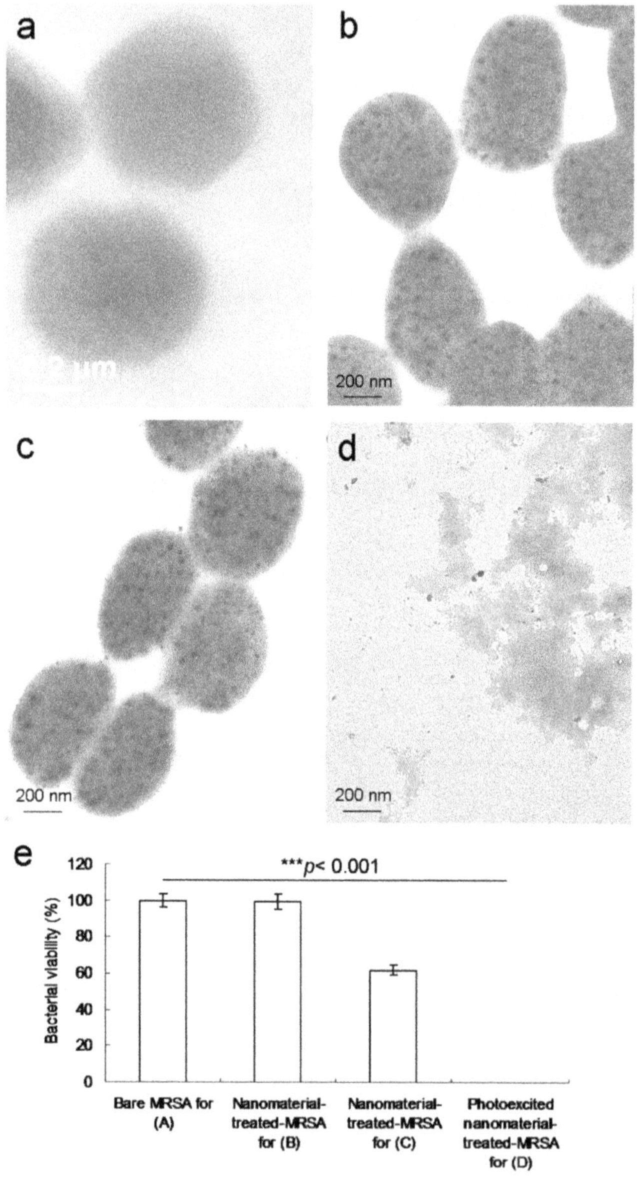

Figure 6. TEM images of (**a**) untreated MRSA; amino-N-GQD–$Ab_{protein\ A}$-treated bacteria after (**b**) 3 h and (**c**) 4 days of incubation; (**d**) material–Ab-treated bacteria after photoexcitation (0.07 W cm^{-2}; Ex: 670 nm) for 60 s followed by 3 h of additional incubation to effectively process the photodynamic therapy (PDT) action. (**e**) MRSA viability (%) based on the viable count of material–$Ab_{protein\ A}$-treated MRSA obtained using the CFU assay in a short excitation period with the same treatment. * p value obtained using the Student's t test (* $p < 0.05$, ** $p < 0.01$, *** $p < 0.001$). Delivered dose: OD_{600} = ~0.05 μg·mL^{-1} for bacteria and 0.5 μg·mL^{-1} for material–$Ab_{protein\ A}$. Data are presented as mean ± SD (n = 6).

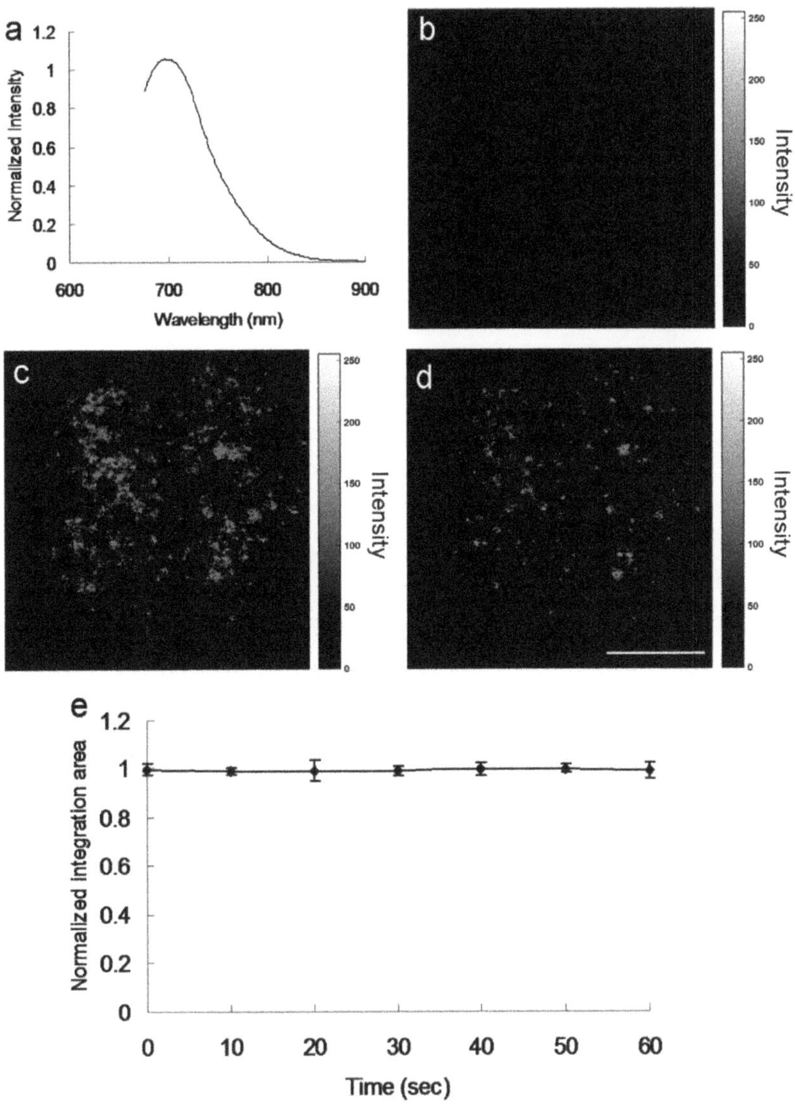

Figure 7. (**a**) Relative PL spectra of amino-N-GQDs exposed to one-photon excitation (Ex/Em: 670/696 nm). Gray-level luminescence images of (**b**) untreated MRSA and (**c**,**d**) amino-N-GQD–$Ab_{protein\ A}$-treated MRSA and material-treated bacteria after photoexcitation for 60 s (0.07 W cm^{-2}; Ex: 670 nm), respectively. On 3 h of additional incubation, the PDT action would be effectively processed. (**e**) Photostability assessment results. For emission wavelengths of 675–900 nm and as a function of exposure time (0−60 s), the relative intensity of the integrated area is almost constant, indicating high photostability. Delivered dose: OD_{600} = ~0.05 μg·mL^{-1} for bacteria and 0.5 μg·mL^{-1} for material–$Ab_{protein\ A}$. Scale bar: 50 μm.

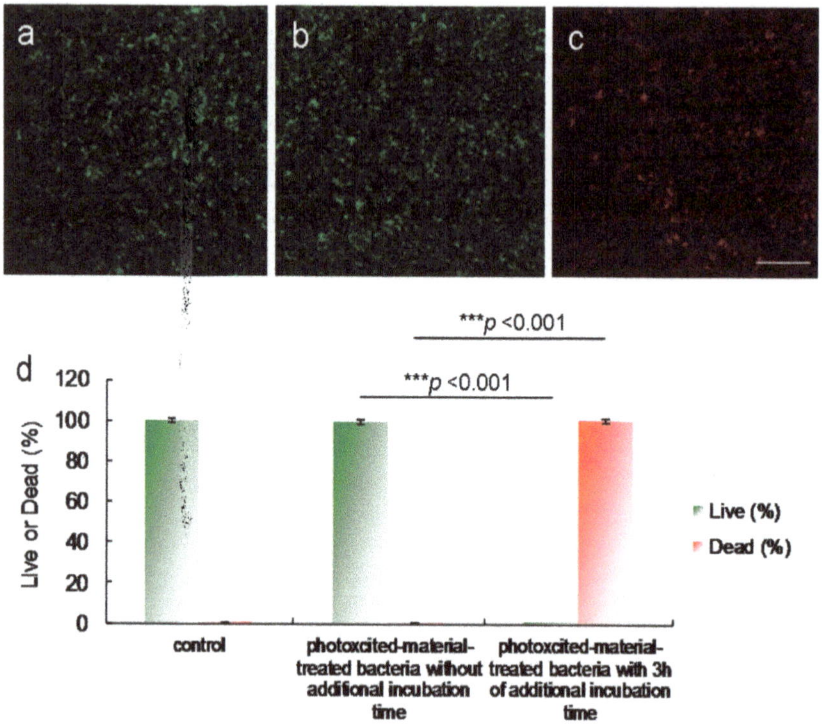

Figure 8. Images obtained after 60-s laser photoexcitation exposure (0.07 W·cm^{-2}; Ex: 670 nm) of (**a**) bacteria alone with 3 h of additional incubation, and amino-N-GQD-Ab$_{\text{protein A}}$-treated MRSA (**b**) without additional incubation, and (**c**) with 3 h of additional incubation. The LIVE/DEAD kit was used to stain bacteria before images were captured. (**d**) Viability (%) determination results. For the percentages of live and dead bacteria, *** $p < 0.001$. * p value was obtained using the Student's t test (* $p < 0.05$, ** $p < 0.01$, *** $p < 0.001$). Delivered dose: OD_{600} = ~0.05 µg·mL^{-1} for bacteria and 0.5 µg·mL^{-1} for material–Ab$_{\text{protein A}}$. Data are presented as mean ± SD (n = 6). Scale bar: 50 µm.

3. Materials and Methods

3.1. Laser Exposure

A continuous-wave diode laser at 670 nm (Coherent, Inc., Santa Clara, CA, USA) was used for the laser irradiation experiments. Material–Ab (0.5 µg·mL^{-1}) was treated with MRSA or *E. coli* (OD_{600} = ~0.05) and placed in a 96-well cell culture plate (Catalog number 174925, Thermo Fisher Scientific, Waltham, MA, USA). Thereafter, the material–Ab-treated-MRSA or -*E. coli* was incubated at 37 °C in the dark. The laser was focused and irradiated on the sample. Relevant experiments were then conducted.

3.2. Singlet Oxygen Quantum Yield (Φ_Δ) Measurement

Φ_Δ was obtained from a previous study. Φ_Δ measurements were conducted in D_2O at 355 nm using *meso*-tetra (4-sulfonatophenyl) porphine dihydrochloride (CAS number 139050-15-0, Sigma Aldrich Co., St. Louis, MO, USA) as a reference (Φ_Δ = 0.64) [18].

3.3. PL QY Measurement

The PL *QY* of the contrast agent is usually the ratio of the emitted photons to the absorbed photons, and is calculated as follows:

$$QY = QY_{ref}\left(\eta^2/\eta^2_{ref}\right)(I/A)\left(A_{ref}/I_{ref}\right) \quad (1)$$

where QY_{ref} is the QY of Cy5.5 (CAS number 2260669-71-2, Thermo Fisher Scientific, Waltham, MA, USA) dissolved in dimethyl sulfoxide (DMSO) as a reference, η is the refractive index of ddH_2O = 1.3333 (η_{ref} of DMSO = 1.479), A is the absorbance at the excitation wavelength, and I is the integrated PL intensity.

All Materials and Methods for this article can be found in the Supplementary Materials.

4. Conclusions

GQDs have been utilized in numerous research fields. However, their use as PSs to directly generate ROS has not received much attention. In this study, we fabricated amino-N-GQDs. The amino-N-GQDs could generate ROS and were effectively used in PDT to eliminate the Gram-positive MDR bacteria, MRSA, at low energy levels within an extremely short photoexcitation period. Amino-N-GQDs have properties, such as high absorption efficiency, strong luminescence, and high stability, making them promising contrast agents for use in biological specimens. Consequently, amino-N-GQDs can be used to perform dual-modality PDT and bioimaging, providing an alternative means for efficiently managing malignant and MDR species.

Supplementary Materials: The following are available online at https://www.mdpi.com/article/10.3390/ijms22189695/s1.

Author Contributions: W.-S.K. initiated and conducted the all experiments. P.-C.W. and C.-Y.H. conducted the all experiments. C.-Y.C. conducted the experiments related to optics & photonics. J.-Y.W., P.-C.C. and M.-H.H. conducted the experiments related to biology and materials. S.-H.L., C.-C.C. and Y.-S.L. designed this work and conducted the all experiments related to biology and materials. The manuscript was written through contributions of all authors. All authors have given approval to the final version of the manuscript, and agree to be accountable for all aspects of the work. All authors have read and agreed to the published version of the manuscript.

Funding: This research was supported by Research Start-Up Fund of Nanjing University of Information Science & Technology, China (2018r047); State Key Laboratory for Chemistry and Molecular Engineering of Medicinal Resources, Guangxi Normal University, China (CMEMR2021-B11); Allergy Immunology and Microbiome Center, China Medical University Children's Hospital, China Medical University, Taiwan (1JA8); An Nan Hospital, China Medical University, Taiwan (ANHRF110-35); E-Da Hospital, Taiwan; Ministry of Science and Technology, Taiwan (110-2221-E-006-013-MY3); Academia Sinica Healthy Longevity Grand Challenge Competition, Taiwan (AS-HLGC-110-07).

Institutional Review Board Statement: Not applicable.

Informed Consent Statement: Not applicable.

Data Availability Statement: Not applicable.

Conflicts of Interest: The authors declare that they have no competing interest.

References

1. Choi, W.; Lahiri, I.; Seelaboyina, R.; Kang, T.S. Synthesis of graphene and its applications: A review. *Crit. Rev. Solid State Mater. Sci.* **2010**, *35*, 52–71. [CrossRef]
2. Heerema, S.J.; Dekker, C. Graphene nanodevices for DNA sequencing. *Nat. Nanotechnol.* **2016**, *11*, 127–1136. [CrossRef]
3. Xu, X.Y.; Ray, R.; Gu, Y.L.; Ploehn, H.J.; Gearheart, L.; Raker, K.; Scrivens, W.A. Electrophoretic analysis and purification of fluorescent single-walled carbon nanotube fragments. *J. Am. Chem. Soc.* **2004**, *126*, 12736–12737. [CrossRef] [PubMed]
4. Wang, Y.; Hu, A. Carbon quantum dots: Synthesis, properties and application. *J. Mater. Chem. C* **2014**, *2*, 6921–6939. [CrossRef]
5. Kuo, W.S.; Shen, X.C.; Chang, C.Y.; Kao, H.F.; Lin, S.H.; Wang, J.Y.; Wu, P.C. Multiplexed graphene quantum dots with excitation-wavelength-independent photoluminescence, as two-photon probes, and in ultraviolet-near infrared bioimaging. *ACS Nano* **2020**, *14*, 11502–11509. [CrossRef] [PubMed]
6. Tabish, T.A.; Scotton, C.J.; Ferguson, D.C.J.; Lin, L.; van der Veen, A.; Lowry, S.; Ali, M.; Jabeen, F.; Ali, M.; Winyard, P.G.; et al. Biocompatibility and toxicity of graphene quantum dots for potential application in photodynamic therapy. *Nanomedicine* **2018**, *13*, 1923–1937. [CrossRef]
7. Li, H.T.; He, X.D.; Liu, Y.; Huang, H.; Lian, S.Y.; Lee, S.T.; Kang, Z.H. One-step ultrasonic synthesis of water-soluble carbon nanoparticles with excellent photoluminescent properties. *Carbon* **2011**, *49*, 605–609. [CrossRef]

8. Hu, C.; Li, M.; Qiu, J.; Sun, Y.P. Design and fabrication of carbon dots for energy conversion and storage. *Chem. Soc. Rev.* **2019**, *48*, 2315–2337. [CrossRef]
9. Zhou, L.; Lin, Y.H.; Huang, Z.Z.; Ren, J.S.; Qu, X.G. Carbon nanodots as fluorescence probes for rapid, sensitive, and lable-free detection of Hg^{2+} and biothiols in complex matrices. *Chem. Commun.* **2012**, *48*, 1147–1149. [CrossRef]
10. He, H.; Kim, K.H.; Danilov, A.; Montemurro, D.; Yu, L.; Park, Y.W.; Lombardi, F.; Bauch, T.; Moth-Poulsen, K.; Iakimov, T.; et al. Uniform doping of graphene close to the Dirac point by polymer-assisted assembly of molecular dopants. *Nat. Commun.* **2018**, *9*, 3956. [CrossRef]
11. Carrero-Sanchez, J.C.; Elias, A.L.; Mancilla, R.; Arrellin, G.; Terrones, H.; Laclette, J.P.; Terrones, M. Biocompatibility and toxicological studies of carbon nanotubes doped with nitrogen. *Nano Lett.* **2006**, *6*, 1609–1616. [CrossRef]
12. Gong, K.; Du, F.; Xia, Z.; Durstock, M.; Dai, L. Nitrogen-doped carbon nanotube arrays with high electrocatalytic activity for oxygen reduction. *Science* **2009**, *5915*, 760–764. [CrossRef]
13. Trauzettel, B.; Bulaev, D.V.; Loss, D.; Burkard, G. Spin qubits in graphene quantum dots. *Nat. Phys.* **2007**, *192*, 192–196. [CrossRef]
14. Wang, X.; Li, X.; Zhang, L.; Yoon, Y.; Weber, P.K.; Wang, H.; Guo, J.; Dai, H. N-doping of graphene through electrothermal reactions with ammonia. *Science* **2009**, *324*, 768–771. [CrossRef] [PubMed]
15. Kuo, W.S.; Yeh, T.S.; Chang, C.Y.; Liu, J.C.; Chen, C.H.; So, E.C.; Wu, P.C. Amino-functionalized nitrogen-doped graphene quantum dots for efficient enhancement of two-photon-excitation photodynamic therapy: Functionalized nitrogen as a bactericidal and contrast agent. *Inter. J. Nanomed.* **2020**, *15*, 6961–6973. [CrossRef]
16. Baptista, M.S.; Cadet, J.; Mascio, P.D.; Ghogare, A.A.; Greer, A.; Hamblin, M.R.; Lorente, C.; Nunez, S.C.; Ribeiro, M.S.; Thomas, A.H.; et al. Type I and type II photosensitized oxidation reactions: Guidelines and mechanistic pathways. *Photochem. Photobiol.* **2017**, *93*, 912–919. [CrossRef] [PubMed]
17. Kuo, W.S.; Chen, H.H.; Chen, S.Y.; Chang, Y.T.; Chen, P.C.; Hou, Y.I.; Shao, Y.T.; Kao, H.F.; Hsu, C.L.L.; Chen, Y.C.; et al. Graphene quantum dots with nitrogen-doped content dependence for highly efficient dual-modality photodynamic antimicrobial therapy and bioimaging. *Biomaterials* **2017**, *120*, 185–194. [CrossRef] [PubMed]
18. Shi, L.; Hernandez, B.; Selke, M. Singlet oxygen generation from water-soluble quantum dot-organic dye nanocomposites. *J. Am. Chem. Soc.* **2006**, *128*, 6278–6279. [CrossRef]
19. Hummers, W.S.; Offeman, R.E. Preparation of graphitic oxide. *J. Am. Chem. Soc.* **1958**, *80*, 1339. [CrossRef]
20. Kuo, W.S.; Shao, Y.T.; Huang, K.S.; Chou, T.M.; Yang, C.H. Antimicrobial amino-functionalized nitrogen-doped graphene quantum dots for eliminating multidrug-resistant species in dual-modality photodynamic therapy and bioimaging under two photon excitation. *ACS Appl. Mater. Interfaces* **2018**, *10*, 14438–14446. [CrossRef]
21. Tang, L.; Ji, R.; Li, X.; Teng, K.S.; Lau, S.P. Energy-level structure of nitrogen-doped graphene quantum dot. *J. Mater. Chem. C* **2013**, *1*, 4908–4915. [CrossRef]
22. Ferrari, A.C.; Basko, D.M. Raman spectroscopy as a versatile tool for studying the properties of graphene. *Nat. Nanotechnol.* **2013**, *8*, 235–246. [CrossRef] [PubMed]
23. Bootsma, M.C.J.; Diekmann, O.; Bonten, M.J.M. Controlling methicillin-resistant *Staphylococcus aureus*: Quantifying effects of interventions and rapid diagnostic testing. *Proc. Natl. Acad. Sci. USA* **2006**, *103*, 5620–5625. [CrossRef] [PubMed]
24. Lyon, D.Y.; Brunet, L.; Hinkal, G.W.; Wiesner, M.R.; Alvarez, P.J.J. Antimicrobial activity of fullerene water suspensions (nC_{60}) is not due to ros-mediated damage. *Nano Lett.* **2008**, *8*, 1539–1543. [CrossRef]
25. Kinen, M.M.; Kamal-Eldin, A.; Lampi, A.M.; Hopia, A. Effect of α- and γ-tocopherols on formation of hydroperoxides and two decomposition products from methyl linoleate. *J. Am. Oil Chem. Soc.* **2000**, *77*, 801–806. [CrossRef]
26. Sleytr, U.B.; Messner, P. Crystalline surface layers on bacteria. *Ann. Rev. Microbiol.* **1983**, *37*, 311–339. [CrossRef]
27. Debabov, V.G. Bacterial and archaeal s-layers as a subject of nanobiotechnology. *Molecular Biol.* **2004**, *38*, 578–591. [CrossRef]
28. Würth, C.; Grabolle, M.; Pauli, J.; Spieles, M.; Resch-Genger, U. Relative and absolute determination of fluorescence quantum yields of transparent Samples. *Nat. Protoc.* **2013**, *8*, 1535–1550. [CrossRef]
29. Thompson, A.; Lever, J.R.; Canella, K.A.; Miura, A.; Posner, G.H.; Seliger, H.H. Chemiluminescence mechanism and quantum yield of synthetic vinylpyrene analogues of benzo[*a*]pyrene-7,8-dihydrodiol. *J. Am. Chem. Soc.* **1986**, *108*, 4498–4504. [CrossRef]

International Journal of
Molecular Sciences

MDPI

Article

Photodynamic Inactivation of an Endodontic Bacteria Using Diode Laser and Indocyanine Green-Loaded Nanosphere

Naoya Higuchi [1], Jun-ichiro Hayashi [2,*], Masanori Fujita [1], Yuki Iwamura [2], Yasuyuki Sasaki [2], Ryoma Goto [2], Tasuku Ohno [2], Eisaku Nishida [2], Genta Yamamoto [2], Takeshi Kikuchi [2], Akio Mitani [2] and Mitsuo Fukuda [2]

1 Department of Endodontics, School of Dentistry, Aichi Gakuin University, Nagoya 464-8651, Aichi, Japan; kinchan@dpc.agu.ac.jp (N.H.); wisteria@dpc.agu.ac.jp (M.F.)
2 Department of Periodontology, School of Dentistry, Aichi Gakuin University, Nagoya 464-8651, Aichi, Japan; yukiwa@dpc.agu.ac.jp (Y.I.); ysasaki@dpc.agu.ac.jp (Y.S.); gryoma@dpc.agu.ac.jp (R.G.); tasuku@dpc.agu.ac.jp (T.O.); enishida@dpc.agu.ac.jp (E.N.); genta@dpc.agu.ac.jp (G.Y.); tkikuchi@dpc.agu.ac.jp (T.K.); minita@dpc.agu.ac.jp (A.M.); fukuda-m@dpc.agu.ac.jp (M.F.)
* Correspondence: jun1row@dpc.agu.ac.jp

Abstract: Apical periodontitis, an inflammatory lesion causing bone resorption around the apex of teeth, is treated by eradicating infectious bacteria from the root canal. However, it has a high recurrence rate and often requires retreatment. We investigated the bactericidal effect of antimicrobial photodynamic therapy (aPDT)/photodynamic antimicrobial chemotherapy (PACT) using indocyanine green (ICG)-loaded nanospheres coated with chitosan and a diode laser on a biofilm of *Enterococcus faecalis*, a pathogen of refractory apical periodontitis. Biofilm of *E. faecalis* was cultured in a porcine infected root canal model. ICG solution was injected into the root canal, which was then irradiated with a laser (810 nm wavelength) from outside the root canal. The bactericidal effect was evaluated by colony counts and scanning electron microscopy. The result of the colony counts showed a maximum 1.89 log reduction after irradiation at 2.1 W for 5 min. The temperature rise during aPDT/PACT was confirmed to be within a safe range. Furthermore, the light energy transmittance through the root was at a peak approximately 1 min after the start of irradiation, indicating that most of the ICG in the root canal was consumed. This study shows that aPDT/PACT can suppress *E. faecalis* in infected root canals with high efficiency.

Keywords: antimicrobial photodynamic therapy; indocyanine green; endodontics; *Enterococcus faecalis*

Citation: Higuchi, N.; Hayashi, J.-i.; Fujita, M.; Iwamura, Y.; Sasaki, Y.; Goto, R.; Ohno, T.; Nishida, E.; Yamamoto, G.; Kikuchi, T.; et al. Photodynamic Inactivation of an Endodontic Bacteria Using Diode Laser and Indocyanine Green-Loaded Nanosphere. *Int. J. Mol. Sci.* **2021**, *22*, 8384. https://doi.org/10.3390/ijms22168384

Academic Editors: Antonino Mazzaglia, Angela Scala and Enrico Caruso

Received: 20 June 2021
Accepted: 30 July 2021
Published: 4 August 2021

1. Introduction

Antimicrobial photodynamic therapy (aPDT)/photodynamic antimicrobial chemotherapy (PACT) has been attracting attention as a promising method to eradicate pathogens from infectious lesions [1–3]. In theory with aPDT/PACT, reactive oxygen species (ROS), which are produced by irradiating light on a photosensitizer, demonstrate a bactericidal effect and suppress tissue infection. More specifically, the light-irradiated photosensitizer absorbs light energy and transitions to an excited singlet state in which the energy is enhanced. Because this state is unstable, most photosensitizers emit fluorescence and return to the ground state, but some transition to a triplet state because of the intersystem crossing of energy. When energy is transferred from a photosensitizer in a triplet state to a nearby triplet oxygen, the triplet oxygen is excited and becomes a singlet oxygen. This singlet oxygen destroys bacteria in infectious lesions [1,4,5]. In recent years, the drug resistance of bacteria caused by the overuse of antibiotics in the treatment of infectious diseases has become a serious problem. Therefore, the clinical application of aPDT/PACT, which can prevent the development of drug-resistant bacteria, has advanced as an alternative to antibacterial agents [6]. Other reported characteristics of aPDT/PACT include the suppression of endotoxin and protease activity, biological effects such as the anti-inflammatory benefits of the light itself, and the absence of side effects such as tissue damage.

Dentistry, along with dermatology, is a field in which aPDT/PACT is actively used [7]. A wide range of research has been conducted on periodontal treatment, treatment of peri-implantitis, endodontic treatment, treatment of tooth extraction sockets, and applications in oral cleaning [8–11]. aPDT/PACT has already been commercialized, and clinical applications continue to be developed in some countries. Methylene Blue (MB) [12,13] and Toluidine Blue O (TBO) [14,15] are currently used as photosensitizers, but there is still room for improvement in the extent of their bactericidal effect and their effect against biofilm, and basic research seeking more effective photosensitizers for dental applications is currently being conducted.

Endodontic disease and periodontal disease both cause destruction of periapical and periodontal tissue, including alveolar bone, and are major causes of tooth loss. Apical periodontitis, which is an endodontic disease that develops with the progress of dental caries, greatly affects the preservation of teeth. For the treatment of this disease, the infected root canal is typically disinfected as much as possible using sodium hypochlorite and calcium hydroxide. However, these medicaments have safety issues because they are strongly alkaline, and therefore an alternative method is needed. Additionally, the success rate of treatment even by specialists is reported to be only 70% or less because of the complexity of the root canal shape and the presence of bacteria that are difficult to eliminate and which may recur [16]. *Enterococcus faecalis* is strongly suspected to be related to refractory apical periodontitis, and this Gram-positive pathogen that remains in the root canal after endodontic treatment is thought to be one of the causes of recurrence [17].

Because sterilization of the root canal greatly affects the success or failure of infected root canal treatment, various medicaments have been studied. Sodium hypochlorite, which has been widely used in the past, is still an indispensable and effective medicament for chemical cleaning of root canals; however, safety problems such as serious medical accidents caused by misuse have been reported [18,19]. Calcium hydroxide, which is used as a disinfectant, is associated with misuse accidents [20], a low bactericidal effect for some bacterial species, and difficulty in use for root canal treatment of teeth requiring pulp regeneration treatment. Therefore, the development of a safe root canal cleaning method that does not use sodium hypochlorite or calcium hydroxide is important, and aPDT/PACT is considered to be a promising alternative treatment.

In endodontic aPDT/PACT, in addition to MB, TBO, Rose Bengal (RB), and curcumin (CUR), indocyanine green (ICG) has been used as a photosensitizer, and many studies have investigated the application of these photosensitizers to cleaning and disinfecting root canals for their bactericidal and antibiofilm effects. As for photosensitizers other than those, aPDT/PACT with porphyrin and chlorin has been reported to be effective against *E. faecalis* biofilm, suppressing the maturation, and phthalocyanine was also found to be as effective as conventional aPDT/PACT at a low concentration against *E. faecalis* biofilms [21–23]. ICG is a photosensitizer that is considered to be safe for use in humans, with relatively good results reported for its use in endodontic treatment [24]. Pourhajibagher et al. reported that ICG has high antimicrobial activity on biofilms compared with other photosensitizers [25]. Furthermore, some studies have reported that the dosage of ICG could be adjusted to further enhance the bactericidal activity for endodontic pathogens or other bacteria [26,27]. However, because the number of studies is still small and the experimental conditions of such studies including the concentration and dosage of the photosensitizer, the type of light source, and the irradiation method differ, the consensus on the effectiveness of ICG is still unclear.

In recent years, we have been developing an aPDT/PACT system using ICG and a diode laser to treat marginal periodontitis [28–30]. One feature of our method is the use of ICG-loaded nanospheres coated with chitosan (ICG-Nano/c) as the original dosage form of a photosensitizer. The theory behind this method is that the chitosan coating would positively charge the photosensitizers and improve their binding to negatively charged bacteria. In this study, we conducted basic research investigating the bactericidal effect of this aPDT/PACT system against *E. faecalis* biofilm, assuming future clinical applications. Other notable features of this study are the creation of an infected root canal model to reproduce the biofilm environment and irradiation of laser light from outside the root canal orifice to improve safety compared with conventional intra-root-canal irradiation. The effect on the biofilm, the temperature change of the root surface, and changes in the intensity of the laser light at the apex were evaluated.

2. Results

2.1. Bactericidal Effects aPDT/PACT with ICG-Nano/c on Planktonic E. faecalis

First, the bactericidal effect of the aPDT/PACT system using ICG-Nano/c on planktonic *E. faecalis* was confirmed by colony counts. Because a bactericidal effect of 99% or more was observed for planktonic *P. gingivalis* with 1 min laser irradiation at a peak output power of 0.7 W (0.49 W/cm^2) in previous studies using the current laser equipment, the same output was used to investigate the bactericidal effect on *E. faecalis*. When the irradiation time was 1 min, only approximately an 80.6% bactericidal effect was observed (0.71 log reduction), but 99.4% of the bacteria were killed in 3 min (2.21 log reduction), and 99.8% were killed in 5 min (2.76 log reduction) (Figure 1A and Table A1a). Next, the bactericidal effect was evaluated under varied irradiation outputs. When the output was doubled or tripled, a significant decrease in the number of detected bacteria was observed in a power-dependent manner (Figure 1B and Table A1b). The effects were 1.99 log reduction (98.9% killing) at 0.7 W, 5.18 (99.9993%) at 1.4 W (0.98 W/cm^2), and 5.96 (99.99989%) at 2.1 W (1.46 W/cm^2).

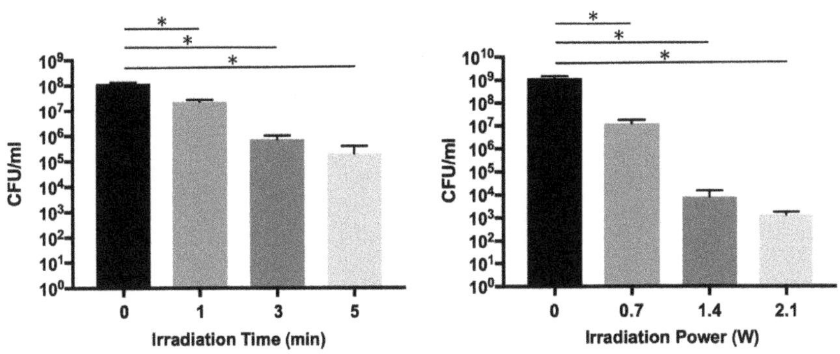

Figure 1. Bactericidal effect of aPDT/PACT using ICG-Nano/c on planktonic *E. faecalis*. (**A**) Bactericidal effect with irradiation of different durations (1, 3, and 5 min) indicated by viable colony counts. The diode laser was applied with a 0.7 W peak power output without air-cooling. (**B**) Comparison of the bactericidal effect with different irradiation powers (0.7, 1.4, and 2.1 W (0.49, 0.98 and 1.46 W/cm^2)). The laser operated for 5 min without air-cooling. Data are presented as the mean ± standard deviation. Asterisks (*) denote $p < 0.05$ ($n = 3$).

To confirm that this bactericidal effect was due to aPDT/PACT, a comparison of the effect was made between groups with and without photosensitizers (Figure 2 and Table A2). While a significant decrease in the number of colony-forming units (CFUs) was observed in the aPDT/PACT group with photosensitizers, there was no bactericidal effect in the laser alone group without photosensitizers (Figure 2A). Additionally, we confirmed if the bactericidal effect was influenced by heat caused by the laser irradiation. As shown in Figure 2B, a sufficient bactericidal effect was observed when the laser was used with air-cooling to reduce the effects of heat. There was a 4.38 log reduction (99.996% killing) with irradiation at 2.1 W for 5 min.

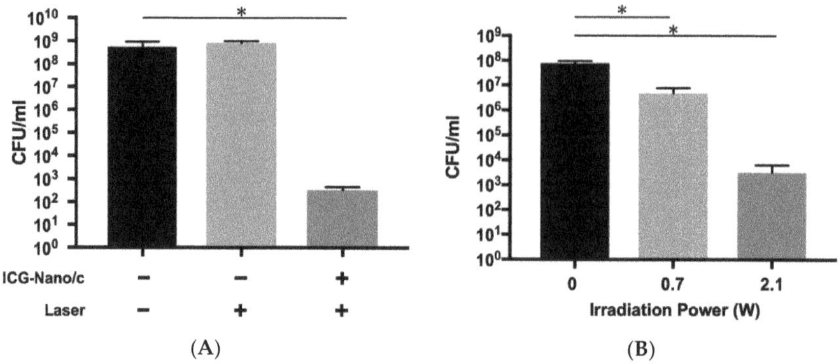

Figure 2. Confirmation that the bactericidal effect is due to aPDT/PACT. (**A**) Comparison of bactericidal effects with and without photosensitizers. The laser was applied at 2.1 W (1.46 W/cm^2) for 5 min without air-cooling. A sufficient bactericidal effect could not be obtained in the absence of photosensitizers. (**B**) Bactericidal effect with air-cooling (2 L/min). A marked bactericidal effect was obtained, even when air-cooling reduced the effect of heat. Data are presented as mean ± standard deviation. Asterisks (*) denote $p < 0.05$ ($n = 3$).

2.2. Bactericidal Effects on *E. faecalis* Biofilm in the Infected Root Canal Model

To verify the effect of aPDT/PACT in a biologically similar environment, an infected root canal model with an *E. faecalis* biofilm was prepared, and the sterilization experiment was performed. As shown in Figure 3 and Table A3, the number of bacteria detected in the root canals of the model was significantly lower in both aPDT/PACT groups than in the control group. A 1.89 log reduction (98.7%) was observed when the power of the irradiated laser was 2.1 W (1.46 W/cm^2), while a 1.21 log reduction (93.9% killing) was obtained at 0.7 W (0.49 W/cm^2) irradiation.

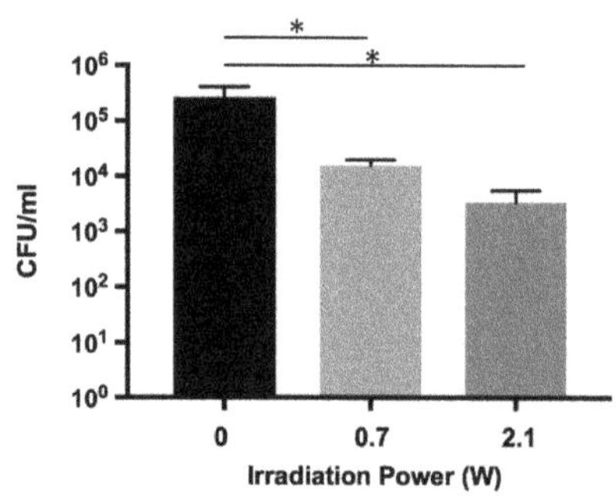

Figure 3. Bactericidal effect on *E. faecalis* biofilm in an infected root canal model. Lasers with different powers (0.7 W and 2.1 W (0.49 and 1.46 W/cm^2)) were applied for 5 min with air-cooling. A marked bactericidal effect was observed at 2.1 W irradiation. Data are presented as the mean ± standard deviation. Asterisks (*) denote $p < 0.05$ ($n = 3$).

2.3. Scanning Electron Microscopy (SEM) Observations of *E. faecalis* Biofilm on Dentin Blocks after Treatment

An SEM image of the *E. faecalis* biofilm formed on a dentin block is shown in Figure 4A. The biofilm, which is composed of bacterial cells and extracellular polymeric substances, appeared to be spread in a thick sheet form covering the dentinal tubules. Figure 4B,C shows the images after aPDT/PACT. The biofilm appeared to be thinner and less abundant, and the dentinal tubules were exposed, but the morphology of the bacterial cells could still be confirmed. Furthermore, the nanoparticles of the photosensitizer were also observed to adhere to and aggregate with the remaining biofilm. The thinnest biofilm was observed in blocks treated with NaOCl, where the dentinal tubules could be clearly visualized in the blocks (Figure 4D).

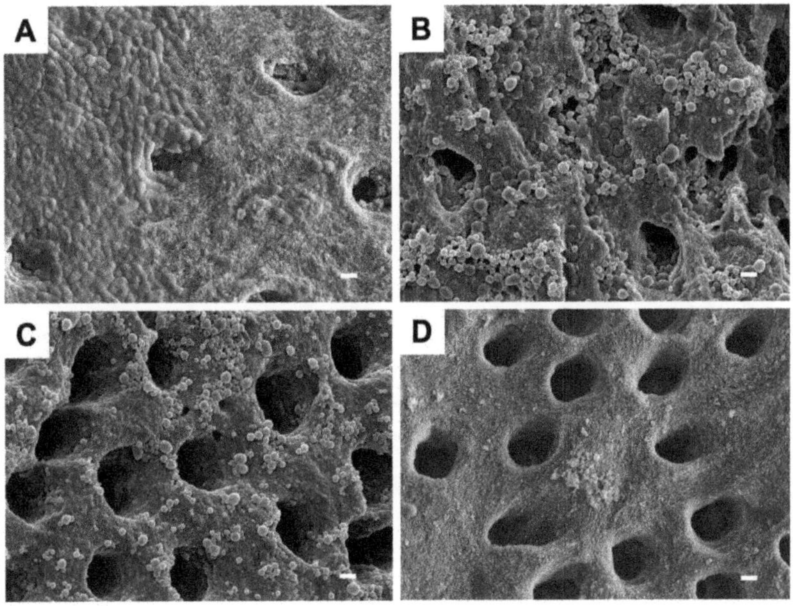

Figure 4. Scanning electron microscope images of *E. faecalis* biofilm after aPDT/PACT and NaOCl treatments. (**A**) Biofilm without irradiation is shown as a control. (**B**) Biofilm after aPDT/PACT with laser irradiation at 0.7 W. (**C**) Biofilm after aPDT/PACT with laser irradiation at 2.1 W (1.46 W/cm^2). (**D**) Biofilm after treatment with NaOCl. All images are displayed at a magnification of $\times$5000. The length of the bar represents 1 µm.

2.4. Effect of Cooling on Temperature Elevation of the Root Surface during Laser Irradiation

For future clinical application of this aPDT/PACT system to human root canals, it is necessary to avoid damage to the tissues around the roots caused by heat generated by the laser irradiation. The temperature elevation of the root surface of extracted teeth was investigated with air-cooling and intermittent irradiation (Figure 5). When aPDT/PACT using a photosensitizer was performed with an output power of 2.1 W (1.46 W/cm^2) without cooling, the temperature of the root surface increased by up to 15 °C. However, with cooling of the root surface at the same output, the maximum elevation in temperature was 8 °C. The maximum elevation in temperature was reached 60 s after the start of irradiation and then gradually decreased. However, when the root canal was filled with physiological saline without a photosensitizer, the temperature rose only 2.5 °C even when irradiated at 2.1 W. The temperature rose to the same level as when saline was used at 0.7 W.

2.5. Measurement of the Light Energy Transmitted through the Tooth Root

In aPDT/PACT, a singlet oxygen is generated by adding light energy to the photosensitizer pigment. However, when the photosensitizer is exposed to light, the pigment is thought to gradually decompose, and its function as a photosensitizer is diminished. In the aPDT/PACT method used in this study, a photosensitizer was injected into the root canal, and then light was irradiated from outside the orifice of the root canal. The light is mostly absorbed by the photosensitizer in the root canal and the dentin of the root canal wall but partly passes through the root. As the pigment is decomposed by light, the light energy absorbed by the photosensitizer decreases, and as a result, energy transmitted through the tooth root increases.

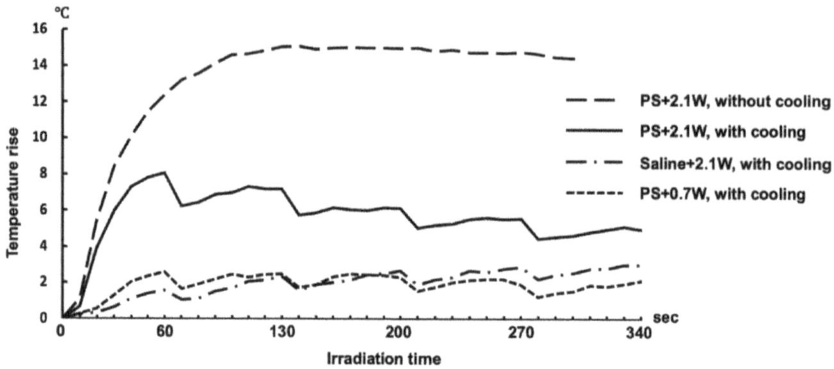

Figure 5. Temperature transitions on the root surface of extracted human teeth during laser irradiation. In the group without air-cooling, the laser was applied continuously for 5 min. In the air-cooled groups, the laser operated intermittently for 5 min with a 10 s rest period every minute. The average of the highest temperatures within a heat distribution captured by thermography is indicated in the polygonal line graph (n = 5).

Therefore, the change in energy transmitted through the tooth root was measured during irradiation for 5 min to investigate whether the pigment decomposed (Figure 6A). Less energy was transmitted through the roots when the root canal was filled with saline than when the root canal was empty. At 5 s after irradiation, the level of energy that was transmitted through the root filled with ICG-Nano/c was approximately half of that transmitted through the root filled with saline. The level of energy that was transmitted gradually increased and was approximately equal to that of the control at 60 s (Figure 6B). Measurements were taken up to 340 s (a total of 5 min irradiation and 40 s rest), and the amount of transmitted energy was almost flat after 140 s (Figure A1).

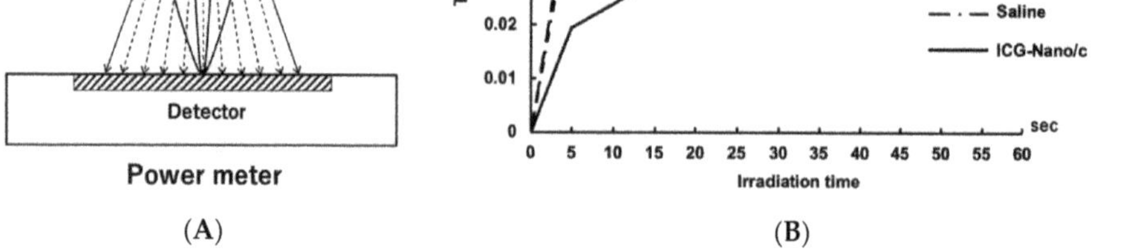

Figure 6. Transition of the light energy transmitted through the tooth root. (**A**) Schematic diagram of laser irradiation and detection. The guided circle of the irradiation range was matched to the outer circumference of the root. (**B**) The transition of the energy transmitted through the root is shown in a polygonal line graph. The value displayed in the power meter was recorded every 5 s (n = 5).

3. Discussion

In this study, we investigated the bactericidal effect of aPDT/PACT using ICG-Nano/c on *E. faecalis*, a widely known pathogen of endodontic lesions such as refractory apical periodontitis. *E. faecalis* is also a pathogen of urinary tract infections [31], and its broad spectrum of resistance to antibiotics such as vancomycin is regarded as a public health problem [32]. Genomic analysis of vancomycin-resistant strains revealed that *E. faecalis* develops drug resistance by horizontal gene transfer mechanisms including transposons [33]. Therefore, sterilization methods that do not rely on antibiotics are required for the extermination of this bacteria. Our aPDT/PACT method showed a sufficient bactericidal effect on *E. faecalis* without dependence on antibacterial agents, which is of clinical significance.

The bactericidal effect on planktonic bacteria was investigated in relation to the irradiation energy and irradiation time. Our previous studies on a pathogen of marginal periodontitis confirmed that the aPDT/PACT method has a bactericidal effect of a 2 log reduction (1 min) to a 5 log reduction (5 min) against planktonic cells of *Porphyromonas gingivalis* [28]. However, when testing the bactericidal effects on planktonic cells of *E. faecalis* with the same output (0.7 W, duty cycle of 50%, 0.49 W/cm^2) in this study, the effect was less than a 1 log reduction at 1 min and approximately a 2 log reduction even at 5 min; this was considered to be weaker than the effect on *P. gingivalis*. Therefore, when the energy output of the laser was increased to three times the original setting (2.1 W, 1.46 W/cm^2), the same level of bactericidal effect as on *P. gingivalis* could be obtained. The reason why higher energy was required to exert a sufficient bactericidal effect on *E. faecalis* is considered to be related to the difference in the antioxidative mechanisms of both bacteria. *E. faecalis* is a Gram-positive facultative anaerobic bacterium, which is more resistant to oxygen than Gram-negative obligate anaerobic bacteria such as *P. gingivalis*, and numerous antioxidative enzymes have been identified in *E. faecalis* [34]. *E. faecalis* possesses manganese-containing superoxide dismutase, which is a classical ROS scavenging enzyme [35], and catalase of *E. faecalis* converts hydrogen peroxide to water and oxygen in a heme-dependent manner [36]. Additionally, *E. faecalis* is known to produce extracellular free radicals and exert pathogenicity [37]. Thus, because *E. faecalis* has various functions to resist oxidative stress, a higher output than that for *P. gingivalis* may have been required to achieve a sufficient bactericidal effect.

Many previous studies have focused on the bactericidal effect of aPDT/PACT against monospecies biofilms of *E. faecalis*, with variable results. Zand et al. reported that no living bacteria were detected in the biofilm-formed dentin block after aPDT/PACT treatment with TB [38]. Kishen et al. showed a bactericidal effect of a 5 log reduction by aPDT/PACT using MB and RB [39], and Akbari et al. showed a 2.81 log reduction with nanographene oxide including ICG [27]. However, a number of studies have reported that the bactericidal effect of aPDT/PACT on *E. faecalis* biofilm was less than 90%. According to Oda et al., aPDT/PACT with MB and CUR resulted in detection rates of living bacteria of 29.80% and 26.20%, respectively [40]. Golmohamadpour et al. applied three types of metal–organic frameworks including ICG for aPDT/PACT, resulting in detection rates of living bacteria from 45.12% to 62.67% [41]. Furthermore, potassium iodide-potentiated PDT with MB and RB increased bactericidal effects to 86.50% and 91.50%, respectively, in a study by Li et al. [42]. In this study, we used aPDT/PACT with ICG-Nano/c in an infected root canal model consisting of an *E. faecalis* monospecies biofilm in root canals. We detected a significant decrease in bacterial count, with a bactericidal effect of 98.7% (1.89 log reduction) (Figure 3). Because of differences in methodologies such as laser conditions, analysis methods, and biofilm production, a simple comparison of studies cannot be easily made, but it is apparent that aPDT/PACT with ICG-Nano/c produces a significant bactericidal effect. However, the bactericidal effect in our study did not exceed 99%. It was previously reported that mature (weeks old) biofilms experience a significantly lower bactericidal effect compared with immature (days old) biofilms [43]. Because we used an infected root canal model that involves a 3-week-old biofilm, the maturation of the biofilm may have impacted the extent of the bactericidal effect of aPDT/PACT.

Although quantitative measurement was not performed, we were able to observe an apparent biofilm reduction on SEM images (Figure 4). Extracellular polymeric substances form at least 90% of the total biofilm mass and enhance the structural complexity and strength of the biofilm [44]. Hence, extracellular polymeric substances are among the first lines of defense against antibiotic and biocide diffusion and thus reduce photosensitizer penetration. Various components of extracellular polymeric substances are subject to attack by ROS [45], which form from aPDT/PACT. The effect of ROS damage to the *E. faecalis* biofilm is evidenced by SEM imaging [46]. Chitosan can also damage the extracellular polymeric substances and disrupt the structure of biofilm [47]. Thus, it can be hypothesized that the chitosan in ICG-Nano/c and the ROS resulting from irradiation cause damage to the extracellular polymeric substances in biofilm.

The laser output used in aPDT/PACT on the *E. faecalis* biofilm was 2.1 W (duty cycle of 50%, 1.46 W/cm^2), which was higher than that of a commercially available general aPDT/PACT system. It is known that ICG generates heat when excited by laser light [48]. Additionally, because laser light barely passes through the root dentin, the energy absorbed by dentin is converted into heat at a high rate. Therefore, the irradiated site was air-cooled to avoid heat injury, in consideration of future clinical applications. Previous research focusing on the temperature increase of the root canal surface caused by the vertical condensation root canal filling technique [49,50] and on the effect of elevated bone temperatures on living organisms [51] has established that a 10 °C increase in the root surface temperature may damage the periodontal tissues, cementum, periodontal ligament, and alveolar bone. In this study, we found that the temperature of the root surface reached a maximum at 60 s after irradiation in aPDT/PACT, but even the maximum output of 2.1 W averaged a temperature increase of only 8 °C. Additionally, after the maximum temperature elevation was observed, the temperature dropped immediately as a result of the 10-s resting time and air-cooling. Furthermore, the temperature of the root surface gradually decreased. These results are within the critical temperature elevation range of 10 °C. Therefore, little damage would have been sustained by the periodontal tissue given that the maximum temperature range lasted only 10–20 s.

When the empty root canal was irradiated with a laser, 0.06 W of energy was transmitted through the root (Figure 6). Because the irradiation energy was set to a peak output of 2.1 W and a duty cycle of 50%, the theoretical average energy irradiated on the root was 1.05 W. Therefore, approximately 94% of the energy was absorbed by the root dentin. In the root canals filled with ICG-Nano/c solution, the transmitted energy of the laser was 0.02 W at 5 s after the start of irradiation but gradually increased thereafter to 0.04 W at 1 min. This was almost the same value as when physiological saline was placed in the root canal. This result indicates that the ICG-Nano/c absorbed laser energy immediately after the start of irradiation, but the absorbance decreased over time. ICG decomposes in light [52]. The polymethine chain of ICG exposed to light forms a dioxane ring producing singlet oxygen, which is further decomposed into a carbonylated molecule. Therefore, the decrease in energy absorption is thought to be caused by the decomposition of ICG-Nano/c in the laser light, and its function as a photosensitizer is gradually lost. If most of the photosensitizer in the root canal is consumed within 1 min of the start of irradiation, 5 min of irradiation is longer than necessary. The reason why the bactericidal effect increased in a time-dependent manner against the planktonic *E. faecalis* in Figure 1 may be that a sufficient amount of photosensitizer was present around the bacteria. To establish a clinical application of this method in the future, it will be necessary to investigate the bactericidal effect under changing conditions, such as a smaller interval for the irradiation time and multiple injections of photosensitizer.

Cleaning the root canal with hypochlorous acid produces high bactericidal activity, and this medicament is still widely used today. In this study, hypochlorous acid showed a high biofilm removal effect (Figure 4D). However, the outflow from the apical foramen is regarded as a problem, and there is a demand for a safer method. Furthermore, conventional aPDT/PACT for bacterial decontamination of root canals recommends inserting the tip of the optical probe into the root canal, which raises safety concerns. The advantage of the method used in this study is that a high bactericidal effect equal to that of previous studies can be obtained without inserting a probe into the root canal. Irradiation was performed from 1 cm above the root canal orifice. It can be concluded that this method of aPDT/PACT functions well without inserting the laser tip into the root canal, as evidenced by the consumption of most of the photosensitizer injected into the root canal. Our findings suggest that this method is safer than conventional methods.

One limitation of this study is that it only validates the bactericidal effect against a single bacterial biofilm. Because biofilm usually consists of a complex bacterial population, it is possible that the reaction in actual root canals may not be the same as in the case of a single bacterial biofilm. Additionally, the effect of this method on apical periodontitis is unclear because no clinical evaluation was conducted. In fact, as can be seen in Figure 4D, our method shows less bactericidal effect than NaOCl. However, there is still room for improvement in this method, such as multiple injections of photosensitizer, and many problems remain to be solved to achieve clinical application. In our future research, we plan to optimize the method and evaluate its effect in clinical studies.

4. Materials and Methods

4.1. Preparation of ICG-Nano/c

ICG-Nano/c was Prepared by the Emulsion Solvent diffusion method in oil [53], following a protocol from Nagahara et al. [28].

4.2. Bacterial Strain

Enterococcus faecalis was used as the test organism because it is often found in persistent and retreatment cases [54]. The *E. faecalis* ATCC 19,433 strain was grown aerobically on brain–heart infusion (BHI) agar plates (Bacto; DIFCO Laboratories, Franklin Lakes, NJ, USA) at 37 °C. Single colonies were then inoculated into BHI broth and cultured aerobically to the mid-log phase at 37 °C. The number of bacteria was adjusted with sterilized saline spectrophotometrically to a cell density of approximately 10^8 CFU per mL (optical density of 0.1 at 600 nm) before subsequent experiments. To summarize, 5 mg of ICG (Ophthagreen; Santen Pharmaceutical, Osaka, Japan), 100 mg of poly (lactic-co-glycolic acid), and 100 mg of Span80 (Kanto Chemical Co., Inc., Tokyo, Japan) solution, which was dissolved in 3 mL acetone and 1 mL of methanol, was added at 2 mL/minute to 100 mL of 2% hexaglycerin-condensed ricinoleate containing triglyceride and n-hexane. The total mixture was stirred at 35 °C for 3 h under a vacuum (400 rpm). It was then centrifuged to remove oil and n-hexane. After centrifugation, nanospheres were mixed with 50 mL of a solution of 2% poly-vinyl alcohol/0.5% chitosan to form a chitosan coating. The mixture was further centrifuged at 26,960$\times$ *g* for 10 min at 4 °C. The nanosphere cluster was suspended in mannitol solution to separate each particle and frozen at −45 °C for 2 h. Preparation of ICG-Nano/c was completed by freeze-drying for 48 h. Properties of the produced ICG-Nano/c were (1) it contained 5 mg/g ICG, (2) the average nanosphere size was 560 nm. ICG-Nano/c solutions were freshly made in sterilized saline at 20 mg/mL and used at a final concentration of 10 mg/mL.

4.3. Laser Application

LIGHTSURGE SQUARE (Osada, Tokyo, Japan) was used as a diode laser with a central wavelength of 810 ± 20 nm that can output up to 3 W. Other features of this laser are: (1) light is distributed through the fiber-optic applicator, (2) the diameter of the fiber core is 600 µm, and (3) the spread angle of emitted light is 20.49°. In this study, we set it to a repeated pulse mode with a pulse width of 100 msec, 50% duty cycle, and used it under various conditions (peak power outputs: 0.7–2.1 W, 0.49–1.46 W/cm^2, irradiation time: 1, 3, 5 min). The light probe was placed 10 mm above the surface of the samples. The irradiation spot area was set to a diameter of 0.956 mm.

4.4. Preparation of the Infected Root Canal Model

An infected root canal model was prepared using extracted porcine teeth (Figure 7). Specifically, the roots of the first and second molars were separated and extracted, and the root length was adjusted to 13 mm. The root canals were then cleaned and shaped up to #60 with a K file (MANI, Inc., Utsunomiya, Japan) using conventional methods, chemically cleaned with sodium hypochlorite solution and EDTA solution, and sterilized in an autoclave. The apical foramen was closed with an immediate polymerization resin (Ortho Crystal; JM Ortho, Tokyo, Japan) and fixed in a polymerase chain reaction tube. The *E. faecalis* bacterial solution was inoculated into the root canal and cultured at 37 °C for 21 days. During the culture period, the medium was changed three times a week.

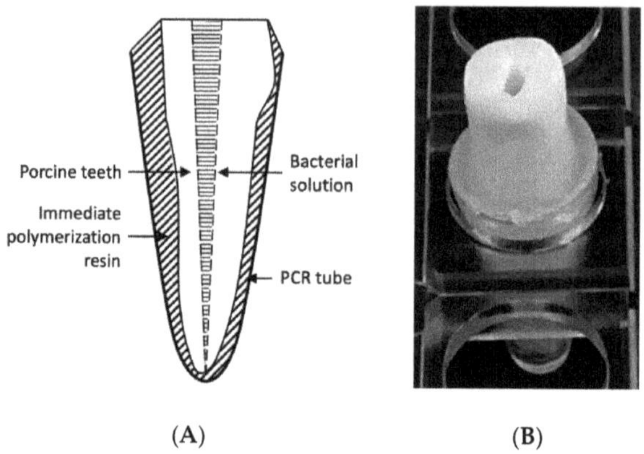

Figure 7. Preparation of the infected root canal model based on porcine teeth. (**A**) Schematic diagram of the infected root canal model. (**B**) Photograph of the infected root canal model. The root was separated from the first or second molar, and the length was adjusted to 13 mm. The root canal was cleaned mechanically and chemically, and then the root was autoclaved before it was fixed in a PCR tube with immediate polymerization resin. A bacterial solution of *E. faecalis* was inoculated in the root canal and incubated for 21 days at 37 °C.

4.5. Bactericidal Assay on Planktonic Cells

The experimental group was treated with aPDT/PACT as follows: 100 µL of the bacterial solution and 100 µL of the ICG-Nano/c solution prepared to 20 mg/mL with sterile saline (ICG-Nano/c final concentration: 10 mg/mL) were added to a microtube and irradiated with the laser at intensities of 0.7, 1.4, and 2.1 W for 1, 3, and 5 min. After serial dilution, the total solution was spread on BHI agar plates, and a colony count was performed. There were three samples per condition. Two control groups were set up as follows: (1) a positive control group in which the same amount of sterile saline was added instead of the ICG-Nano/c solution and no laser irradiation was performed and (2) a laser alone group in which the same amount of sterile saline was added instead of the ICG-Nano/c solution and laser irradiation was performed.

4.6. Bactericidal Assay on the Biofilm in the Infected Root Canal Model

Using the completed infected root canal model, the components of the medium were washed away with 2 mL of sterile saline using a syringe (Terumo Syringe 2.5 mL SS-02SZ; Terumo, Tokyo, Japan) and an irrigation needle (Nishika Rootclin Needle 23G; Nishika, Yamaguchi, Japan). The root canals were then dried with paper points (#60) (Morita Paper Point; J. MORITA CORP., Suita, Japan). ICG-Nano/c solution was added to the experimental group (final concentration 10 mg/mL), and laser irradiation was performed while blowing air (2 L/min) from a nozzle built into the tip of the light probe. The irradiation intensities were 0.7 and 2.1 W. The irradiation time was 5 min, and irradiation was stopped for 10 s every minute. The two control groups were set up in the same way as the assay for planktonic cells. Paper points (#55) were placed for 1 min in the root canals. After the paper points were removed, serial dilution and colony counting were performed. There were three samples per condition.

4.7. SEM Observations of Biofilm on Dentin Blocks after aPDT/PACT Treatment

A dentin block (5 mm × 5 mm × 2 mm) was prepared from extracted human teeth. The block was chemically cleaned with sodium hypochlorite solution and EDTA solution, sterilized in an autoclave, immersed in the *E. faecalis* bacterial solution, and cultured aerobically at 37 °C for 21 days. After the block was washed with 5 mL of sterile saline using a syringe, aPDT/PACT treatment was performed on the experimental group. Irradiation was performed under the same conditions described in Section 4.6. Control groups were also set up with the same conditions described in Section 4.6. As NaOCl treatment, dentin blocks were placed in 3% NaOCl solution (Dental Antiformin; Nishika, Yamaguchi, Japan) for 1 min.

The samples were washed with 5 mL of sterile saline and fixed with a mixture of 4% paraformaldehyde and 5% glutaraldehyde for 24 h at 4 °C. The samples were washed again in the same manner, dehydrated with an ascending ethanol series (50, 60, 70, 80, 90, 95, 100, 100%), replaced with t-butyl alcohol, and freeze-dried for 4 days using a freeze-dryer (EYE4 FDS-1000 type; Tokyo Rika Kikai Co. Ltd., Tokyo, Japan). Subsequently, platinum was vapor-deposited using a sputtering device (JUC-5000; JEOL, Tokyo, Japan), subjected to conductive treatment, and observed under SEM (JXA-8530FA; JEOL).

4.8. Confirmation of the Cooling Effect on the Root Surface during Laser Irradiation

Two-thirds of the crowns of extracted human mandibular premolars were removed, the pulp chamber was opened, and the root canals were cleaned and shaped up to #60 with a K file using conventional methods. The root canals were chemically cleaned with sodium hypochlorite solution and EDTA solution and sterilized in an autoclave as described above. ICG-Nano/c solution (10 mg/mL) was injected into the root canals of the experimental group, and sterile saline was injected into the root canals of the control group. Irradiation was performed with or without air-cooling under the conditions described in Section 4.6. During laser irradiation, the temperature of the tooth root surface was measured with a thermographic camera (InfReC Thermo GEAR G100; Nippon Avionics, Yokohama, Japan) at a shooting distance of 15 cm. The minimum detection limit was 270 μm at this shooting distance. The frame time was 60 Hz and thermal images were recorded every 10 s.

4.9. Measurement of the Light Energy Transmitted through the Tooth Root

Tooth root samples were prepared as described in Section 4.8. A laser irradiated the samples from the top of the roots, and the transmitted light that passed through the root was measured with a power meter (Nova; Ophir Optronics Solutions Ltd., Jerusalem, Israel) set at the apex. ICG-Nano/c solution (10 mg/mL) was injected into the root canals of the experimental group, and sterile saline was injected into the root canals of the control group. The tip of the laser probe was placed above the upper edge of the root canal so that the guide circle was matched to the outer circumference of the root. The irradiation conditions were the same as described in Section 4.6. The value displayed on the power meter was recorded every 5 s.

4.10. Statistical Analysis

The data were tested for normality (Shapiro–Wilk and Kolmogorov–Smirnov), and a normal distribution was confirmed. Therefore, the data were further analyzed with a parametric test (Tukey's test) to compare irradiation conditions and CFU counts, using SPSS v15.0 (IBM, Armonk, NY) with significance accepted at $p < 0.05$.

5. Conclusions

In this in vitro study, we present an aPDT/PACT method using ICG-loaded nanospheres coated with chitosan for the treatment of infected root canals. We produced a biofilm of the endodontic pathogen *E. faecalis* in an infected root canal model and examined the bactericidal effect on the biofilm. Our results showed that the viable cell counts of *E. faecalis* were reduced by more than 98% without an unsafe temperature rise in the root. Morphological observation by SEM confirmed a clear reduction of biofilm on the dentin block, but the removal was not complete. The laser light transmitted through the roots increased over time and had almost reached a maximum at 1 min, likely because of the consumption of photosensitizer in the root canal. For future clinical application of this method, it will be necessary to improve the bactericidal activity, perhaps by injecting the photosensitizer multiple times.

Author Contributions: Conceptualization, N.H., A.M. and M.F. (Mitsuo Fukuda); methodology, N.H. and J.-i.H.; formal analysis, N.H. and M.F. (Masanori Fujita); investigation, N.H. and M.F. (Masanori Fujita); resources, Y.I., Y.S. and M.F. (Mitsuo Fukuda); data curation, N.H., J.-i.H., R.G., T.O., E.N. and G.Y.; writing—original draft preparation, N.H. and J.-i.H.; writing—review and editing, Y.I., T.O., E.N. and G.Y.; visualization, N.H. and J.-i.H.; supervision, T.K., A.M. and M.F. (Mitsuo Fukuda); funding acquisition, N.H., J.-i.H. and M.F. (Mitsuo Fukuda). All authors have read and agreed to the published version of the manuscript.

Funding: This study was supported by Grants-in-Aid for Scientific Research from the Ministry of Education, Culture, Sports, Science and Technology (MEXT) of Japan, Grant Numbers 16K11572 for N.H. and 19K10140 for J.-i.H. This study was also supported by donation from The Hori Science and Arts Foundation for M.F. (Mitsui Fukuda).

Institutional Review Board Statement: The study was conducted according to the guidelines of the Declaration of Helsinki, and approved by the Ethics Committee of Aichi Gakuin University (Approval Number: 550; Date of Approval: 10 January 2019).

Informed Consent Statement: Not applicable.

Data Availability Statement: The data presented in this study are available on request from the corresponding author.

Acknowledgments: We would like to thank Shigemi Goto for his administrative support in the department, and we are also grateful to Kazuhiko Nakata for his cooperation in launching this project.

Conflicts of Interest: The authors declare no conflict of interest.

Appendix A

The tables below show the actual values of data indicated in Figures 1–3.

Table A1. Bactericidal effect of aPDT/PACT using ICG-Nano/c on planktonic *E. faecalis*. Actual values of data indicated in Figure 1 are shown. (**a**) Bactericidal effect with irradiation of different durations. The diode laser was applied with 0.7 W peak power output without air-cooling. (**b**) Bactericidal effect with different irradiation powers. The laser operated for 5 min without air-cooling.

a	Control	1 min, 0.7 W	3 min, 0.7 W	5 min, 0.7 W
CFU/mL Average	1.19×10^8	2.31×10^7	7.33×10^5	2.00×10^5
SD	0.18×10^8	0.41×10^7	3.05×10^5	2.00×10^5
log10	8.07	7.36	5.87	5.3
log reduction		0.71	2.21	2.77
% reduction		80.59	99.38	99.83
b	**Control**	**0.7 W, 5 min**	**1.4 W, 5 min**	**2.1 W, 5 min**
CFU/mL Average	1.21×10^9	2.31×10^7	8.00×10^3	1.33×10^3
SD	0.26×10^9	0.41×10^7	7.21×10^3	0.41×10^3
log10	9.09	7.10	3.90	3.12
log reduction		1.99	5.18	5.96
% reduction		98.98	99.9993	99.99989

Table A2. Confirmation that the bactericidal effect is due to aPDT/PACT. Actual values of data indicated in Figure 2 are shown. (**a**) Comparison of bactericidal effects with and without photosensitizers. The laser was applied at 2.1 W for 5 min without air-cooling. (**b**) Bactericidal effect with air-cooling (2 L/min). The laser was applied for 5 min.

a	Control	Laser Only	ICG-Nano/C with Laser
CFU/mL Average	5.58×10^8	8.29×10^8	3.33×10^2
SD	3.50×10^8	1.65×10^8	1.15×10^2
log10	8.74	8.92	2.52
log reduction		−0.17	6.22
% reduction		−48.4	99.99994
b	**Control**	**0.7 W, 5 min**	**2.1 W, 5 min**
CFU/mL Average	8.00×10^7	4.80×10^6	3.27×10^3
SD	1.69×10^7	2.98×10^6	3.12×10^3
log10	7.9	6.88	3.51
log reduction		1.22	4.38
% reduction		94.0	99.9959

Table A3. Bactericidal effect of aPDT/PACT with IGC-Nano/c on *E. faecalis* biofilm in an infected root canal model. Actual values of data indicated in Figure 3 are shown. Lasers with different powers were applied for 5 min with air-cooling.

	Control	0.7 W, 5 min	2.1 W, 5 min
CFU/mL Average	2.63×10^5	1.60×10^4	3.40×10^3
SD	1.46×10^5	0.4×10^8	2.15×10^3
log10	5.42	4.2	3.53
log reduction		1.21	1.88
% reduction		93.93	98.70

Appendix B

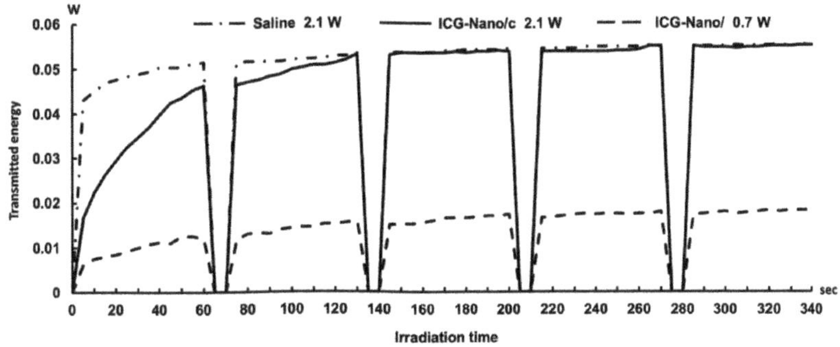

Figure A1. Transition of transmitted energy over the total irradiation time. The laser was irradiated on the tooth root intermittently for 5 min with a rest period of 10 s every minute. Canals of the root were filled with ICG-Nano/c, saline or none. The light energy transmitted through the tooth root was measured during full time irradiation and recorded every 5 s.

References

1. Cieplik, F.; Deng, D.; Crielaard, W.; Buchalla, W.; Hellwig, E.; Al-Ahmad, A.; Maisch, T. Antimicrobial photodynamic therapy—What we know and what we don't. *Crit. Rev. Microbiol.* **2018**, *44*, 571–589. [CrossRef]
2. Sperandio, F.F.; Huang, Y.Y.; Hamblin, M.R. Antimicrobial photodynamic therapy to kill Gram-negative bacteria. *Recent Pat. Antiinfect. Drug Discov.* **2013**, *8*, 108–120. [CrossRef]

3. Wainwright, M. Photodynamic antimicrobial chemotherapy (PACT). *J. Antimicrob. Chemother.* **1998**, *42*, 13–28. [CrossRef] [PubMed]
4. Hu, X.; Huang, Y.Y.; Wang, Y.; Wang, X.; Hamblin, M.R. Antimicrobial Photodynamic Therapy to Control Clinically Relevant Biofilm Infections. *Front. Microbiol.* **2018**, *9*, 1299. [CrossRef] [PubMed]
5. Ragàs, X.; He, X.; Agut, M.; Roxo-Rosa, M.; Gonsalves, A.R.; Serra, A.C.; Nonell, S. Singlet oxygen in antimicrobial photodynamic therapy: Photosensitizer-dependent production and decay in *E. coli*. *Molecules* **2013**, *18*, 2712–2725. [CrossRef]
6. Akhtar, F.; Khan, A.U.; Misba, L.; Akhtar, K.; Ali, A. Antimicrobial and antibiofilm photodynamic therapy against vancomycin resistant Staphylococcus aureus (VRSA) induced infection in vitro and in vivo. *Eur. J. Pharm. Biopharm.* **2021**, *160*, 65–76. [CrossRef]
7. Carrera, E.T.; Dias, H.B.; Corbi, S.C.T.; Marcantonio, R.A.C.; Bernardi, A.C.A.; Bagnato, V.S.; Hamblin, M.R.; Rastelli, A.N.S. The application of antimicrobial photodynamic therapy (aPDT) in dentistry: A critical review. *Laser Phys.* **2016**, *26*, 123001. [CrossRef] [PubMed]
8. Moro, M.G.; de Carvalho, V.F.; Godoy-Miranda, B.A.; Kassa, C.T.; Horliana, A.C.R.T.; Prates, R.A. Efficacy of antimicrobial photodynamic therapy (aPDT) for nonsurgical treatment of periodontal disease: A systematic review. *Lasers Med. Sci.* **2021**. [CrossRef] [PubMed]
9. Chambrone, L.; Wang, H.L.; Romanos, G.E. Antimicrobial photodynamic therapy for the treatment of periodontitis and peri-implantitis: An American Academy of Periodontology best evidence review. *J. Periodontol.* **2018**, *89*, 783–803. [CrossRef]
10. Chiniforush, N.; Pourhajibagher, M.; Shahabi, S.; Kosarieh, E.; Bahador, A. Can Antimicrobial Photodynamic Therapy (aPDT) Enhance the Endodontic Treatment? *J. Lasers Med. Sci.* **2016**, *7*, 76–85. [CrossRef]
11. Ervolino, E.; Statkievicz, C.; Toro, L.F.; de Mello-Neto, J.M.; Cavazana, T.P.; Issa, J.P.M.; Dornelles, R.C.M.; de Almeida, J.M.; Nagata, M.J.H.; Okamoto, R.; et al. Antimicrobial photodynamic therapy improves the alveolar repair process and prevents the occurrence of osteonecrosis of the jaws after tooth extraction in senile rats treated with zoledronate. *Bone* **2019**, *120*, 101–113. [CrossRef] [PubMed]
12. Tortamano, A.C.A.; Anselmo, G.G.; Kassa, C.T.; Godoy-Miranda, B.; Pavani, C.; Kato, I.T.; Wainwright, M.; Prates, R.A. Antimicrobial photodynamic therapy mediated by methylene blue in surfactant vehicle on periodontopathogens. *Photodiagn. Photodyn. Ther.* **2020**, *31*, 101784. [CrossRef]
13. Nagai, Y.; Suzuki, A.; Katsuragi, H.; Shinkai, K. Effect of antimicrobial photodynamic therapy (aPDT) on the sterilization of infected dentin in vitro. *Odontology* **2018**, *106*, 154–161. [CrossRef] [PubMed]
14. Nuernberg, M.A.A.; Wainwright, M.; Miessi, D.M.J.; Scalet, V.; Olivo, M.B.; Ervolino, E.; Garcia, V.G.; Theodoro, L.H. Effects of butyl toluidine blue photosensitizer on antimicrobial photodynamic therapy for experimental periodontitis treatment in rats. *Photodiagn. Photodyn. Ther.* **2020**, *31*, 101868. [CrossRef] [PubMed]
15. Wiench, R.; Skaba, D.; Matys, J.; Grzech-Leśniak, K. Efficacy of Toluidine Blue-Mediated Antimicrobial Photodynamic Therapy on. *Antibiotics* **2021**, *10*, 349. [CrossRef]
16. Wu, M.K.; Shemesh, H.; Wesselink, P.R. Limitations of previously published systematic reviews evaluating the outcome of endodontic treatment. *Int. Endod. J.* **2009**, *42*, 656–666. [CrossRef] [PubMed]
17. Seneviratne, C.J.; Suriyanarayanan, T.; Swarup, S.; Chia, K.H.B.; Nagarajan, N.; Zhang, C. Transcriptomics Analysis Reveals Putative Genes Involved in Biofilm Formation and Biofilm-associated Drug Resistance of Enterococcus faecalis. *J. Endod.* **2017**, *43*, 949–955. [CrossRef] [PubMed]
18. Gernhardt, C.R.; Eppendorf, K.; Kozlowski, A.; Brandt, M. Toxicity of concentrated sodium hypochlorite used as an endodontic irrigant. *Int. Endod. J.* **2004**, *37*, 272–280. [CrossRef] [PubMed]
19. Bowden, J.R.; Ethunandan, M.; Brennan, P.A. Life-threatening airway obstruction secondary to hypochlorite extrusion during root canal treatment. *Oral Surg. Oral Med. Oral Pathol. Oral Radiol. Endod.* **2006**, *101*, 402–404. [CrossRef]
20. Sharma, S.; Hackett, R.; Webb, R.; Macpherson, D.; Wilson, A. Severe tissue necrosis following intra-arterial injection of endodontic calcium hydroxide: A case series. *Oral Surg. Oral Med. Oral Pathol. Oral Radiol. Endod.* **2008**, *105*, 666–669. [CrossRef]
21. Diogo, P.; Fernandes, C.; Caramelo, F.; Mota, M.; Miranda, I.M.; Faustino, M.A.F.; Neves, M.G.P.M.; Uliana, M.P.; de Oliveira, K.T.; Santos, J.M.; et al. Antimicrobial Photodynamic Therapy against Endodontic *Enterococcus faecalis* and *Candida albicans* Mono and Mixed Biofilms in the Presence of Photosensitizers: A Comparative Study with Classical Endodontic Irrigants. *Front. Microbiol.* **2017**, *8*, 498. [CrossRef]
22. De Freitas, L.M.; Lorenzón, E.N.; Cilli, E.M.; de Oliveira, K.T.; Fontana, C.R.; Mang, T.S. Photodynamic and peptide-based strategy to inhibit Gram-positive bacterial biofilm formation. *Biofouling* **2019**, *35*, 742–757. [CrossRef]
23. Er Karaoğlu, G.; Uğur Ydın, Z.; Erdönmez, D.; Göl, C.; Durmuş, M. Efficacy of antimicrobial photodynamic therapy administered using methylene blue, toluidine blue and tetra 2-mercaptopyridine substituted zinc phthalocyanine in root canals contaminated with Enterococcusaecalis. *Photodiagn. Photodyn. Ther.* **2020**, *32*, 102038. [CrossRef]
24. Beltes, C.; Sakkas, H.; Economides, N.; Papadopoulou, C. Antimicrobial photodynamic therapy using Indocyanine green and near-infrared diode laser in reducing Entrerococcus faecalis. *Photodiagn. Photodyn. Ther.* **2017**, *17*, 5–8. [CrossRef] [PubMed]
25. Pourhajibagher, M.; Kazemian, H.; Chiniforush, N.; Hosseini, N.; Pourakbari, B.; Azizollahi, A.; Rezaei, F.; Bahador, A. Exploring different photosensitizers to optimize elimination of planktonic and biofilm forms of Enterococcus faecalis from infected root canal during antimicrobial photodynamic therapy. *Photodiagn. Photodyn. Ther.* **2018**, *24*, 206–211. [CrossRef] [PubMed]

26. Bilici, K.; Atac, N.; Muti, A.; Baylam, I.; Dogan, O.; Sennaroglu, A.; Can, F.; Yagci Acar, H. Broad spectrum antibacterial photodynamic and photothermal therapy achieved with indocyanine green loaded SPIONs under near infrared irradiation. *Biomater. Sci.* **2020**, *8*, 4616–4625. [CrossRef] [PubMed]
27. Akbari, T.; Pourhajibagher, M.; Hosseini, F.; Chiniforush, N.; Gholibegloo, E.; Khoobi, M.; Shahabi, S.; Bahador, A. The effect of indocyanine green loaded on a novel nano-graphene oxide for high performance of photodynamic therapy against Enterococcus faecalis. *Photodiagn. Photodyn. Ther.* **2017**, *20*, 148–153. [CrossRef]
28. Nagahara, A.; Mitani, A.; Fukuda, M.; Yamamoto, H.; Tahara, K.; Morita, I.; Ting, C.C.; Watanabe, T.; Fujimura, T.; Osawa, K.; et al. Antimicrobial photodynamic therapy using a diode laser with a potential new photosensitizer, indocyanine green-loaded nanospheres, may be effective for the clearance of Porphyromonas gingivalis. *J. Periodontal Res.* **2013**, *48*, 591–599. [CrossRef]
29. Fujimura, T.; Mitani, A.; Fukuda, M.; Mogi, M.; Osawa, K.; Takahashi, S.; Aino, M.; Iwamura, Y.; Miyajima, S.; Yamamoto, H.; et al. Irradiation with a low-level diode laser induces the developmental endothelial locus-1 gene and reduces proinflammatory cytokines in epithelial cells. *Lasers Med. Sci.* **2014**, *29*, 987–994. [CrossRef]
30. Sasaki, Y.; Hayashi, J.I.; Fujimura, T.; Iwamura, Y.; Yamamoto, G.; Nishida, E.; Ohno, T.; Okada, K.; Yamamoto, H.; Kikuchi, T.; et al. New Irradiation Method with Indocyanine Green-Loaded Nanospheres for Inactivating Periodontal Pathogens. *Int. J. Mol. Sci.* **2017**, *18*, 154. [CrossRef]
31. Abat, C.; Huart, M.; Garcia, V.; Dubourg, G.; Raoult, D. Enterococcus faecalis urinary-tract infections: Do they have a zoonotic origin? *J. Infect.* **2016**, *73*, 305–313. [CrossRef]
32. Sparo, M.; Delpech, G.; García Allende, N. Impact on Public Health of the Spread of High-Level Resistance to Gentamicin and Vancomycin in Enterococci. *Front. Microbiol.* **2018**, *9*, 3073. [CrossRef] [PubMed]
33. Paulsen, I.T.; Banerjei, L.; Myers, G.S.; Nelson, K.E.; Seshadri, R.; Read, T.D.; Fouts, D.E.; Eisen, J.A.; Gill, S.R.; Heidelberg, J.F.; et al. Role of mobile DNA in the evolution of vancomycin-resistant Enterococcus faecalis. *Science* **2003**, *299*, 2071–2074. [CrossRef]
34. Szemes, T.; Vlkova, B.; Minarik, G.; Tothova, L.; Drahovska, H.; Turna, J.; Celec, P. On the origin of reactive oxygen species and antioxidative mechanisms in Enterococcus faecalis. *Redox Rep.* **2010**, *15*, 202–206. [CrossRef] [PubMed]
35. Peppoloni, S.; Posteraro, B.; Colombari, B.; Manca, L.; Hartke, A.; Giard, J.C.; Sanguinetti, M.; Fadda, G.; Blasi, E. Role of the (Mn)superoxide dismutase of Enterococcus faecalis in the in vitro interaction with microglia. *Microbiology* **2011**, *157*, 1816–1822. [CrossRef]
36. Frankenberg, L.; Brugna, M.; Hederstedt, L. Enterococcus faecalis heme-dependent catalase. *J. Bacteriol.* **2002**, *184*, 6351–6356. [CrossRef]
37. Vlková, B.; Celec, P. Does Enterococcus faecalis contribute to salivary thiobarbituric acid-reacting substances? *In Vivo* **2009**, *23*, 343–345. [PubMed]
38. Zand, V.; Milani, A.S.; Amini, M.; Barhaghi, M.H.; Lotfi, M.; Rikhtegaran, S.; Sohrabi, A. Antimicrobial efficacy of photodynamic therapy and sodium hypochlorite on monoculture biofilms of Enterococcus faecalis at different stages of development. *Photomed. Laser Surg.* **2014**, *32*, 245–251. [CrossRef] [PubMed]
39. Kishen, T.J.; Mohapatra, B.; Diwan, A.D.; Etherington, G. Post-traumatic thoracic scoliosis with rib head dislocation and intrusion into the spinal canal: A case report and review of literature. *Eur. Spine J.* **2010**, *19* (Suppl. 2), S183–S186. [CrossRef] [PubMed]
40. Oda, D.F.; Duarte, M.A.H.; Andrade, F.B.; Moriyama, L.T.; Bagnato, V.S.; de Moraes, I.G. Antimicrobial action of photodynamic therapy in root canals using LED curing light, curcumin and carbopol gel. *Int. Endod. J.* **2019**, *52*, 1010–1019. [CrossRef]
41. Golmohamadpour, A.; Bahramian, B.; Khoobi, M.; Pourhajibagher, M.; Barikani, H.R.; Bahador, A. Antimicrobial photodynamic therapy assessment of three indocyanine green-loaded metal-organic frameworks against Enterococcus faecalis. *Photodiagn. Photodyn. Ther.* **2018**, *23*, 331–338. [CrossRef]
42. Li, R.; Yuan, L.; Jia, W.; Qin, M.; Wang, Y. Effects of Rose Bengal- and Methylene Blue-Mediated Potassium Iodide-Potentiated Photodynamic Therapy on Enterococcus faecalis: A Comparative Study. *Lasers Surg. Med.* **2021**, *53*, 400–410. [CrossRef]
43. Ghorbanzadeh, A.; Bahador, A.; Sarraf, P.; Ayar, R.; Fekrazad, R.; Asefi, S. Ex vivo comparison of antibacterial efficacy of conventional chemomechanical debridement alone and in combination with light-activated disinfection and laser irradiation against Enterococcus faecalis biofilm. *Photodiagn. Photodyn. Ther.* **2020**, *29*, 101648. [CrossRef]
44. Chandra, J.; Kuhn, D.M.; Mukherjee, P.K.; Hoyer, L.L.; McCormick, T.; Ghannoum, M.A. Biofilm formation by the fungal pathogen Candida albicans: Development, architecture, and drug resistance. *J. Bacteriol.* **2001**, *183*, 5385–5394. [CrossRef]
45. Beirão, S.; Fernandes, S.; Coelho, J.; Faustino, M.A.; Tomé, J.P.; Neves, M.G.; Tomé, A.C.; Almeida, A.; Cunha, A. Photodynamic inactivation of bacterial and yeast biofilms with a cationic porphyrin. *Photochem. Photobiol.* **2014**, *90*, 1387–1396. [CrossRef]
46. Garcez, A.S.; Núñez, S.C.; Azambuja, N.; Fregnani, E.R.; Rodriguez, H.M.; Hamblin, M.R.; Suzuki, H.; Ribeiro, M.S. Effects of photodynamic therapy on Gram-positive and Gram-negative bacterial biofilms by bioluminescence imaging and scanning electron microscopic analysis. *Photomed. Laser Surg.* **2013**, *31*, 519–525. [CrossRef] [PubMed]
47. Shrestha, A.; Kishen, A. Antibacterial Nanoparticles in Endodontics: A Review. *J. Endod.* **2016**, *42*, 1417–1426. [CrossRef] [PubMed]
48. Fernandez-Fernandez, A.; Manchanda, R.; Lei, T.; Carvajal, D.A.; Tang, Y.; Kazmi, S.Z.; McGoron, A.J. Comparative study of the optical and heat generation properties of IR820 and indocyanine green. *Mol. Imaging* **2012**, *11*, 99–113. [CrossRef] [PubMed]
49. Zhou, X.; Chen, Y.; Wei, X.; Liu, L.; Zhang, F.; Shi, Y.; Wu, W. Heat transfers to periodontal tissues and gutta-percha during thermoplasticized root canal obturation in a finite element analysis model. *Oral Surg. Oral Med. Oral Pathol. Oral Radiol. Endod.* **2010**, *110*, 257–263. [CrossRef]

50. Lipski, M.; Woźniak, K. In vitro infrared thermographic assessment of root surface temperature rises during thermafil retreatment using system B. *J. Endod.* **2003**, *29*, 413–415. [CrossRef] [PubMed]
51. Yoshida, K.; Uoshima, K.; Oda, K.; Maeda, T. Influence of heat stress to matrix on bone formation. *Clin. Oral Implant. Res.* **2009**, *20*, 782–790. [CrossRef]
52. Saxena, V.; Sadoqi, M.; Shao, J. Degradation kinetics of indocyanine green in aqueous solution. *J. Pharm. Sci.* **2003**, *92*, 2090–2097. [CrossRef] [PubMed]
53. Yamamoto, H.; Kuno, Y.; Sugimoto, S.; Takeuchi, H.; Kawashima, Y. Surface-modified PLGA nanosphere with chitosan improved pulmonary delivery of calcitonin by mucoadhesion and opening of the intercellular tight junctions. *J. Control. Release* **2005**, *102*, 373–381. [CrossRef] [PubMed]
54. Hancock, H.H.; Sigurdsson, A.; Trope, M.; Moiseiwitsch, J. Bacteria isolated after unsuccessful endodontic treatment in a North American population. *Oral Surg. Oral Med. Oral Pathol. Oral Radiol. Endod.* **2001**, *91*, 579–586. [CrossRef] [PubMed]

International Journal of
Molecular Sciences

MDPI

Article

Improving Consistency of Photobiomodulation Therapy: A Novel Flat-Top Beam Hand-Piece versus Standard Gaussian Probes on Mitochondrial Activity

Andrea Amaroli [1,2,*], Praveen Arany [3], Claudio Pasquale [2], Stefano Benedicenti [2], Alessandro Bosco [4] and Silvia Ravera [5]

1 Department of Orthopedic Dentistry, Faculty of Dentistry, First Moscow State Medical University (Sechenov University), 119991 Moscow, Russia
2 Department of Surgical and Diagnostic Sciences, University of Genoa, 16132 Genoa, Italy; clodent@gmail.com (C.P.); stefano.benedicenti@unige.it (S.B.)
3 Departments of Oral Biology, Surgery and Biomedical Engineering, University at Buffalo, Buffalo, NY 14260, USA; prarany@buffalo.edu
4 Bosco Ottica S.r.l., Castel Rozzone, 24040 Bergamo, Italy; alessandro.bosco@boscoottica.it
5 Department of Experimental Medicine, University of Genoa, 16132 Genoa, Italy; silvia.ravera@unige.it
* Correspondence: andrea.amaroli.71@gmail.com; Tel.: +39-010-3537309

Citation: Amaroli, A.; Arany, P.; Pasquale, C.; Benedicenti, S.; Bosco, A.; Ravera, S. Improving Consistency of Photobiomodulation Therapy: A Novel Flat-Top Beam Hand-Piece versus Standard Gaussian Probes on Mitochondrial Activity. *Int. J. Mol. Sci.* **2021**, *22*, 7788. https://doi.org/10.3390/ijms22157788

Academic Editors: Antonino Mazzaglia, Angela Scala and Enrico Caruso

Received: 30 June 2021
Accepted: 18 July 2021
Published: 21 July 2021

Abstract: The tremendous therapeutic potential of photobiomodulation therapy in different branches of medicine has been described in the literature. One of the molecular mechanisms for this treatment implicates the mitochondrial enzyme, cytochrome C oxidase. However, the efficacy and consistency of clinical outcomes with photobiomodulation treatments has been fiercely debated. This work was motivated by this need to improve photobiomodulation devices and delivery approaches. We designed a novel hand-piece with a flat-top beam profile of irradiation. We compared the beam profile versus a standard hand-piece and a fibre probe. We utilized isolated mitochondria and performed treatments at various spots within the beam, namely, the centre, left and right edge. We examined mitochondrial activity by assessing ATP synthesis with the luciferin/luciferase chemiluminescent method as a primary endpoint, while mitochondrial damage was assessed as the secondary endpoint. We observed a uniform distribution of the power density with the flat-top prototype compared to a wide Gaussian beam profile with the standard fibre and standard hand-piece. We noted increased production of ATP in the centre of all three beams with respect to the non-treated controls ($p < 0.05$). Both the fibre and standard hand-piece demonstrated less increase in ATP synthesis at the edges than the centre ($p < 0.05$). In contrast, ATP synthesis was increased homogenously in the flat-top handpiece, both in the centre and the edges of the beam. Fibre, standard hand-piece and the flat-top hand-piece prototype have discrete beam distribution characteristics. This significantly affected the mitochondrial activity with respect to their position within the treated areas. Flat-top hand-piece enhances the uniformity of photobiomodulation treatments and can improve the rigour and reproducibility of PBM clinical outcomes.

Keywords: phototherapy; light therapy; low-level laser therapy; ATP; mitochondria respiratory chain; light-emitting diode; near-infrared light; energetic metabolism

1. Introduction

The ability of visible and near-infrared light (NIR) to influence body healing has been described by several ancient civilizations [1]. The first scientific description by Prof. Endre Mester about fifty years ago outlined the effects of low dose laser interaction with tissues describing the non-thermal therapeutic benefits of biophotonics energy [2–4]. Since then, growing evidence has demonstrated that visible and near-infrared light can modulate metabolism in various life forms, from bacteria and protozoa to animals and humans [5]. While all life forms need energy for survival, unlike plant-cell, the animal cell does not appear to directly employ light as a metabolic source of energy [6]. Nonetheless, there are several specialized cells in the human body capable of photoreception that harness light to enable vision and circadian rhythm [7]. Additionally, several biological molecules are capable of interacting with visible and NIR light wavelengths. These include oxyhaemoglobin, melanin, cytochrome and metalloproteins, sulphur-protein, water and lipids [8].

A major site for light interaction has been noted to involve the mitochondria [8]. These interactions can result via direct interaction of cytochromes belonging to the respiratory chain. Alternatively, they can indirectly impact biophysical properties of water, lipids and voltage-gated ion channels as well as changes in calcium homeostasis and membrane fluidity [8–10]. In other words, the photon can transfer its energy to the photoacceptor in the mitochondria that leads to an electronically excited state and production of energy substrates such as ATP [11,12]. Thus, this ability to modulate the cellular metabolism and functions via non-ionizing and non-thermal light treatments is termed photobiomodulation (PBM) therapy; popularly termed low-level light/laser therapy or cold laser treatments.

A growing literature has noted the popularity of photobiomodulation therapy in discrete branches of medicine from performance enhancement, supportive cancer care to neurorehabilitation. There has been significant recent progress in our understanding of PBM mechanisms, extending beyond the initial effects on the mitochondria to cell membrane receptors and ion transporters as well as an extracellular growth factor, TGF-β1 [13]. However, the standardization and the repeatability of the photobiomodulation process is currently under debate [1,9–11,13–15]. Two major parameters that can influence this have been identified. First, the variation in tissue optical properties such as scattering due to their microstructure, wavelength-dependent absorption of photon-energy, skin colour and thickness affects the transmittance and reflectance of laser light [16–18]. These parameters are not assessed directly in vitro lab studies. Therefore, a coherent translation of the results from in vitro to clinical patients is limited. Second, the photobiomodulation parameters do not directly correlate with precise light and cellular photoacceptor interaction. These can drastically vary with even small changes in the photon-energy delivered [11].

Hanna et al., [19] recently demonstrated that by moving the hand-piece from contact to many centimetres away from the target, the treatment power varies with the use of a standard gaussian hand-piece with respect to a flat-top beam delivery system. The same authors also reported improved cell growth and differentiation with treatments using the flat-top hand-piece than the standard probe. They attributed this response to the more homogenous power distribution within the treatment area through the flat-top probe. However, this has not yet been experimentally demonstrated. This study was designed to address this issue. We examined the beam profiles of a novel hand-piece with a flat-top beam profile compared to a standard hand-piece and fibre probe. Next, we investigated the effects of the three probes on isolated mitochondria by assessing ATP generation in various areas of the laser beam, namely in the centre and at the edges on either side. Finally, we also examined the effects of various treatment doses with the three probes on potential mitochondrial damage.

2. Materials and Methods

2.1. Experimental Design and Purpose

The experiments followed standardized methods for mitochondria isolation, PBM treatments and ATP synthesis evaluation as noted in our prior works with a few modifications [11,20,21]. We aimed to compare the effectiveness of laser treatments on the entire treatment spot-size area. The primary endpoint was the impact of the beam power distribution on mitochondrial activity. The secondary endpoint was the induction of mitochondrial damage. An iLux, Real-Time MecOS 2.0, (Mectronic Medicale S.R.L., Grassobbio, Bergamo, Italy) was employed to perform treatments through probes such as fibre, standard hand-piece and the novel flat-top hand-piece prototype.

As the flat-top hand-piece keeps the spot-size area constant from contact to many centimetres [19], this was kept constant at 1 cm^2 with all probes set up. A 635-nm red light pointer (>0.5 mW) was used to measure probe spot-size areas on a graph, and the correct distance to obtain it was evaluated with a spacer (2 cm) (Figure 1A). Therefore, a circular area of 1 cm^2 was outlined on a square microscope coverslip for transmitted light microscopy (2.4 cm on each side). The coverslip had a thickness of 0.13 mm and optical properties in accordance with ISO 8255 and ISO 8255-1. The two sides of the coverslip were suspended at 3 cm from an absorbing material mat positioned at the bottom.

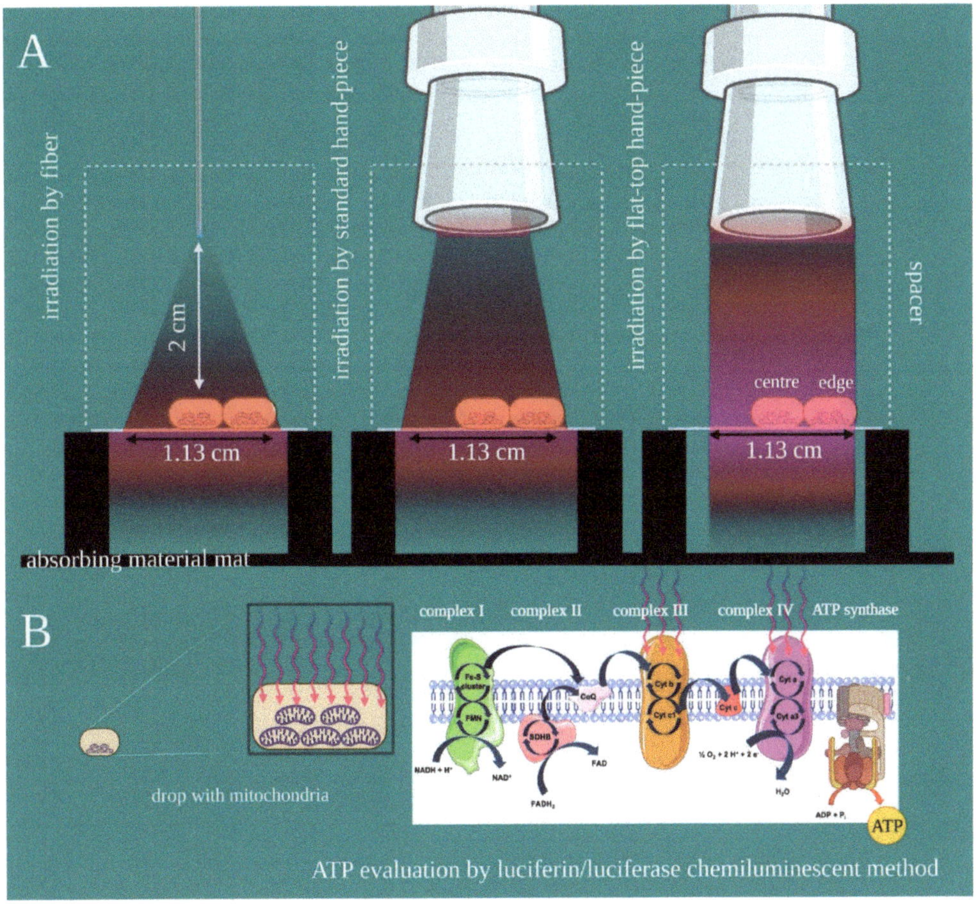

Figure 1. Experimental design. (**A**) A drop of isolated mitochondria enriched-fraction was pipetted on a coverslip, in the centre or the edge of the laser beam circular area, according to the experimental purpose. The drop was then irradiated with an 808-nm laser light (diameter 1.13 cm; spot-size 1 cm^2) through a fibre, a standard hand-piece, or the flat-top hand-piece. (**B**) Primary and secondary endpoints were evaluated as ATP synthesis through the luciferin/luciferase chemiluminescent method. Image created with BioRender.com.

A 50 µL drop, approximatively 1/3rd of laser beam, of isolated mitochondria enriched-fraction, was pipetted in the centre or on edge (left and right) (Figure 1). The circular area was then illuminated with an 808-nm laser light (spot size 1 cm^2) through a fibre, a standard hand-piece, or the flat-top hand-piece prototype. A power meter Pronto-250 (Gentec electro-optics, Inc. Canada) was employed to monitor the power at the target and 2 cm from away. Then, the laser with power set at 1 W, which allowed for the generation of the power density of 1 W/cm^2 and energy and fluence of 60 J and 60 J/cm^2, respectively. The treatments were performed in continuous wave mode for 60 s. For the controls, treatments were performed with the same experimental set-up, but with the device switched off (0 W, 60 s). The laser parameter was chosen in accordance with our prior work on isolated mitochondria [21]. Due to possible undesirable thermal effects, adverse events were avoided by monitoring the irradiation with a thermal camera FLIR ONE Pro-iOS (FLIR Systems, Inc. designs, Portland, OR, USA, dynamic range: −20 °C/+400 °C; resolution 0.1 °C) during treatments. Our previous study showed the power of 1 W treatments with similar parameters stimulates the mitochondria complex III and IV [21]. This study examined ATP production with the luciferin/luciferase chemiluminescent method (Figure 1B). The reliability of the experimental set-up was evaluated with the dosimetry for the isolated mitochondria model, described by our team in previous work [22]. To avoid operator bias, the treatments and the data analyses were performed by different operators in a blinded manner.

2.2. Design of the Flat-Top Hand-Piece

The prototype of the hand-piece with flat-top beam profile relies on the probe with an international industrial patent (n.00001425863). Basically, fibre and hand-pieces employed in laser treatments have a non-uniform distorted beam profile resulting in their divergent power density distribution [21,23]. Improvements may be obtained through beam collimation with an alignment between the fibre's spot delivering light (Figure 2A) and the focal point of a positive lens (Figure 2a,b). However, this approach does not ensure true collimation due to two issues. The first issue is that a large optical fibre size has a longer focal length of the lens resulting in a more diffractive effect. Hence, the beam is not truly collimated, but tends to diverge (partially collimated). The second issue is the fibres routinely used in devices for phototherapy. The optical fibre may transmit light radiation through a series of continuous internal reflections across its length. This feature may support the homogeneous distribution of the power density on the spot size enabling reasonable non-distortive magnification to expand the spot size. However, these critical reflections require fibres parameters incompatible with current clinical devices.

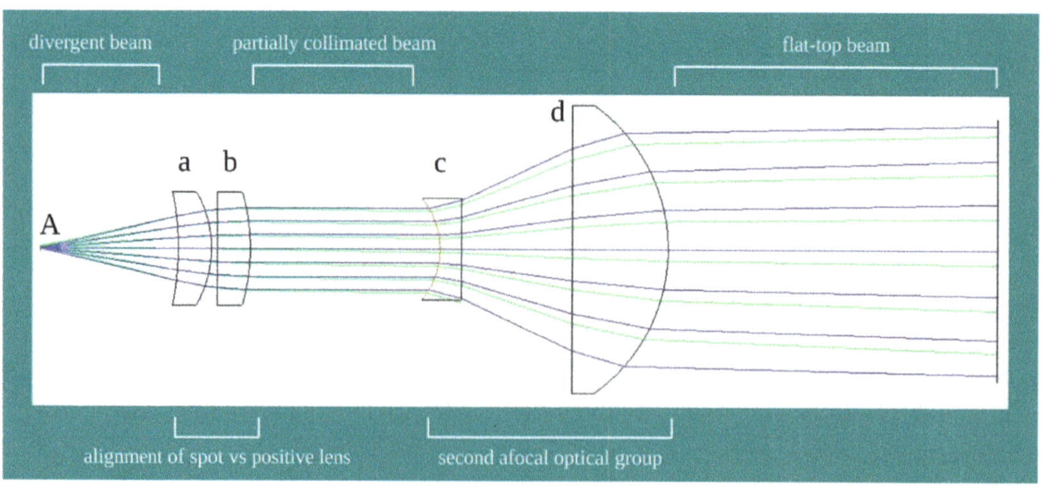

Figure 2. Schematic design of the flat-top hand-piece. (**A**) = fibre's spot delivering laser light; (**a**,**b**) = collimating lens; (**c**,**d**) = couple of lens to generate an afocal optical group.

To improve the distribution of the photons on the spot-size area and meet the features of the iLux, real-time MecOS 2.0, a novel integrated optical system was introduced

(Figure 2c,d). This system does not use filters or artifices to modify photon intensity and density, and generates a standard Gaussian beam energy profile. A simplified scheme of the novel hand-piece with a flat-top beam profile, project PRG004.20.02 (Bosco Ottica srl, Castel Rozzone, Bergamo, Italy), is shown in Figure 2. First, the alignment of the fibre's spot delivering light (Figure 2A) with the focal point of the positive lens (Figure 2a,b) generates partial collimation. Then, a second afocal optical group was introduced (Figure 2c,d), which reduces the residual divergence and improves the collimation of the beam and, consequently, the flatness of the wavefront.

2.3. Characterization of the Probes Beam Profile

The power density distribution of the beam was characterized with a camera Spiricon SP928 equipped with the software BeamGage Professional (Ophir Spiricon Europe GmbH, Darmstadt, Germany). The measuring tool shows a palette of colours; the power density was measured in a point reflected as purple = 0.1 W/cm^2; fuchsia = 0.3 W/cm^2; blue = 0.6 W/cm^2; light blue = 0.9 W/cm^2; green = 1 W/cm^2; yellow = 1.6 W/cm^2; orange = 2.0 W/cm^2; Red = 2.2 W/cm^2.

2.4. Mitochondria Enriched-Fraction Isolation

Bovine liver from two males and females was acquired at the slaughterhouse, Ceva, Torino, Italy. These specimens were less than 1 year old and were bred for human consumption following the directives of the Italian Ministry of Agricultural, Food and Forestry Policies. As the animals were not bred or sacrificed at the University of Genoa, ethical committee approval was deemed unnecessary. Specimens were collected and immediately processed after slaughter, following all safety rules. To isolate mitochondria enriched fraction, the bovine livers were washed in PBS and homogenized in a buffer solution containing: 0.25 M sucrose, 0.15 M KCl, 10 mM Tris-HCl pH 7.4 and 1 mM EDTA. The homogenate was then centrifuged at 800× *g* for 10 min. The supernatant sample was filtered and centrifuged at 12,000× *g* for 15 min. The pellet was resuspended in another buffer containing: 0.25 M sucrose, 75 mM mannitol, 10 mM Tris-HCl pH 7.4, 1 mM EDTA. Lastly, the supernatant was centrifuged at 12,000× *g* for 15 min and the mitochondrial pellet was resuspended in the same buffer [11,20,21,24].

2.5. Evaluation of Mitochondrial ATP Synthesis

To evaluate the ATP production through the ATP synthase (Fo-F1 ATP synthase), mitochondria enriched-fraction treated with or without the 808-nm laser were diluted in a solution containing: 100 mM Tris-HCl pH 7.4, 100 mM KCl, 1 mM EGTA, 2.5 mM EDTA, 5 mM $MgCl_2$, 0.2 mM di(adenosine-5′) penta-phosphate, 0.6 mM ouabain, ampicillin (25 μg/mL), 5 mM KH2PO4 and 5 mM pyruvate + 2.5 mM malate, used as respiratory substrates. The ATP synthesis started after the addition of 0.1 mM ADP and was monitored for 2 min, in a luminometer (Glomax 20/20, Promega, 20126 Milan, Italy) by the luciferin/luciferase chemiluminescent method [11,20,21,23]. An ATP standard solution between 10^{-9} and 10^{-7} M was used for calibration.

2.6. Statistical Analysis

Statistical analyses were performed with GraphPad Prism software version 7 (GraphPad Software, La Jolla, CA, USA). All parameters were tested by one-way ANOVA followed by the Bonferroni test. Data are expressed as mean ± standard deviation (SD) from 3 to 5 independent determinations performed in duplicate. In the figures, SD is shown as error bars. An error probability with $p < 0.05$ was selected as significant.

3. Results

3.1. Characterization of the Probes Beam Profile and Irradiation

Power density is differentially distributed when 1 W of power (1 W/cm^2 power density) is irradiated through a fibre (A), a standard hand-piece (B) and the flat-top hand-piece (C) within the 1 cm^2 treatment spot area (Figure 3). As indicated in the material and methods section and on the figure, different colours indicate differences in power density distribution. The green colour (1 W/cm^2) is distributed on only the 27–30% and 39–41% of the irradiated area when a fibre or a standard hand-piece was employed (Figures 3 and 4). The remaining areas demonstrated power density in the range of 0.6–0.1 W/cm^2. Conversely, a wider area of ~90% of green colour was described during irradiation through the flat-top hand-piece. Additionally, in the centre of both the spot sizes generated through fibre and standard hand-piece irradiations, higher energy distribution was pointed out by red, orange and yellow colours. Therefore, a distribution of the power density such as a wide-Gaussian beam profile can be considered for fibre and standard hand-piece, compared with a more uniform profile observed with the flat-top prototype.

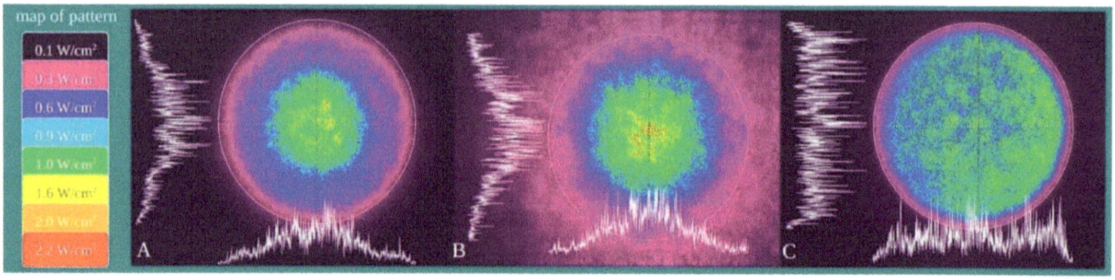

Figure 3. Characterization of the fibre (**A**), standard hand-piece (**B**) and the novel flat-top hand-piece prototype; (**C**) beam profile through a camera Spiricon SP928 equipped with the software BeamGage Professional. Setting the instrument for an irradiation with 1 W the main colours displayed are: purple = 0.1 W/cm^2; fuchsia = 0.3 W/cm^2; blue = 0.6 W/cm^2; light blue = 0.9 W/cm^2; green = 1 W/cm^2; yellow = 1.6 W/cm^2; orange = 2.0 W/cm^2; red = 2.2 W/cm^2.

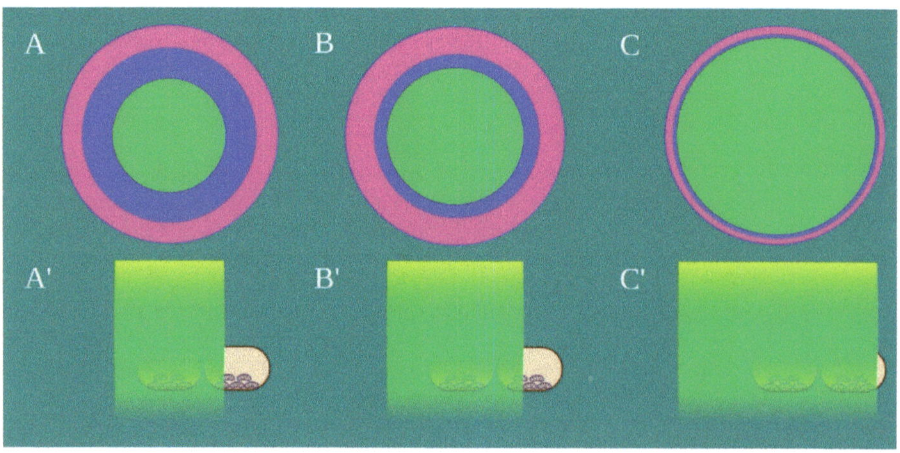

Figure 4. Schematic representation of the laser beam profile (**A–C**) and the drop's area irradiated by the power density of 1 W/cm^2, green colour, (**A'–C'**) Image created with BioRender.com.

According to Hanna et al. [19] and also in our experimental set-up, the irradiation with the flat-top hand-piece keeps power constant in contact mode and at 2 cm from the target (1.03 ± 0.02 W vs. 1.01 ± 0.03 W; $p > 0.05$). Irradiation with fibre or the standard hand-piece experienced a statistically significant decrease of 0.23 and 0.22 W, $p < 0.05$. A statistically significant increment of the drop temperature was observed after irradiation (~2.1 °C), but the thermal increase was similar with the three probes ($p > 0.05$).

3.2. Evaluation of Mitochondrial ATP Synthesis

ATP synthesis in isolated mitochondria was assessed following PBM treatments with various probes. The mitochondria in the centre of the laser beam with all three probes demonstrated increased ATP production compared to the non-treated control ($p < 0.05$) (Figure 5). However, PBM treatments performed with fibre or standard hand-piece on the mitochondria placed at the edges of the beam had a significantly less increase in ATP synthesis compared to the centre of the beam ($p < 0.05$). This difference was more evident in the fibre group compared to the standard hand-piece ($p < 0.05$). In contrast, the flat-top hand-piece group demonstrated increased ATP synthesis in both the centre and edges uniformly throughout the beam area.

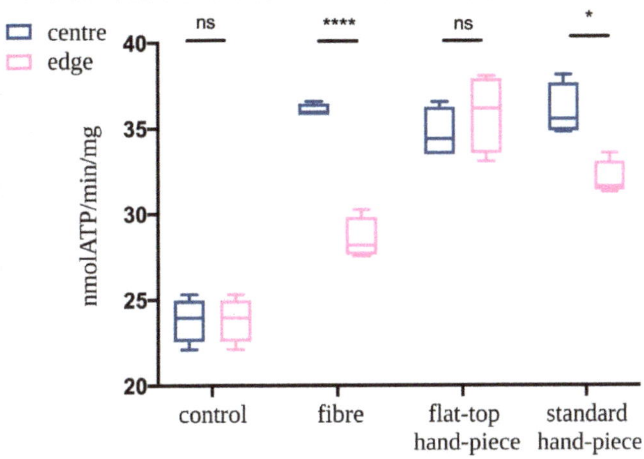

Figure 5. Effect of the photobiomodulation on mitochondrial ATP synthase activity. The isolated mitochondria were placed at the centre or edges of the laser treatment spot (1 cm^2) and the treatments were performed using a fibre, flat-top hand-piece or a standard hand-piece. All samples were treated with PBM 808 nm laser in continuous wave mode with 1 W, 1 W/cm^2, 60 J and 60 J/cm^2 for 60 s. Untreated control samples were placed in a similar set-up with laser at 0 W for 60 s. Data are expressed as mean ± SD. A significant difference between the ATP production of the mitochondria in the centre vs. the edge is indicated by the symbol * and ****, respectively $p < 0.05$ and 0.0001, ns indicates a no-significant difference.

4. Discussion

Photobiomodulation involves changes in cellular metabolism through the transfer of energy from photons to its photoacceptors [8]. The pivotal role of mitochondria in the translation of biophotonic energy to biochemical changes has been previously demonstrated [11,20,21,25–27]. However, the rigour of the photobiomodulation clinical outcomes has been questioned. The device technology used for photobiomodulation delivery plays a critical role in improving treatment standardization [24]. Indeed, our data point out that power density delivery is affected by the probes used for these treatments. In this study, we noted that despite both the fibre and standard hand-piece delivering the correct power density, only a surface area less than 50% of the laser spot size was effectively illuminated. Further, the distribution of the different power densities on the treatment spot size was more non-uniform for the fibre compared to a standard hand-piece. Strikingly, flat-top prototype delivered the most consistent power distribution over 90% of the treatment area. Furthermore, the flat-top hand-piece was able to maintain constant power from contact to a couple of centimetres away allowing for improved clinical consistency during PBM treatment delivery as previously reported by Hanna et al. [19]. This evidence supports the notion that the effectiveness of photobiomodulation therapy could be significantly influenced by both the probes employed and the delivery technique of the operator. The significant improvement in mitochondrial activity with the flat-top hand-piece prototype demonstrated a homogenous treatment spot size in the centre and on either sides. This contrasted with the significant differences in the fibre and standard hand-piece groups indicating that the biological responses were most uniformly achieved with the current flat-top prototype. Improvements in the laser device, fibre and collimating procedure with the lens and probe design could further significantly improve the rigour and consistency of PBM clinical treatments.

As noted in our previous work, PBM responses involve complex events resulting from the absorption and scattering of the photons and the generation of an electromagnetic field [22]. This work was based on three-dimensional modelling of photon waves interacting within the mitochondrial droplet generated between the laser to the coverslip. The differences in the beam power density distribution during treatment with the three probes may impact these interactions affecting the overall photobiomodulation responses. Additionally, recent work from our group observed a limited dose range of a 980 nm diode laser, which affects the complexes III and IV as well as ATP production and oxygen consumption of mitochondria [11]. Slight variations (0.1 W) in the treatment power appeared to drastically modulate the photobiomodulation outcomes. Basically, 0.8–1.1 W kept mitochondria coupled and induced increments of ATP production by increments of complex III and IV activities. In contrast, 0.1–0.2 W uncoupled the mitochondria and had an inhibitory effect of ATP synthesis and increment of oxidative stress, while 0.3–0.7 and 1.2–1.4 W did not appear to affect these responses. These data suggest a major limitation in the reproducibility of photobiomodulation responses lie in a non-homogeneous distribution of the laser energy. Variations in an area of only 1 cm^2 appeared to induce drastically different photobiomodulatory responses (positive, null or negative) in these in vitro studies. Therefore, we would expect significant differences in groups of neighbouring cells and overall tissues when this therapy is used in clinical in vivo scenarios.

This work has some strengths and limitations. A single spot size of 1.13 cm in diameter was used allowing for a limited drop volume of isolated mitochondria sufficient to assess ATP. Further reduction of the drop diameter or an increment of the laser spot-size as well as reducing treatment power could highlight further, perhaps more dramatic, differences. However, the use of three-dimensional dose modelling [22], the temperature monitoring during treatments [11], the standardised ATP synthesis evaluation of treated mitochondria [11,20,21] and the careful assessment of the treatment power at the target surface with a power meter [19], enabled comparisons of the three discrete probe designs and were clear strengths.

5. Conclusions

To summarize, our data demonstrated that the fibre, standard hand-piece and the flat-top hand-piece prototype have different beam energy distribution features. These differences significantly affected our primary endpoint, which was mitochondria activity with respect to their position in the treatment spot size. Our second endpoint showed that even at a power of 1 W, there was no damage to mitochondrial function. These results provide evidence that flat-top hand-piece allows improved photobiomodulation treatment reproducibility, especially in clinical scenarios where the distance from the target surface may vary during treatments and the wide affected area needs uniform irradiation to better experience the photobiomodulation effects.

Author Contributions: Conceptualization, A.A., A.B. and S.R.; methodology, A.A., A.B. and S.R.; software, A.A., A.B. and S.R.; validation, A.A. and A.B.; formal analysis, A.A., A.B. and S.R.; investigation, A.A., A.B., C.P. and S.R.; resources, S.B.; data curation, A.A., A.B. and S.R.; writing—original draft preparation, A.A., A.B., P.A. and S.R.; writing—review and editing, A.A. and P.A.; supervision, P.A., A.A., and S.B.; project administration, A.A. and S.B.; funding acquisition, S.B. All authors have read and agreed to the published version of the manuscript.

Funding: This research received no external funding.

Institutional Review Board Statement: Not applicable.

Informed Consent Statement: Not applicable.

Data Availability Statement: Data available on request from the authors.

Acknowledgments: Authors would like to express special appreciation and thanks to Alberico Benedicenti and Mirco Raffetto, for their guidance on our work.

Conflicts of Interest: The authors declare no conflict of interest.

References

1. Mosca, R.C.; Ong, A.A.; Albasha, O.; Bass, K.; Arany, P. Photobiomodulation Therapy for Wound Care: A Potent, Noninvasive, Photoceutical Approach. *Adv. Skin Wound Care* **2019**, *32*, 157–167. [CrossRef]
2. Mester, E.; Spiry, T.; Szende, B.; Tota, J.G. Effect of laser rays on wound healing. *Am. J. Surg.* **1971**, *122*, 532–535. [CrossRef]
3. Mester, E.; Szende, B.; Gartner, P. The effect of laser beams on the growth of hair in mice. *Radiobiol. Radiother.* **1968**, *9*, 621–626.
4. Mester, E.; Szende, B.; Spiry, T.; Scher, A. Stimulation of wound healing by laser rays. *Acta Chir. Acad. Sci. Hung.* **1972**, *13*, 315–324.
5. Amaroli, A.; Ferrando, S.; Benedicenti, S. Photobiomodulation Affects Key Cellular Pathways of all Life-Forms: Considerations on Old and New Laser Light Targets and the Calcium Issue. *Photochem. Photobiol.* **2019**, *95*, 455–459. [CrossRef]
6. Niklas, K.J. *The Evolutionary Biology of Plant*; University of Chicago: Chicago, IL, USA, 1997; p. 449.
7. Ruban, A.V. Evolution under the sun: Optimizing light harvesting in photosynthesis. *J. Exp. Bot.* **2015**, *66*, 7–23. [CrossRef]
8. Ravera, S.; Colombo, E.; Pasquale, C.; Benedicenti, S.; Solimei, L.; Signore, A.; Amaroli, A. Mitochondrial Bioenergetic, Photobiomodulation and Trigeminal Branches Nerve Damage, What's the Connection? A Review. *Int. J. Mol. Sci.* **2021**, *22*, 4347. [CrossRef] [PubMed]
9. Amaroli, A.; Colombo, E.; Zekiy, A.; Aicardi, S.; Benedicenti, S.; De Angelis, N. Interaction between Laser Light and Osteoblasts: Photobiomodulation as a Trend in the Management of Socket Bone Preservation-A Review. *Biology* **2020**, *9*, 409. [CrossRef] [PubMed]
10. Colombo, E.; Signore, A.; Aicardi, S.; Zekiy, A.; Utyuzh, A.; Benedicenti, S.; Amaroli, A. Experimental and Clinical Applications of Red and Near-Infrared Photobiomodulation on Endothelial Dysfunction: A Review. *Biomedicines* **2021**, *9*, 274. [CrossRef]
11. Amaroli, A.; Pasquale, C.; Zekiy, A.; Utyuzh, A.; Benedicenti, S.; Signore, A.; Ravera, S. Photobiomodulation and Oxidative Stress: 980 nm Diode Laser Light Regulates Mitochondrial Activity and Reactive Oxygen Species Production. *Oxid. Med. Cell. Longev.* **2021**, *3*, 6626286.
12. Chen, A.C.; Arany, P.R.; Huang, Y.Y.; Tomkinson, E.M.; Sharma, S.K.; Kharkwal, G.B.; Saleem, T.; Mooney, D.; Yull, F.E.; Blackwell, T.S.; et al. Low-level laser therapy activates NF-kB via generation of reactive oxygen species in mouse embryonic fibroblasts. *PLoS ONE* **2011**, *6*, e22453. [CrossRef]
13. Arany, P.R. Craniofacial Wound Healing with Photobiomodulation Therapy: New Insights and Current Challenges. *J. Dent. Res.* **2016**, *95*, 977–984. [CrossRef]
14. Fornaini, C.; Arany, P.; Rocca, J.P.; Merigo, E. Photobiomodulation in Pediatric Dentistry: A Current State-of-the-Art. *Photobiomodul. Photomed. Laser Surg.* **2019**, *3*, 7798–7813. [CrossRef]
15. Pasquale, C.; Colombo, E.; Benedicenti, S.; Signore, A.; Amaroli, A. 808-Nm Near-Infrared Laser Photobiomodulation versus Switched-Off Laser Placebo in Major Aphthae Management: A Randomized Double-Blind Controlled Trial. *Appl. Sci.* **2021**, *11*, 4717. [CrossRef]
16. Salehpour, F.; Cassano, P.; Rouhi, N.; Hamblin, M.R.; De Taboada, L.; Farajdokht, F.; Mahmoudi, J. Penetration Profiles of Visible and Near-Infrared Lasers and Light-Emitting Diode Light Through the Head Tissues in Animal and Human Species: A Review of Literature. *Photobiomodul. Photomed. Laser Surg.* **2019**, *37*, 581–595. [CrossRef] [PubMed]
17. Hochman-Elam, L.N.; Heidel, R.E.; Shmalberg, J.W. Effects of laser power, wavelength, coat length, and coat color on tissue penetration using photobiomodulation in healthy dogs. *Can. J. Vet. Res.* **2020**, *84*, 131–137.
18. Souza-Barros, L.; Dhaidan, G.; Maunula, M.; Solomon, V.; Gabison, S.; Lilge, L.; Nussbaum, E.L. Skin color and tissue thickness effects on transmittance, reflectance, and skin temperature when using 635 and 808 nm lasers in low intensity therapeutics. *Lasers Surg. Med.* **2018**, *50*, 291–301. [CrossRef] [PubMed]
19. Hanna, R.; Agas, D.; Benedicenti, S.; Ferrando, S.; Laus, F.; Cuteri, V.; Lacava, G.; Sabbieti, M.G.; Amaroli, A. A Comparative Study Between the Effectiveness of 980 nm Photobiomodulation Delivered by Hand-Piece with Gaussian vs. Flat-Top Profiles on Osteoblasts Maturation. *Front Endocrinol.* **2019**, *20*, 92. [CrossRef]
20. Ravera, S.; Ferrando, S.; Agas, D.; De Angelis, N.; Raffetto, M.; Sabbieti, M.G.; Signore, A.; Benedicenti, S.; Amaroli, A. 1064 nm Nd:YAG laser light affects transmembrane mitochondria respiratory chain complexes. *J. Biophotonics* **2019**, *12*, 201900101. [CrossRef]
21. Amaroli, A.; Ravera, S.; Parker, S.; Panfoli, I.; Benedicenti, A.; Benedicenti, S. An 808-nm Diode Laser with a Flat-Top Handpiece Positively Photobiomodulates Mitochondria Activities. *Photomed. Laser Surg.* **2016**, *34*, 564–571. [CrossRef]

22. Amaroli, A.; Benedicenti, S.; Bianco, B.; Bosco, A.; Clemente Vargas, M.R.; Hanna, R.; Kalarickel Ramakrishnan, P.; Raffetto, M.; Ravera, S. Electromagnetic Dosimetry for Isolated Mitochondria Exposed to Near-Infrared Continuous-Wave Illumination in Photobiomodulation Experiments. *Bioelectromagnetics* **2021**, *18*, 22342.
23. Selting, W. *Atlas of Laser Therapy: State of the Art*, 4th ed.; Teamwork Media Srl: Villa Carcina, Italy, 2016; pp. 225–236.
24. Lescuyer, P. Progress in the Definition of a Reference Human Mitochondrial Proteome. *Proteomics* **2003**, *3*, 157–167. [CrossRef]
25. Passarella, S.; Karu, T. Absorption of monochromatic and narrow band radiation in the visible and near IR by both mitochondrial and non-mitochondrial photoacceptors results in photobiomodulation. *J. Photochem. Photobiol. B* **2014**, *140*, 344–358. [CrossRef] [PubMed]
26. Karu, T.I. Mitochondrial signaling in mammalian cells activated by red and near-IR radiation. *Photochem. Photobiol.* **2008**, *84*, 1091–1099. [CrossRef] [PubMed]
27. Manteĭfel', V.M.; Andreĭchuk, T.N.; Karu, T.I. Reaktsiia mitokhondrial'nogo appparata limfotsitov na obluchenie He-Ne-lazerom i na mitogen fitohemaggliutinin [The effect of irradiation by a He-Ne laser and phytohemagglutinin on lymphocyte mitochondria]. *Mol. Biol.* **1991**, *25*, 273–280.

International Journal of *Molecular Sciences*

MDPI

Article

Photoinactivation of *Pseudomonas aeruginosa* Biofilm by Dicationic Diaryl-Porphyrin

Viviana Teresa Orlandi [1,*], Eleonora Martegani [1], Fabrizio Bolognese [1], Nicola Trivellin [2], Francesco Garzotto [3] and Enrico Caruso [1]

[1] Department of Biotechnologies and Life Sciences, University of Insubria, Via JH Dunant 3, 21100 Varese, Italy; e.martegani@uninsubria.it (E.M.); fabrizio.bolognese@uninsubria.it (F.B.); enrico.caruso@uninsubria.it (E.C.)
[2] Department of Industrial Engineering, University of Padova, Via Gradenigo 6A, 35131 Padova, Italy; nicola.trivellin@unipd.it
[3] Institute of Oncology IOV—IRCCS, 35128 Padova, Italy; f.garzotto@gmail.com
* Correspondence: viviana.orlandi@uninsubria.it

Abstract: In recent years, antimicrobial photodynamic therapy (aPDT) has received increasing attention as a promising tool aimed at both treating microbial infections and sanitizing environments. Since biofilm formation on biological and inert surfaces makes difficult the eradication of bacterial communities, further studies are needed to investigate such tricky issue. In this work, a panel of 13 diaryl-porphyrins (neutral, mono- and di-cationic) was taken in consideration to photoinactivate *Pseudomonas aeruginosa*. Among cationic photosensitizers (PSs) able to efficiently bind cells, in this study two dicationic showed to be intrinsically toxic and were ruled out by further investigations. In particular, the dicationic porphyrin (P11) that was not toxic, showed a better photoinactivation rate than monocationic in suspended cells. Furthermore, it was very efficient in inhibiting the biofilms produced by the model microorganism *Pseudomonas aeruginosa* PAO1 and by clinical strains derived from urinary tract infection and cystic fibrosis patients. Since *P. aeruginosa* represents a target very difficult to inactivate, this study confirms the potential of dicationic diaryl-porphyrins as photo-activated antimicrobials in different applicative fields, from clinical to environmental ones.

Keywords: diaryl-porphyrins; antimicrobial photodynamic therapy; aPDT; *Pseudomonas aeruginosa*; biofilm

Citation: Orlandi, V.T.; Martegani, E.; Bolognese, F.; Trivellin, N.; Garzotto, F.; Caruso, E. Photoinactivation of *Pseudomonas aeruginosa* Biofilm by Dicationic Diaryl-Porphyrin. *Int. J. Mol. Sci.* **2021**, *22*, 6808. https://doi.org/10.3390/ijms22136808

Academic Editor: Rustam I. Aminov

Received: 7 June 2021
Accepted: 22 June 2021
Published: 24 June 2021

Publisher's Note: MDPI stays neutral with regard to jurisdictional claims in published maps and institutional affiliations.

1. Introduction

In recent years, in clinical field, the attention was focused on "ESKAPE" pathogens for their ability to "escape" from antimicrobials' action. The term "ESKAPE" arises from the following six name species: *Enterococcus faecium*, *Staphylococcus aureus*, *Klebsiella pneumoniae*, *Acinetobacter baumannii*, *Pseudomonas aeruginosa* and *Enterobacter* spp [1]. These bacterial species are associated with most of nosocomial infections and the highest risk of mortality [2]. All of them have been recently reported by the World Health Organization (WHO) in the list of the 12 bacterial species against which new antibiotics are urgently needed [3]. In particular, *P. aeruginosa*, thanks to its wide genome (5–7 Mbp), displays a wide capacity to use various carbon sources and adapts to several environments including soils, waters, sewages and is a common part of the microflora of different animals. In addition, the highest part of its genome is dedicated to regulatory genes and networks that are fundamental for the response and adaptation to different and changing environments [4]. In humans, it causes severe acute or chronic infections in a variety of tissues and body sites, including skin, middle-ear, eyes and urinary tract, especially in immunocompromised patients [5,6]. *P. aeruginosa* is a leading cause of nosocomial infections, including, i.e., urinary tract catheter-associated infections (CAUTIs), central-line associated bloodstream infections (CLABSIs) and ventilator-associated pneumonia (VAP), as well as chronic lung infections in cystic fibrosis patients [7,8]. The major cause of persistent *P. aeruginosa* infections is the presence of biofilm, formed on tissues or on the surface of surgical implants or medical devices. *P. aeruginosa* cells attach on a surface through twitching motility driven by type IV pili and develop microcolonies. At this step, they secrete exopolysaccharides (Psl, Pel and alginate) and other components such as polypeptides and extracellular DNA. During biofilm maturation, the community acquires its typical three-dimensional structure that planktonic cells can leave for colonizing other surfaces [9]. As a whole, biofilm formation and production of many virulence factors (i.e., pyocyanin, pyoverdine, elastases, proteases, rhamnolipids, exotoxin A) contribute to pathogenicity in *P. aeruginosa* [10]. Furthermore, *P. aeruginosa* is characterized by an outer membrane that acts as a selective barrier to prevent the antibiotic entrance, in addition to several non-specific porins that govern membrane permeability [11]. In addition, a wide variety of efflux pumps (i.e., mexAB-OprM, MexCD-OprJ, MexEF-OprN) are responsible for the resistance to different classes of antibiotics (β-lactams, quinolones and aminoglycosides) [12]. The production of extended-spectrum β-lactamases and enzymes modifying aminoglycosides worsen the spread of multidrug resistant strains [13].

In this alarming scenario, the visible-light based techniques are gathering attention. The main approach exploiting visible light as an anti-infective agent is antimicrobial photodynamic therapy (aPDT) [14]. The photodynamic process is based on the simultaneous presence of three components: a source of light energy, a photosensitive compound and molecular oxygen. Upon the photoexcitation of a photosensitizer (PS) with an appropriate light wavelength, several reactive oxygen species (ROSs) are released: hydroxyl radical ($OH^{\cdot}$), hydrogen peroxide (H_2O_2), superoxide anion (O_2^{-}) and singlet oxygen ($^1O_2$). The elicited oxidative stress compromises the integrity of macromolecules, including lipids, proteins and nucleic acids, and cellular structures leading to microbial cell death [15,16]. An interesting advantage of this technique is based on the efficacy observed against both sensitive and antibiotic resistant strains [17]. In addition, several studies highlight the potential of aPDT on microbial biofilms [18–21]. Until now, several photosensitizers have been taken into consideration for the inhibition and/or eradication of *P. aeruginosa* biofilms. Methylene blue, belonging to phenothiazine compounds, was successful as antibiofilm PS [22]. Curcumin, a well-known natural PS, inhibited the biofilm formation of *P. aeruginosa*, reducing the EPS (extracellular polymeric substance) production by 94% [19]. A PS belonging to the boron-dipyrro-methene (BODIPY) class, upon activation with green light, was efficient not only in inhibiting, but also in eradicating biofilms of *P. aeruginosa* PAO1 [23]. Among the most investigated dyes in the photodynamic field, the family of porphyrins showed a promising potential in biofilm treatment. For example, the 5,10,15,20-tetrakis [4 -(3-N,N-dimethylammoniumpropoxy)phenyl]porphyrin (TAPP) inhibited the biofilm formation of *P. aeruginosa* [24]. A recent study showed the efficacy of cationic zinc-porphyrins in disrupting and detach the matrix of 16–18 h-old biofilms of *P. aeruginosa* [25]. Porphyrins are widely distributed in nature in both prokaryotic and eukaryotic organisms as components of cytochromes, heme groups and chlorophylls, and are involved in many biological processes, such as photosynthesis and oxygen or electron transport [26]. The extensive electron delocalization on the macrocycle ring is responsible for the intense absorption of porphyrins in the visible range. Their typical spectrum is characterized by a higher absorption band around 420 nm, known as Soret band, and weaker bands between 500 and 600 nm (Q bands), making porphyrins suitable PSs to be activated by different light sources, including wide-spectrum emission lamps, sunlight, light-emitting diodes (LEDs) and lasers [27]. Along with this aspect, porphyrins show strong photosensitizing abilities, due to their long-lived triplet state and notable yield of singlet oxygen production, making these compounds almost ideal photosensitizers [28]. Further, the chemical synthesis of porphyrins is relatively simple and cost-efficient and generally involves the condensation of pyrroles with suitable aldehydes. The resulting tetra-pyrrolic ring is a versatile skeleton bearing four different substituents in meso-positions, meaning that high number of combinations could lead to the production of different molecules with desired chemico-physical features [27]. Photosensitizers should not be toxic and mutagenic in the dark towards both eukaryotic and prokaryotic cells. In literature, conflicting observations have been reported on dark toxicity of the most investigated porphyrin, tetracationic TMPyP (5,10,15,20-Tetrakis(1-methyl-4-pyridinio)-porphyrin tetra(p-toluenesulfonate). Eckl assumed that the dark toxicity of TMPyP, observed in several bacterial species, could be attributed to photoinactivation effects caused by any residual light in a laboratory. When keeping the bacteria under dark conditions (<10 nW cm^2), no dark toxicity in *Escherichia coli* was detected for a very high concentration of TMPyP (250 μM) and incubation times up to 24 h [29]. On the other hand, *P. aeruginosa* strains isolated from Fibrosis Cystic patients were sensitive in the dark to porphyrins [30].

Among synthetic porphyrins, diaryl-porphyrins, bearing two substituents in two meso-positions, were shown to be efficient PSs both in antitumoral and in antifungal applications [31,32]. In the present study a panel of 5,15 meso-substituted diaryl-porphyrins was assayed for anti-biofilm activity. The compounds were chosen due to their different degree of amphiphilicity, molecular symmetry and charge and were tested against *P. aeruginosa*.

2. Results

2.1. Panel of Diaryl-Porphyrins

A panel of novel diaryl-porphyrins, synthetized by our group and previously tested as antimicrobials [33] antifungals [32] and antitumorals [31,34], was investigated for anti-Pseudomonas activity (Table 1). The neutral and asymmetrical P1 and P2 bear in meso-positions (positions 5 and 15) a pentafluorophenyl group, associated with a C4 or C8 para-bromoalkyloxy-phenyl group, respectively. In compounds P3 and P5, the C4 or C8 bromoalkyloxy-phenyl chain is associated with a phenyl group in position 5. The symmetrical P4 and P6 bear two phenyl groups with para-bromobutoxy and two para-bromooctanoxy chains, respectively. The positive charge of cationic diaryl-porphyrins derives from a pyridinium group. All monocationic PSs are asymmetrical molecules bearing a phenyl (P7, P8) or a pentafluorophenyl group (P9, P10) in position 5 and a pyridinobutoxy-phenyl (P7, P9) or pyridinooctanoxy-phenyl group (P8, P10) in position 15. The three dicationic symmetric porphyrins are characterised by benzyl group as alkylating group of the pyridyl substituent (P11), or alkoxy-linked pyridinium at the end of four (P12) or eight (P13) carbon chains.

Table 1. List of diaryl-porphyrins (P1-P13) used in this study.

	PS	Chemical Structure	Chemical Denomination	Ref
Non-ionic (ø)	P1	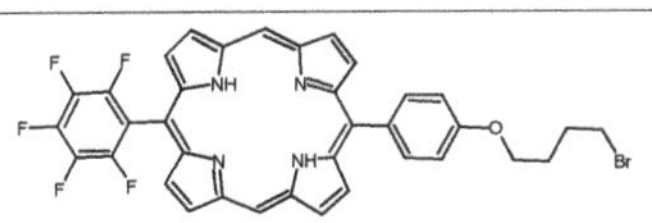	5-Pentafluorophenyl-15-[4-(4-Bromobutoxy)Phenyl]-21H,23H-porphyrin	[34]
	P2	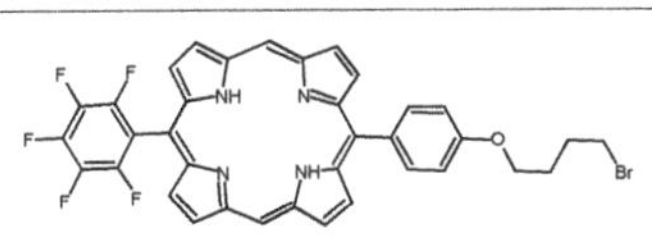	5-Pentafluorophenyl-15-[4-(8-Bromooctanoxy)Phenyl]-21H,23H-porphyrin	[34]
	P3	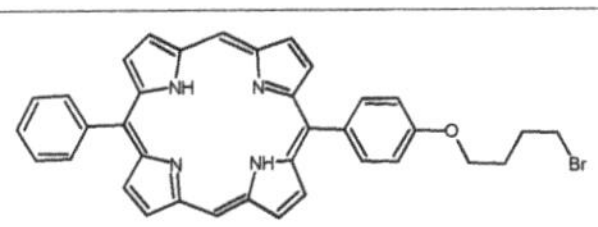	5-Phenyl-15-[4-(4-bromobutoxy)phenyl]-21H,23H-porphyrin	[31]
	P4	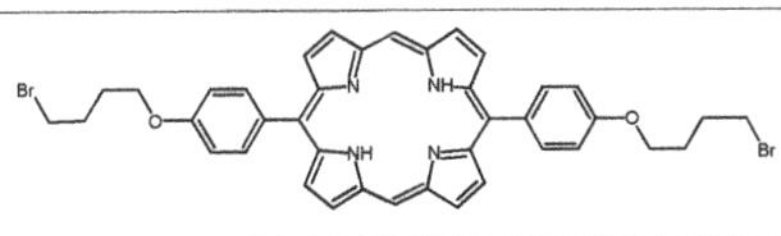	5,15-Di[4-(4-bromobutoxy)phenyl]-21H,23H-porphyrin	[31]

Table 1. *Cont.*

	PS	Chemical Structure	Chemical Denomination	Ref
	P5	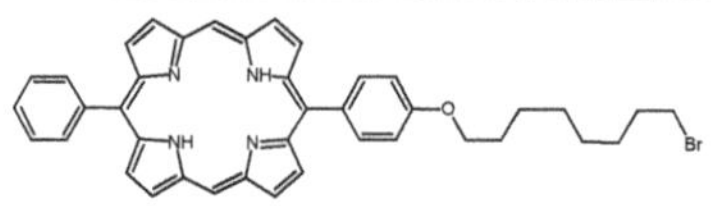	5-Phenyl-15-[4-(8-bromooctanoxy)phenyl]-21H,23H-porphyrin	[31]
	P6	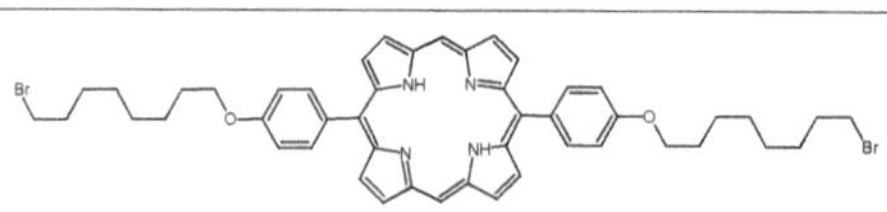	5,15-Di[4-(8-bromooctanoxy)phenyl]-21H,23H-porphyrin	[31]
Monocationic (+)	P7	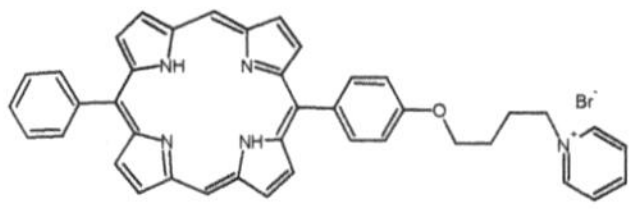	5-Phenyl-15-[4-(4-pyridinobutoxy)phenyl]-21H,23H-porphyrin	[31]
	P8	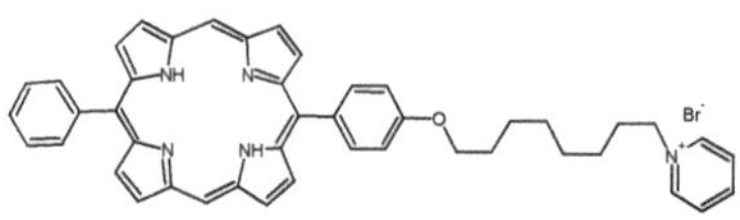	5-Phenyl-15-[4-(4-pyridinooctanoxy)phenyl]-21H,23H-porphyrin	[31]
	P9	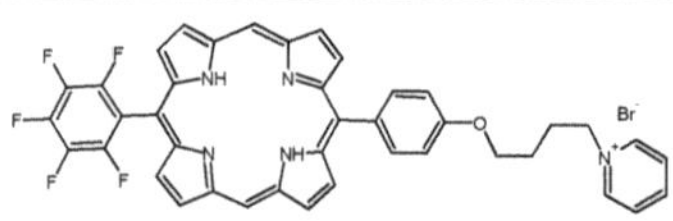	5-Pentafluorophenyl-15-[4-(4-Pyridinobutoxy)Phenyl]-21H,23H-porphyrin	[34]
	P10	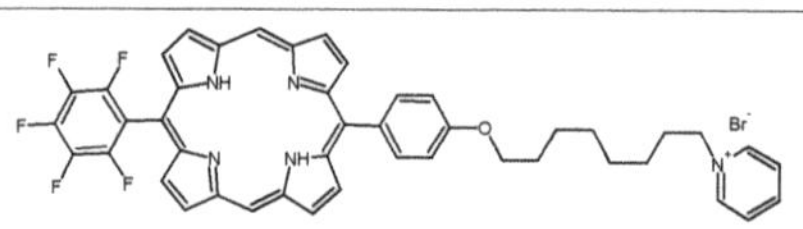	5-Pentafluorophenyl-15-[4-(4-Pyridinooctanoxy)Phenyl]-21H,23H-porphyrin	[34]
Dicationic (++)	P11	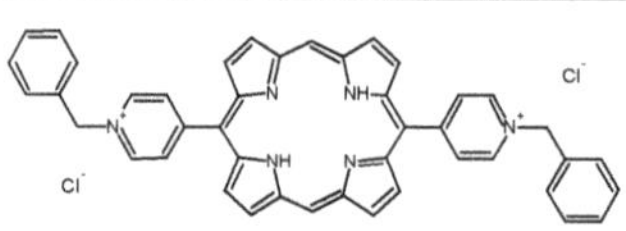	5,15-di(N-benzyl-4-pyridyl)-21H,23H-porphyrin	[33]
	P12	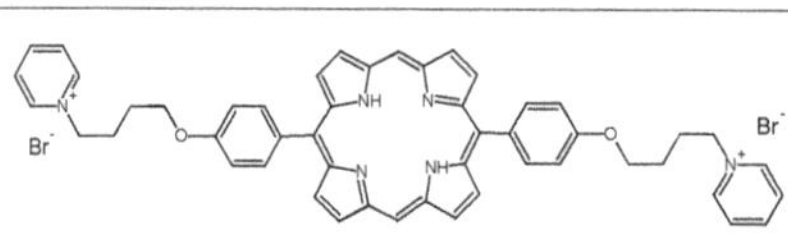	5,15-Di[4-(4-pyridinobutoxy)phenyl]-21H,23H-porphyrin	[31]
	P13	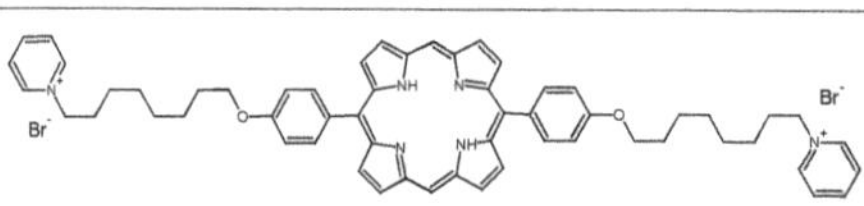	5,15-Di[4-(4-pyridinooctanoxy)phenyl]-21H,23H-porphyrin	[31]

Intrinsic Toxicity of Diaryl-Porphyrins

The effect of diaryl-porphyrins under dark incubation was investigated because the "ideal" PS should not display intrinsic toxicity. The reduction of microbial viability in the presence of photosensitizer without irradiation should be avoided [35]. Since PSs are dissolved in DMSO, it was evaluated that, under the tested conditions, the solvent did not impair microbial growth (Figure 1). Indeed, *P. aeruginosa* was almost insensitive to neutral and monocationic porphyrins up to the longest tested dark incubation (6 h). On the other hand, two of the three dicationic PSs were intrinsically toxic. The observed killing effects increase with longer incubation times: after 6 h of dark incubation, P12 impaired the growth of sample at ~10^4 CFU/spot and P13 at ~10^3 CFU/spot (Figure 1).

PS [10 μM]		Log_{10} reduction			Spot test images (6 h incubation)
Incubation time		10′	1 h	6 h	CFU/spot 10^7 10^6 10^5 10^4 10^3 10^2
Untreated		0	0	0	
DMSO		0	0	0	
Ø	P1	0	0	0	
	P2	0	0	0	
	P3	0	0	0	
	P4	0	0	0	
	P5	0	0	0	
	P6	0	0	0	
+	P7	0	0	0	
	P8	0	0	0	
	P9	0	0	0	
	P10	0	0	0	
++	P11	0	0	0	
	P12	1.5	2	3.5	
	P13	1	1.5	1.5	

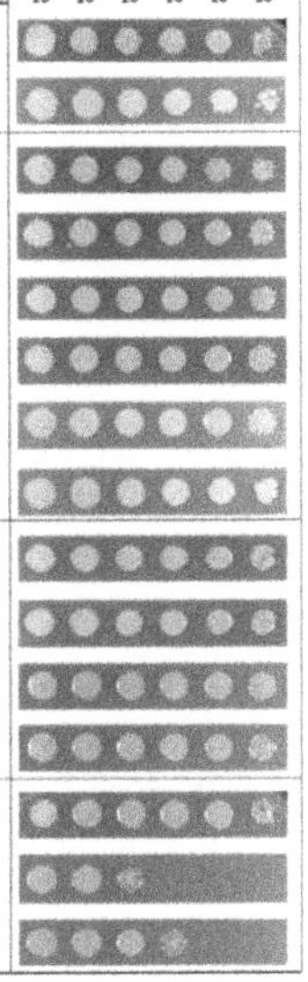

Figure 1. Analysis of intrinsic toxicity of diaryl-porphyrins (P1-P13). PSs were administered at a concentration of 10 μM to samples of *P. aeruginosa* PAO1 at decreasing concentrations (from 10^9 up 10^4 CFU/mL). After dark incubation for 10 min, 1 and 6 h, volumes of ~5 μL of each sample were replica plated on LB agar. After overnight incubation at 37 °C, the growth spots were checked. In the second column, Log10 reduction values represent the mean of three independent experiments for each dark incubation. Representative images reported in the last column refer to growth spots at the corresponding bacterial denisities (from 10^7 to 10^2 CFU/spot) upon 6 h of dark incubation with the tested PS. Each experiment has been repeated at least three times with independent cultures.

2.2. Diaryl-Porphyrins Binding and Photoinactivation Rates

As an optimal interaction between PS and microorganism is required for the following oxidative stress elicited by irradiation [36], the binding rate of porphyrins to microbial cells was evaluated. The yield of PS binding was strictly related to diaryl-porphyrin charge. After 1 h of dark incubation, neutral PSs (P1–P6) showed a very low affinity, less than ~8% of PS binding except P2 with a binding yield of ~15% (Figure 2). On the other hand, positively charged compounds (P7–P13), both mono- and dicationic, were able to strongly interact with *P. aeruginosa* cells (Figure 2).

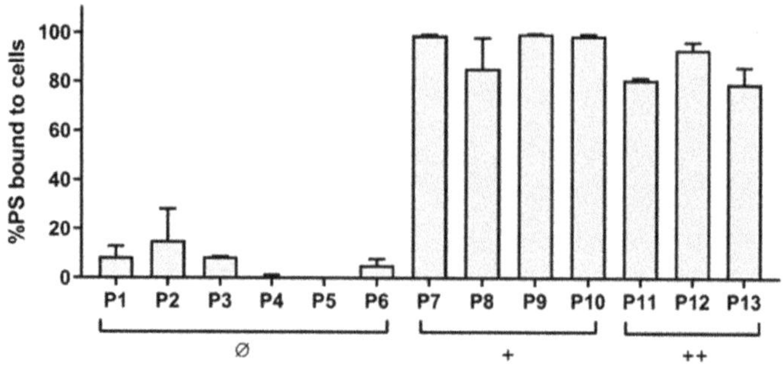

Figure 2. Binding assay of diaryl-porphyrins (P1–P13) to *P. aeruginosa* PAO1. Bacterial cells at 10^8 CFU/mL were dark incubated for 1 h with neutral (Ø), monocationic (+) and dicationic (++) porphyrins administered at a concentration of 30 μM. Samples were centrifuged (10,000× *g*, 5 min) and the visible spectra of the supernatants were recorded. The rate of bound PS was inferred. Three independent experiments have been performed and the mean ± standard deviation of percentage of PS bound to cells is reported for each diaryl-porphyrin.

Since the dark toxicity is not desired in aPDT applications, P12 and P13 were excluded from the following investigations. The antimicrobial potential of the remaining diaryl-porphyrins was investigated under irradiation of a LED emitting at 410 nm and fitting with typical Soret band of porphyrins. Since *P. aeruginosa* is sensitive to light at 410 nm [37], a light dose without toxic effect (20 J/cm^2) was chosen for the activation of porphyrins. After 1 h of dark incubation with diaryl-porphyrins to favor the interaction between PS and cell, bacteria were irradiated. The administration of neutral porphyrins (P1–P6) to *P. aeruginosa* did not affect cell viability upon activation by blue light, neither at the lowest cell concentration (10^2 CFU/spot) (Figure 3).

PS		Log_{10} reduction	CFU/spot 10^7 10^6 10^5 10^4 10^3 10^2
Untreated		0	
DMSO		0	
Ø	P1	0	
	P2	0	
	P3	0	
	P4	0	
	P5	0	
	P6	0	
+	P7	2.5	
	P8	0	
	P9	0	
	P10	0	
++	P11	5.5	

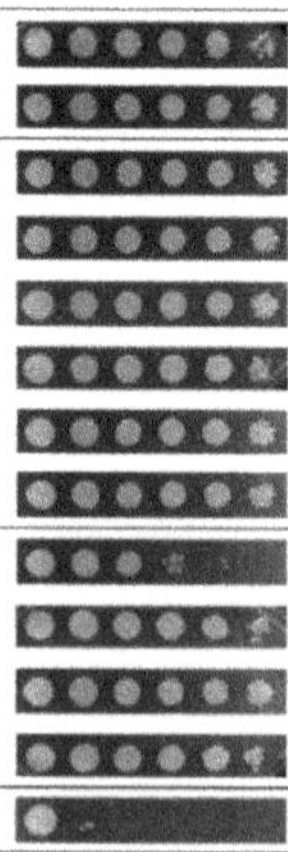

Figure 3. Photodynamic activity of diaryl-porphyrins on *P. aeruginosa* PAO1 evaluated by the photo-spot test. The neutral (Ø), monocationic (+) and dicationic (++) porphyrins were administered at a final concentration of 10 μM to samples of *P. aeruginosa* PAO1 at decreasing concentrations (from 10^9 up 10^4 CFU/mL). After 1 h dark incubation, volumes of ~5 μL of each sample were replica plated on LB agar and irradiated under 410 nm light (20 J/cm^2). Cells were incubated at 37 °C O/N and growth spots were checked and representative images are reported in the last column. Log10 reduction values represent the mean of at least three independent experiments.

Higher concentrations (up 30 μM) of neutral PS failed in photoactivation (data not shown). Among the four monocationic porphyrins, P7 10 μM caused a Log10 reduction of ~2.5 unit. It is noteworthy that longer chain (8 carbon) in P8 or the presence of 5 Fluoro atoms on phenyl residue in 5 positions in P9, compromised completely the photoactivation. The dicationic P11 displayed an activity twofold higher than monocationic P7 (Figure 3). Thus, at the end of this screening, the dicationic diaryl-porphyrin P11 resulted the best candidate to photoinactivate the suspended form of *P. aeruginosa*: a significant decrease of more than 4 Log units (from 10^8 to 10^4 CFU/mL) was obtained upon PDT treatment (Figure 4).

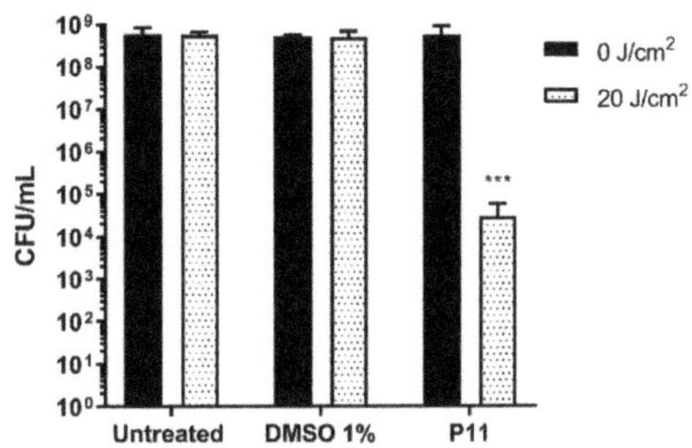

Figure 4. Photodynamic inactivation of *P. aeruginosa* PAO1 by P11. Bacterial samples at ~10^8 CFU/mL suspended in water were incubated in the dark for 1 h with P11 10 μM. After dark incubation, cells were irradiated under light at 410 nm (20 J/cm^2) and cellular viability was checked. Values, presented as CFU/mL, are the mean of at least three independent experiments and the bars represent standard deviations. Statistical analyses were performed by one-way ANOVA *** $p < 0.0001$).

2.3. Photodynamic-Inhibition of Biofilm Formation

The ability to form structured communities, both on inert surfaces and biological tissues, renders *P. aeruginosa* particularly tolerant to conventional antibiotic therapies. Photodynamic therapy (PDT) is a promising approach to tackle bacterial infections in biofilm lifestyle, both in inhibiting biofilm formation and/or eradicating formed biofilms. Notwithstanding, few studies have been carried out employing porphyrins for the photodynamic treatment of microbial biofilms, and moreover no reports describe the inhibition of *P. aeruginosa* biofilms by photoactivation of porphyrins. Thus, the potential of P11 in inhibiting the formation of biofilms was tested on PAO1 strain and two clinical isolates, *P. aeruginosa* UR48 isolated from a patient with catheter-associated urinary tract infections (CAUTI), and BT1 from the sputum of a cystic fibrosis (CF) patient [38,39]. Among the chosen strains, BT1 formed a biofilm with the highest biomass value (OD590~27), while PAO1 and UR48, 7 and 12, respectively (Figure 5). Since cells forming the adherent phase of all the considered strains showed a comparable density (~10^8 CFU/well), it can be inferred that CF isolate was able to hyperproduce extracellular components of matrix biofilm. Upon administration of DMSO 2.5%, and even more P11 30 μM dissolved in DMSO, an increase of crystal violet staining was observed in PAO1. Since no changes in cellular concentrations of planktonic and adherent phases were observed, DMSO could induce the formation of extracellular matrix. However, in all the strains, the combination of P11 and blue light irradiation (30 J/cm^2) caused a relevant and statistically significant inhibition of biomass adhesion. Similarly, the cellular concentrations of planktonic and adherent subpopulations were significantly lower than control samples (Figure 5).

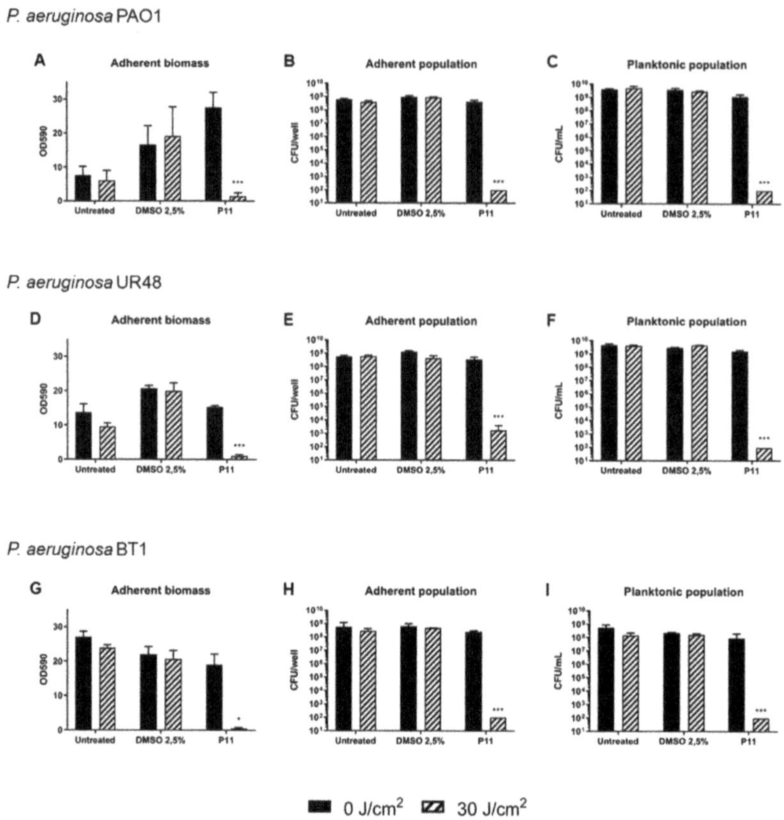

Figure 5. Inhibition of biofilm formation of *P. aeruginosa* PAO1 (**A–C**), UR48 (**D–F**) and BT1 (**G–I**) upon photodynamic treatment with diaryl-porphyrin P11. Overnight cultures of *P. aeruginosa* PAO1, UR48 and BT1 strains were diluted 500-fold in M9 minimal medium added with glucose (10 mM) and casamino acids (0.2% *V*/*V*) reaching a concentration of ~10^7 CFU/mL and inoculated in 12-well microplate. P11 was added at a final concentration of 30 μM and incubated in the dark for 1 h. Upon irradiation with a final dose of 30 J/cm^2 (100 mW/cm^2, 300 s), bacteria were grown O/N at 37 °C to form biofilm. The graphs report values of the optical density at 590 nm (OD 590) after biofilm staining with crystal violet (**A**,**D**,**G**), values of adherent population density (CFU/well) (**B**,**E**,**H**) and planktonic population concentration (CFU/mL) (**C**,**F**,**I**). Dark control samples are represented as black bars and light-treated samples as striped bars. Data represent the mean of at least three independent experiments ± the standard deviation. Statistical analyses were performed by one-way ANOVA (* $p < 0.05$; *** $p < 0.0001$).

The photoinactivation protocol was also applied to *P. aeruginosa* PAO1 strain expressing the green fluorescent protein (GFP) under arabinose induction. Before confocal laser scanning microscope (CLSM) analysis, GFP expression was induced by arabinose to highlight viable cells with functional and active protein machinery. In the control biofilms the detection of GFP signal can be appreciated and no signal was observed under the combination of photosensitizer and light at 410 nm (Figure 6). The obtained results support the effectiveness of PDI mediated by diaryl-porphyrins in inhibiting biofilm formation.

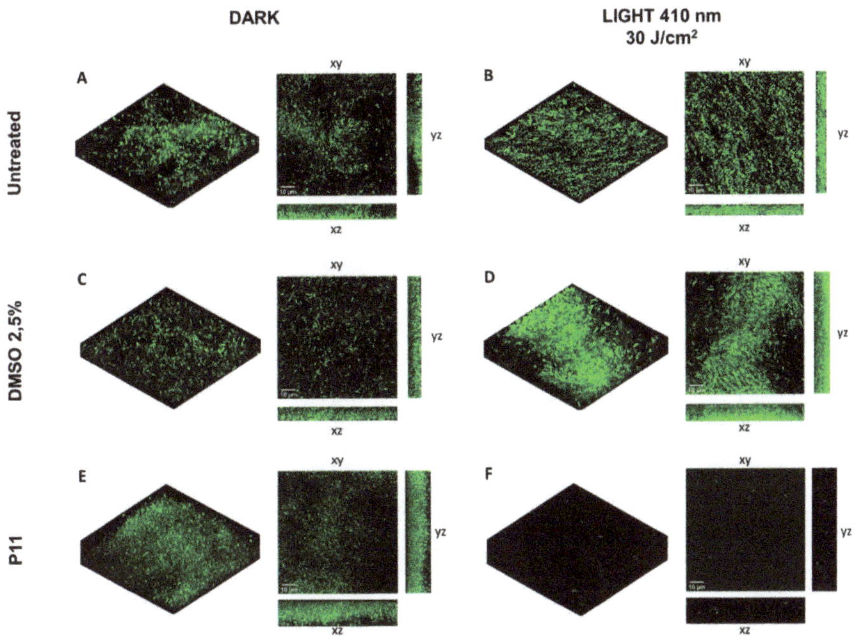

Figure 6. Inhibition of biofilm formation by *P. aeruginosa* PAO1-GFP by dicationic diaryl-porphyrin P11. GFP-tagged PAO1 cells were inoculated on coverslip glass and treated with P11 30 µM for 1 h in the dark. Cells were irradiated with light at 410 nm (30 J/cm^2) and incubated at 37 °C in static to let form biofilm. After overnight incubation, GFP expression was induced and biofilm formed on glasses was analyzed by confocal microscopy. Untreated and DMSO-treated biofilms are included as controls. Confocal images of dark controls are shown in panels (**A**,**C**,**E**), while irradiated samples are shown in panels (**B**,**D**,**F**). Images of biofilms are shown in volume view, and in xy, xz and yz projections (scale bar = 10 µm).

2.4. Photodynamic-Eradication of Formed Biofilms

If biofilm inhibition is a crucial step in infection prevention, the control of infection through eradication of mature biofilm remains the most arduous challenge. Therefore, the effect of porphyrin-mediated photodynamic treatment was evaluated on 24 h-old biofilms formed by the model strain PAO1. PDT experimental conditions were set as follows: upon 24 h of biofilm growth, porphyrin P11 (30 µM final concentration) was gently administered to the samples, without modifying the biofilm environment and, after 1 h of dark incubation, samples were irradiated with light at 410 nm (30 J/cm^2). DMSO-treated samples and dark controls were included in each experiment. The total adherent biomass did not significantly change upon DMSO or P11 administration, in both light and dark conditions, as compared to the untreated dark sample. Interestingly, activation of P11 by blue light caused a significant decrease of 2 Log units in both adherent and planktonic populations (Figure 7B,C). These results suggested a mild anti-biofilm effect on the cellular component of biofilm, both sessile and planktonic, of the dicationic diaryl-poprhyrin P11.

On the other hand, the photoinactivation protocol (P11 30 µM, 30 J/cm^2) applied to a 24 h grown biofilm of *P. aeruginosa* PAO1 expressing GFP showed a clear antimicrobial effect. In the control biofilms, a comparable fluorescent signal was detected (Figure 8A–E), and in the photoinactivated biofilm the fluorescent signal almost disappeared (Figure 8F). It can be hypothesized that the treatment impaired cell functions, including the activity of the cellular protein synthetic machinery. Even if the viability of sessile bacteria was slightly compromised, most of the cells seemed to be damaged immediately after PDT treatment.

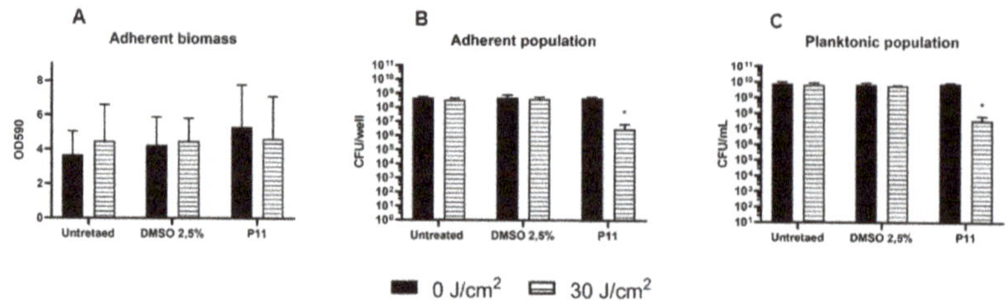

Figure 7. Assay of eradication of *P. aeruginosa* PAO1 biofilm by porphyrin P11. Overnight cultures of *P. aeruginosa* PAO1 were diluted 500-fold in M9 minimal medium added with glucose (10 mM) and casamino acids (0.2% *V*/*V*) reaching a concentration of ~10^7 CFU/mL and inoculated in 12-well microplate to let form biofilm. 24 h-old biofilm was treated with PS at a concentration of 30 μM. After dark incubation for 1 h, biofilm was irradiated (30 J/cm^2). After irradiation, the adherent biomass (OD_{590}) (**A**) and viable counts from sessile (CFU/well) (**B**) and planktonic phases (CFU/mL) (**C**), respectively, have been evaluated. Data represent the mean of at least three independent experiments ± the standard deviation. Statistical analyses were performed by one-way ANOVA (* $p < 0.05$).

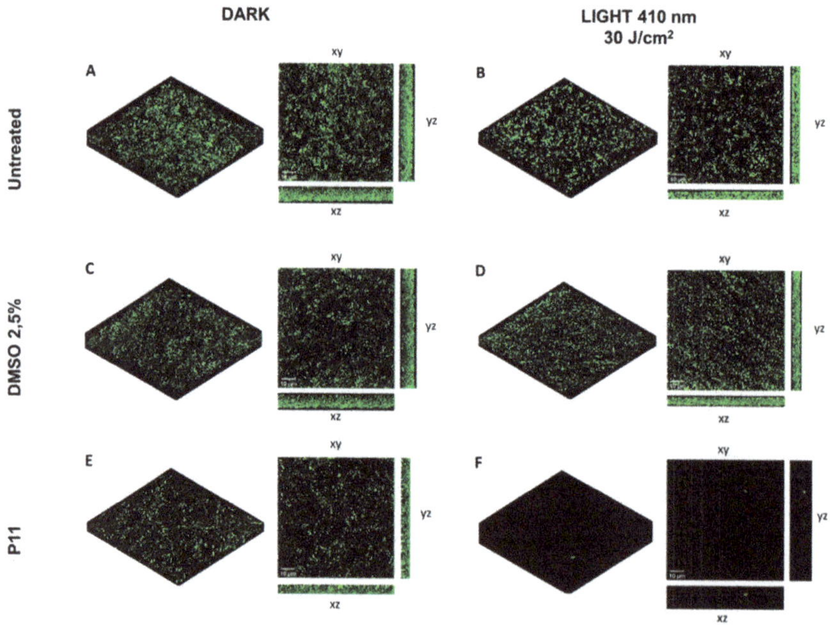

Figure 8. Assay of *P. aeruginosa* PAO1-GFP biofilm eradication with dicationic diaryl-porphyrin P11. A 24 h-old PAO1-GFP biofilm grown on coverslip glass was treated with P11 (30 μM final concentration) and irradiated with 410 nm blue light at 30 J/cm^2. Upon induction of GFP expression, confocal analyses have been performed. Biofilms (24 h) of untreated and DMSO-treated samples were included in the experiment. Dark controls are shown in panels (**A**,**C**,**E**), while irradiated samples are depicted in panels (**B**,**D**,**F**). Images of biofilms are shown in volume view, and in xy, xz and yz projections (scale bar = 10 μm).

3. Discussion

Since *P. aeruginosa* is a pathogen difficult to eradicate for its resistance to antibiotics and tolerance to antimicrobial treatments, it is very interesting to acquire information on its sensitivity to novel drugs, independently from photodynamic applications. The dicationic compounds showed different activities: the porphyrins, P12 and P13, bearing substituents with longer chains (4 carbon and 8 carbon) in 5 and 15 positions were toxic, while P11 bearing a benzyl chain in the same positions was not intrinsically toxic. Therefore, the presence of an alkyl chain that increases the degree of lipophilicity can favour the cross and perhaps the injury of the outer membrane that represents the ideal target for new approaches [40].

In literature, conflicting observations have been reported on dark toxicity of the most investigated porphyrins [29,30]. It is noteworthy that P12 and P13 show a higher Soret band than the other diaryl-porphyrins under investigation in this study (Figure 9), and it cannot be ruled out that the occasional exposure to wide spectrum daylight for experimental set-up could be sufficient to elicit a mild photo-oxidative stress.

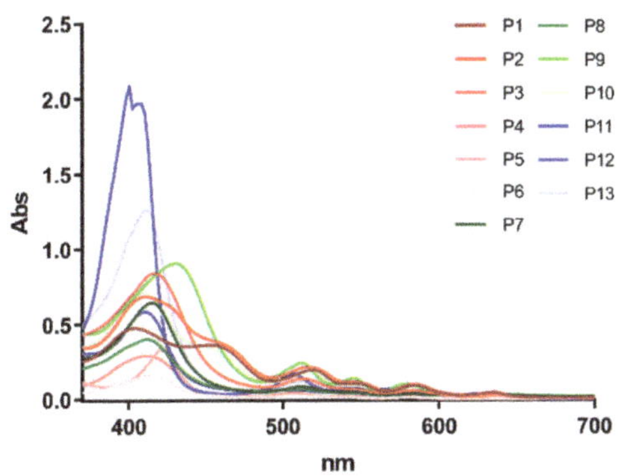

Figure 9. Visible light absorption spectra of diaryl-porphyrins (P1-P13).

It can be hypothesized that the positive charge on P11 is strongly delocalized, in fact the pyridine nucleus is directly coordinated with the tetrapirrole system of porphyrin. The greater charge distribution should disadvantage a strong interaction of diaryl-porphyrins with anionic counterparts of the cell wall of both Gram-negative and Gram-positive bacteria. Instead, P12 and P13 show a strongly localized charge only in the pyridine positioned far from the tetrapyrrole nucleus, this may suggest a strong electrostatic interaction with the bacterial cell wall. The electrostatic interaction may be sufficient to disassemble cell wall integrity, impairing bacterial viability, independently from irradiation.

Moreover, the eukaryotic microorganism *Candida albicans* showed to be sensitive in the dark to most cationic and dicationic diaryl-porphyrins tested in this study, irrespective of Soret band height [32]. Cormick reported that tri- and tetracationic porphyrins were more tightly bound to *C. albicans* cells than anionic porphyrins, supporting the requirement of positive charge on porphyrins to promote the electrostatic interaction with the yeast cell wall [41]. Indeed, the net negative charge of yeast cell wall is conferred by a robust polysaccharide skeleton linked to mannoproteins and chitin [42]. Since diaryl-porphyrins were intrinsically toxic to *P. aeruginosa* and *C. albicans* in a different rate, irrespective of their absorbance spectrum, their potential activation by indoor light cannot be considered as the only factor appointed for intrinsic toxicity.

On the basis of the obtained results, several considerations may be carried out about efficiency of binding and photo-inactivation yields. Neutral diaryl-porphyrins were not able to bind tightly to *P. aeruginosa* cells, as the most part of PSs was recovered in the supernatant, and furthermore, upon irradiation, no killing was observed. On the other hand, cationic compounds (P8, P9, P10), even if tightly bound to the cell wall, did not elicit any photooxidative stress upon irradiation. Thus, a good binding is not necessarily the only requirement for a successful photoactivation process in *P. aeruginosa*. The factor that could affect the potential activity of PSs is the capacity of PSs to bind and penetrate the cell wall and, possibly, reach cytoplasmic targets. In this regard, Sulek reported that cationic TMPyP was not efficient in photoinactivating *E. coli* cells, and it was necessary to administer verapamil, an efflux pump inhibitor, to potentiate the antibacterial effect [43]. The inhibitor could increase PS's accumulation in the bacteria cell of molecules attached to cell wall via electrostatic interactions. A damage of the outer surface of the outer membrane of Gram-negative bacteria could be less dangerous than that of the inner part of the same envelope or the cytoplasmic environment. The photoactivation of P11 impaired the cell wall of *E. coli*: the outer membrane of approx. 90% of the treated cells appeared fuzzier and lacked the pronounced margin of the envelope of control cells, and 60–70% of the cells underwent envelope and cytoplasmic changes with the arising of honeycomb-like structures [33]. Thus, the impairment of the cell wall seems relevant and necessary upon photodynamic treatment. The outer membrane forms a barrier that impairs neutral PS binding and penetration. The high yield of binding of cationic diaryl-porphyrins (80–100%) could be ascribable to the electrostatic force displayed between negative-charged lipopolysaccharides on the outer layer of the outer membrane and positively charged PSs. As previously reported by other authors, one or more positive charges are required on the PS structure for a good interaction with Gram-negative bacteria [43]. The mechanism of binding of cationic PSs with bacteria is the so-called "self-promoted uptake" pathway. This pathway involves the binding of the cationic molecules to LPS that results in the progressive displacement of divalent cations (Ca^{2+}, Mg^{2+}) electrostatically bound to the LPS, thereby weakening the outer membrane. The destabilization of the LPS coat results in the formation of "cracks" in the permeability barrier, and divalent cations neutralize the negative character of cell membrane and eliminate electrostatic repulsion between PS and the bacterial envelope [43]. Furthermore, a partial hydrophobic feature associated to C4 and C8 substituents of P12 and P13 could facilitate the cross of the cytoplasmic membrane. A certain degree of amphiphilicity was reported as peculiar and advantageous for porphyrins in photodynamic applications [43]. Neutral PSs that do not interact with *P. aeruginosa*, neither affect its viability upon dark incubation.

As reported in literature, it was necessary to employ high concentration of TMPyP (225 µM) to induce 4-fold reduction in *P. aeruginosa* biofilms and cause detachment of the biofilm from the substratum [44]. Another PDI study using a lower concentration of TMPyP (20 µM) under 64.8 J/cm^2 demonstrated that in *P. aeruginosa* polysaccharides of the biofilm matrix may be a primary target of photodynamic damage [45]. Patel et al. reported that a cationic zinc (II) porphyrin, ZnPor, at low concentration (~20 µM), resulted in the extensive disruption and detachment of the matrix of 16–18 h-old biofilms of *P. aeruginosa* [25]. However, it has been recently hypothesized that the presence of negative charges in the EPS matrix could protect bacteria from the interaction with positively charged PSs. Furthermore, larger molecules are disadvantaged in the penetration through the biofilm matrix if compared to smaller ones [46]. Since no standard procedures are yet available for biofilm studies as guidelines for MIC evaluation value (minimal inhibitory concentration) [47], comparisons can not be made between different experimental approaches. In our case, stringent conditions were applied in antibiofilm PDT and no changes in biofilm environment were made upon biofilm growth. The diaryl-porphyrin P11 (30 µM) caused a 2-Log unit depletion of both adherent and planktonic populations of PAO1 biofilm. In addition, confocal analyses allowed to observe the biofilm adherent population immediately upon photo-treatment. PDT provoked a certain damage to embedded PAO1 cells, suggesting that they could be more sensitive to other antimicrobial agents delivered upon photodynamic treatment. The embedding of porphyrins in nanoparticles or polymeric coatings could be exploited both for skin disinfection and surface sanitization [48,49].

In conclusion, this study confirmed that the dicationic diarylic porphyrin P11 that previously showed to be efficient in inhibiting both Gram-negative *E. coli* and *P. aeruginosa* and Gram-positive *Enterococcus faecalis* and *S. aureus*, is optimal to inhibit the formation of *P. aeruginosa* biofilm. Since a mild effect on formed biofilm was obtained, these results could pave the way through the development of combined antibiofilm strategies where P11-

mediated PDT in addition to other antimicrobial approaches could successfully eradicate *P. aeruginosa* biofilms.

4. Materials and Methods

4.1. Photosensitizers

A panel of 13 diaryl-porphyrins previously described [31,33,34] has been used in this study (Table 1). As shown in Figure 1, P1–P6 porphyrins are neutral molecules, P7–P10 are monocationic and P11–P13 di-cationic, respectively. PSs were dissolved in DMSO (Sigma Aldrich, Milano, Italy) at a final concentration of 1 or 0.5 mM, as requested, and stored at 4 °C until needed.

4.2. Microbial Strains and Culture Conditions

Three different strains of *P. aeruginosa* were considered for this study: *P. aeruginosa* PAO1 as model pathogen to photoinactivate [4] and two clinical strains previously considered: UR48 from urinary tract infection [39] and BT1 from cystic fibrosis patient [38]. *P. aeruginosa* strains were grown in Luria Bertani (LB) (Sigma Aldrich, Milano, Italy) on an orbital shaker at 200 rpm, or in solid media (15 g/L agar) at 37 °C. For biofilm formation, *P. aeruginosa* strains were grown in M9 minimal medium added with glucose (10 mM) (Sigma Aldrich, Milano, Italy) and casamino acids (0.2% *V*/*V*) (Sigma Aldrich, Milano, Italy) at 37 °C in static conditions. When necessary, the bacterial concentration was determined by viable count technique. Briefly, an aliquot of each sample was ten-fold serially diluted and a 10 μL of each diluted and undiluted sample was inoculated on LB Agar. After overnight incubation at 37 °C, the colony count was performed and the corresponding cellular concentration expressed as CFU/mL was calculated.

4.3. Light Source

The lighting unit device is equipped with a head composed by 25 high power LEDs with maximum emission peak at 410 nm blue light, suitable for the activation of porphyrins and allows the uniform irradiation of a square area of 75 mm × 75 mm. The system is powered by a specific PC based control system, which allows the setting of irradiation time and irradiance values for a precise evaluation of the radiation fluence rate.

4.4. Photo-Spot Test Assay

The spot test previously optimized [50] was used to screen the intrinsic toxicity and the photoactivity of diaryl-porphyrins. Upon overnight growth in LB, *P. aeruginosa* PAO1 culture (~10^9 CFU/mL) was suspended in phosphate buffer saline (PBS-KH_2PO_4/K_2HPO_4 10 mM, pH 7.4) and 10-fold serially diluted from ~10^9 to ~10^4 CFU/mL in 96-well plates.

To investigate the intrinsic toxicity of PSs, undiluted and diluted bacterial suspensions were incubated in the dark with PSs 10 μM. After 10 min, 1 h or 6 h of dark incubation, volumes of ~5 μL of each sample were replica plated on LB agar. Untreated samples and DMSO treated samples were included as controls. After O/N incubation at 37 °C, the growth of treated samples was compared to control growth spots of decreasing cell density (from ~10^7 to ~10^2 CFU/spot, respectively). For example, if the growth spot at 10^2 CFU/spot was not observed, a 2-log unit decrease was recorded. Similarly, a 3-log unit was recorded if the spot at 10^3 CFU/spot was not observed. Thus, higher values correspond to higher dark toxicity. The experiments have been repeated at last three times with independent cultures.

To investigate the photo-inactivation rates of diaryl-porphyrins, the PSs were administered at a final concentration of 10 μM to samples of *P. aeruginosa* PAO1 at decreasing concentrations (from 10^9 up 10^4 CFU/mL), as previously described. After 1 h dark incubation, volumes of ~5 μL of each sample were replica plated on LB agar and irradiated under 410 nm light (20 J/cm^2). After O/N incubation at 37 °C, the growth spot was checked and compared to untreated control of decreasing cell density (from ~10^7 to ~10^2 CFU/spot, respectively). For example, if the growth spot at 10^2 CFU/spot was not observed, a 2-log unit decrease was recorded. Similarly, a 3-log unit was recorded if the spot at 10^3 CFU/spot was not observed. Thus, higher values correspond to higher antimicrobial efficiency. Photo-spot tests were performed at least in triplicate.

4.5. Photoinactivation of Suspended Cells

Upon overnight growth of *P. aeruginosa* PAO1, cells were ten-fold diluted in sterile deionized water, to reach approximate concentrations of 10^8 CFU/mL. Porphyrins were added to cell suspension at a final concentration of 10 μM. Untreated cells, DMSO-treated cells and not irradiated controls were also included. Cells were incubated in the dark for 60 min and then irradiated (20 J/cm^2). Soon after irradiation, the number of viable cells was evaluated by viability count, as previously described. Photoinactivation experiments were performed at least in triplicate.

4.6. Photodynamic Treatment of Biofilms

The effect of diaryl-porphyrins in inhibiting the biofilm formation of *P. aeruginosa* was evaluated as follows. Overnight cultures of *P. aeruginosa* PAO1, UR48 and BT1 strains were diluted 500-fold in M9 minimal medium added with glucose (10 mM) and casamino acids (0.2% *V*/*V*) reaching a concentration of ~10^7 CFU/mL and inoculated in 12-well microplate. Porphyrins were added at a final concentration of 30 μM and incubated in the dark for 1 h. Upon irradiation with a final dose of 30 J/cm^2 (100 mW/cm^2, 300 s), bacteria were grown O/N at 37 °C to form biofilm. In order to evaluate the effect of the different treatments on the cellular viability of suspended and adherent populations, the planktonic phase was axenically collected and adherent cells were recovered by scraping and suspended in 1 mL of PBS. Viable counts—expressed as CFU/mL in cell suspensions and as CFU/well in adherent biomass were estimated by a plate count technique, as previously described. The total adherent biomass was quantified by crystal violet (CV) staining. Briefly, planktonic biomass was removed and wells were washed once with 1 mL PBS. One millilitre of 0.1% (*W*/*V*) CV was added to each well for approximately 20 min to stain the biofilm, after which the CV was removed and each well was gently washed with 1 mL PBS. The remaining CV, which indicated the amount of biofilm present, was dissolved in acetic acid 30% for 10 min. The amount of solubilized dye was spectrophotometrically measured at 590 nm.

To evaluate the eradication of biofilm, overnight cultures of *P. aeruginosa* PAO1 were diluted 500-fold in M9 minimal medium added with glucose (10 mM) and casamino acids (0.2% *V*/*V*) reaching a concentration of ~10^7 CFU/mL and inoculated in 12-well microplate to let form biofilm. 24 h-old biofilms were treated with PS at a concentration of 30 μM, dark incubated for 1 h and irradiated (30 J/cm^2). After irradiation, the adherent biomass (OD_{590}) and viable counts from sessile (CFU/well) and planktonic phases (CFU/mL), respectively, have been evaluated as previously described.

In both experimental setups (inhibition of biofilm formation and eradication of formed biofilm), a panel of the following controls was included: DMSO treated and not irradiated biofilm (+DMSO; -light), DMSO treated and irradiated biofilm (+DMSO; +light), PS treated and not irradiated biofilm (+PS; -light), untreated and irradiated biofilm (-PS; +light) and untreated and not irradiated biofilm (-PS; -light). All experiments were independently repeated at least three times.

4.7. Confocal Microscopy Analyses

Anti-biofilm activity of porphyrins on *Pseudomonas aeruginosa* PAO1 was analyzed using PAO1_pVOGFP recombinant strain, in which GFP fluorescent protein is expressed under the control of pBAD arabinose inducible promoter [51]. Overnight culture of *P. aeruginosa* PAO1_pVOGFP was diluted 500-fold in M9 minimal medium added with glucose (10 mM) and casamino acids (0.2% *V*/*V*) reaching a concentration of ~10^7 CFU/mL and inoculated on coverslip glass positioned in 35 mm Petri dish. P11 30 μM was added at a final concentration of 30 μM and incubated in the dark for 1 h. Upon irradiation with a final dose of 30 J/cm^2 (100 mW/cm^2, 300 s), bacteria were grown O/N at 37 °C to form biofilm. Planktonic phase was removed and GFP expression was induced for 1 h at 37 °C by the addition of fresh medium containing arabinose 0.1% *W/V*. Finally, the coverslip was placed on a microscope glass slide for the acquisition of the adherent biofilm images. To evaluate the eradication, the recombinant strain was inoculated as previously described in this section and let form biofilm on coverslip glass. A 24 h old biofilm, after 1 h dark incubation with P11 at 30 μM, was irradiated (30 J/cm^2). After irradiation, GFP expression was induced for confocal analysis.

All microscopic image acquisitions were performed on a Leica TCS SP5 CLSM (Leica Microsystems, Wetzlar, Germany) equipped for GFP visualization (excitation laser at 488 nm). Images were obtained using a x63 objective lens. Simulated 3D images of *P. aeruginosa* biofilm were generated using the free open-source software ImageJ (National Institute of Health, Bethesda, MD, USA).

4.8. Photosensitizer Binding Assay

All the photosensitizers were tested for their ability to bind bacterial cells. Upon overnight growth of *P. aeruginosa* PAO1, cultures were centrifuged at 5000× *g* for 10 min and the supernatants were removed. Pellets were resuspended and 10-fold diluted in sterile deionized water to obtain samples at 10^8 CFU/mL. The bacterial concentration was evaluated by a plate count technique. Porphyrins at the concentration of 30 μM were added to the cells and samples were incubated for 1 h at 37 °C in the dark. This concentration was optimal to detect absorbance spectrum of all the tested porphyrins. Untreated cells, PSs treated cells and cells added with DMSO 4% (*V/V*) were included as controls. After dark incubation, samples were centrifuged (10,000× *g* for 5 min) and the visible spectra of the supernatants were recorded (k = 380–700 nm). A calibration plot (μM vs. OD) was obtained for each PS. The amount of PS not bound to bacterial cells was inferred interpolating the data on the calibration plot. The percentage of each PS bound to *P. aeruginosa* cells is represented as the mean ± standard deviation of at least three independent experiments.

4.9. Statistical Analyses

Photoinactivation experiments on suspended cells and biofilm formation by each microbial strain were performed at least three times with independent cultures, and statistical analyses were assessed by one-way ANOVA. If homogeneity of variance was not observed, post hoc test was performed.

Author Contributions: Conceptualization, V.T.O. and E.C.; methodology, V.T.O., E.M. and F.B.; irradiating apparatus, LED parameters check and LED software, N.T. and F.G.; experimental investigation, E.M. and F.B.; data curation, V.T.O. and E.C.; writing—original draft preparation, E.C. and V.T.O. All authors have read and agreed to the published version of the manuscript.

Funding: This research received no external funding.

Conflicts of Interest: The authors declare no conflict of interest.

References

1. Rice, L.B. Federal Funding for the Study of Antimicrobial Resistance in Nosocomial Pathogens: No ESKAPE. *J. Infect. Dis.* **2008**, *197*, 1079–1081. [CrossRef]
2. Ma, Y.; Wang, C.; Li, Y.; Li, J.; Wan, Q.; Chen, J.; Tay, F.R.; Niu, L. Considerations and Caveats in Combating ESKAPE Pathogens against Nosocomial Infections. *Adv. Sci.* **2019**, *7*, 1901872. [CrossRef]
3. Mulani, M.S.; Kamble, E.; Kumkar, S.N.; Tawre, M.S.; Pardesi, K.R. Emerging Strategies to Combat ESKAPE Pathogens in the Era of Antimicrobial Resistance: A Review. *Front. Microbiol.* **2019**, *10*, 539. [CrossRef]
4. Stover, C.K.; Pham, X.Q.; Erwin, A.L.; Mizoguchi, S.D.; Warrener, P.; Hickey, M.J.; Brinkman, F.S.L.; Hufnagle, W.O.; Kowalik, D.J.; Lagrou, M.; et al. Complete genome sequence of *Pseudomonas aeruginosa* PAO1, an opportunistic pathogen. *Nature* **2000**, *406*, 959–964. [CrossRef] [PubMed]
5. Nordmann, P.; Naas, T.; Fortineau, N.; Poirel, L. Superbugs in the coming new decade; multidrug resistance and prospects for treatment of *Staphylococcus aureus*, *Enterococcus* spp. and *Pseudomonas aeruginosa* in 2010. *Curr. Opin. Microbiol.* **2007**, *10*, 436–440. [CrossRef] [PubMed]

6. Migiyama, Y.; Yanagihara, K.; Kaku, N.; Harada, Y.; Yamada, K.; Nagaoka, K.; Morinaga, Y.; Akamatsu, N.; Matsuda, J.; Izumikawa, K.; et al. *Pseudomonas aeruginosa* Bacteremia among Immunocompetent and Immunocompromised Patients: Relation to Initial Antibiotic Therapy and Survival. *Jpn. J. Infect. Dis.* **2016**, *69*, 91–96. [CrossRef] [PubMed]
7. Driscoll, J.A.; Brody, S.L.; Kollef, M.H. The Epidemiology, Pathogenesis and Treatment of *Pseudomonas aeruginosa* Infections. *Drugs* **2007**, *67*, 351–368. [CrossRef] [PubMed]
8. El Zowalaty, A.; Al Thani, A.A.; Webster, T.J.; Schweizer, H.P.; Nasrallah, G.; Marei, H.; Ashour, H. *Pseudomonas aeruginosa*: Arsenal of resistance mechanisms, decades of changing resistance profiles, and future antimicrobial therapies. *Future Microbiol.* **2015**, *10*, 1683–1706. [CrossRef]
9. Rasamiravaka, T.; Labtani, Q.; Duez, P.; El Jaziri, M. The Formation of Biofilms by *Pseudomonas aeruginosa*: A Review of the Natural and Synthetic Compounds Interfering with Control Mechanisms. *BioMed Res. Int.* **2015**, *2015*, 759348. [CrossRef]
10. Moradali, M.F.; Ghods, S.; Rehm, B.H.A. *Pseudomonas aeruginosa* Lifestyle: A Paradigm for Adaptation, Survival, and Persistence. *Front. Cell. Infect. Microbiol.* **2017**, *7*, 39. [CrossRef]
11. Hancock, R.E.W.; Brinkman, F.S.L. Function of *Pseudomonas Porins* in Uptake and Efflux. *Annu. Rev. Microbiol.* **2002**, *56*, 17–38. [CrossRef]
12. Li, X.-Z.; Nikaido, H. Efflux-Mediated Drug Resistance in Bacteria. *Drugs* **2004**, *64*, 159–204. [CrossRef]
13. Wright, G.D. Bacterial resistance to antibiotics: Enzymatic degradation and modification. *Adv. Drug Deliv. Rev.* **2005**, *57*, 1451–1470. [CrossRef] [PubMed]
14. Hamblin, M.R.; Abrahamse, H. Can light-based approaches overcome antimicrobial resistance? *Drug Dev. Res.* **2019**, *80*, 48–67. [CrossRef] [PubMed]
15. Allison, R.R.; Moghissi, K. Photodynamic Therapy (PDT): PDT Mechanisms. *Clin. Endosc.* **2013**, *46*, 24–29. [CrossRef]
16. Saint-Denis, T.; Dai, T.; Izikson, L.; Astrakas, C.; Anderson, R.R.; Hamblin, M.R.; Tegos, G.P. All you need is light. *Virulence* **2011**, *2*, 509–520. [CrossRef] [PubMed]
17. Hamblin, M.R. Antimicrobial photodynamic inactivation: A bright new technique to kill resistant microbes. *Curr. Opin. Microbiol.* **2016**, *33*, 67–73. [CrossRef]
18. Prochnow, E.P.; Martins, M.R.; Campagnolo, C.B.; Santos, R.C.; Villetti, M.A.; Kantorski, K.Z. Antimicrobial photodynamic effect of phenothiazinic photosensitizers in formulations with ethanol on *Pseudomonas aeruginosa* biofilms. *Photodiagnosis Photodyn. Ther.* **2016**, *13*, 291–296. [CrossRef] [PubMed]
19. Abdulrahman, H.; Misba, L.; Ahmad, S.; Khan, A.U. Curcumin induced photodynamic therapy mediated suppression of quorum sensing pathway of *Pseudomonas aeruginosa*: An approach to inhibit biofilm in vitro. *Photodiagnosis Photodyn. Ther.* **2020**, *30*, 101645. [CrossRef]
20. Sarker, R.R.; Tsunoi, Y.; Haruyama, Y.; Ichiki, Y.; Sato, S.; Nishidate, I. Combined Addition of Ethanol and Ethylenediaminetetraacetic Acid Enhances Antibacterial and Antibiofilm Effects in Methylene Blue-Mediated Photodynamic Treatment against *Pseudomonas aeruginosa* In Vitro. *Photochem. Photobiol.* **2021**, *97*, 600–606. [CrossRef]
21. Anju, V.T.; Paramanantham, P.; Siddhardha, B.; Sruthil Lal, S.B.; Sharan, A.; Alyousef, A.A.; Arshad, M.; Syed, A. Malachite green-conjugated multi-walled carbon nanotubes potentiate antimicrobial photodynamic inactivation of planktonic cells and biofilms of *Pseudomonas aeruginosa* and *Staphylococcus aureus*. *Int. J. Nanomed.* **2019**, *14*, 3861–3874. [CrossRef] [PubMed]
22. Pérez-Laguna, V.; García-Luque, I.; Ballesta, S.; Pérez-Artiaga, L.; Lampaya-Pérez, V.; Rezusta, A.; Gilaberte, Y. Photodynamic therapy using methylene blue, combined or not with gentamicin, against *Staphylococcus aureus* and *Pseudomonas aeruginosa*. *Photodiagnosis Photodyn. Ther.* **2020**, *31*, 101810. [CrossRef] [PubMed]
23. Orlandi, V.T.; Rybtke, M.; Caruso, E.; Banfi, S.; Tolker-Nielsen, T.; Barbieri, P. Antimicrobial and anti-biofilm effect of a novel BODIPY photosensitizer against *Pseudomonas aeruginosa* PAO1. *Biofouling* **2014**, *30*, 883–891. [CrossRef]
24. Mamone, L.; Ferreyra, D.; Gándara, L.; Di Venosa, G.; Vallecorsa, P.; Sáenz, D.; Calvo, G.; Batlle, A.; Buzzola, F.; Durantini, E.N.; et al. Photodynamic inactivation of planktonic and biofilm growing bacteria mediated by a meso-substituted porphyrin bearing four basic amino groups. *J. Photochem. Photobiol. B Biol.* **2016**, *161*, 222–229. [CrossRef] [PubMed]
25. Patel, N.; Swavey, S.; Robinson, J. A Cationic Porphyrin, ZnPor, Disassembles *Pseudomonas aeruginosa* Biofilm Matrix, Kills Cells Directly, and Enhances Antibiotic Activity of Tobramycin. *Antibiotics* **2020**, *9*, 875. [CrossRef]
26. Phoenix, D.A.; Dennison, S.R.; Harris, F. Photodynamic antimicrobial chemotherapy. In *Novel Antimicrobial Agents Strategies*; Wiley: Hoboken, NJ, USA, 2014; Volume 42, pp. 295–330. [CrossRef]
27. Amos-Tautua, B.M.; Songca, S.P.; Oluwafemi, O.S. Application of Porphyrins in Antibacterial Photodynamic Therapy. *Molecules* **2019**, *24*, 2456. [CrossRef]
28. Malatesti, N.; Munitic, I.; Jurak, I. Porphyrin-based cationic amphiphilic photosensitisers as potential anticancer, antimicrobial and immunosuppressive agents. *Biophys. Rev.* **2017**, *9*, 149–168. [CrossRef]
29. Eckl, D.B.; Dengler, L.; Nemmert, M.; Eichner, A.; Bäumler, W.; Huber, H. A Closer Look at Dark Toxicity of the Photosensitizer TMPyP in Bacteria. *Photochem. Photobiol.* **2017**, *94*, 165–172. [CrossRef] [PubMed]
30. Donnelly, R.F.; McCarron, P.A.; Cassidy, C.M.; Elborn, J.S.; Tunney, M.M. Delivery of photosensitisers and light through mucus: Investigations into the potential use of photodynamic therapy for treatment of *Pseudomonas aeruginosa* cystic fibrosis pulmonary infection. *J. Control. Release* **2007**, *117*, 217–226. [CrossRef]
31. Caruso, E.; Malacarne, M.C.; Banfi, S.; Gariboldi, M.B.; Orlandi, V.T. Cationic diarylporphyrins: In vitro versatile anticancer and antibacterial photosensitizers. *J. Photochem. Photobiol. B Biol.* **2019**, *197*, 111548. [CrossRef] [PubMed]

32. Orlandi, V.T.; Martegani, E.; Bolognese, F.; Trivellin, N.; Mat'átková, O.; Paldrychová, M.; Baj, A.; Caruso, E. Photodynamic Therapy by Diaryl-Porphyrins to Control the Growth of Candida albicans. *Cosmetics* **2020**, *7*, 31. [CrossRef]
33. Orlandi, V.T.; Caruso, E.; Tettamanti, G.; Banfi, S.; Barbieri, P. Photoinduced antibacterial activity of two dicationic 5,15-diarylporphyrins. *J. Photochem. Photobiol. B Biol.* **2013**, *127*, 123–132. [CrossRef]
34. Caruso, E.; Cerbara, M.; Malacarne, M.C.; Marras, E.; Monti, E.; Gariboldi, M.B. Synthesis and photodynamic activity of novel non-symmetrical diaryl porphyrins against cancer cell lines. *J. Photochem. Photobiol. B Biol.* **2019**, *195*, 39–50. [CrossRef]
35. Cieplik, F.; Deng, D.; Crielaard, W.; Buchalla, W.; Hellwig, E.; Al-Ahmad, A.; Maisch, T. Antimicrobial photodynamic therapy—What we know and what we don't. *Crit. Rev. Microbiol.* **2018**, *44*, 571–589. [CrossRef]
36. Demidova, T.N.; Hamblin, M.R. Effect of Cell-Photosensitizer Binding and Cell Density on Microbial Photoinactivation. *Antimicrob. Agents Chemother.* **2005**, *49*, 2329–2335. [CrossRef] [PubMed]
37. Martegani, E.; Bolognese, F.; Trivellin, N.; Orlandi, V.T. Effect of blue light at 410 and 455 nm on *Pseudomonas aeruginosa* biofilm. *J. Photochem. Photobiol. B Biol.* **2020**, *204*, 111790. [CrossRef]
38. Bragonzi, A.; Wiehlmann, L.; Klockgether, J.; Cramer, N.; Worlitzsch, D.; Döring, G.; Tümmler, B. Sequence diversity of the *mucABD* locus in *Pseudomonas aeruginosa* isolates from patients with cystic fibrosis. *Microbiology* **2006**, *152*, 3261–3269. [CrossRef] [PubMed]
39. Orlandi, V.T.; Villa, F.; Cavallari, S.; Barbieri, P.; Banfi, S.; Caruso, E.; Clerici, P. Photodynamic therapy for the eradication of biofilms formed by catheter associated *Pseudomonas aeruginosa* strains. *Microbiol. Med.* **2011**, *26*. [CrossRef]
40. Motta, S.; Vecchietti, D.; Martorana, A.; Brunetti, P.; Bertoni, G.; Polissi, A.; Mauri, P.; Di Silvestre, D. The Landscape of *Pseudomonas aeruginosa* Membrane-Associated Proteins. *Cells* **2020**, *9*, 2421. [CrossRef] [PubMed]
41. Cormick, M.P.; Alvarez, M.G.; Rovera, M.; Durantini, E.N. Photodynamic inactivation of *Candida albicans* sensitized by tri- and tetra-cationic porphyrin derivatives. *Eur. J. Med. Chem.* **2009**, *44*, 1592–1599. [CrossRef] [PubMed]
42. Kapteyn, J.C.; Montijn, R.C.; Dijkgraaf, G.J.; Ende, H.V.D.; Klis, F.M. Covalent association of beta-1,3-glucan with beta-1,6-glucosylated mannoproteins in cell walls of *Candida albicans*. *J. Bacteriol.* **1995**, *177*, 3788–3792. [CrossRef] [PubMed]
43. Sułek, A.; Pucelik, B.; Kobielusz, M.; Barzowska, A.; Dąbrowski, J.M. Photodynamic Inactivation of Bacteria with Porphyrin Derivatives: Effect of Charge, Lipophilicity, ROS Generation, and Cellular Uptake on Their Biological Activity In Vitro. *Int. J. Mol. Sci.* **2020**, *21*, 8716. [CrossRef]
44. Collins, T.L.; Markus, E.A.; Hassett, D.J.; Robinson, J.B. The Effect of a Cationic Porphyrin on *Pseudomonas aeruginosa* Biofilms. *Curr. Microbiol.* **2010**, *61*, 411–416. [CrossRef] [PubMed]
45. Beirão, S.; Fernandes, S.; Coelho, J.; Faustino, M.A.F.; Tomé, J.P.C.; Neves, M.D.G.P.M.S.; Tomé, A.; Almeida, A.; Cunha, A. Photodynamic Inactivation of Bacterial and Yeast Biofilms With a Cationic Porphyrin. *Photochem. Photobiol.* **2014**, *90*, 1387–1396. [CrossRef]
46. Maisch, T. Resistance in antimicrobial photodynamic inactivation of bacteria. *Photochem. Photobiol. Sci.* **2015**, *14*, 1518–1526. [CrossRef] [PubMed]
47. Wiegand, I.; Hilpert, K.; Hancock, R. Agar and broth dilution methods to determine the minimal inhibitory concentration (MIC) of antimicrobial substances. *Nat. Protoc.* **2008**, *3*, 163–175. [CrossRef]
48. Caruso, E.; Orlandi, V.; Malacarne, M.; Martegani, E.; Scanferla, C.; Pappalardo, D.; Vigliotta, G.; Izzo, L. Bodipy-Loaded Micelles Based on Polylactide as Surface Coating for Photodynamic Control of *Staphylococcus aureus*. *Coatings* **2021**, *11*, 223. [CrossRef]
49. Caruso, E.; Ferrara, S.; Ferruti, P.; Manfredi, A.; Ranucci, E.; Orlandi, V.T. Enhanced photoinduced antibacterial activity of a BODIPY photosensitizer in the presence of polyamidoamines. *Lasers Med. Sci.* **2018**, *33*, 1401–1407. [CrossRef] [PubMed]
50. Orlandi, V.T.; Martegani, E.; Bolognese, F. Catalase A is involved in the response to photooxidative stress in *Pseudomonas aeruginosa*. *Photodiagnosis Photodyn. Ther.* **2018**, *22*, 233–240. [CrossRef] [PubMed]
51. Orlandi, V.T.; Bolognese, F.; Rolando, B.; Guglielmo, S.; Lazzarato, L.; Fruttero, R. Anti-Pseudomonas activity of 3-nitro-4-phenylfuroxan. *Microbiology* **2018**, *164*, 1557–1566. [CrossRef] [PubMed]

International Journal of *Molecular Sciences*

MDPI

Article

Antibacterial Photodynamic Inactivation of Fagopyrin F from Tartary Buckwheat (*Fagopyrum tataricum*) Flower against *Streptococcus mutans* and Its Biofilm

Jaecheol Kim [1,2], Suna Kim [3], Kiuk Lee [1], Ryun Hee Kim [1,2] and Keum Taek Hwang [1,2,*]

[1] Department of Food and Nutrition, Research Institute of Human Ecology, Seoul National University, Seoul 08826, Korea; ddeol@snu.ac.kr (J.K.); leku@snu.ac.kr (K.L.); ryunheekim@snu.ac.kr (R.H.K.)
[2] BK21 FOUR Education and Research Team for Sustainable Food & Nutrition, Seoul National University, Seoul 08826, Korea
[3] Division of Human Ecology, College of Natural Science, Korea National Open University, Seoul 03078, Korea; ksuna7@mail.knou.ac.kr
* Correspondence: keum@snu.ac.kr

Abstract: The objective of this study was to determine reactive oxygen species (ROS) produced by fagopyrin F-rich fraction (FFF) separated from Tartary buckwheat flower extract exposed to lights and to investigate its antibacterial photodynamic inactivation (PDI) against *Streptococcus mutans* and its biofilm. ROS producing mechanisms involving FFF with light exposure were determined using a spectrophotometer and a fluorometer. *S. mutans* and its biofilm inactivation after PDI treatment of FFF using blue light (BL; 450 nm) were determined by plate count method and crystal violet assay, respectively. The biofilm destruction by ROS produced from FFF after exposure to BL was visualized using confocal laser scanning microscopy (CLSM) and field emission scanning electron microscope (FE-SEM). BL among 3 light sources produced type 1 ROS the most when applying FFF as a photosensitizer. FFF exposed to BL (5 and 10 J/cm^2) significantly more inhibited *S. mutans* viability and biofilm formation than FFF without the light exposure ($p < 0.05$). In the PDI of FFF exposed to BL (10 J/cm^2), an apparent destruction of *S. mutans* and its biofilm were observed by the CLSM and FE-SEM. Antibacterial PDI effect of FFF was determined for the first time in this study.

Keywords: Tartary buckwheat; fagopyrin F; fagopyrin; photosensitizer; ROS; photodynamic therapy

Citation: Kim, J.; Kim, S.; Lee, K.; Kim, R.H.; Hwang, K.T. Antibacterial Photodynamic Inactivation of Fagopyrin F from Tartary Buckwheat (*Fagopyrum tataricum*) Flower against *Streptococcus mutans* and Its Biofilm. *Int. J. Mol. Sci.* **2021**, *22*, 6205. https://doi.org/10.3390/ijms22126205

Academic Editors: Antonino Mazzaglia, Angela Scala and Enrico Caruso

Received: 17 May 2021
Accepted: 3 June 2021
Published: 8 June 2021

1. Introduction

Buckwheat plants contain protofagopyrins, which are converted to fagopyrins when the plant extract is exposed to light [1]. Both protofagopyrins and fagopyrins belong to naphthodianthrone and act as photosensitizers (PS) [1,2]. Protofagopyrins were converted to fagopyrins most rapidly when exposed to blue light (BL) and fluorescence light, and fagopyrins maintained a stable structure even when exposed to various light sources for 8 h [3]. Hypericin, one of naphthodianthrones structurally similar to fagopyrins and protofagopyrins, has been studied as a PS for various photodynamic effects including antibacterial effects [4–6]. Benković et al. reported the structural characteristics of fagopyrins, which are similar to hypericin [1]. Fagopyrins and hypericin have the same absorbance spectra (maximum absorbance: 590 nm) [1,3]. Also, fluorescence of fagopyrins and hypericin was detected at an excitation wavelength of 330 nm and emission wavelength of 590 nm [1,7–9]. However, to the best of our knowledge, photodynamic effects of fagopyrins and protofagopyrins have not yet been reported. Tartary buckwheat flowers (TBF) are richer in fagopyrins (protofagopyrins) than the other parts of the buckwheat plant (e.g., stems, leaves, flowers, groats, and hulls), regardless of the cultivar type [1,3,7,9]. Six fagopyrins (A–F) were reported to exist in buckwheat plants with fagopyrins A, E, and F identified [1,8]. In addition, Kim and Hwang reported that fagopyrin F (FF) accounted for more than 93% of total fagopyrins in TBF extracts exposed to light [9]. Given these results, in this study, we prepared FF-rich fraction (FFF) separated from the TBF extract exposed to lights and further determined photodynamic effect of the FFF.

Photodynamic therapy (PDT) is a clinical treatment against harmful microorganisms, tumors, viruses, and parasites and is based on the photochemical reaction of PS [10,11]. PS molecules absorb light of a specific wavelength initiating reactive oxygen species (ROS) production, which leads to selective cellular or tissue destruction (Figure 1) [10–12]. After exposure to light, energy from the triplet excited state of the PS is transferred to two ROS production mechanisms [10–13]. In the type 1 ROS photodynamic mechanism, the PS transfers hydrogen or electrons to biomolecules from its surroundings. This process initially produces ROS in the form of superoxide anion radicals ($O_2^{\bullet-}$), which further generates other ROS molecules, such as hydrogen peroxide and hydroxyl radicals, inside the cell [10–13]. In the type 2 ROS photodynamic mechanism, energy from the triplet excited state of the PS is directly transferred to oxygen molecules in the ground energy state ($^3O_2$). This process produces singlet oxygen ($^1O_2$), which exhibits strong oxidative properties [10–13].

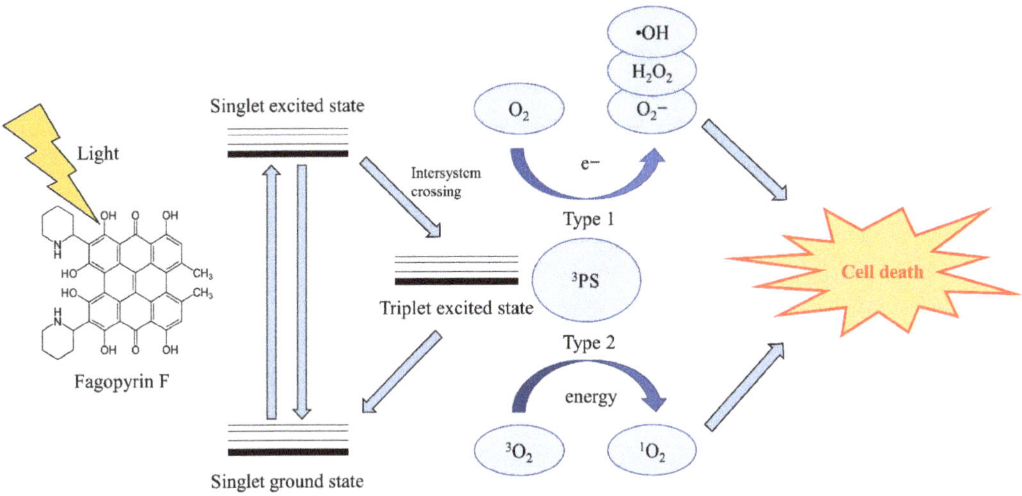

Figure 1. Structure of fagopyrin F and schematic illustration of photodynamic therapy.

One of the most widely studied PDT applications is antibacterial photodynamic inactivation (PDI), which can be applied to several bacterial species. Among possible treatment sites for PDI, oral cavity is exposed outside, not inside the human body; thus, PDI can be directly applied to the oral cavity to inactivate various harmful bacteria. Since the PDI method for the oral cavity is simple, and PDI does not cause antibiotic resistance to harmful bacteria in the oral cavity, PDI has been studied fairly extensively for eliminating harmful bacteria in the oral cavity [14–16]. *Streptococcus mutans* is a gram-positive bacterium that causes dental erosion by forming biofilm [16]. Biofilm, a complex of bacteria and their secondary metabolites (e.g., sugars, acids, and glucans), forms a layer on the surface of the teeth [16]. Various PS have been used to inactivate harmful oral bacteria. Among them, curcumin and hypericin originated from food sources, are typical PS against harmful oral bacteria [4,17–19]. PDI of curcumin or hypericin as a PS destroyed most *S. mutans* (over 99%) and inhibited their biofilm formation [4,17,18]. Application of PDI to *S. mutans* has been studied in terms of biofilm inactivation and destruction, and the treatment effect of PDI has been mainly confirmed using spectrometer and microscope techniques such as confocal laser scanning microscopy (CLSM) and field emission scanning electron microscope (FE-SEM) [14,16,20–26].

The purpose of this study was to evaluate photosensitizing ability of FFF separated from TBF extract by measuring the amount of ROS produced after exposure to lights and to investigate the PDI effect of FFF against *S. mutans* and its biofilm.

2. Results and Discussion

2.1. ROS Production by FFF

Intracellular ROS productions in the *S. mutans* suspensions added with FFF when exposed to BL (450 nm), yellow light (YL; 590 nm), and red light (RL; 660 nm) were shown in Figure 2a. The fluorescence intensity of the *S. mutans* suspension with FFF exposed to BL (5 or 10 J/cm^2) was significantly higher than that of YL or RL (Figure 2a), indicating that BL was an appropriate light source to produce ROS by FFF. The fluorescence intensity of the *S. mutans* suspension added with FFF along with D-mannitol when exposed to BL (5 or 10 J/cm^2) was significantly lower than that with FFF only (Figure 2b), suggesting that FFF exposed to BL might produce hydroxyl radicals, which might be then scavenged by D-mannitol.

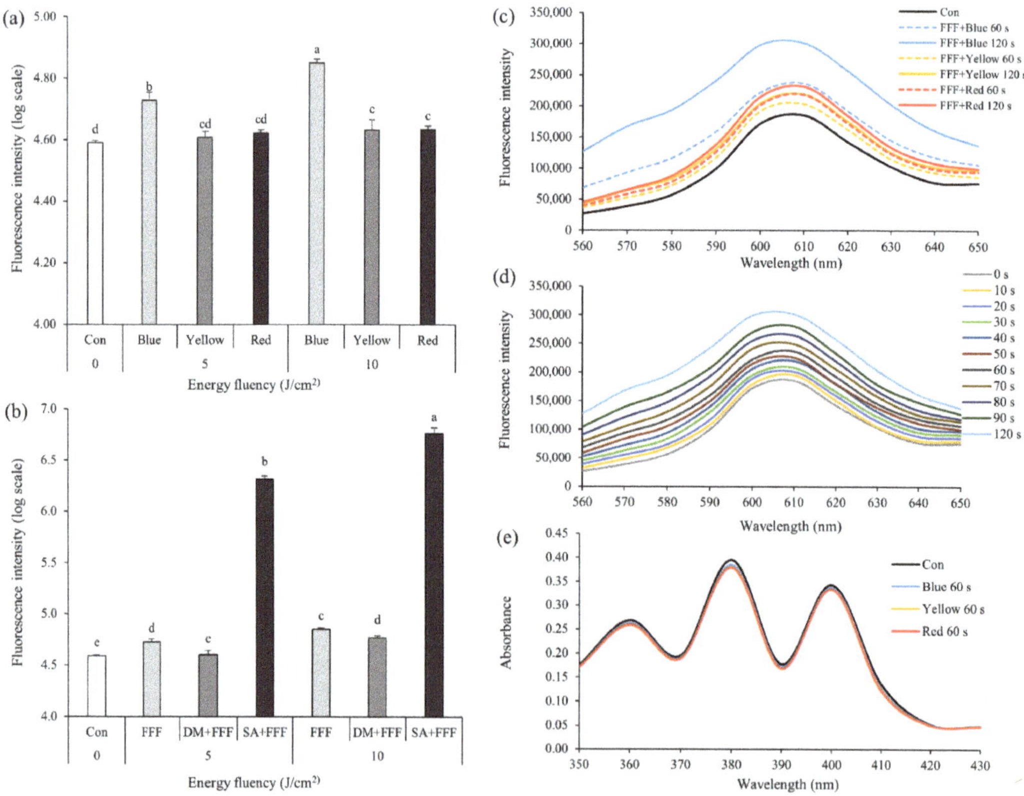

Figure 2. Production of reactive oxygen species (ROS) by fagopyrin F-rich fraction (FFF; final concentration of 5 μg/mL) from Tartary buckwheat flower extract after exposure to lights. (**a**) Intracellular ROS production in *Streptococcus mutans* treated by FFF with different lights. (**b**) Intracellular ROS production in *S. mutans* treated by FFF with blue light (5 or 10 J/cm^2) with different radical scavengers. DM: D-mannitol (final concentration of 100 mM). SA: sodium azide (final concentration of 100 mM). (**c**) Superoxide production in *S. mutans* treated by FFF with different lights. (**d**) Superoxide production in *S. mutans* treated by FFF with blue light. (**e**) Singlet oxygen production treated by FFF with different lights. (**a–e**) Con: control with FFF without irradiation. (**a**,**b**) Different small letters indicate significant differences ($p < 0.05$; one-way ANOVA and Duncan's multiple range test). (**a**,**b**) Bars are means ± standard deviations (n = 3). (**c–e**) Data lines are means (n = 3).

On the other hand, when the *S. mutans* suspension with FFF along with sodium azide (NaN_3) was exposed to BL (5 or 10 J/cm^2), the fluorescence intensity did not decrease as the singlet oxygen might be scavenged by sodium azide, and the fluorescence intensity rapidly increased during exposure to BL (Figure 2b). This result implies that the fluorescence intensity increases by binding 2′,7′-dichlorofluorescin diacetate (DCFH-DA) to azide radical ($N_3^\bullet$) derived from NaN_3 rather than singlet oxygen produced by FFF [20,27]. That is, when FFF was exposed to BL, singlet oxygen might not be generated or the amount of singlet oxygen might not be enough to be combined with DCFH-DA.

Intracellular superoxide productions in the *S. mutans* suspensions added with FFF when exposed to BL, YL, and RL were shown in Figure 2c. The fluorescence intensity of the *S. mutans* suspension with FFF when exposed to YL or RL for 120 s was slightly higher than the control (Figure 2c), whereas the intensity of the FFF suspension exposed to BL for 120 s increased in a time-dependent manner (Figure 2d). Superoxide production by FFF exposed to BL was higher than that exposed to YL or RL. These results were consistent with those measured by DCFH-DA.

When 9,10-anthracenediyl-bis(methylene)dimalonic acid (ABDA) added with FFF was exposed to BL, YL, or RL, its absorbance, whose reduction indicates more singlet oxygen production, differed little from that of the control, regardless of the wavelength (Figure 2e). However, the previous study observed a rapid decrease in the absorbance when ABDA added with PS such as methylene blue, toluidine blue, and rose bengal was exposed to light [21,28], suggesting these PS may very effectively produce singlet oxygen following the type 2 ROS production mechanism. Thus, our result suggests that FFF exposed to light little produces singlet oxygen.

In summary, ROS production mechanism of FFF exposed to lights might be type 1. Also, type 1 ROS were more produced when FFF was exposed to BL than YL and RL. Therefore, subsequent PDI experiments were conducted using BL.

2.2. Photodynamic Inactivation of Planktonic S. mutans

The colony forming unit (CFU) of *S. mutans* did not significantly decrease after FT (*S. mutans* suspension treated with FFF (final concentration of 5 µg/mL) without irradiation), MT (*S. mutans* suspension in 2.5% methanol (the same concentration of methanol in the suspension treated with FFF at the final concentration of 5 µg/mL) with no FFF treatment nor irradiation), or MIT (*S. mutans* suspension in 2.5% methanol without FFF, which was exposed to BL of 10 J/cm^2) treatment compared to NT (*S. mutans* suspension with no FFF treatment nor irradiation) (Figure 3a), suggesting 5 µg/mL FFF or 2.5% methanol (final concentration) in the suspension without BL exposure might not kill *S. mutans*. The BL exposure (10 J/cm^2) of the suspension without FFF (MIT) did not affect *S. mutans* viability in this study, although BL (400–450 nm) is known to have antibacterial effects [20]. Paschoal et al. reported that *S. mutans* was not killed when exposed to BL (450 ± 30 nm; 24, 48, or 72 J/cm^2) without PS treatment [17]. Moreover, *S. mutans* viability did not decrease when exposed to BL (405 nm; 25.4 J/cm^2) without a PS [18]. In this study, *S. mutans* was exposed to BL (5 or 10 J/cm^2), in which energy fluences were lower than in the previous studies [17,18]. Therefore, it is certain that BL (5 and 10 J/cm^2) tested in this study does not affect *S. mutans* viability.

When the *S. mutans* suspension added with FFF was exposed to BL (5 and 10 J/cm^2), CFU in the PDI treatment groups were significantly lower than the NT and FT (Figure 3a). CFU in the *S. mutans* suspension added with FFF when exposed to BL decreased in a dose-dependent and energy fluence-dependent manners. In this study, when *S. mutans* suspension added with FFF (5 µg/mL) was incubated for 10 min and then exposed to BL (450 nm; 10 J/cm^2), *S. mutans* viability decreased by 97.6% compared to the FT. Lüthi et al. reported that when *S. mutans* was incubated with hypericin (10 µg/mL) for 30 min and then exposed to BL (400–505 nm) with energy fluence of 128.4 and 256.8 J/cm^2, *S. mutans* viability decreased by 99.2 and 99.998%, respectively, compared to the control [4]. Paschoal et al. also reported that when *S. mutans* was incubated with curcumin (2 mM; 736.8 µg/mL) for 1 min and then exposed to BL (450 ± 30 nm; 24 J/cm^2), *S. mutans* viability decreased by 99.13% compared to the control [17]. However, considering that PS concentrations and energy fluences in the previous studies were higher than those in this study, FFF may have comparable potency to kill *S. mutans* to other PS. On the other hand, Ribeiro et al. reported that when *S. mutans* was incubated with riboflavin (40 μg/mL) for 10 min and exposed to BL (455 ± 20 nm; 32.4 J/cm^2), *S. mutans* viability decreased by 77.5% compared to the control [19], in which a smaller decrease in viability of *S. mutans* was observed although a higher PS concentration and a higher energy fluence were applied than in this study.

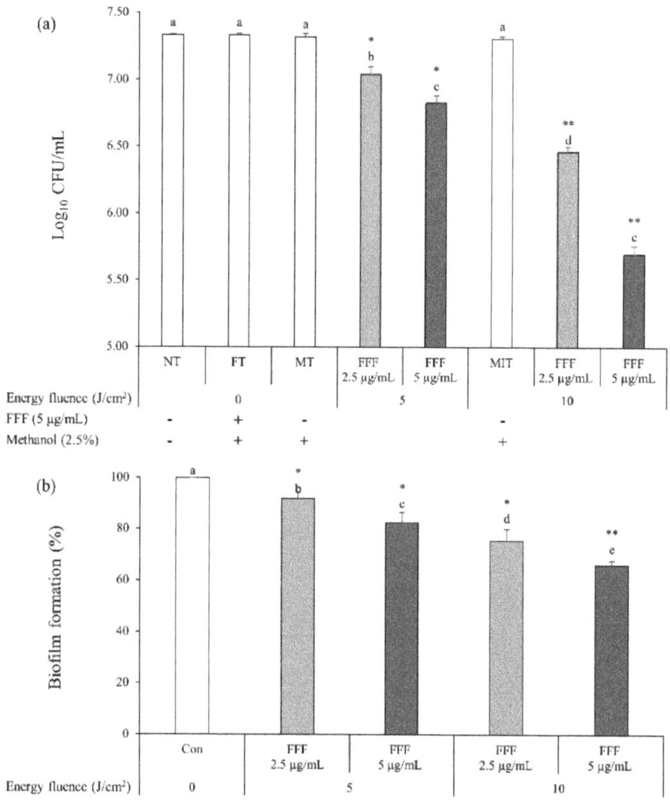

Figure 3. Photodynamic effect of fagopyrin F-rich fraction (FFF) from Tartary buckwheat flower extract against *Streptococcus mutans* and its biofilm. (**a**) Inactivation of *S.mutans* treated with FFF exposed to blue light (BL). NT: *S. mutans* suspension with no FFF treatment nor irradiation. FT: *S. mutans* suspension treated with FFF (final concentration of 5 μg/mL) without irradiation. MT: *S. mutans* suspension in 2.5% methanol (the same concentration of methanol in the suspension treated with FFF at the final concentration of 5 μg/mL) with no FFF treatment nor irradiation. MIT: *S. mutans* suspension in 2.5% methanol without FFF, which was exposed to BL (10 J/cm^2). (**b**) Biofilm formation of *S. mutans* treated with FFF exposed to BL. Con: control with FFF (final concentration of 5 μg/mL) without irradiation. (**a**,**b**) Different small letters indicate significant differences ($p < 0.05$; one-way ANOVA and Duncan's multiple range test). *, ** Significant difference compared to FT (**a**) or Con (**b**) ($p < 0.05$, $p < 0.01$; independent t-test). Bars are means ± standard deviations (n = 3).

2.3. Effects of PDI Treatment of FFF on S. mutans Biofilm Formation

S. mutans biofilm formation in the PDI treatment groups was significantly lower than that in the control (Figure 3b). The *S. mutans* suspension added with FFF (5 μg/mL) exposed to BL (10 J/cm^2) had the highest inhibition in the biofilm formation among the

PDI treatment groups. Inhibition rates of biofilm formation with the PDI treatments of FFF at 2.5 μg/mL exposed to BL of 5 J/cm^2 and FFF at 5 μg/mL exposed to BL of 10 J/cm^2 were the lowest (11.6%) and the highest (34.0%), respectively, compared to the control (Figure 3b). In this study, the PDI treatment with a higher concentration of FFF formed less biofilm at the same energy fluence (Figure 3b). However, the PDI treatment with FFF at 2.5 μg/mL exposed to BL of 10 J/cm^2 formed a significantly less biofilm than the PDI treatment with FFF at 5 μg/mL exposed to BL of 5 J/cm^2 ($p < 0.05$; Figure 3b). These results suggest that increased energy fluence might be more efficient than increased FFF concentration in inhibiting *S. mutans* biofilm formation.

In this study, we confirmed for the first time that FF in the FFF, when exposed to BL, exerts antibacterial and antibiofilm effects in *S. mutans*. As previously stated, inactivation of *S. mutans* and biofilm formation by PDI using hypericin was reported [15]. It has been also reported that hypericin has a PDI effect on several gram-positive bacteria [4,15]. Thus, further studies are warranted to confirm the PDI effects of FFF on other harmful gram-positive bacteria.

2.4. Visualization of Antibiofilm Effects of FFF by CLSM

CLSM was performed to visualize the PDI effects of FFF against *S. mutans* biofilms (Figure 4). SYTO 9 and propidium iodide dyes were used for staining living cells green and dead cells red, respectively, and confirmed whether bacterial death occurred after the PDI treatment of FFF [29]. The fluorescence images of the NT and FT showed dense green staining indicating live bacteria (Figure 4). However, when the energy fluence increased, this green staining gradually dimmed and red staining gradually increased, indicating bacterial death. These results suggest that the PDI treatment of FFF may be effective in destructuring *S. mutans* biofilms. These observations were similar to CLSM image color changes from previous PDI studies using other PS [21,30].

2.5. Visualization of Antibiofilm Effect of FFF by FE-SEM

FE-SEM was used to visualize the PDI effect of FFF against *S. mutans* biofilms (Figure 5). Cell division in *S. mutans* occurs toward poles of cells and tends not to completely separate; hence the bacteria form chain shapes as they grow [31]. Before the PDI treatment as in the NT and FT, most *S. mutans* looked like short-chains in the biofilm (Figure 5). Namely, *S. mutans* formed an intact original shape without cell damage. However, after the PDI treatment of FFF, *S. mutans* cell membranes were destroyed by ROS produced from FFF (Figure 5). These FE-SEM images indicate that the cell membrane, not the intracellular cytoplasm, might be the major site of the damage mediated by ROS derived from FFF. Cell membrane destruction and cytoplasm leakage were clearly observed in the FE-SEM images at 15,000× and 100,000× magnifications, respectively (Figure 5). Also, *S. mutans* chain structures collapsed, resulting from cell membrane destruction by ROS from the PDI treatment of FFF. The FE-SEM images also showed a vast *S. mutans* biofilm destruction, as well as significant morphological changes in *S. mutans* chains after the PDI of FFF. This *S. mutans* biofilm destruction was energy fluence-dependent. Briefly, biofilms after the PDI treatment of FFF exposed to BL of 10 J/cm^2 were more morphologically destroyed than BL of 5 J/cm^2 (Figure 5). These data were consistent with the results described in Section 2.3. Also, the FE-SEM images of *S. mutans* biofilm destroyed by the PDI in this study were similar to those presented in the previous studies [21,23].

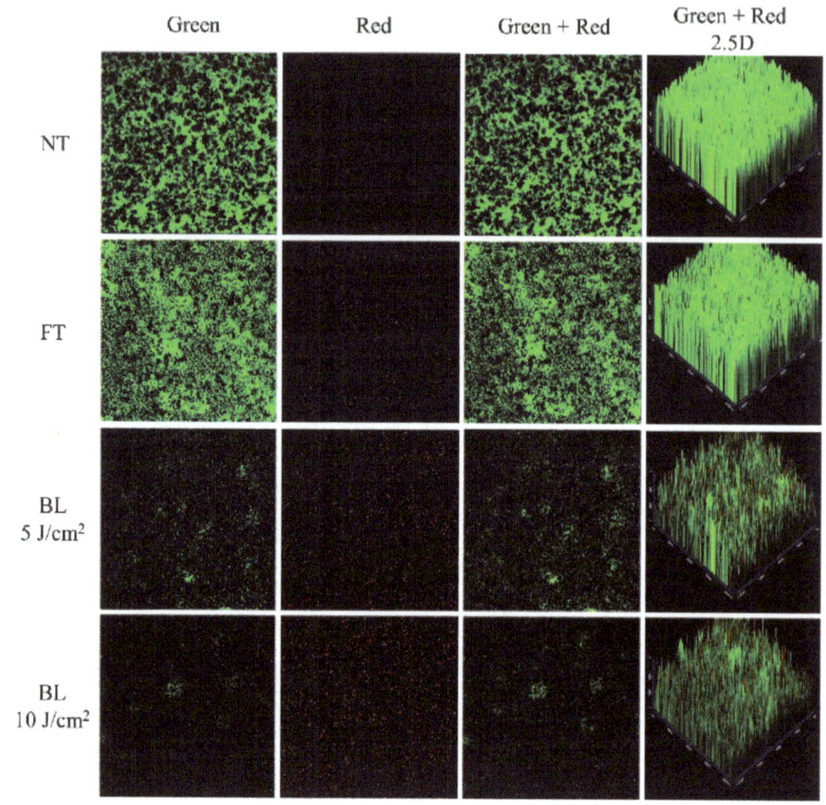

Figure 4. Confocal laser scanning microscopy images of *Streptococcus mutans* biofilm treated with photodynamic therapy of fagopyrin F-rich fraction (FFF; final concentration of 5 μg/mL) exposed to blue light (BL). NT: *S. mutans* biofilm with no FFF treatment nor irradiation. FT: *S. mutans* biofilm treated with FFF without irradiation.

In general, ROS produced from PDI destroys cell membranes and denatures cellular DNA. Takasaki et al. reported that a PS destroys cell membranes via ROS production after attaching to the cell membrane rather than destroying DNA by intracellular ROS production [32]. ROS produced by PS membrane attachment inactivates the membrane transport system and associated enzymes, thereby inducing lipid peroxidation, which damages the cell membrane structure [32,33]. Esmatabadi et al. reported that after PDI-mediated cell membrane destruction, cytoplasm contents and metabolites were released from cells, and DNA was damaged [34]. In this study, we observed the destruction and cytoplasmic release of *S. mutans* cell membranes via the PDI treatment of FFF. Similar to other PS, FFF is believed to selectively bind to cytoplasmic membrane components and cause direct cell death by ROS via destruction of the membrane.

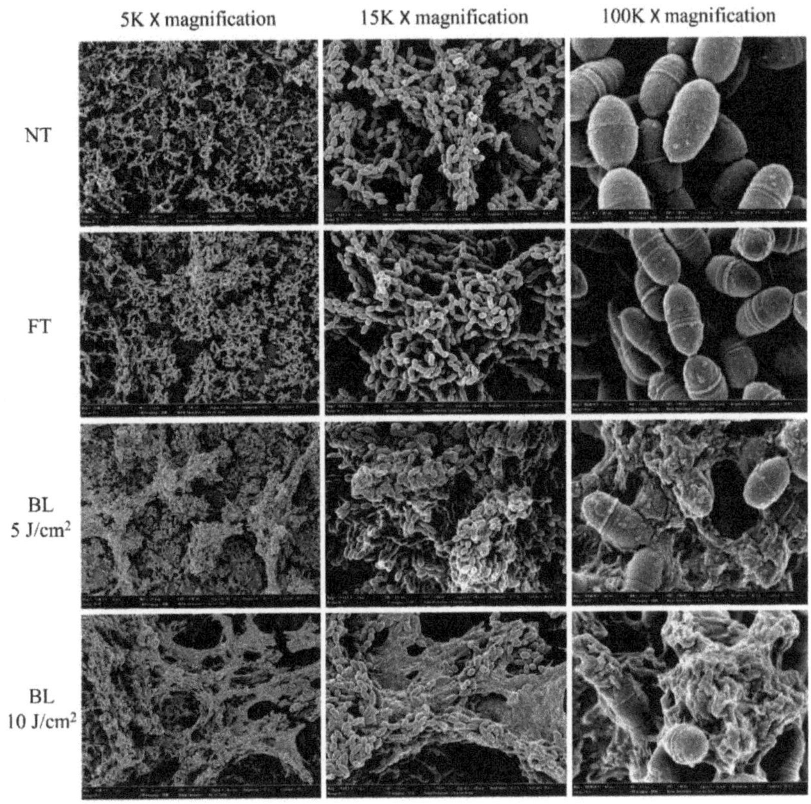

Figure 5. Field emission scanning electron microscope images of *Streptococcus mutans* biofilm treated with photodynamic therapy of fagopyrin F-rich fraction (FFF; final concentration of 5 μg/mL) exposed to blue light (BL). NT: *S. mutans* biofilm with no FFF treatment nor irradiation. FT: *S. mutans* biofilm treated with FFF without irradiation.

3. Materials and Methods

3.1. Chemicals and Reagents

Brain heart infusion (BHI) broth was purchased from BD (Becton, Dickinson and Company, Franklin Lakes, NJ, USA). ABDA, acetic acid, agar, crystal violet solution, DCFH-DA, D-mannitol, dihydroethidium (DHE), ethanol, formic acid, glycerol, and sodium azide were purchased from Sigma-Aldrich Co. (St. Louis, MO, USA). Glutaraldehyde, hexamethyldisilazane, osmium tetroxide, paraformaldehyde, and sodium cacodylate buffer (SCB) were purchased from Electron Microscopy Science (EMS) (Hatfield, PA, USA). Methanol and acetonitrile were purchased from J.T. Baker (Phillipsburg, NJ, USA). Phosphate buffer saline (PBS) and LIVE/DEAD™ Bacterial Viability Kit (L-7012) were purchased from Thermo Fisher Scientific (Waltham, MA, USA).

3.2. Bacterial Strain and Culture

S. mutans KCTC 3298 from the Korean Collection for Type Cultures (KCTC, Jeongeup, Korea) was cultured in the BHI broth at 37 °C. *S. mutans* stock solution was prepared by inoculating a single colony of bacteria from BHI agar plate into BHI broth (10 mL) and incubating for 24 h. After incubation, equal volumes of aliquoted bacterial broth of *S. mutans* and 50% glycerol were mixed, and the mixture was stored at −80 °C as stock. For subsequent experiments, the thawed *S. mutans* stock was inoculated into BHI broth

(1%) and was cultured for 24 h. The cultured suspension of *S. mutans* was inoculated into fresh BHI broth (1%) and cultured for 24 h. The activated *S. mutans* (1.1×10^9 CFU/mL) was used in subsequent experiments.

3.3. Preparation of FFF and Light Sources

TBF was collected as described in the previous study [9]. Freeze-dried TBF powder (500 mg) was extracted with methanol (20 mL) in a water bath (BS-11, Lab Companion, Seoul, Korea) at 60 °C for 60 min. The extract was centrifuged at 1500× *g* for 10 min. The supernatant was filtered through a 0.2 µm syringe filter (DISMIC-13JP, Advantec, Tokyo, Japan) and the filtrate was exposed to fluorescent light (400–700 nm; LED Bulb 12 W, Philips Korea, Seoul, Korea) for 2 h to convert protofagopyrins to fagopyrins. An HPLC system (e2695, Waters, Milford, MA, USA) equipped with a photodiode array detector (2998, Waters, USA) and an Agilent Zorbax Eclipse XDB (250 mm × 4.6 mm, 5 µm; Agilent, Palo Alto, CA, USA) column was used to separate FFF from the TBF extract. Mobile phases were 0.1% formic acid in distilled water (A) and 0.1% formic acid in acetonitrile (B). Flow rate was 1.2 mL/min. Gradient elution was set as follows: 60% B in 0–2 min, 60–100% B in 2–8 min, 100% B in 8–9.5 min, 100–60% B in 9.5–10 min, and 60% B in 10–11 min. Column temperature was 50 °C. Injection volume was 100 µL. The fraction (FFF) eluted between 9.0 and 10.0 min of retention time, at which fagopyrins detected at 590 nm were collected mostly, were dried using a centrifugal vacuum concentrator (VC 2200, Labogene, Seoul, Korea). The FFF powder was dissolved in methanol and stored at −80 °C for subsequent experiments. FF accounted for 92.33 ± 0.08% (n = 3) in the FFF, when calculated based on the peak areas obtained by HPLC-PDA and UPLC-MS/MS [3].

BL (450 nm; ABI 12 W Blue LED, ABI, Indianapolis, IN, USA), YL (590 nm; 15 W PI200, Bissol LED, Seoul, Korea), and RL (660 nm; ABI 12 W DEEP RED LED, ABI) were used in this study to find a suitable wavelength to activate FFF as a PS. Output powers of light sources were expressed as power density (W/cm^2) and energy fluence (J/cm^2), which were calculated as follows [23]:

Power density (W/cm^2) = output power (W)/area (cm^2)
Energy fluence (J/cm^2) = power density (W/cm^2) × exposure time (s)

3.4. ROS Production by FFF

3.4.1. Intracellular ROS Production

Intracellular ROS produced in FFF-incorporated *S. mutans* suspension exposed to light was measured by fluorescence spectroscopy using DCFH-DA as described in previous studies with some modifications [20,27]. The cultured *S. mutans* suspension was diluted with PBS to 10^8 CFU/mL. The suspension was incubated with DCFH-DA (final concentration of 5 µM) at 37 °C for 30 min in the dark. FFF (final concentration of 5 µg/mL) was added to the suspension, which was then treated with D-mannitol (type 1 ROS scavenger; final concentration of 100 mM) or sodium azide (type 2 ROS scavenger; final concentration of 100 mM) or not treated. The suspension was exposed to BL, YL, or RL (5 or 10 J/cm^2). After the irradiation, fluorescence intensity was measured by SpectraMax iD3 (Molecular Devices, San Jose, CA, USA) at an excitation wavelength of 485 nm and an emission wavelength of 525 nm. Control was treated with the same final concentration of FFF without exposure to light.

3.4.2. Superoxide Production

Intracellular superoxide produced in FFF-incorporated *S. mutans* suspension exposed to light was measured by fluorescence spectroscopy using DHE as described in previous studies with some modifications [35,36]. The cultured *S. mutans* suspension was diluted with PBS to 10^8 CFU/mL. DHE (final concentration of 5 mM) was added to the suspension. The suspension was incubated at 37 °C for 30 min in the dark. FFF (final concentration of 5 µg/mL) was added to the suspension, which was then incubated at 37 °C for 15 min in the dark. After incubation, fluorescence intensity (excitation wavelength: 510 nm; emission spectrum: 560–650 nm) of the suspension was measured before exposure to light and every 10 s afterwards for 120 s while exposing the suspension to light (BL, YL, or RL) of 0.16 W/cm^2. Fluorescence intensity was measured using SpectraMax iD3 (Molecular Devices). Control was treated with the same final concentration of FFF without exposure to light.

3.4.3. Singlet Oxygen Detection

Singlet oxygen produced when FFF was exposed to light was measured by spectrophotometry using ABDA as described in a previous study with some modification [37].

ABDA (final concentration of 20 μM) and FFF (final concentration of 5 μg/mL) in PBS were added in 96-well plate (SPL, Pocheon, Korea). The absorbance spectrum (350–430 nm) was measured using SpectraMax iD3 (Molecular Devices). After measurement of the initial absorbance spectrum without exposure to light, it was measured every 10 s for 60 s while exposing the wells to light (BL, YL, or RL) of 0.16 W/cm^2. Control was treated with the same final concentration of FFF without exposure to light.

3.5. Effect of S. mutans Inactivation with PDI Treatment of FFF

The effect of *S. mutans* inactivation with PDI treatment of FFF was performed as described in a previous study with some modification [21]. The PDI effect of FFF against *S. mutans* was determined using the plate count method. The cultured *S. mutans* suspension was diluted with PBS to 10^8 CFU/mL. The suspension with FFF (final concentration of 2.5 or 5 μg/mL) was incubated at 37 °C for 10 min in the dark. Then suspensions were exposed to BL (5 or 10 J/cm^2). After exposure, 10-fold serial dilutions in PBS were performed, and 100 μL of the suspensions were spread on BHI agar plates. After incubating for 48 h, the number of single colonies was counted. To figure out any intervention other than the PDI the following treatments were also tested: NT, FT, MT, and MIT.

3.6. Inhibition of S. mutans Biofilm Formation with PDI Treatment of FFF

Inhibition of *S. mutans* biofilm formation with PDI treatment of FFF was spectrophotometrically determined using crystal violet dye as described in the previous study with some modification [14]. The cultured *S. mutans* suspension was diluted to 10^8 CFU/mL with fresh BHI broth containing 5% sucrose. The suspension (1 mL) was added to each well of a sterile and flat 24-well plate (SPL), and FFF (final concentration of 2.5 or 5 μg/mL) was added into the suspension. The suspension was incubated at 37 °C for 15 min in the dark. After incubation, the suspension was exposed to BL (5 or 10 J/cm^2). The PDI-treated suspension was incubated to form biofilm at 37 °C for 24 h in the dark. The medium was decanted and gently washed twice with 1 mL PBS to remove loosely bound biofilm and unbound planktonic *S. mutans*. Biofilm was stained with 1 mL 0.1% crystal violet for 15 min on a well plate shaker (MX-M, DLAB, Riverside, CA, USA) at 300 rpm and room temperature (RT). The crystal violet dye was removed, and the biofilm was gently washed twice with 1 mL PBS. The stained biofilm was air-dried at RT for 15 min. After drying, 600 μL 33% acetic acid was added to dissolve the stained biofilm. The suspension was dissolved for 10 min on a well-plate shaker (DLAB, USA) at 300 rpm and RT. The absorbance of the dissolved biofilm suspension was measured using a SpectraMax iD3 (Molecular Devices) at 570 nm. Control was treated with the same final concentrations of FFF without exposure to BL. Biofilm formation level was expressed as the percentage of the control.

3.7. Visualization in PDI Effects of FFF against S. mutans Biofilm

3.7.1. CLSM

The *S. mutans* suspension was diluted with fresh BHI broth containing 5% sucrose to 10^6 CFU/mL. The suspension (3 mL) in a sterile confocal dish (SPL) was incubated at 37 °C for 24 h to form *S. mutans* biofilm. The medium was decanted and gently washed twice with 1 mL PBS to remove loosely bound biofilm and unbound planktonic *S. mutans*.

One mL FFF (final concentration of 5 μg/mL) in PBS was added to the biofilm in the confocal dish. The suspension was exposed to BL (5 or 10 J/cm^2). The suspension was removed to obtain the biofilm, which was then gently washed twice with 1 mL PBS. Biofilm was stained using LIVE/DEAD™ Bacterial Viability Kit (SYTO 9 and propidium iodide dye) according to the manufacturer's instruction. The biofilm with staining solution was incubated in the dark at 37 °C for 1 h. The staining solution was removed and 100 μL PBS was added to prevent drying of biofilm. Fluorescence images were observed using CLSM (LSM710, Carl Zeiss, Oberkochen, Germany) at fluorescences of green (excitation wavelength: 488 nm; emission wavelength: 516 nm) and red (excitation wavelength: 543 nm; emission wavelength: 589 nm) under 40 times magnification. Control was treated with the same final concentrations of FFF without exposure to BL.

3.7.2. FE-SEM

The *S. mutans* suspension was diluted to 10^6 CFU/mL in fresh BHI broth containing 5% sucrose, and the suspension (3 mL) was incubated to form biofilm of *S. mutans* on glass coverslips (24 × 24 mm, Paul Marienfield, Lauda-Königshofen, Germany) in 6-well plates (SPL) at 37 °C for 36 h. After biofilm formation on the coverslip, the medium was decanted, and the well was gently washed twice with 1 mL of PBS to remove loosely bound biofilm and unbound planktonic *S. mutans*. One mL FFF (final concentration of 5 μg/mL) in PBS was added into each well, and the suspension was incubated at 37 °C for 5 min in the dark. After incubation, the suspension was exposed to BL (5 or 10 J/cm^2). The suspension with FFF was decanted, and the biofilm on the coverslip was gently washed twice with 1 mL PBS. For primary fixation, the biofilm on the coverslip was soaked for 4 h in Karnovsky fixative containing 2% glutaraldehyde and 2% paraformaldehyde in 0.05 M SCB. The coverslip was washed three times with 0.05 M SCB for 5 min at each time. Post fixation was conducted using 1% osmium tetroxide in 0.1 M SCB at 4 °C for 1 h. After fixation, the coverslip was washed three times with distilled water for 5 min at each time and dehydrated in a series of ethanol solutions (30%, 50%, 70%, 80%, 90%, and 100%) for 10 min each. The coverslip was soaked in hexamethyldisilazane for 10 min for specimen drying, and then the coverslip was put in a vacuum desiccator for 24 h. After drying, the coverslip was mounted on stubs (EMS) and was coated with platinum by EM ACE 200 (Leica, Wetzlar, Germany). Biofilm images on the coverslip were obtained using FE-SEM (AURIGA, Carl Zeiss, Oberkochen, Germany). Control was treated with the same final concentrations of FFF without exposure to BL.

3.8. Statistics

All experiments were conducted in triplicate except for CLSM and FE-SEM. Statistical analysis was conducted by Student's *t*-test and one-way analysis of variance (ANOVA) with Duncan's multiple range test ($p < 0.05$) using SPSS 23.0 software (IBM, Armonk, NY, USA).

4. Conclusions

This study demonstrated that FFF when exposed to BL (450 nm) produced ROS, and the ROS production of FFF was on the basis of the type 1 mechanism. Furthermore, *S. mutans* and its biofilm were destroyed by ROS produced by PDI of FFF. It was elucidated that PDI with FFF (5 μg/mL) exposed to BL (10 J/cm^2) destroyed *S. mutans* biofilms, which were visually confirmed by CLSM and FE-SEM. This study demonstrated for the first time that FFF, present in buckwheat plants, is a potent PS and can be applied to PDI. Although the PDI treatment of FFF was conducted at lower energy fluences of BL and lower concentrations than other PS, PDI effect of FFF against *S. mutans* was similar to curcumin and hypericin and was stronger than riboflavin. Thus, FFF might be a more effective PS for PDI against *S. mutans* than commonly used PS. Further studies would be needed to investigate whether the PDI of FFF can be applied to other gram-positive bacteria and tumors.

Author Contributions: Conceptualization, J.K., S.K., K.L., R.H.K. and K.T.H.; methodology, J.K., S.K. and K.L.; validation, J.K.; formal analysis, J.K.; investigation, J.K.; data curation, J.K. and K.L.; writing—original draft preparation, J.K.; writing—review and editing, S.K., R.H.K. and K.T.H.; visualization, J.K.; supervision, K.T.H. All authors have read and agreed to the published version of the manuscript.

Funding: This research received no external funding.

Institutional Review Board Statement: Not applicable.

Informed Consent Statement: Not applicable.

Data Availability Statement: Not applicable.

Conflicts of Interest: The authors declare no conflict of interest.

References

1. Benković, E.T.; Kreft, S. Fagopyrins and protofagopyrins: Detection, analysis, and potential phototoxicity in buckwheat. *J. Agric. Food. Chem.* **2015**, *63*, 5715–5724. [CrossRef]
2. Ahmed, A.; Khalid, N.; Ahmad, A.; Abbasi, N.A.; Latif, M.S.Z.; Randhawa, M.A. Phytochemicals and biofunctional properties of buckwheat: A review. *J. Agric. Sci.* **2014**, *152*, 349–369. [CrossRef]
3. Kim, J.; Kim, S.; Hwang, K.T. Determination and photochemical conversion of protofagopyrins and fagopyrins in buckwheat plants. *J. Food Compos. Anal.* **2021**, *100*, 103894. [CrossRef]
4. Lüthi, M.; Gyenge, E.B.; Engstrüm, M.; Bredell, M.; Grätz, K.; Walt, H.; Gmür, R.; Maake, C. Hypericin-and mTHPC-mediated photodynamic therapy for the treatment of cariogenic bacteria. *Med. Laser Appl.* **2009**, *24*, 227–236. [CrossRef]
5. Bhatta, A.K.; Keyal, U.; Wang, X.L. Photodynamic therapy for onychomycosis: A systematic review. *Photodiagn. Photodyn. Ther.* **2016**, *15*, 228–235. [CrossRef] [PubMed]
6. Jendželovská, Z.; Jendželovský, R.; Kuchárová, B.; Fedoročko, P. Hypericin in the light and in the dark: Two sides of the same coin. *Front. Plant Sci.* **2016**, *7*, 560. [CrossRef]
7. Stojilkovski, K.; Glavač, N.K.; Kreft, S.; Kreft, I. Fagopyrin and flavonoid contents in common, Tartary, and cymosum buckwheat. *J. Food Compos. Anal.* **2013**, *32*, 126–130. [CrossRef]
8. Benković, E.T.; Žigon, D.; Friedrich, M.; Plavec, J.; Kreft, S. Isolation, analysis and structures of phototoxic fagopyrins from buckwheat. *Food Chem.* **2014**, *143*, 432–439. [CrossRef] [PubMed]
9. Kim, J.; Hwang, K.T. Fagopyrins in different parts of common buckwheat (*Fagopyrum esculentum*) and Tartary buckwheat (*F. tataricum*) during growth. *J. Food Compos. Anal.* **2020**, *86*, 103354. [CrossRef]
10. Kwiatkowski, S.; Knap, B.; Przystupski, D.; Saczko, J.; Kędzierska, E.; Knap-Czop, K.; Kotlińska, J.; Michel, O.; Kotowski, K.; Kulbacka, J. Photodynamic therapy–mechanisms, photosensitizers and combinations. *Biomed. Pharmacother.* **2018**, *106*, 1098–1107. [CrossRef]
11. Oniszczuk, A.; Wojtunik-Kulesza, K.A.; Oniszczuk, T.; Kasprzak, K. The potential of photodynamic therapy (PDT)—Experimental investigations and clinical use. *Biomed. Pharmacother.* **2016**, *83*, 912–929. [CrossRef]
12. Plaetzer, K.; Kiesslich, T.; Verwanger, T.; Krammer, B. The modes of cell death induced by PDT: An overview. *Med. Laser Appl.* **2003**, *18*, 7–19. [CrossRef]
13. Bacellar, I.O.; Tsubone, T.M.; Pavani, C.; Baptista, M.S. Photodynamic efficiency: From molecular photochemistry to cell death. *Int. J. Mol. Sci.* **2015**, *16*, 20523–20559. [CrossRef] [PubMed]
14. Beytollahi, L.; Pourhajibagher, M.; Chiniforush, N.; Ghorbanzadeh, R.; Raoofian, R.; Pourakbari, B.; Bahador, A. The efficacy of photodynamic and photothermal therapy on biofilm formation of *Streptococcus mutans*: An in vitro study. *Photodiagn. Photodyn. Ther.* **2017**, *17*, 56–60. [CrossRef]
15. Ghorbani, J.; Rahban, D.; Aghamiri, S.; Teymouri, A.; Bahador, A. Photosensitizers in antibacterial photodynamic therapy: An overview. *Laser Ther.* **2018**, *27*, 293–302. [CrossRef]
16. Garcia, M.T.; Pereira, A.H.C.; Figueiredo-Godoi, L.M.A.; Jorge, A.O.C.; Strixino, J.F.; Junqueira, J.C. Photodynamic therapy mediated by chlorin-type photosensitizers against *Streptococcus mutans* biofilms. *Photodiagn. Photodyn. Ther.* **2018**, *24*, 256–261. [CrossRef] [PubMed]
17. Paschoal, M.A.; Tonon, C.C.; Spolidório, D.M.; Bagnato, V.S.; Giusti, J.S.; Santos-Pinto, L. Photodynamic potential of curcumin and blue LED against *Streptococcus mutans* in a planktonic culture. *Photodiagn. Photodyn. Ther.* **2013**, *10*, 313–319. [CrossRef]
18. Lee, H.J.; Kang, S.M.; Jeong, S.H.; Chung, K.H.; Kim, B.I. Antibacterial photodynamic therapy with curcumin and *Curcuma xanthorrhiza* extract against *Streptococcus mutans*. *Photodiagn. Photodyn. Ther.* **2017**, *20*, 116–119. [CrossRef] [PubMed]
19. Ribeiro, M.S.; de Melo, L.S.; Farooq, S.; Baptista, A.; Kato, I.T.; Núñez, S.C.; de Araujo, R.E. Photodynamic inactivation assisted by localized surface plasmon resonance of silver nanoparticles: In vitro evaluation on *Escherichia coli* and *Streptococcus mutans*. *Photodiagn. Photodyn. Ther.* **2018**, *22*, 191–196. [CrossRef]
20. Misba, L.; Kulshrestha, S.; Khan, A.U. Antibiofilm action of a toluidine blue O-silver nanoparticle conjugate on *Streptococcus mutans*: A mechanism of type I photodynamic therapy. *Biofouling* **2016**, *32*, 313–328. [CrossRef] [PubMed]

21. Misba, L.; Zaidi, S.; Khan, A.U. Efficacy of photodynamic therapy against *Streptococcus mutans* biofilm: Role of singlet oxygen. *J. Photochem. Photobiol. B* **2018**, *183*, 16–21. [CrossRef]
22. Fumes, A.C.; da Silva Telles, P.D.; Corona, S.A.M.; Borsatto, M.C. Effect of aPDT on *Streptococcus mutans* and *Candida albicans* present in the dental biofilm: Systematic review. *Photodiagn. Photodyn. Ther.* **2018**, *21*, 363–366. [CrossRef]
23. Wang, S.; Huang, Q.; Liu, X.; Li, Z.; Yang, H.; Lu, Z. Rapid antibiofilm effect of Ag/ZnO nanocomposites assisted by dental LED curing light against facultative anaerobic oral pathogen *Streptococcus mutans*. *ACS Biomater. Sci. Eng.* **2019**, *5*, 2030–2040. [CrossRef]
24. Zanin, I.C.J.; Goncalves, R.B.; Junior, A.B.; Hope, C.K.; Pratten, J. Susceptibility of *Streptococcus mutans* biofilms to photodynamic therapy: An in vitro study. *J. Antimicrob. Chemother.* **2005**, *56*, 324–330. [CrossRef]
25. Metcalf, D.; Robinson, C.; Devine, D.; Wood, S. Enhancement of erythrosine-mediated photodynamic therapy of *Streptococcus mutans* biofilms by light fractionation. *J. Antimicrob. Chemother.* **2006**, *58*, 190–192. [CrossRef]
26. Wood, S.; Metcalf, D.; Devine, D.; Robinson, C. Erythrosine is a potential photosensitizer for the photodynamic therapy of oral plaque biofilms. *J. Antimicrob. Chemother.* **2006**, *57*, 680–684. [CrossRef] [PubMed]
27. Bilski, P.; Belanger, A.G.; Chignell, C.F. Photosensitized oxidation of 2′, 7′-dichlorofluorescin: Singlet oxygen does not contribute to the formation of fluorescent oxidation product 2′, 7′-dichlorofluorescein. *Free Radic. Biol. Med.* **2002**, *33*, 938–946. [CrossRef]
28. Entradas, T.; Waldron, S.; Volk, M. The detection sensitivity of commonly used singlet oxygen probes in aqueous environments. *J. Photochem. Photobiol. B* **2020**, *204*, 111787. [CrossRef] [PubMed]
29. Klinger-Strobel, M.; Ernst, J.; Lautenschläger, C.; Pletz, M.W.; Fischer, D.; Makarewicz, O. A blue fluorescent labeling technique utilizing micro-and nanoparticles for tracking in LIVE/DEAD® stained pathogenic biofilms of *Staphylococcus aureus* and *Burkholderia cepacia*. *Int. J. Nanomed.* **2016**, *11*, 575.
30. Lee, Y.H.; Park, H.W.; Lee, J.H.; Seo, H.W.; Lee, S.Y. The photodynamic therapy on *Streptococcus mutans* biofilms using erythrosine and dental halogen curing unit. *Int. J. Oral Sci.* **2012**, *4*, 196–201. [CrossRef]
31. Li, Y.; Liu, Z.; Zhang, Y.; Su, Q.P.; Xue, B.; Shao, S.; Zhu, Y.; Xu, X.; Wei, S.; Sun, Y. Live-cell and super-resolution imaging reveal that the distribution of wall-associated protein A is correlated with the cell chain integrity of *Streptococcus mutans*. *Mol. Oral Microbiol.* **2015**, *30*, 376–383. [CrossRef]
32. Takasaki, A.A.; Aoki, A.; Mizutani, K.; Schwarz, F.; Sculean, A.; Wang, C.Y.; Koshy, G.; Romanos, G.; Ishikawa, I.; Izumi, Y. Application of antimicrobial photodynamic therapy in periodontal and peri-implant diseases. *Periodontology 2000* **2009**, *51*, 109–140. [CrossRef]
33. Gursoy, H.; Ozcakir-Tomruk, C.; Tanalp, J.; Yılmaz, S. Photodynamic therapy in dentistry: A literature review. *Clin. Oral Investig.* **2013**, *17*, 1113–1125. [CrossRef]
34. Esmatabadi, M.D.; Bozorgmehr, A.; Hajjari, S.N.; Sombolestani, A.S.; Malekshahi, Z.V.; Sadeghizadeh, M. Review of new insights into antimicrobial agents. *Cell. Mol. Biol. (Noisy-le-grand)* **2017**, *63*, 40–48. [CrossRef] [PubMed]
35. Zhang, Y.; Dai, M.; Yuan, Z. Methods for the detection of reactive oxygen species. *Anal. Methods* **2018**, *10*, 4625–4638. [CrossRef]
36. Li, M.; Mai, B.; Wang, A.; Gao, Y.; Wang, X.; Liu, X.; Song, S.; Liu, Q.; Wei, S.; Wang, P. Photodynamic antimicrobial chemotherapy with cationic phthalocyanines against *Escherichia coli* planktonic and biofilm cultures. *RSC Adv.* **2017**, *7*, 40734–40744. [CrossRef]
37. Qian, H.S.; Guo, H.C.; Ho, P.C.L.; Mahendran, R.; Zhang, Y. Mesoporous-silica-coated up-conversion fluorescent nanoparticles for photodynamic therapy. *Small* **2009**, *5*, 2285–2290. [CrossRef]

International Journal of *Molecular Sciences*

MDPI

Article

FLTX2: A Novel Tamoxifen Derivative Endowed with Antiestrogenic, Fluorescent, and Photosensitizer Properties

Mario Díaz [1,2,*], Fernando Lobo [3], Dácil Hernández [4], Ángel Amesty [3,5], Catalina Valdés-Baizabal [3,6], Ana Canerina-Amaro [6], Fátima Mesa-Herrera [1], Kevin Soler [7], Alicia Boto [2,4], Raquel Marín [2,6], Ana Estévez-Braun [5,8] and Fernando Lahoz [2,7]

1 Departamento Biología Animal, Edafología y Geología, Universidad de La Laguna, 38200 Tenerife, Spain; fatimamesaherrera@gmail.com
2 Unidad Asociada ULL-CSIC "Fisiología y Biofísica de la Membrana Celular en Enfermedades Neurodegenerativas y Tumorales", 38200 Tenerife, Spain; alicia@ipna.csic.es (A.B.); rmarin@ull.edu.es (R.M.); flahoz@ull.es (F.L.)
3 Programa Agustín de Betancourt, Universidad de la Laguna, 38200 Tenerife, Spain; flobopal@ull.edu.es (F.L.); angelamesty@yahoo.es (Á.A.); cvaldesb@ull.edu.es (C.V.-B.)
4 Instituto de Productos Naturales y Agrobiología del CSIC, Avda. Astrofísico F. Sánchez, 38206 Tenerife, Spain; dacil@ipna.csic.es
5 Instituto Universitario de Bioorgánica "Antonio González", Universidad de La Laguna, 38200 Tenerife, Spain; aestebra@ull.edu.es
6 Departamento Ciencias Médicas Básicas, Universidad de La Laguna, 38200 Tenerife, Spain; anacanerinaamaro@gmail.com
7 Departamento Física, IUdEA, Universidad de La Laguna, 38200 Tenerife, Spain; kscarracedo@gmail.com
8 Departamento Química Orgánica, Universidad de La Laguna, 38200 Tenerife, Spain
* Correspondence: madiaz@ull.es

Citation: Díaz, M.; Lobo, F.; Hernández, D.; Amesty, Á.; Valdés-Baizabal, C.; Canerina-Amaro, A.; Mesa-Herrera, F.; Soler, K.; Boto, A.; Marín, R.; et al. *FLTX2*: A Novel Tamoxifen Derivative Endowed with Antiestrogenic, Fluorescent, and Photosensitizer Properties. *Int. J. Mol. Sci.* **2021**, *22*, 5339. https://doi.org/10.3390/ijms22105339

Academic Editors: Antonino Mazzaglia, Angela Scala and Enrico Caruso

Received: 26 March 2021
Accepted: 8 May 2021
Published: 19 May 2021

Abstract: Tamoxifen is the most widely used selective modulator of estrogen receptors (SERM) and the first strategy as coadjuvant therapy for the treatment of estrogen-receptor (ER) positive breast cancer worldwide. In spite of such success, tamoxifen is not devoid of undesirable effects, the most life-threatening reported so far affecting uterine tissues. Indeed, tamoxifen treatment is discouraged in women under risk of uterine cancers. Recent molecular design efforts have endeavoured the development of tamoxifen derivatives with antiestrogen properties but lacking agonistic uterine tropism. One of this is FLTX2, formed by the covalent binding of tamoxifen as ER binding core, 7-nitrobenzofurazan (NBD) as the florescent dye, and Rose Bengal (RB) as source for reactive oxygen species. Our analyses demonstrate (1) FLTX2 is endowed with similar antiestrogen potency as tamoxifen and its predecessor FLTX1, (2) shows a strong absorption in the blue spectral range, associated to the NBD moiety, which efficiently transfers the excitation energy to RB through intramolecular FRET mechanism, (3) generates superoxide anions in a concentration- and irradiation time-dependent process, and (4) Induces concentration- and time-dependent MCF7 apoptotic cell death. These properties make FLTX2 a very promising candidate to lead a novel generation of SERMs with the endogenous capacity to promote breast tumour cell death in situ by photosensitization.

Keywords: tamoxifen; estrogen receptors; SERM; fluorescence; FRET; reactive oxygen species; superoxide anions; photosensitization; FLTX1; breast cancer; laser dye; molecular dynamics

1. Introduction

Tamoxifen (TX) is a selective modulator of estrogen receptors (SERMs) widely used as coadjuvant therapy in the treatment of breast cancer. Tamoxifen has been prescribed for decades to millions of women as the first therapeutic strategy for prevention or treatment hormone-dependent breast cancer [1–4]. TX is a triphenylethylene molecule that competes the binding of 17β-estradiol to estrogen receptors (ER) and prevents the transcriptional activation of estrogen-dependent genes, which control the proliferation of mammary gland cells. However, tamoxifen is known to significantly enhance the risk of developing endometrial lesions, including hyperplasia, polyps, carcinomas, sarcoma, as well as womb cancer and thromboembolism [5–10]. Notably, tamoxifen-associated endometrial cancer often has a poor clinical outcome, and serious concerns exist on its use for cancer prevention or long-term palliative treatment [9,11]. The current view of these undesired effects is that in most cases they are due to the interaction of TX with either canonical and/or non-canonical molecular targets [11], whose proper identification has been hampered because the lack of appropriate pharmacological and experimental tools. The existence of these adverse secondary effects has boosted many laboratories to search for either modifications on the tamoxifen molecule or novel SERMs structurally unrelated to tamoxifen, which could circumvent the undesirable effects during long-term therapies [1,12–15]. In recent years, we have developed a novel fluorescent tamoxifen derivative, FLTX1, which not only allows tracking canonical (i.e., ER) and non-canonical intracellular targets, but also retains most pharmacological properties of tamoxifen (specific ER binding, antagonism of ER-mediated transcriptional activation and cell proliferation), while lacks the trophic worrisome effects of tamoxifen in the uterus [16]. Of note, beyond its pharmacological properties, FLTX1 is endowed with emergent opto-chemical properties of which tamoxifen lacks [17–20]. In the present study we have designed and developed a new tamoxifen derivative, named FLTX2, formed by the covalent linkage of FLTX1 to a well-known photosentizer, Rose Bengal (RB), which takes advantages of both pharmacological and optochemical properties of FLTX1, and the incorporation of a novel functionality to generate reactive oxygen species (ROS) with high efficiency. Thus, positive gain media based on FLTX1 optical properties were prepared and laser emission was demonstrated under optical pulsed excitation, which is transferred to RB through Förster resonance energy transfer (FRET). In this article we have used a multidisciplinary approach to assess the main properties of FLTX2 with special focus on: (1) ER antagonistic properties, (2) Docking and Molecular Dynamics of the FLTX2-ER interaction (3) Optochemical properties, (4) Irradiation-induced ROS generation, (5) Intracellular binding, and (6) Ability to photosensitize and to induce irradiation-mediated cell death. The outcomes pinpoint to a highly promising pharmacophore with potential applications as specific site-directed photosensitizer in ER-dependent breast cancer.

2. Results and Discussion

2.1. Synthesis of FLTX2 (FLTX1-RB)

In the present study we have designed and synthetized a novel fluorescent tamoxifen derivative, FLTX2, following a synthetic strategy (Figure 1A) which avoided the need of high-cost commercial tamoxifen. The tamoxifen core of dimethyl acetal precursor **1** was synthesized in a few steps from inexpensive materials, according to a reported procedure [21]. As it happens with commercial tamoxifen, a mixture of the E and Z isomers was obtained. Once compound **1** was available, the aldehyde was deprotected in acid media and treated with ethanolamine under reductive conditions to introduce the FLTX1-Rose Bengal (FLTX1-RB) linker. This linker was attached first to 7-nitrobenzofurazan (NBD) and then to RB. Thus, the FLTX1-linker intermediate was treated with 4-chloro-7-nitrobenzofurazan in the presence of a base, providing the desired compound **2**. The **1**→**2** conversion took place in a satisfactory overall yield (29% for the four steps), since the procedure avoids lengthy protection-deprotection steps and is operationally simple. However, we expect to further optimize the yields in a future. In the final synthetic step (conversion **2**→**3**), substrate **2** underwent an esterification reaction with Rose Bengal under mild conditions, affording FLTX2 (**3**). A 3D representation of FLTX2 in a minimum-energy conformation is shown in Figure 1B, where it can be seen that the NBD and Rose Bengal moieties align closely allowing FRET exchange. With respect to the NBD-BR approach, it should be pointed out that the linker length (a two-carbon chain) was chosen as optimal according to our former study [22] aimed at assessing FRET efficiencies between the NBD-moiety of FLTX1 (donor) and RB (acceptor) dipole pairs (see below).

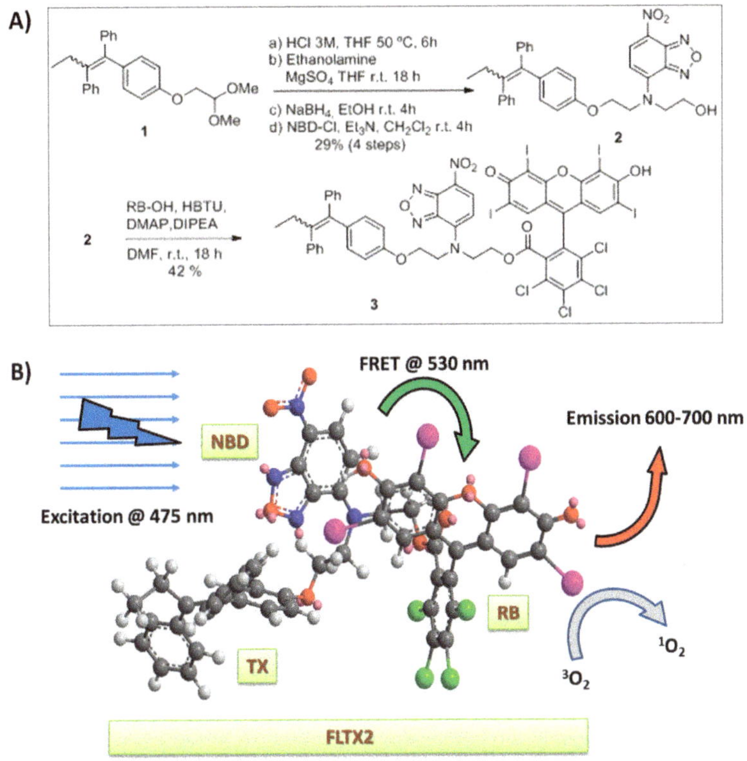

Figure 1. (**A**) Synthesis of FLTX2 (compound 3). (**B**) Minimized 3D representation of FLTX2 (3) highlighting the sequence: excitation wavelength, intramolecular FRET, red emission, and singlet oxygen generation. For details see optical measurements DIPEA: Diisopropylethylamine; DMAP: Dimethylaminopyridine; DMF: Dimethylformamide; HBTU: O-(Benzotriazol-1-yl)- *N,N,N′,N′*-tetramethyluronium hexafluorophosphate; NBD: 7-nitrobenzofurazan; THF: tetrahydrofuran.

2.2. *Affinity Studies of FLTX2 on ER. Comparison with FLTX1*

Competition experiments were performed using rat uterus cytosol. This extract rich in ERα was saturated with 5 nM of labelled E2 in the presence of increasing concentrations of unlabelled tamoxifen or FLTX2 (Figure 2).

As expected, TX was able to competitively displace the [^{3}H]E2 from rat uterine ER in a dose-dependent manner. Similarly, FLTX2 competed off the radiolabelled estradiol binding to ER. The estimated EC_{50} values for TX and FLTX2 were 173.3 $\pm$ 6.14 nM ($p < 0.005$) and 114.4 $\pm$ 3.07 nM ($p < 0.05$), respectively. Hill coefficients (n_H) were close to 1 in both cases. Thus, assuming an RBA (Relative Binding Affinity) value of 100 for TX, the RBA value for FLTX2 was 151.5% indicating a slightly increased affinity of FLTX2 for ER than TX. Thus, it may be concluded that despite significant changes in the molecular complexity and size of FLTX2 compared to TX, the novel derivative preserves the ability to compete E2 for ERα binding, behaving as antiestrogen and likely with a slightly higher affinity than TX and FLTX1 [16]. These properties of fluorescent derivatives FLTX1, and especially FLTX2, were remarkable since other studies have shown that, in general, changes in the side chain of TX lead to compounds with decreased affinity for ER [23–25]. Nonetheless, from present results we cannot ascertain if FLTX2 behaves as a pure antiestrogen and whether it lacks the agonist functionality demonstrated for TX, in particular, in uterine tissues. However, in our previous study on FLTX1 pharmacological properties, we demonstrated that this pharmacophore was completely devoid of uterotrophic and proliferative effects [16]. In addition, FLTX1 was unable to induce ER-mediated transcriptional activity in transfected MCF7 and T47D-KBluc cells expressing a genetic construct containing a luciferase reporter gene under the control of canonical estrogen response elements [16]. These findings were interpreted in terms of a physical hindrance of FLTX1-bound ER to recruit coactivators required for transcriptional activation. Based on the chemical similarities between bulky side chains of FLTX1 and FLTX2, and especially from the results of docking and MD detailed below, it is likely that FLTX2 might also lack agonistic effects exhibited by TX.

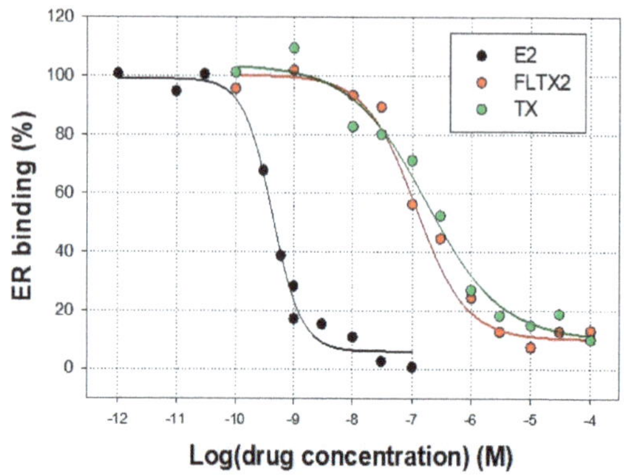

Figure 2. Effects of FLTX2 and TX on [^{3}H]-estradiol competitive ER binding assay. Uterine cytosolic extracts were saturated with 5 nM of labelled estradiol in the presence of increasing concentrations of unlabelled estradiol (E2), tamoxifen (TX), or FLTX2. Data are presented as mean ± SEM of four different assays.

2.3. *Docking of FLTX2 on ERα. Comparison with FLTX1*

In order to explain the increased affinity of FLTX2 for ERα compared to FLTX1 and TX, we carried out a molecular docking study on the reported crystal structure of human ERα ligand binding domain (LBD) in complex with 4-hydroxytamoxifen (PDB code 3ERT), which is an active metabolite of TX and a well-known as selective estrogen receptor modulator (SERM). The ligand-binding domain (LBD) is located in the middle of the carboxy-terminal region of the receptor, the final portion of which is critical and responsible for ligand binding, receptor dimerization, and nuclear translocation among other roles [26–28].

The LBD consists of twelve α-helices (H1 to H12) and a beta sheet/hairpin. The amino acid residues that line the ligand-binding cavity or interact with bound ligand span from helix 3 (H3) to helix 12 (H12). When the receptor binds to the ligand, a change in its three-dimensional structure is produced, and the LBD forms a bag-shaped structure, hydrophobic in nature, that lodges the ligand [29].

The H12 plays an important role as a molecular switch by adopting different conformations that allow ligand-dependent receptor activation [30]. When bound to an agonist, the LBD adopts an active conformation where H12 rests across H3 and H11, forming a groove to accommodate co-regulator binding and facilitate downstream activation process [31]. When bound to an antagonist, H12 is displaced from this position resulting in the distortion of the co-regulator binding groove and the inhibition of receptor activation [26,28,32,33].

Hence, in order to explore the binding mode of FLTX2 derivative, we have docked FLTX2, FLTX1, and TX into the hydrophobic binding pocket of LBD to understand their possible binding modes and key active site interactions (Figure 3).

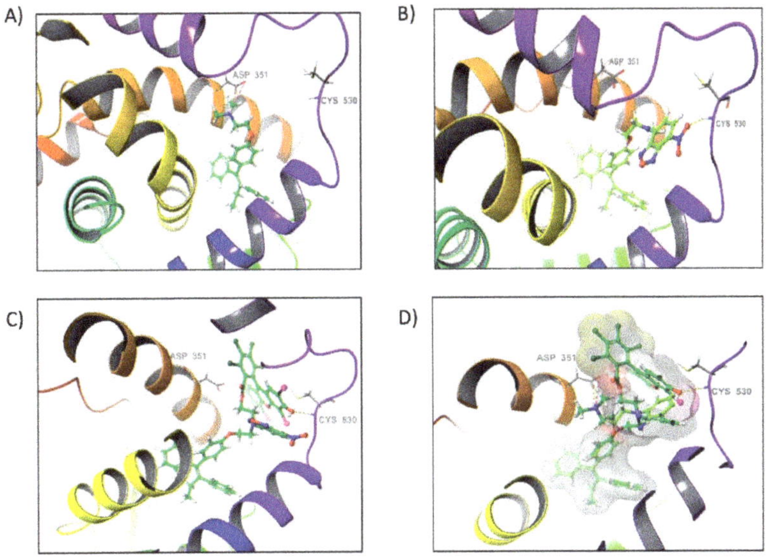

Figure 3. Best Docking poses for TX (**A**), FLTX1 (**B**), and FLTX2 (**C**) in the human ERα ligand binding domains (LBD). (**D**) Overlapped docking poses of TX and FLTX2 within the human ERα LBD.

The analyses of the docking results showed that the moieties of both derivatives (FLTX1 and FLTX2) share a similar pose at the ligand binding site and that it is also located towards the end of the pocket where the ligand is usually found in the crystal structure (Figure 3B,C). These results strongly suggested that these compounds share a common binding mode into the hydrophobic binding pocket of ERα (Figure 3D).

According to the predicted binding modes, FLTX1 as well as FLTX2 display the same type of hydrophobic interactions as tamoxifen does, therefore they probably play a dominant role in protein–ligand interaction. In the favored docking conformations, there are multiple potential hydrophobic interactions involving residues Leu391, Leu349, Leu346, Leu428, Leu387, Leu384, Leu525, Ile424, Glu353, Arg394, Met421, Met388, Met343, Gly420, Ala350, Thr347, and His524, whose side chains are in close proximity to the tamoxifen moiety (see Figure 3A–C).

Additionally, our results also show that the introduction of the NBD moiety to tamoxifen to form the derivative FLTX1, as well as the preparation of the derivative FLTX2 by modifying the derivative FLTX1 to form the derivative Rose Bengal-conjugated FLTX1, gives rise to a new hydrogen bond interaction with the residue Cys530 in H12. However, in the case of FLTX1 this interaction is produced by the nitro group (-NO_2) of NBD while in the case of the derivative FLTX2 the interaction is produced by the hydroxyl group (-OH) present in Rose Bengal (Figure 3D).

The calculations revealed that the affinity for the receptor binding site. It is as high for both derivatives (XPGlide score of $\times$11.55 kcal mol^{-1} for FLTX1 and $-$11.34 kcal mol^{-1} for FLTX2) as for tamoxifen (XPGlide score $-$11.72 kcal mol^{-1}), which was used as internal reference. These findings suggest that novel interactions observed for FLTX1 and FLTX2 with the Cys530 residue in H12 are as effective as that established by tamoxifen between the protonated nitrogen and Asp351 which provides a stabilizing effect on the active antiestrogenic conformation of ER-LBD [34].

In addition, due to the presence of both NBD and RB moieties in FLTX2, additional hydrophobic interactions arise with residues Leu539, Leu536, Val534, Val533, Asp351, Trp383, and Tyr526, which likely stabilize H12 in its typical antagonist conformation, thereby hampering coactivators to interact with the receptor, and more efficiently binding of FLTX2 to the RBS [28]. This issue is further explored in the next section.

2.4. Molecular Dynamic (MD) Simulation

To further evaluate the reliability of the docking study result, we combined it with more accurate molecular dynamics (MD) simulation techniques in order to predict more reliable structures as well as to confirm the stability of the system and study the conformational change in the ligand-receptor complex throughout the course of the MD simulation. The simulations provide every minute of detail on every atom movement with respect to time. This will help in answering the questions arising about the binding mode, stability, deviation, and fluctuation pattern of the protein.

The MD simulation was performed on the best Docking pose of the FLTX2 derivative for 50 ns in an explicit aqueous solution environment with periodic boundary conditions, the OPL-2005 force field and the TIP3P solvent model employing the Desmond simulation package seamlessly integrated into Maestro software. The simulation was also carried out to confirm the orientation of the FLTX2 derivative into the binding site.

The analysis of the trajectory obtained during the simulation as well as the protein–ligand contacts analysis revealed that the compound does not leave the binding site of the protein and remained in the similar orientation during the entire simulation. However, in the graphical snapshot of the production phase the disappearance of the hydrogen bond interaction formed between the hydroxyl group present in Rose Bengal with the residue Cys530 could be observed due to the flexibility of movement of H12, and it could also be detected by the appearance of two new hydrogen bonds mediated by two water molecules formed between the carbonyl group of RB and the residues Gly344 and Thr347. In addition, a previously undetected π–π stacking interaction between TX with the residue Phe404 could also be observed. This is indicative that FLTX2 is further stabilized by other hydrogen bonds, which could explain a significant increase in the FLTX2 affinity of this derivative (Figure 4A).

Additionally, the interaction observed during the simulation between the TX core of FLTX2 with Phe404 (Figure 4B) is quite significant since it is observed around more than 50% of the simulation time in the selected trajectory (Supplementary Figure S7).

In addition, it could also be observed that the FLTX2 interacted with other essential amino acids such as Ala350, Leu525, and Ser527, which are part of the hydrophobic pocket by means of hydrophobic interactions (Figure 4B). These interactions of the protein with the ligand may be quantified as the 'interaction fraction' throughout the whole molecular dynamic simulation (Figure 4C and Supplementary Figure S7). The stacked bar charts are normalized over the course of the trajectory (i.e., a value of 0.7 indicates that along 70% of the simulation time the specific interaction is maintained). This allowed us to know the role of each of the particular bonds, with the amino acid residues responsible for the stabilization of the complex.

This model and subsequent dynamical observation has notably expanded our knowledge on the potential antiestrogenic activity of FLTX2 on the ERα. Thus, FLTX2 retains the ability of TX to get into the ERα ligand binding groove and to establish similar interactions with the LBD though and with similar affinity likely through π–π stacking interaction with Phe404. However, the large moiety formed by NBD and RB expand outside the ligand groove and stabilizes H12 in the antiestrogen conformation, by the hydroxyl group present in Rose Bengal with the residue Cys530, and two hydrogen bonds with the carbonyl group of RB and the residues Gly344 and Thr347. In this sense, FLTX2 is expected to exhibit a ligand kinetics that tends to the formation of the ERα-FLTX2 complex, thereby behaving as an antagonist as effective as TX and FLTX1.

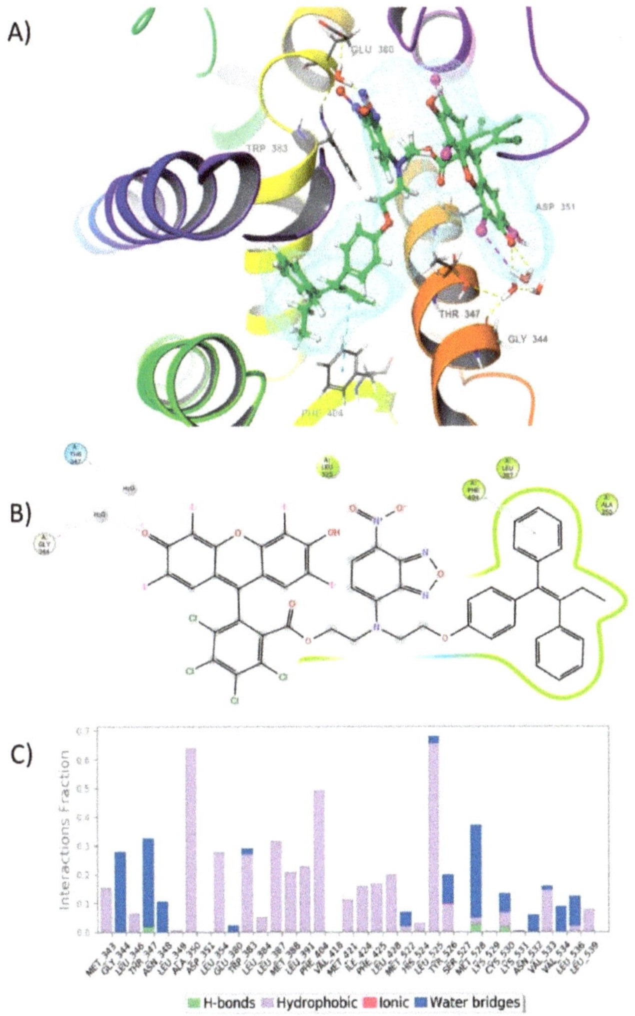

Figure 4. Molecular Dynamics study for FLTX2 on the human ERα ligand binding domain. (**A**) Interactions of FLTX2 with key amino acid residues at the hydrophobic binding pocket of human ERα LBD. (**B**) 2D Ligand-protein interaction diagrams and interaction strength quantified by the frequency of occurrences in the trajectory when a minimum percentage of 50% is achieved. Colors indicate type of residue: green are lipophilic residues; blue are polar residues; purple are basic residues. Ligand atoms that are exposed to the solvent are marked with grey spheres. (**C**) Stacked bar charts of human ERα LBD interaction with FLTX2.

In our previous study on the pharmacological properties of FLTX1, we demonstrated the total absence of estrogen-like effects both in vitro and in vivo as one of the most outstanding properties of this pharmacophore [16]. The antiestrogenic ability of FLTX1 and FLTX2 is mechanistically similar to that of TX, i.e., by binding the LBD domain through their triphenylethylene cores and the lateral side moieties protruding out the LDB pocket and displacing helix 12 from adopting its agonist conformation [12,35]. Further, the compact volume of NDB-RB moiety and its spatial conformation respect to the triphenyethylene core as indicated by the minimal-energy conformations and molecular dynamics simulation

(Figures 3 and 4) strongly suggest a conformational hindrance for helixes 3 and 11 to reallocate helix 12 [12,35,36]. It is known that for agonist ER ligands, helix 12 is stabilized in a conformation that allows it to form one side of the coactivator-binding site [12,37–39]. The geometry of FLTX2-bound ER structure would hamper the sequential recruitment of tissue-specific coregulators, including steroid receptor co-activator-1 (SRC-1), amplified in breast cancer-1 (AIB1) and CREB-binding protein (CBP) which are required in endometrial cells for estrogenic agonism and carcinogenesis [11,28,37,40,41]. Future studies will be needed to confirm the pure antiestrogenic nature of FLTX2.

2.5. Fluorescence Properties and FRET Mechanism

The visible optical absorption of FLTX1, RB, and FLTX2 solutions at a concentration of 10 μM in DMSO are given in Figure 5A.

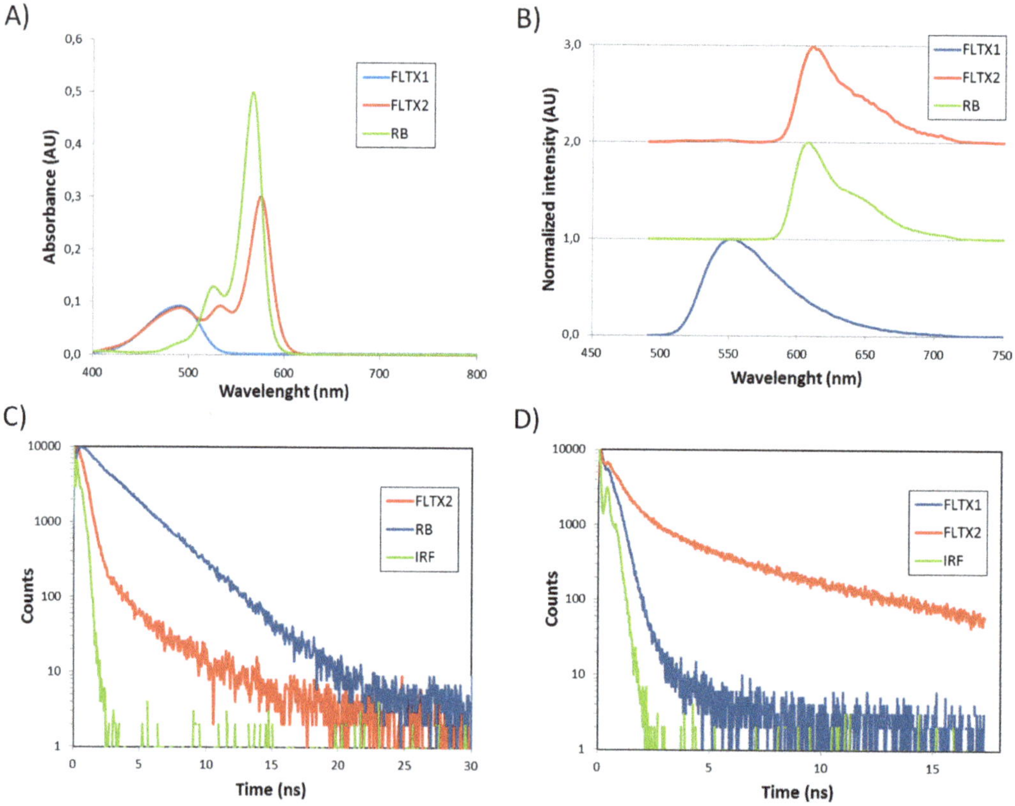

Figure 5. Optical properties of FLTX2. (**A**) Absorption spectra of FLTX1, RB, and FLTX2. (**B**) Normalized emission spectra of FLTX1, RB, and FLTX2. (**C**) Decay of the fluorescence of RB and of FLTX2 with detection tuned at 608 nm. (**D**) Decay of the fluorescence of FLTX1 and of FLTX2 with detection tuned at 550 nm.

The characteristic broad band centered at about 488 nm is observed for FLTX1 [18]. The absorption spectrum of RB shows an intense band at about 565 nm and a shoulder at 525 nm. The FLTX2 complex shows the absorption features of their constituent moieties. The absorption spectrum has a broad band at 488 nm, associated with the FLTX1 moiety, and also the absorption bands of RB but slightly red shifted (main band at 574 nm and shoulder at 532 nm). The red shift observed in the bands related to the RB moiety indicates that the covalent bonding between FLTX1 and RB changes the environment of the RB electronic density and has an influence on its energy levels. Similar effects have been reported for variations of the RB molecular environment [42].

The normalized fluorescence emission spectra of the FLTX1, RB, and FLTX2 solutions under excitation at 475 nm are given in Figure 5B. An intense and green emission band with a maximum about 550 nm is observed for the FLTX1. The RB emission is characterized by a broad red band with its maximum at about 608 nm. When FLTX2 is excited at 475 nm, the emission of the FLTX1 moiety is almost completely quenched and essentially the emission is only due to the RB moiety. However, upon 475 nm excitation the FLTX1 moiety of the FLTX2 complex is excited. Therefore, the fluorescence results confirm that the FLTX1 moiety efficiently transfers its energy to the RB part of the complex, which may then relax radiatively, giving place to the observed emission spectrum of FLTX2 through a Förster Resonance Energy transfer (FRET) mechanism.

2.6. Absorption and Fluorescence Analysis

A detailed comparison of the absorption spectra of RB and of FLTX2 reveals a change in the ratio of the main RB absorption band (longer wavelength) to its shoulder (shorter wavelength). Interestingly, the relative intensity of the main absorption band of RB to its shoulder has been considered as a measure of the aggregation of RB [43]. Shorter ratio values indicate more aggregation. In the case of RB, the ratio value is 3.85, while it decreases to 3.19 for the FLTX2 complex. This result could point to the formation of more dimmers in the FLTX2 solutions than in pure RB preparations.

In order to better understand the FRET mechanism, which underlies the fluorescent properties of FLTX2, the decays of the luminescence have been recorded. The temporal evolution of the RB emission band decay is given in Figure 5C. The decay can be fitted to an exponential decay curve:

$$I(t) = A\,e^{-t/\tau} \tag{1}$$

where τ represents the lifetime of the emission band, and A is the pre-exponential factor. The fitting was performed taking into account the reconvolution to the IRF. The best fitting was found for a lifetime constant of 2.6 ns. The decay of the fluorescence of FLTX2 when the detection is tuned at 608 nm, which corresponds to the maximum of the emission band of the RB moiety, shows a two exponential decay behavior and can be fitted to a curve:

$$I(t) = A_1 e^{-t/\tau_1} + A_2 e^{-t/\tau_2} \tag{2}$$

where, τ_1 and τ_2 are the decay constants of the fast and slow components, respectively. A_1 and A_2 represent the pre-exponential factors, which are related to the weight of the fast and slow components, respectively. An average life-time can be defined from the fitting parameters as [44]:

$$\tau_{av} = \frac{A_1\tau_1^2 + A_2\tau_2^2}{A_1\tau_1 + A_2\tau_2} \tag{3}$$

The decay constant parameters obtained from the fitting are 0.31 and 2.14 ns, for the fast and slow components, respectively. The main contribution to the decay curve corresponds to the fast component, and the average life-time is 0.51 ns, which is notably shorter than that found for the pristine RB molecules in the same solvent. There are several reasons to explain the shorter average life-time found for the RB moiety of FLTX2 as compared to the pristine RB molecules. First, non-radiative vibrational relaxation of the FLTX2 molecule might be higher than in RB molecules, as the chemical environment of the RB moiety in FLTX2 has changed. Second, the presence of dimers in FLTX2 could be higher than in pristine RB solutions, as already indicated by the absorption measurements. Since shorter decays are expected in dimers as compared to monomers, these aggregates could be partially responsible of the shortening of the average life-time. Finally, back-transfer from the RB moiety to the FLTX1 compound might take place in the FLTX2 complex. This would represent a further relaxation mechanism of the RB unit and would contribute to the shortening of its life-time. The decay of the fluorescence of FLTX1 was also measured at 550 nm upon 475 nm excitation. The decay could be fitted to a double-exponential decay, and the following fitting parameters were obtained, τ_1 = 0.15 ns and τ_2 = 0.54 ns. The main contribution to the decay of the fluorescence corresponds to the fast decay constant contribution. The average life-time obtained was 0.2 ns. Although the efficiency of the FRET mechanism between the FLTX1 and the RB moieties in the FLTX2 complex is high and the emission of FLTX1 moiety is almost completely quenched, a small fluorescence signal can still be detected at 550 nm, which corresponds to the FLTX1 moiety of FLTX2 (Figure 5D). Moreover, the decay of the fluorescence of this signal could be measured when FLTX2 was excited at 475 nm and the detection was tuned at 550 nm. One would expect a shortening of the life-time, as compared to pristine FLTX1, because of the new and efficient FRET relaxation mechanism to RB moiety. Surprisingly, the decay of the fluorescence of FLTX2 at 550 nm is slower than for FLTX1 [22]. It could be fitted to a double-exponential decay with τ_1 = 0.35 ns and τ_2 = 4.35 ns, which provide an average life-time of 2.35 ns. The fact that a longer average life-time was found for the FLTX1 moiety of FLTX2 as compared to isolated FLTX1 molecules can be understood in terms of back-transfer from the RB moiety. Indeed, the population of the excited state of the FLTX1 moiety is regulated by the longer life-time of the RB moiety, which may partially back-transfer its excited energy to populate the excited state of the FLTX1 moiety and, consequently regulates the decay rate of fluorescence.

Finally, the intramolecular FRET efficiency, E_{FRET}, between the NBD and RB moieties of the FLTX2 complex can be estimated using the equation:

$$E_{FRET} = 1 - \frac{I_{550\text{nm}-FLTX2}}{I_{550\text{nm}-FLTX1}} \quad (4)$$

where $I_{550\text{nm-FLTX2}}$ and $I_{550\text{nm-FLTX1}}$ represent the emission intensities of the green emission band centered at about 550 nm (due to the NBD moieties), in the FLTX2 and FLTX1 molecules, respectively. From the experimental emission spectra of FLTX1 and FLTX2 solutions under excitation at 475 nm, the estimated efficiency (E_{FRET}) was around 0.99, which indicates an extremely highly efficient energy transfer within the FLTX2 NBD-RB moiety. This observation strongly agrees with the negligible green fluorescence detected in the confocal images shown further on in Figure 7.

In summary, the optical spectroscopic properties found in FLTX2, as compared to those of their constituent moieties, i.e., FLTX1 and RB, allow us to conclude that an efficient energy transfer process occurs in FLTX2 complexes. Indeed, under blue excitation at 475 nm, the FLTX1 moiety of FLTX2, promotes to the excited state and transfers its energy to the RB moiety. Therefore, these optical results suggest the possible use of FLTX2 to produce ROS under blue excitation, as it is shown in the next section.

2.7. Laser-Stimulated ROS Generation In Vitro

In order to evaluate the potential capacity of FLTX2 as a photosensitizer, we first addressed the generation of ROS (mainly superoxide anions) using the Nitroblue tetrazolium (NBT)/formazan method depicted in Figure 6A [45].

Under this approach, the amount of oxidized formazan will depend on the photosensitizer activity of FLTX2. In our assays, we initially checked for the effect of irradiation at 473 nm, during 5 min, on the spectral response of NBT plus FLTX2 solutions in comparison with NBT plus Rose Bengal (RB) mixtures at the same concentration. In the presence of ROS, NBT transforms into formazan, which shows a characteristic broad absorption band in the 500–600 nm spectral range. The absorption spectra were recorded before and after the irradiation process. Figure 6B shows the absorption spectra of the mixtures after irradiation, after subtracting the spectra before irradiation. An absorption band in the 500–600 nm range is observed for the FLTX2 solution upon irradiation, which is attributed to formazan formation, while a negligible signal is detected in the RB alone solution. Further, the response was found to be concentration-dependent (Figure 6C). We chose the smallest concentration of FLTX2 (50 μM) to assess the effect of irradiation time on the spectral features of FLTX2. Thus, when FLTX2 samples were irradiated at 473 nm for 0, 2, 4, 6, 8, and 10 min, and the absorbance recorded between 400 and 800 nm changes in formazan generation were detected at all times above 0 min, indicating ROS generation in an irradiation time-dependent manner (Figure 6D). The increase in absorbance was totally attributable to FLTX2 molecule, since NBT, RB, or TX used alone failed to produce significant changes of absorbance in the 500–600 nm range (not shown). Further, plotting of maximal absorbance at 574 nm versus irradiation time (not shown) reveals a time-dependent increase which exhibits saturation after 8 min irradiation time.

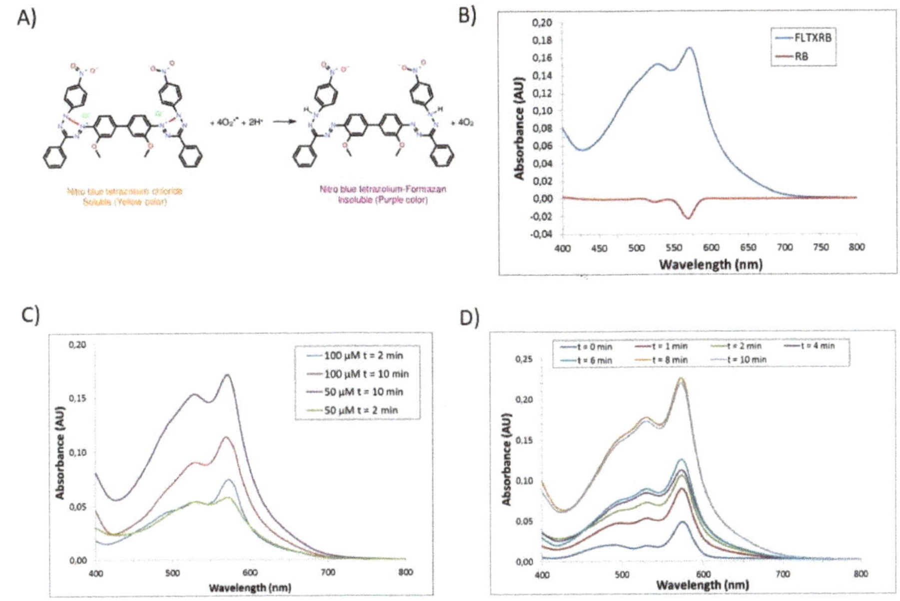

Figure 6. FLTX2-induced ROS generation. (**A**) Superoxide anion-induced formation of nitro blue tetrazolium-formazan. (**B**) Absorption spectra of equimolar FLTX2 and RB solutions after laser irradiation for 10 min. (**C**) Concentration-dependence of FLTX2-induced formazan formation irradiated for 2 or 10 min. (**D**) Time-dependence for FLTX2-induced (50 µM) formazan generation.

Overall, these results indicate that, at least in vitro, irradiation of FLTX2 at the excitation wavelength of the FLTX1 moiety causes an intramolecular FRET to RB, which, in turn, undergoes formation of oxygen singlets to produce superoxide anions. These outcomes demonstrate that FLTX2 is an efficient source of ROS given proper laser stimulation. Generation of ROS may be finely tuned by adjusting irradiation times and occurs with a high FRET efficiency even in aqueous solutions. This finding is relevant from a biological point of view, and also for potential photodynamic applications.

2.8. Cellular Labeling of FLTX2 and Occurrence of FRET on MCF-7 Cells

MCF-7 cells were incubated with different concentrations of FLTX2 and visualized by confocal microscopy. Preparations were excited at 450 nm and the fluorescence recorded at 600 nm. Results in Figure 7A show a concentration-dependent fluorescence at the emission band of RB (600 nm). A significant fluorescence was detected in the nucleus, but most of the fluorescent signal originates at the cytoplasmic perinuclear space, both being increased as the FLTX2 concentration did. On average, cytoplasmic labeling was 3.8 times higher than that in the nucleus (Figure 7B). The evidence that FRET mechanism was responsible for the red fluorescence comes from the fact that the expected emission at 530 due to NBD moiety excitation was almost undetectable (Figure 7C), well below 5% of total cellular fluorescence

(Figure 7D). The fact that most fluorescence comes from outside the nucleus disagree with the classical notion that ERs (both α and β) are essentially nuclear receptors. However, it is now widely accepted that their natural location under unstimulated conditions (i.e., in the absence of estradiol or agonist ligands) is extranuclear, being that they are found even in intracellular organelles such mitochondria and cell membranes [16,26,27,46,47].

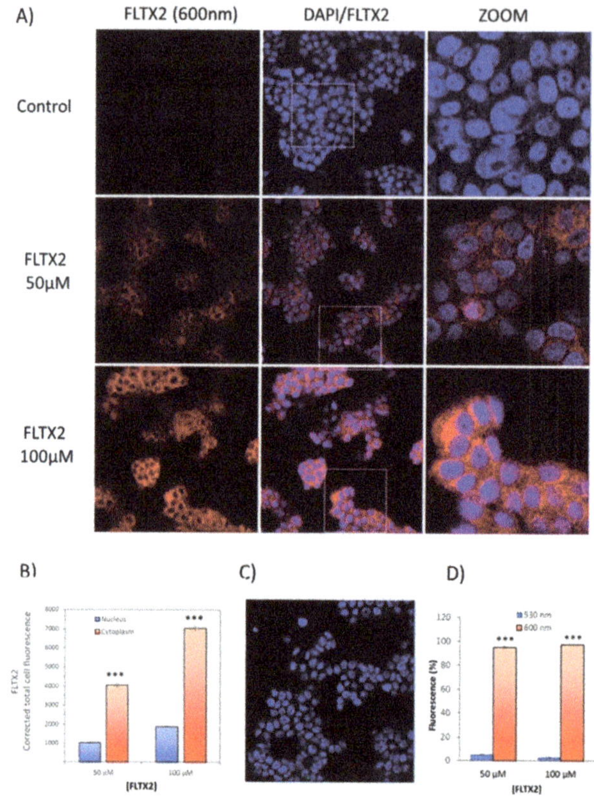

Figure 7. Cellular labeling of FLTX2 and occurrence of FRET in MCF-7 cells. (**A**) Representative images of irradiated preparations (450 nm, 30 min) incubated with different concentrations of FLTX2 (0, 50, and 100 µM). Fluorescent signals were recorded at the emission band of RB (600 nm). DAPI was used as nuclear marker. (**B**) Concentration-dependent fluorescent signals in cytoplasmic and nuclear compartments. (**C**) Demonstration of FRET efficiency in FLTX2 irradiation. Fluorescent signal was virtually absent at the emission wavelength of the NBD moiety of FLTX1. (**D**) bar chart comparing fluorescence signals recorded at 530 nm (NBD moiety) and 600 nm (NBD to RB transfer). *** Statistically significant at $p < 0.005$.

2.9. Photodynamic Effects of FLTX2 on Cell Cultures

MCF-7 cells were used as a human cellular breast cancer model. We initially explored the effect of irradiation itself on untreated cells. The results shown in Figure 8A show that irradiation per se under 240 mW/cm^2 pump power density does not affect cell viability even at exposures as long as 30 min.

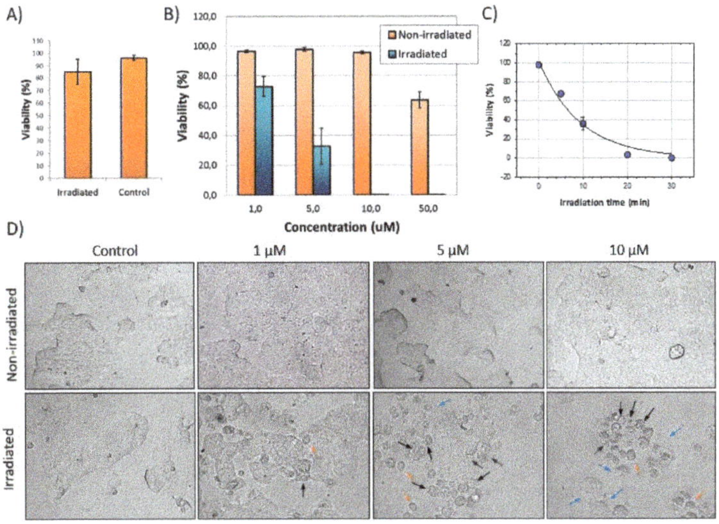

Figure 8. Photodynamic effects of FLTX2. (**A**) Effects of laser irradiation (475 nm) for 30 min on MCF-7 cells viability. (**B**) Concentration-dependence effects on MCF-7 cells viability irradiated for 30 min. (**C**) Effects of irradiation time on FLTX2-induced toxicity. (**D**) Phase contrast transmission representative images of irradiated and non-irradiated MCF-7 cells at different concentrations of FLTX2. Arrows indicate different apoptotic traits: intracellular vacuolation (black arrows), membrane blebbing (blue arrows), cell shrinkage (red arrows), cellular fragmentation (green arrows).

In the next experiment, cells were submitted to treatment with either FLTX2 or DMSO for 2 h before irradiation using an ad hoc laser device built in our group to allow simultaneous irradiation of all wells in the plate. After 30 min irradiation, cells were incubated for 24 h under culture conditions before being processed for cell viability. Results demonstrate significant FLTX2-induced toxicity upon irradiation even at the lowest concentration assayed (1 µM), 10 µM FLTX2 being the lowest dose causing the highest mortality Figure 8B. We next assessed the effect of irradiation time on FLTX2-induced toxicity on MCF-7 cells. In these experiments we explored the effect of 10 µM FLTX2 at different irradiation times on MCF-7 cell viability (Figure 8C). We could fit the experimental data to a two-parameter exponential decay function ($p = 0.0007$, $R^2 = 0.98$) which yielded a t_{50} equal 6,63 min. Thus, it is evident that irradiation time is as determinant as concentration in producing FLTX2-dependent toxicity. Finally, judging from morphological changes shown in Figure 8D, FLTX2-induced cell death likely occurs though induction of apoptosis as suggested by the signs of chromatin condensation, intracellular vacuolation (black arrows), extensive membrane blebbing (blue arrows), cell shrinkage (red arrows), altered membrane integrity, aberrant morphology, and cellular fragmentation (green arrows), which collectively represent typical damage-associated patterns of extrinsic apoptosis or type I PCD (programmed cell death) [48,49]. In agreement with our findings, previous studies in cultured cells have shown that Rose Bengal acetate photodynamic therapy (RBAc-PDT) induces exposure and release of damage-associated molecular patterns (DAMPs) [50,51]. The interpretation of these findings is that irradiated FLTX2 leads to uncontrolled generation of ROS to levels where the intracellular antioxidant systems become overwhelmed and provoke irreversible oxidative stress.

3. Materials and Methods

3.1. Synthesis of FLTX2

The synthesis of FLTX2 is schematized in Figure 1. The dimethyl acetal precursor 1 was synthesized according to a previously described method [21]. After aldehyde deprotection in acid medium, the FLTX-RB linker was introduced using a reductive amination with ethanolamine. Then, the 7-nitrobenzofurazan (NBD) moiety was attached to the amino group to yield compound 2. The esterification of alcohol 2 with Rose Bengal provided FLTX2 (3).

3.1.1. Synthesis of Compound **2**

To a solution of acetal 1 (200 mg, 0.515 mmol) in THF (2.5 mL) was added a 3M HCl aqueous solution (2.5 mL). The reaction was stirred for 6 h at 50 °C. The mixture was then allowed to cool to room temperature, diluted with water (50 mL) and extracted with EtOAc (3 × 50 mL). The joint organic phases were dried over $MgSO_4$ and concentrated under vacuum. The resulting crude aldehyde (not shown) was dissolved in THF (2 mL) under nitrogen atmosphere and stirred with magnesium sulphate (1 g) and ethanolamine (60 μL, 1.0 mmol) for 18 h. The reaction was then filtered and concentrated under vacuum. The crude oily product was dissolved in ethanol (5 mL), treated with $NaBH_4$ (46 mg, 1.2 mmol) and stirred at room temperature for 4 h. The reaction was quenched with a 5% HCl aqueous solution, extracted with EtOAc, dried over $MgSO_4$, and concentrated under vacuum.

The resulting amine was dissolved in CH_2Cl_2 (5 mL) and treated with triethylamine (0.25 mL) and 4-chloro-7-nitrobenzofurazan (NBD-Cl, 100 mg, 0.5 mmol). The reaction was stirred for 4 h at room temperature and then concentrated under vacuum. The residue was purified by column chromatography (hexane/EtOAc, from 9:1 to 1:1) to give compound 2 as a yellow solid (81 mg, 29% for the four steps).

^{1}H NMR (500 MHz, $CDCl_3$, isomer mixture) δ 8.40–8.27 (m, 1H), 7.50–7.07 (m, 8H), 7.05–6.98 (m, 2H), 6.98–6.96 (m, 2H), 6.86–6.82 (m, 1H), 6.77 (d, J = 8.8 Hz, 1H), 6.50 (d, J = 8.8 Hz, 1H), 6.35/6.27 ([d, J = 9.1 Hz/ d, J = 9.0 Hz], 1H), 4.55–4.50 (m, 1H), 4.44–4.38 (m, 1H), 4.38 (t, J = 4.9 Hz, 1H), 4.31–4.25 (m, 1H), 4.22 (t, J = 5.0 Hz, 1H), 4.21–4.16 (m, 1H), 4.13 (t, J = 5.2 Hz, 1H), 4.06 (t, J = 5.2 Hz, 1H), 2.45/2.44 ([ddd, J = 8.6, 7.9, 7.6 Hz/ ddd, J = 8.6, 7.6, 7.4 Hz], 2H), 0.92/0.91 ([t, J = 7.5 Hz/ t, J = 7.5 Hz], 3H).

^{13}C NMR (126 MHz, $CDCl_3$, isomer mixture). The isomer signals corresponding to the same carbon are separated by a dash. Some key signals are described) δ 156.6/155.7 (=C-O), 145.4 (C), 144.7/144.6 (C), 143.6/143.1 (C), 142.31/142.26 (C), 142.1/141.9 (C), 138.0/137.9 (C), 137.1/136.6 (C), 135.2 (C), 132.1, 130.8, 130.7, 129.7, 129.6, 129.4, 128.2, 127.9, 127.8, 127.4, 126.6, 126.2, 126.1, 125.8, 122.8 (C, C-NO_2), 114.0/113.2 (CH, CH=C-O), 102.3/102.2 (CH, NBD), 65.9 (CH_2O), 63.4/60.3 (CH_2O), 56.9/56.8 (CH_2N), 54.2/54.0 (CH_2N), 31.9/29.0 (CH_2), 13.5/7.7 (CH_3).

HRMS-ESI: Calcd for $C_{32}H_{31}N_4O_5$ $[M + H]^+$: 551.2294, found: 551.2273.

Elemental Analysis calcd for $C_{32}H_{30}N_4O_5$: C, 69.80%; H, 5.49%; N, 10.18%; found: C: 69.81%, H: 5.52%, N: 9.98%.

^{1}H NMR and ^{13}C NMR spectra can be found in the Supplementary material (Figures S1 and S2).

3.1.2. Synthesis of FLTX2 (Compound **3**)

Compound 2 (81 mg, 0.15 mmol) was dissolved in dry DMF (5 mL) under nitrogen atmosphere, and the solution was treated with HBTU (62 mg, 0.16 mmol), DMAP (20 μL, 20 mg, 0.16 mmol), Rose Bengal (167 mg, 0.16 mmol), and DIPEA (0.78 mL, 4.5 mmol). The reaction mixture was stirred at room temperature for 18 h, and then was concentrated under vacuum. The residue was purified by column chromatography (EtOAc) to give FLTX2 (3) (93 mg, 42%) as a red solid.

^{1}H NMR (500 MHz, acetone-d_6, isomer mixture) δ 8.54/8.50 ([d, J = 9.0 Hz/ d, J = 9.0 Hz], 1H), 7.69/7.66 (s/s, 1H), 7.38 (t, J = 7.7 Hz), 7.30–7.08 (m, 8H), 7.04–6.95 (m, 1H),

6.89 (d, J = 9.0 Hz, 3H), 6.77 (d, J = 8.8 Hz), 6.54 (d, J = 8.8 Hz), 6.49/6.41 ([d, J = 9.0 Hz/ d, J = 9.0 Hz], 1H), 4.48–4.44 (m, 1H), 4.44 (t, J = 6.0 Hz), 4.42–4.36 (m, 2H), 4.37–4.32 (m, 1H), 4.27–4.20 (m, 2H), 4.19–4.13 (m, 1H), 2.49–2.37 (m, 2H), 0.90–0.77 (m, 3H).

^{13}C NMR (126 MHz, acetone-d$_6$, mixture of isomers. Some isomer signals corresponding to the same carbon are separated by a dash). δ 171.9 (C, CO/COCH=C(O)), 163.1 (C, CO_2), 157.72 (C), 157.70 (C), 157.2 (C), 156.4 (C), 145.10/144.97 (C), 143.7, 143.4, 142.3, 141.3, 139.4, 138.6, 138.4, 136.7, 136.6, 135.7, 135.5, 134.7, 134.1, 132.6, 131.7, 130.6, 130.4, 129.7, 129.6, 129.2, 128.1, 127.9, 127.3, 126.5, 126.0, 125.6, 114.2 (CH), 113.4 (CH), 110.80/110.77 (=C), 103.11/103.03 (CH, NBD), 96.4 (C, =C-I), 75.6 (2×C, =C-I), 65.7/65.5 (CH_2O), 63.3 (CH_2O), 54.1 (C, =C-I), 53.7/53.6 (CH_2N), 51.9/51.8 (CH_2N), 12.92/12.86 (CH_3). An aliphatic CH_2 signal is overlapped with the solvent.

HRMS-ESI: Calcd for $C_{52}H_{31}Cl_4I_4N_4O_9$ $[M-H]^-$: 1502.7024; found: 1502.6992.

Elemental Analysis: calcd for $C_{52}H_{32}Cl_4I_4N_4O_9$: C, 41.46%; H, 2.14%; N, 3.72%; found: C, 41.65%; H, 1.93%; N, 3.45%.

^{1}H NMR and ^{13}C NMR spectra can be found in the Supplementary material (Figures S3 and S4), as well as DEPT (Figure S5) and HSQC (Figure S6) experiments.

3.2. Estrogen Receptor Competitive Binding Assay

Enriched preparations of Estrogen receptor were obtained from uterine cytosol fraction from mature female Sprague-Dawley rats following the procedure described elsewhere [16]. Aliquots of 100 μL cytosol were incubated with 5 nM [^{3}H]E2 and increasing concentrations of non-radioactive FLTX2 or TX (0.1 nM–100 μM) for 18 h at 4 °C. Then, 200 μL of dextran (0.08%)-coated charcoal (0.8%) suspension prepared in TRIS-EDTA-Glycerol-Mg buffer was added to each tube and incubated for 10 min. Suspensions were then centrifuged for 10 min at 3000 g. Radioactivity was then measured in the supernatant in 4 mL scintillation cocktail Optiphase Hisafe 2 (PerkinElmer) by LKB Rackbeta counter (LKB Instrument). Corrections were made for non-specific binding. Relative binding affinity (RBA) was calculated as the ratio of FLTX2 and TX IC_{50} values obtained from dose-response curves.

3.3. Docking Studies

In order to get a deeper insight about the higher affinity of FLTX2 over tamoxifen on ERα, we initially performed in silico docking studies. The X-ray coordinates of human ERα ligand binding domains (LBD) were extracted from the Protein Data Bank (PDB code 3ERT). The PDB structures were prepared for docking using the Protein Preparation Workflow (Schrodinger, LLC, New York, NY, USA, 2020) accessible from the Maestro program (Maestro, version 12.3; Schrodinger, LLC: New York, NY, USA, 2020). The substrate and water molecules were removed beyond 5 Å, bond corrections were applied to the cocrystallized ligands and an exhaustive sampling of the orientations of groups was performed. Finally, the receptors were optimized in Maestro 12.3 by using OPLS3e force field before docking study. In the final stage the optimization and minimization on the ligand-protein complexes were carried out with the OPLS3e force field and the default value for RMSD of 0.30 Å for non-hydrogen atoms were used. The receptor grids were generated using the prepared proteins, with the docking grids centered at the bound ligand for each receptor. A receptor grid was generated using a 1.00 van der Waals (vdW) radius scaling factor and 0.25 partial charge cutoff. The binding sites were enclosed in a grid box of 20 Å^3 without constrains. The three-dimensional structures of the ligands to be docked were generated and prepared using LigPrep as implemented in Maestro 12.3 (LigPrep, Schrodinger, LLC: New York, NY, USA, 2020) to generate the most probable ionization states at pH 7 ± 1 (retaining the original ionization state). These conformations were used as the initial input structures for the docking. In this stage a series of treatments were applied to the structures. Finally, the geometries were optimized using OPLS3e force field. These conformations were used as the initial input structures for the docking. The ligands were docked using the extra precision mode (XP) [52] without using any constraints and a 0.80 van der Waals (vdW) radius scaling factor and 0.15 partial charge cutoff. The dockings were carried out with flexibility of the residues of the pocket near to the ligand. The generated ligand poses were evaluated with empirical scoring function implemented in Glide, GlideScore, which was used to estimate binding affinity and rank ligands [53]. The XP Pose Rank was used to select the best-docked pose for each ligand.

3.4. Molecular Dynamics Simulation

Optimized Potentials for Liquid Simulations-2005 (OPLS2005) [54] force field in Desmond Molecular Dynamic System was used in order to study the behavior of the ligand-target complex. The docking resulting complexes were solvated with an orthorhombic box of TIP3P (Transferable Intermolecular Potential 3-Point) water [55] and counter ions were added, creating an overall neutral system simulating approximately 0.15 M NaCl. The ions were equally distributed in a water box. The final system was subjected to a MD simulation up to 50 ns using Desmond program [56]. The method selected was NPT (Noose-Hover chain thermostat at 300 K, Martyna-Tobias-Klein barostat method at 1.01325 bar with a relaxation time of 2 ps, isotropic coupling, and a 9 Å radius cut-off was used for coulombic short-range interaction) constraints were not applied. During the simulations process, smooth particle Mesh-Ewald method was used to calculate long-range electrostatic interactions. For multiple time step integration, RESPA (Reversible reference System Propagator Algorithm) was applied to integrate the equation of motion with Fourier-space electrostatics computed every 6 fs, and all remaining interactions computed every 2 fs [57]. MD simulations were carried out on these equilibrated systems for a time period of 50 ns, frames of energy and trajectory were captured after every 1.2 ps and 9.6 ps, respectively. The quality of MD simulations was assessed by the Simulation Event Analysis tool. Ligand-receptor interactions were identified using the Simulation Interaction Diagram tool.

3.5. Optical Measurements

A continuous wave (CW) 473 nm diode laser was used in the laser irradiation studies. The collimated beam of the laser was expanded using two convergence lenses to obtain a 1 cm diameter collimated beam. The irradiation density was 240 mW/cm^2. In the in vitro experiments, the frontal face of a quartz cuvette was irradiated, providing a homogenous excitation of the solutions. In the case of the in vivo studies, a mirror was utilized to obtain a homogeneous illumination from the top of the cell culture, which completely covered the cell culture well.

Optical absorption in the visible spectral range was performed in an Agilent Cary 5000 spectrophotometer equipped with double beam cuvette holders. For fluorescence studies, an Edinburgh Instruments LifeSpec II fluorescence spectrometer was used. An Edinburgh Instruments EPL-475 picosecond pulsed diode laser (typical temporal pulse width 80 ps) was selected as the excitation source to pump samples at 475 nm. Fitting of decay curves to mono- or double-exponential equations was made using instrumental response function (IRF) reconvolution analysis with Edinburgh Instruments FAST software. All measurements were conducted at room temperature.

3.6. ROS Generation

The potential capacity of FLTX2 as a photosensitizer, and its ability to produce reactive oxygen species, was determined using a colorimetric assay based on the nitroblue tetrazolium (NBT)/formazan method [45]. NBT is a yellow, water-soluble compound that is transformed into dark-blue formazan upon reaction with superoxide ions and other reactive oxygen species. NBT-formazan displays a broad absorption band at 500–600 nm whose intensity can be measured [20,58]. The amount of formazan is thus directly related to photosensitizer activity of FLTX2.

NBT tablets were purchased from Sigma Aldrich and dissolved in deionized water. One mL solutions of 50 μM FLTX2 (in DMSO) and 100 μM NBT (final concentrations) were prepared and irradiated with a CW flux of laser radiation at 473 nm at a power density of about 240 mW/cm^2 during 0, 4, 8, 12, 16, and 20 min. Control solutions without FLTX2 were also prepared and irradiated at the same times and control mixtures. Samples were assessed spectrophotometrically in the range 400–800 nm and the absorbance, after subtraction at t = 0, was plotted as a function of wavelength and irradiation times.

3.7. Cellular Culture

MCF-7 cells were grown in standard DMEM culture medium containing 10% FBS and maintained at 37 °C under 95% air/5% CO_2 atmosphere. Upon reaching 90% confluency, cells were detached from the flasks with trypsin-EDTA and seeded depending on the subsequent use, i.e., viability/toxicity assays or fluorescence analyses.

3.8. FLTX2-Induced Cellular Toxicity

For viability assays, MCF-7 cells were seeded at a density of 25.000 cells per well in chambered cell culture slides (Falcon). Cells were grown for 72 h and after this time, media was replaced with a fresh one containing the corresponding dose of FLTX2 and incubated for 2 h. After incubation, cells were irradiated for 5, 10, 15, or 30 min with the 473 nm diode laser and irradiation power density was 240 mW/cm^2 as previously described. Corresponding non-irradiated control was performed using the opposite well to the irradiated one in the same slide. After 24 h irradiation, 30 μL of Trypan Blue 0.4% was added and pictures from 5 random fields per well were taken. Experiments were performed by triplicated. Dead cells are easily recognizable by their trypan blue staining and their tendency to be floating. Viability was automatically measured using the cell counter application from ImageJ software.

3.9. Confocal Microscopy Studies

Cellular fluorescence studies were performed following the procedure described in Morales et al., (2016) [46]. Briefly, MCF7 cells were seeded in 8-well chamber slides (40,000 cells/well) and maintained for 24 h at 37 °C in a 95% air/5% CO_2 atmosphere. For fixation, cells were incubated for 1 min with the fixative solution (2% paraformaldehyde, 0.1% glutaraldehyde, 150 mM sucrose), followed by 2 min in 0.5% Nonidet P-40 solution at room temperature. After fixation, cells were washed three times with PBS and incubated in BSA 5% (30 min) to block background noise.

Once washed, cells were incubated for 2 h with FLTX2 (50 or 100 μM), dissolved in DMSO (<2%), and washed again with PBS. A drop of mounting solution containing glycerol and DAPI was applied to each well and the slide was covered with a coverslip and analysed by confocal microscopy (Leica SP8 with software LAS X from Leica). Preparations were excited at 450 nm and the fluorescence recorded at 530 nm and 600 nm. Fluorescence quantification was done using ImageJ software (Rasband, W.S., ImageJ 1997–2018. https://imagej.nih.gov/ij/ accessed on 2–15 July 2020). Fluorescence from the nucleus and cytoplasm were measured independently. Corrected Total Cell Fluorescence (CTCF) was calculated for each compartment as:

$$\mathrm{CTCF} = \mathrm{Id} - (\mathrm{CA} * \mathrm{Bkg}) \tag{5}$$

where CTCF is the corrected total cell fluorescence, Id is the integrated fluorescence density, CA is the cell area, and Bkg is the mean background fluorescence [46].

3.10. Statistical Analysis

When required, data were submitted to one-way ANOVA test followed by Tukey's post hoc test, or by non-parametric Kruskal Wallis test followed by Games-Howell's post hoc. Student-Newman-Keuls *t*-test was also used to determine differences between treatments or times. Dose-response curves were fitted to four parameters logistic equation using nonlinear regression analysis tools included in the statistical software.

$$\%\text{Binding} = \text{min} + \frac{\text{max} - \text{min}}{1 + \left(\frac{\text{x}}{\text{EC50}}\right)^{\text{nH}}} \tag{6}$$

where EC50 is the concentration producing 50% of total binding, nH is the Hill coefficient, and x is the experimental binding.

4. Conclusions

The newly developed FLTX2 is a compound that is pharmacologically and optochemically active. FLTX2 molecule maintains the triphenylethylene core of TX essential to bind the LBD pocket of ER but the lateral side formed by the NBD-linker-RB moiety protrudes out the LBD and displaces helix 12 from adopting its agonist conformation. Experimentally, FLTX2 retains the antiestrogen potency of TX in breast cancer cell lines, and is likely to be devoid of estrogenic effects as it was demonstrated for its predecessor FLTX1. Fluorescence studies demonstrate an efficient intramolecular FRET mechanism in FLTX2. Moreover, under blue light excitation, the FLTX1 moiety of the complex efficiently transfers its excited energy to the RB moiety through a FRET mechanism. RB excited states can then relax, transferring its energy to oxygen ground state triplets to induce ROS. Indeed, FLTX2 exhibits a highly efficient ability to generate superoxide radicals and to induce cell death, likely through ROS-induced apoptosis. Consistent with the high intramolecular FRET efficiency of FLTX2, nearly all fluorescence in fixed MCF7 cells is observed in the emission band of RB. Further, most fluorescent signals originate at the cytoplasmic perinuclear space, as expected for unstimulated MCF-7 cells.

Overall, these properties of FLTX2 make this derivative a potential heir of TX, as a pharmacophore endowed with the ability to specifically bind ER and to antagonize ER activation in estrogen-dependent breast cancer cells, which upon appropriate irradiation would undergo uncontrolled oxidative stress.

Supplementary Materials: The following are available online at https://www.mdpi.com/article/10.3390/ijms22105339/s1, Figures S1 and S2: ^{1}H NMR and ^{13}C NMR spectra of compound **2**. Figures S3 and S4: ^{1}H NMR and ^{13}C NMR spectra of compound **3**. Figures S5 and S6: DEPT and HSQC, respectively. Figure S7: Molecular Dynamics study for FLTX2 on the human ERα LBD: Interactions diagram.

Author Contributions: M.D. did the conceptualization, structuration and supervision of the whole study. M.D. also performed the radioactive ligand-binding experiments, statistical analyses and wrote the manuscript. A.B. and F.L. (Fernando Lobo) synthesized and purified FLTX2. A.E.-B. and Á.A. performed the docking and molecular dynamics studies. D.H. and F.M.-H. were in charge of the ROS generation experiments. A.C.-A. and K.S. performed the MCF7 cells irradiation and FLTX2-induced toxicity assays. C.V.-B. and R.M. did all confocal microscopy studies. F.L. (Fernando Lahoz) was responsible for the optical measurements and FRET analyses. M.D., A.B., F.L. (Fernando Lahoz) and A.E.-B. secured funds for the study. All authors have read and agreed to the published version of the manuscript.

Funding: This work was mainly supported by research grant SAF2014-61644-EXP from MINECO (Spain) to MD and project SAF-2013-48399-R (MINECO-MCIU)-European Social Funds (ESF) to AB. Partial financial support was provided by Agencia Estatal de Investigación (AEI-MICINN and MAT2016-79866-R, Spain), EU-FEDER (PID2019-110430GB-C21 and PID2019-107335RA-I00, European Union) and Gobierno de Canarias (ProID2020010067, Spain) to FL, Grant RTI2018-094356-B-C21 from MICIU (Spain) to AEB. F. Lobo, C. Valdez and A. Amesty were awarded "Agustín de Bethancourt"-Tenerife 2030 postdoctoral grants (Universidad de la Laguna-Cabildo de Tenerife, cofinanced by MEDI & FDCAN Funds).

Institutional Review Board Statement: Not applicable.

Informed Consent Statement: Not applicable.

Data Availability Statement: The datasets generated during and/or analysed during the current study are available from the corresponding author on reasonable request.

Acknowledgments: We acknowledge the financial contribution of Vicerrectorado de Investigacion (Universidad de La Laguna) for the open access publication of this article. ACA and FM-H. were supported by ULL-LaCaixa predoctorate fellowships. D.H. acknowledges her current contract (TRANSALUDAGRO) financed by Cabildo de Tenerife (Program TF INNOVA 2016-21) with MEDI and FDCAN funds.

Conflicts of Interest: The authors declare no conflict of interest.

References

1. Jordan, V.C. Tamoxifen as the first targeted long-term adjuvant therapy for breast cancer. *Endocr. Relat. Cancer* **2014**, *21*, 235–246. [CrossRef]
2. Jordan, V.C.; Obiorah, I.; Fan, P.; Kim, H.R.; Ariazi, E.; Cunliffe, H.; Brauch, H. The St. Gallen Prize Lecture 2011: Evolution of long-term adjuvant anti-hormone therapy: Consequences and opportunities. *Breast* **2011**, *20*, 1–11. [CrossRef]
3. Cersosimo, R.J.; Hernández, L.; Gagnon, L. Tamoxifen for prevention of breast cancer. *Ann. Pharmacother.* **2003**, *37*, 268–273. [CrossRef] [PubMed]
4. Waks, A.G.; Winer, E.P. Breast Cancer Treatment: A Review. *JAMA* **2019**, *321*, 288–300. [CrossRef]
5. Senkus-Konefka, E.; Konefka, T.; Jassem, J. The effects of tamoxifen on the female geniatal tract. *Cancer Treat. Rev.* **2004**, *30*, 291–301. [CrossRef] [PubMed]
6. Mourits, M.J.E.; De Vries, E.G.E.; Willemse, P.H.B.; Ten Hoor, K.A.; Hollema, H.; Van Der Zee, A.G.J. Tamoxifen treatment and gynecologic side effects: A review. *Obstet. Gynecol.* **2001**, *97*, 855–866. [CrossRef] [PubMed]
7. Iqbal, J.; Ginsburg, O.M.; Wijeratne, T.D.; Howell, A.; Evans, G.; Sestak, I.; Narod, S.A. Endometrial cancer and venous thromboembolism in women under age 50 who take tamoxifen for prevention of breast cancer: A systematic review. *Cancer Treat. Rev.* **2012**, *38*, 318–328. [CrossRef]
8. Díaz, M.; Marrero-Alonso, J.; García Marrero, B.; Marín, R.; Gomez, T.; Alonso, R. Cellular and molecular basis for acute nongenomically mediated actions of SERMs. In *Selective Estrogen Receptor Modulators: A New Brand of Multitarget Drugs*; Springer: Berlin/Heidelberg, Germany, 2006; pp. 79–102; ISBN 3540242279.
9. Dong, C.; Chen, L. Second malignancies after breast cancer: The impact of adjuvant therapy. *Mol. Clin. Oncol.* **2014**, *2*, 331–336. [CrossRef]
10. Early Breast Cancer Trialists' Collaborative Group (EBCTCG). Aromatase inhibitors versus tamoxifen in early breast cancer: Patient-level meta-analysis of the randomised trials. *Lancet* **2015**, *386*, 1341–1352. [CrossRef]
11. Hu, R.; Hilakivi-Clarke, L.; Clarke, R. Molecular mechanisms of tamoxifen-associated endometrial cancer (Review). *Oncol. Lett.* **2015**, *9*, 1495–1501. [CrossRef]
12. McDonnell, D.P.; Wardell, S.E. The molecular mechanisms underlying the pharmacological actions of ER modulators: Implications for new drug discovery in breast cancer. *Curr. Opin. Pharmacol.* **2010**, *10*, 620–628. [CrossRef]
13. Rivera-Guevara, C.; Camacho, J. Tamoxifen and its New Derivatives in Cancer Research. *Recent Pat. Anticancer Drug Discov.* **2011**, *6*, 237–245. [CrossRef]
14. Davies, C.; Godwin, J.; Gray, R.; Clarke, M.; Cutter, D.; Darby, S.; McGale, P.; Pan, H.C.; Taylor, C.; Wang, Y.C.; et al. Relevance of breast cancer hormone receptors and other factors to the efficacy of adjuvant tamoxifen: Patient-level meta-analysis of randomised trials. *Lancet* **2011**, *378*, 771–784. [CrossRef]
15. May, F.E. Novel drugs that target the estrogen-related receptor alpha: Their therapeutic potential in breast cancer. *Cancer Manag. Res.* **2014**, *6*, 225–252. [CrossRef] [PubMed]
16. Marrero-Alonso, J.; Morales, A.; García Marrero, B.; Boto, A.; Marín, R.; Cury, D.; Gómez, T.; Fernández-Pérez, L.; Lahoz, F.; Díaz, M. Unique SERM-like properties of the novel fluorescent tamoxifen derivative FLTX1. *Eur. J. Pharm. Biopharm.* **2013**, *85*, 898–910. [CrossRef] [PubMed]
17. Lahoz, F.; Oton, C.J.; López, D.; Marrero-Alonso, J.; Boto, A.; Díaz, M. Whispering gallery mode laser based on antitumor drug–dye complex gain medium. *Opt. Lett.* **2012**, *37*, 4756. [CrossRef] [PubMed]
18. Lahoz, F.; Oton, C.J.; López, D.; Marrero-Alonso, J.; Boto, A.; Díaz, M. High efficiency amplified spontaneous emission from a fluorescent anticancer drug-dye complex. *Org. Electron.* **2013**, *14*, 1225–1230. [CrossRef]
19. Lahoz, F.; Martín, I.R.; Urgellés, M.; Marrero-Alonso, J.; Marín, R.; Saavedra, C.J.; Boto, A.; Díaz, M. Random laser in biological tissues impregnated with a fluorescent anticancer drug. *Laser Phys. Lett.* **2015**, *12*, 045805. [CrossRef]
20. Díaz, M.; Scholz, L.E.; Marrero-Alonso, J.; Boto, A.; Marín, R.; Lobo, F.; Hernández, D.; Amesty, A.; Estévez-Braun, A.; Quinto-Alemany, D.; et al. Opto-chemical and laser properties of FLTX1, a novel fluorescent tamoxifen derivative, and its potential applications in breast cancer photodynamic chemotherapy. *Opt. Mater. (Amst)* **2018**, *84*, 442–446. [CrossRef]
21. Kushner, P.; Cyrus, H.; Davis, M. Subsituted Triphenyl Buthenes. U.S. Patent 8063249, 22 November 2011.
22. Lahoz, F.; Scholz, L.E.; Boto, A.; Díaz, M. FRET mechanism between a fluorescent breast-cancer drug and photodynamic therapy sensitizers. *Spectrochim. Acta Part A Mol. Biomol. Spectrosc.* **2020**, *239*, 1–5. [CrossRef]

23. Katzenellenbogen, B.S.; Choi, I.; Delage-Mourroux, R.; Ediger, T.R.; Martini, P.G.V.; Montano, M.; Sun, J.; Weis, K.; Katzenellenbogen, J.A. Molecular mechanisms of estrogen action: Selective ligands and receptor pharmacology. *J. Steroid Biochem. Mol. Biol.* **2000**, *74*, 279–285. [CrossRef]
24. Wilson, S.; Ruenitz, P.C.; Ruzicka, J.A. Estrogen receptor affinity and effects on MCF-7 cell growth of triarylethylene carboxylic acids related to tamoxifen. *J. Steroid Biochem. Mol. Biol.* **1992**, *42*, 613–616. [CrossRef]
25. Rickert, E.L.; Oriana, S.; Hartman-Frey, C.; Long, X.; Webb, T.T.; Nephew, K.P.; Weatherman, R.V. Synthesis and characterization of fluorescent 4-hydroxytamoxifen conjugates with unique antiestrogenic properties. *Bioconjug. Chem.* **2010**, *21*, 903–910. [CrossRef] [PubMed]
26. Yaşar, P.; Ayaz, G.; User, S.D.; Güpür, G.; Muyan, M. Molecular mechanism of estrogen–estrogen receptor signaling. *Reprod. Med. Biol.* **2017**, *16*, 4–20. [CrossRef] [PubMed]
27. Tecalco-Cruz, A.C.; Pérez-Alvarado, I.A.; Ramírez-Jarquín, J.O.; Rocha-Zavaleta, L. Nucleo-cytoplasmic transport of estrogen receptor alpha in breast cancer cells. *Cell. Signal.* **2017**, *34*, 121–132. [CrossRef]
28. Arnal, J.-F.; Lenfant, F.; Metivier, R.; Flouriot, G.; Henrion, D.; Adlanmerini, M.; Fontaine, C.; Gourdy, P.; Chambon, P.; Katzenellenbogen, B.; et al. Membrane and Nuclear Estrogen Receptor Alpha Actions: From Tissue Specificity to Medical Implications. *Physiol. Rev.* **2017**, *97*, 1045–1087. [CrossRef]
29. Gu, X. Helix 12 in the human estrogen receptor (hER) is essential for the hER function by overcoming nucleosome repression in yeast. *J. Cell. Biochem.* **2002**, *86*, 224–238. [CrossRef]
30. Souza, P.C.T.; Textor, L.C.; Melo, D.C.; Nascimento, A.S.; Skaf, M.S.; Polikarpov, I. An alternative conformation of ERβ bound to estradiol reveals H12 in a stable antagonist position. *Sci. Rep.* **2017**, *7*, 1–11. [CrossRef]
31. Bruning, J.B.; Parent, A.A.; Gil, G.; Zhao, M.; Nowak, J.; Pace, M.C.; Smith, C.L.; Afonine, P.V.; Adams, P.D.; Katzenellenbogen, J.A.; et al. Coupling of receptor conformation and ligand orientation determine graded activity. *Nat. Chem. Biol.* **2010**, *6*, 837–843. [CrossRef]
32. Heldring, N.; Pawson, T.; McDonnell, D.; Treuter, E.; Gustafsson, J.Å.; Pike, A.C.W. Structural insights into corepressor recognition by antagonist-bound estrogen receptors. *J. Biol. Chem.* **2007**, *282*, 10449–10455. [CrossRef]
33. Foulds, C.E.; Feng, Q.; Ding, C.; Bailey, S.; Hunsaker, T.L.; Malovannaya, A.; Hamilton, R.A.; Gates, L.A.; Zhang, Z.; Li, C.; et al. Proteomic analysis of coregulators bound to ERα on DNA and nucleosomes reveals coregulator dynamics. *Mol. Cell* **2013**, *51*, 185–199. [CrossRef] [PubMed]
34. Jordan, V.C. Antiestrogens and selective estrogen receptor modulators as multifunctional medicines. 1. Receptor interactions. *J. Med. Chem.* **2003**, *46*, 883–908. [CrossRef] [PubMed]
35. Shiau, A.K.; Barstad, D.; Loria, P.M.; Cheng, L.; Kushner, P.J.; Agard, D.A.; Greene, G.L. The structural basis of estrogen receptor/coactivator recognition and the antagonism of this interaction by tamoxifen. *Cell* **1998**, *95*, 927–937. [CrossRef]
36. Brzozowski, A.M.; Pike, A.C.W.; Dauter, Z.; Hubbard, R.E.; Bonn, T.; Engström, O.; Öhman, L.; Greene, G.L.; Gustafsson, J.Å.; Carlquist, M. Molecular basis of agonism and antagonism in the oestrogen receptor. *Nature* **1997**, *389*, 753–758. [CrossRef]
37. Shang, Y. Molecular mechanisms of oestrogen and SERMs in endometrial carcinogenesis. *Nat. Rev. Cancer* **2006**, *6*, 360–368. [CrossRef]
38. Hall, J.M.; McDonnell, D.P. Coregulators in nuclear estrogen receptor action: From concept to therapeutic targeting. *Mol. Interv.* **2005**, *5*, 343–357. [CrossRef]
39. Nettles, K.W.; Bruning, J.B.; Gil, G.; O'Neill, E.E.; Nowak, J.; Hughs, A.; Kim, Y.; DeSombre, E.R.; Dilis, R.; Hanson, R.N.; et al. Structural plasticity in the oestrogen receptor ligand-binding domain. *EMBO Rep.* **2007**, *8*, 563–568. [CrossRef]
40. Shang, Y.; Brown, M. Molecular determinants for the tissue specificity of SERMs. *Science* **2002**, *295*, 2465–2468. [CrossRef]
41. Manavathi, B.; Dey, O.; Gajulapalli, V.N.R.; Bhatia, R.S.; Bugide, S.; Kumar, R. Derailed estrogen signaling and breast cancer: An authentic couple. *Endocr. Rev.* **2013**, *34*, 1–32. [CrossRef]
42. Chang, C.C.; Yang, Y.T.; Yang, J.C.; Da Wu, H.; Tsai, T. Absorption and emission spectral shifts of rose bengal associated with DMPC liposomes. *Dye. Pigment.* **2008**, *79*, 170–175. [CrossRef]
43. Islam, S.D.M.; Ito, O. Solvent effects on rates of photochemical reactions of rose bengal triplet state studied by nanosecond laser photolysis. *J. Photochem. Photobiol. A Chem.* **1999**, *123*, 53–59. [CrossRef]
44. Lakowicz, J.R. *Principles of Fluorescence Spectroscopy*, 3rd ed.; Lakowicz, J.R., Ed.; Springer: Berlin/Heidelberg, Germany, 2006.
45. Zhang, Y.; Dai, M.; Yuan, Z. Methods for the detection of reactive oxygen species. *Anal. Methods* **2018**, *10*, 4625–4638. [CrossRef]
46. Morales, A.; Marín, R.; Marrero-Alonso, J.; Boto, A.; Díaz, M. Colocalization of estrogen receptors with the fluorescent tamoxifen derivative, FLTX1, analyzed by confocal microscopy. In *Methods in Molecular Biology*; Humana Press Inc.: Totowa, NJ, USA, 2016; Volume 1366, pp. 163–173.
47. Ford, C.H.J.; Al-Bader, M.; Al-Ayadhi, B.; Francis, I. Reassessment of estrogen receptor expression in human breast cancer cell lines. *Anticancer Res.* **2011**, *31*, 521–527. [PubMed]
48. Elmore, S. Apoptosis: A Review of Programmed Cell Death. *Toxicol. Pathol.* **2007**, *35*, 495–516. [CrossRef]
49. Tower, J. Programmed cell death in aging. *Ageing Res. Rev.* **2015**, *23*, 90–100. [CrossRef]
50. Panzarini, E.; Inguscio, V.; Fimia, G.M.; Dini, L. Rose Bengal Acetate PhotoDynamic Therapy (RBAc-PDT) Induces Exposure and Release of Damage-Associated Molecular Patterns (DAMPs) in Human HeLa Cells. *PLoS ONE* **2014**, *9*, e105778. [CrossRef]
51. Atenco-Cuautle, J.C.; Delgado-López, M.G.; Ramos-García, R.; Ramírez-San-Juan, J.C.; Ramirez-Ramirez, J.; Spezzia-Mazzocco, T. Rose bengal as a photosensitizer in the photodynamic therapy of breast cancer cell lines. *Proc. SPIE* **2019**, *11070*. [CrossRef]

52. Friesner, R.A.; Murphy, R.B.; Repasky, M.P.; Frye, L.L.; Greenwood, J.R.; Halgren, T.A.; Sanschagrin, P.C.; Mainz, D.T. Extra precision glide: Docking and scoring incorporating a model of hydrophobic enclosure for protein-ligand complexes. *J. Med. Chem.* **2006**, *49*, 6177–6196. [CrossRef]
53. Friesner, R.A.; Banks, J.L.; Murphy, R.B.; Halgren, T.A.; Klicic, J.J.; Mainz, D.T.; Repasky, M.P.; Knoll, E.H.; Shelley, M.; Perry, J.K.; et al. Glide: A New Approach for Rapid, Accurate Docking and Scoring. 1. Method and Assessment of Docking Accuracy. *J. Med. Chem.* **2004**, *47*, 1739–1749. [CrossRef]
54. Banks, J.L.; Beard, H.S.; Cao, Y.; Cho, A.E.; Damm, W.; Farid, R.; Felts, A.K.; Halgren, T.A.; Mainz, D.T.; Maple, J.R.; et al. Integrated Modeling Program, Applied Chemical Theory (IMPACT). *J. Comput. Chem.* **2005**, *26*, 1752–1780. [CrossRef]
55. Jorgensen, W.L.; Chandrasekhar, J.; Madura, J.D.; Impey, R.W.; Klein, M.L. Comparison of simple potential functions for simulating liquid water. *J. Chem. Phys.* **1983**, *79*, 926–935. [CrossRef]
56. Bowers, K.J.; Dror, R.O.; Shaw, D.E. The midpoint method for parallelization of particle simulations. *J. Chem. Phys.* **2006**, *124*, 184109. [CrossRef]
57. Gibson, D.A.; Carter, E.A. Time-reversible multiple time scale ab initio molecular dynamics. *J. Phys. Chem.* **1993**, *97*, 13429–13434. [CrossRef]
58. Halliwell, B.; Gutteridge, J.M.C. *Free Radicals in Biology and Medicine*; Halliwell, B., Gutteridge, J.M.C., Eds.; Oxford University Press: New York, NY, USA, 2015; ISBN 9780198717485.

International Journal of
Molecular Sciences

MDPI

Article

Nitrogen-Doped Titanium Dioxide Mixed with Calcium Peroxide and Methylcellulose for Dental Bleaching under Visible Light Activation

Minal Thacker [1], Yi-Ning Chen [1], Chun-Pin Lin [2,3] and Feng-Huei Lin [1,4,*]

1 Graduate Institute of Biomedical Engineering, National Taiwan University, Taipei 10051, Taiwan; d05548019@ntu.edu.tw (M.T.); d2825065d@gmail.com (Y.-N.C.)
2 Graduate Institute of Clinical Dentistry, School of Dentistry, National Taiwan University, Taipei 10617, Taiwan; chunpinlin@gmail.com
3 National Taiwan University Hospital, College of Medicine, National Taiwan University, Taipei 10617, Taiwan
4 Institute of Biomedical Engineering and Nanomedicine, National Health Research Institutes, Miaoli County 35053, Taiwan
* Correspondence: double@ntu.edu.tw; Tel.: +886-928-260-400

Abstract: The available tooth whitening products in the market contain high concentrations of hydrogen peroxide (H_2O_2) as an active ingredient. Therefore, in order to curb the high H_2O_2 concentration and instability of liquid H_2O_2, this study evaluated the efficacy and cytotoxicity of the bleaching gel composed of 10% calcium peroxide (CaO_2) and visible-light-activating nitrogen-doped titanium dioxide (N-TiO_2) with methyl cellulose as a thickener. Extracted bovine teeth were discolored using coffee and black tea stain solution and were divided into two groups (n = 6). Bleaching was performed thrice on each tooth specimen in both the groups, with one minute of visible light irradiation during each bleaching time. The CIELAB L*a*b* values were measured pre- and post-bleaching. The N-TiO_2 calcinated at 350 °C demonstrated a shift towards the visible light region by narrowing the band gap energy from 3.23 eV to 2.85 eV. The brightness (ΔL) and color difference (ΔE) increased as bleaching progressed each time in both the groups. ANOVA results showed that the number of bleaching significantly affected ΔE ($p < 0.05$). The formulated bleaching gel exhibits good biocompatibility and non-toxicity upon exposure to 3T3 cells. Our findings showed that CaO_2-based bleaching gel at neutral pH could be a stable, safe, and effective substitute for tooth whitening products currently available in the market.

Keywords: bleaching; calcium peroxide; titanium dioxide; nitrogen doping; visible light

Citation: Thacker, M.; Chen, Y.-N.; Lin, C.-P.; Lin, F.-H. Nitrogen-Doped Titanium Dioxide Mixed with Calcium Peroxide and Methylcellulose for Dental Bleaching under Visible Light Activation. *Int. J. Mol. Sci.* **2021**, *22*, 3759. https://doi.org/10.3390/ijms22073759

Academic Editor: Antonino Mazzaglia

Received: 10 March 2021
Accepted: 31 March 2021
Published: 4 April 2021

1. Introduction

The desire to flaunt a beautiful smile has turned into an aesthetic necessity today. Hence, tooth whitening has become rampant in esthetic dentistry [1]. The demand for that near-perfect smile has triggered the production of several whitening products and the success of each such product is dependent on the type of tooth discoloration [2]. Tooth discoloration can be classified into two main categories: external staining, which is mainly due to poor oral hygiene, smoking, chlorhexidine in mouthwashes, pigmented food, dental caries, or beverages; whereas internal staining is attributed to factors like age, antibiotics, or excessive fluoride content [3,4].

Hydrogen peroxide is the most common active ingredient in dental bleaching products, due to its ability to release free radicals [5]. These free radicals trigger the oxidation and splitting of pigmented organic molecules into smaller molecules. In turn, these smaller molecules reflect more light, making the tooth appear brighter and whiter, achieving a successful bleaching [6–8]. For vital tooth bleaching, both in-office and home bleaching techniques are widely used [9]. Office bleaching products usually contain a high concentration of H_2O_2 (35–40%) to generate high levels of free radicals for a short treatment time in one appointment, while home bleaching products contain 6–10% H_2O_2 and take more than one month to achieve desired results [10,11]. The higher H_2O_2 concentration or longer exposure time in contact with bleaching gel is effective in whitening. However, side effects might be generated. The most common complications include tooth sensitivity and gingival irritation [9]. With the increasing concentration of H_2O_2, adverse side effects, such as dental hypersensitivity, soft tissue irritation, and cytotoxicity, of dental bleach may occur. Moreover, 0.1 to 6.0% hydrogen peroxide or equivalent for hydrogen peroxide releasing from dental bleaching products is reported safe as per the European Scientific Committee on Consumer Products (SCCP) [12]. However, dental bleaching gel consisting of varying concentrations of H_2O_2 showed toxicity in different cell lines, such as fibroblasts and dental pulp cells (DPCs), in previous reports [13–16]. Therefore, a safer yet efficient bleaching product is long overdue.

Calcium peroxide (CaO_2) can be a potential substitute of liquid H_2O_2 in dental bleaching products because of its characteristics, such as a more effective source of H_2O_2, and relatively stable nature than liquid H_2O_2. CaO_2 is considered to be the safest form of solid peroxy compounds. Upon dissolution in water, CaO_2 is capable of reacting in the medium for a longer time and releases H_2O_2 in a controlled manner. The maximum H_2O_2 released per gram of CaO_2 is 0.47 g, which is considered safe in dentistry [17,18]. According to previous studies, CaO_2 is capable of releasing H_2O_2 and O_2 independently. Moreover, the releasing rates can be controlled by various factors, such as pH and temperature; that is, an increase in pH leads to a decrease in the release of H_2O_2 and increases the O_2 yield, while the increased temperature improves the release of O_2 [19].

In the past, there have been studies showing the use of high-energy ultraviolet (UV)-light-assisted catalysts to reduce the concentration of hydrogen peroxide in dental bleaching products [20]. However, UV-light imposed greater risks on the eyes and skin of the patients as well as that of the dentists [21]. To address the challenges imposed by the harmful UV light, a visible light-activating photocatalyst titanium dioxide was recently introduced. The feasibility of TiO_2 can be attributed to its nontoxicity, low price, and photostability [22]. Titanium dioxide (TiO_2), in its crude form, is a well-known photocatalytic material activated by UV light but can be modified to achieve photo response at visible light. Doping TiO_2 with nitrogen turns out to be the most effective way to reduce the band gap, and transition its photo response from UV light to visible light [23–25].

Given this background, the focus of this study was to develop a safe and effective dental bleaching product for clinical use. This product comprises calcium peroxide and visible light-activating nitrogen-doped TiO_2 (N-TiO_2) as a photocatalyst to increase the rate of bleaching. The cytotoxicity and efficacy of the fabricated dental bleach were examined in vitro by co-culturing with 3T3 cells, while the effectiveness of the prepared dental bleach was further evaluated using stained bovine teeth model in vitro.

2. Results

2.1. X-ray Diffraction (XRD) Analysis

XRD analysis was used to characterize the crystal phase structure of the prepared photocatalytic TiO_2. The XRD pattern of N-TiO_2, along with the pattern of pure TiO_2 used as a reference, is presented in Figure 1. The crystal phase composition of pure TiO_2 and N-TiO_2 contained a mixture of major anatase phase and minor rutile phase. The XRD profile of N-TiO_2 calcined at 350 °C exhibited anatase peaks at 25.25°, 37.68°, 47.94°, 53.87°, and 55.01° and were in good accordance with the (101), (004), (200), (105), and (211) peak position of anatase TiO_2 (JCPDS 86-1157) and rutile peaks at 27.4° and 36.03°, which were in conformity with the (110) and (101) peak position of rutile TiO_2 (JCPDS 21-1276).

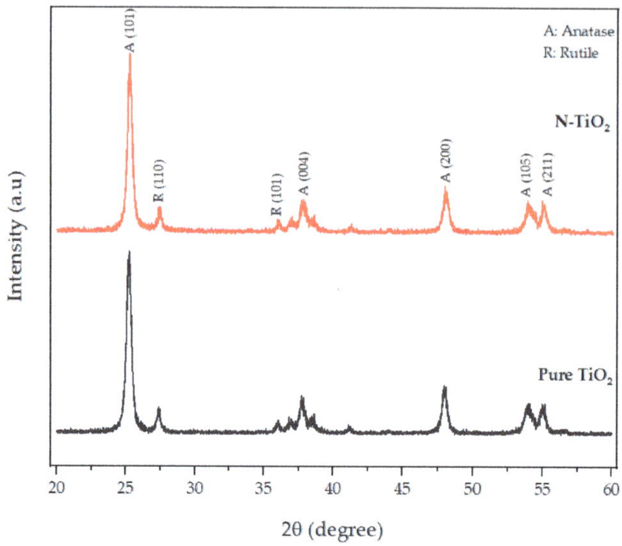

Figure 1. XRD patterns of TiO_2 and nitrogen-doped TiO_2.

2.2. X-ray Photoelectron Spectroscopy (XPS) Analysis

XPS was performed in order to characterize the chemical state and surface composition of the prepared photocatalytic TiO_2. The spectrum of pure TiO_2 and N-TiO_2 is presented in Figure 2a,b, respectively. TiO_2 predominantly contained Ti, O, and C elements, while TiO_2 upon heat treatment indicated the presence of Ti, O, N, and C elements. Amongst these elements in N-TiO_2, the C1s peak was located at 288.6 eV, and it represented the contaminated residual precursor, which was not completely removed during heat treatment. Moreover, accidental carbon during the process may cause the presence of the C element. The N1s peak was found to be at 400.6 eV, while Ti 2p doublets were located at 464.4 and 458.4 eV corresponding to Ti $2p_{3/2}$ and Ti $2p_{1/2}$, which were approximately in accordance with the actual values. Subsequently, the O1s peak could be fitted into two peaks located at binding energies of 529.8 and 532.05 eV.

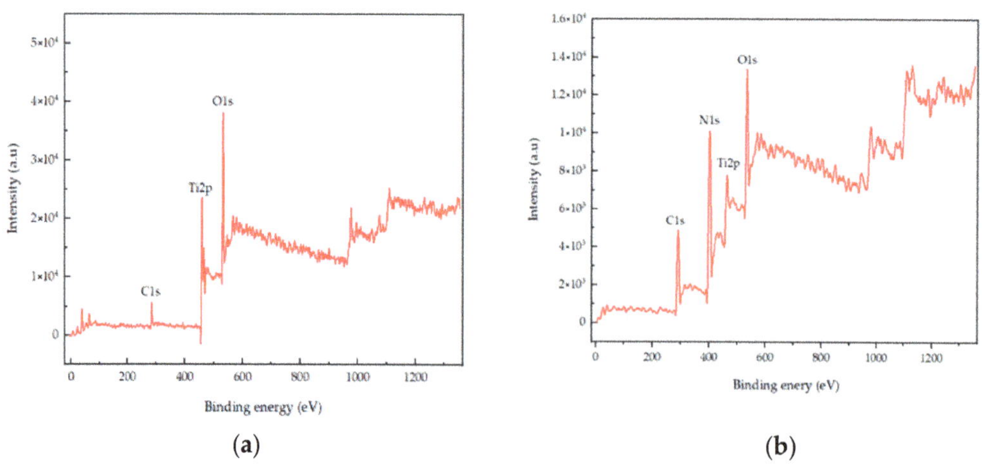

Figure 2. (**a**) XPS survey spectra of TiO_2 and (**b**) XPS survey spectra of N-TiO_2.

2.3. Morphology Analysis

The morphology of doped TiO_2 was observed using SEM and is shown in Figure 3. According to the results, N-TiO_2 calcinated at 350 °C showed agglomerated clusters with spherical morphology. At high temperatures, particles tend to aggregate due to the particle growth process.

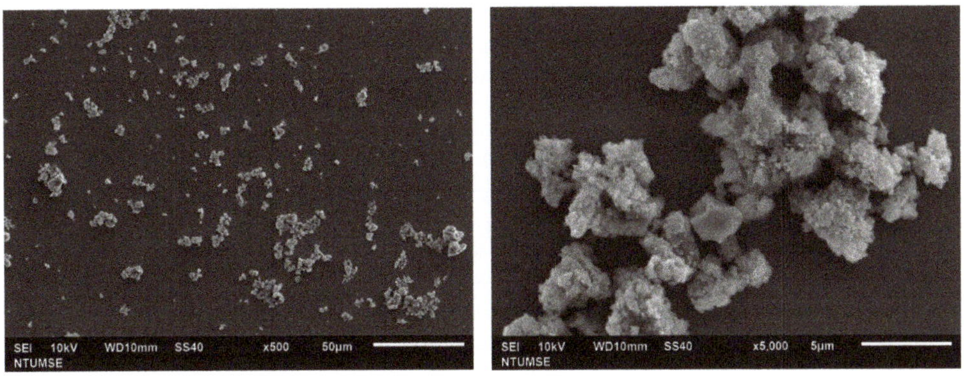

Figure 3. SEM micrograph of N-TiO_2.

2.4. UV-Vis Absorption Spectra

The optical property of the prepared TiO_2 photocatalyst was measured using UV-vis spectroscopy. Figure 4 shows the UV-Vis absorption spectra of pure TiO_2 (undoped) and N-TiO_2. According to the results, pure TiO_2 had an absorption edge around 390 nm, while the absorption edge of N-TiO_2 was found to be around 450 nm. This shift toward the visible light in the absorption spectra of N-TiO_2 was due to the incorporation of nitrogen into the TiO_2 lattice, eventually leading to band gap narrowing.

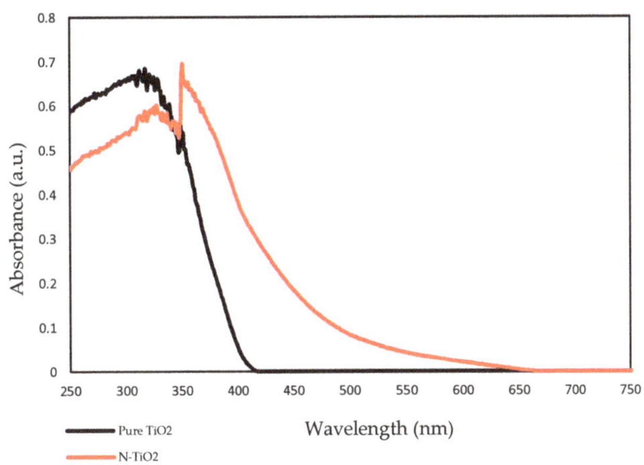

Figure 4. UV-Vis absorption spectrum of pure TiO_2 and N-TiO_2.

2.5. Color Analysis of Tooth Bleaching

Figure 5 shows the ΔL, Δa, Δb, and ΔE values of each bleaching time in the coffee- and black tea-stained bovine teeth groups, respectively. ΔL, Δa, and Δb stand for the difference of L, a, and b between baseline and each bleaching time while the color difference (ΔE) was calculated according to the formula $\Delta E = [(\Delta L)^2 + (\Delta a)^2 + (\Delta b)^2]^{1/2}$. After the bleaching treatment, the values of ΔL and ΔE gradually increased in both the groups, while the values of Δa and Δb exhibited a decrease. According to the statistical analysis, ΔE in both the groups showed significant differences ($p < 0.05$) between bleaching times (Table 1). The photographs of a representative image of tooth bleaching from black tea and coffee stained groups at the baseline and 1, 2, 3 times of bleaching is shown in Figures S1 and S2 respectively.

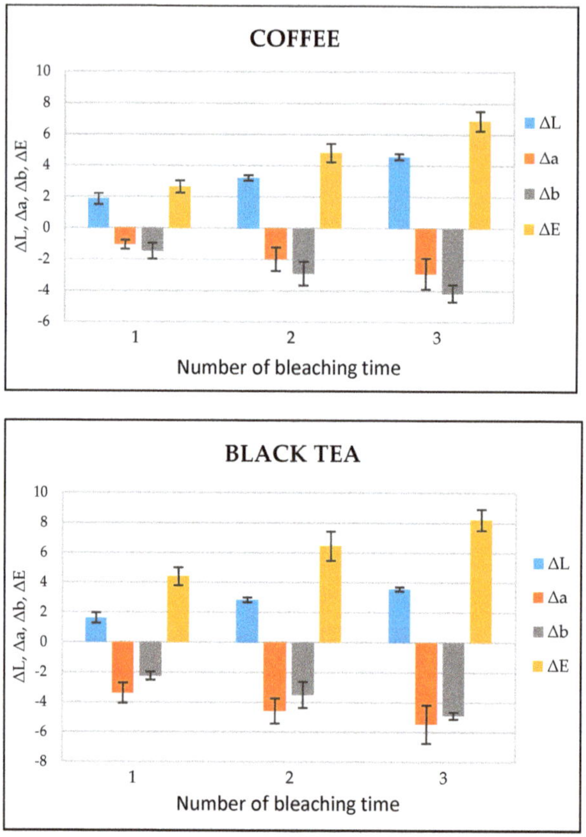

Figure 5. ΔL, Δa, Δb, and ΔE values of coffee- and black tea-stained bovine tooth samples treated with dental bleaching gel ($n = 6$).

Table 1. Mean ΔE and standard deviation.

	Mean ΔE (SD)		
Solution	**Times**		
	1	**2**	**3**
Coffee	2.638 (0.392) [a]	4.822 (0.601) [b]	6.874 (0.629) [c]
Black Tea	4.362 (0.589) [a]	6.460 (0.973) [b]	8.213 (0.705) [c]

Different superscript lowercase letters in coffee and black tea rows indicate statistically significant differences among evaluation times in the same group.

2.6. Biological Assays

The cytotoxicity and biocompatibility of the bleaching gel were determined via LDH and WST-1 assay, respectively, and the results are presented in Figures 6 and 7, respectively.

WST-1 is a colorimetric cell proliferation assay used to measure the biocompatibility while the LDH assay determines the cellular cytotoxicity. 3T3 cells were used in this study. Conditioned medium containing 0.2 g of bleaching gel (10% CaO_2 + 1% N-TiO_2 + 1.5% methylcellulose) per mL was prepared and used in this study. The WST-1 result showed a cell viability of 93.7% at 24 h while the LDH assay showed very low cytotoxicity of 7.3% when exposed to dental bleach-conditioned medium for 24 h. No significant toxicity was observed in 3T3 cells when cultured with the experimental bleaching gel.

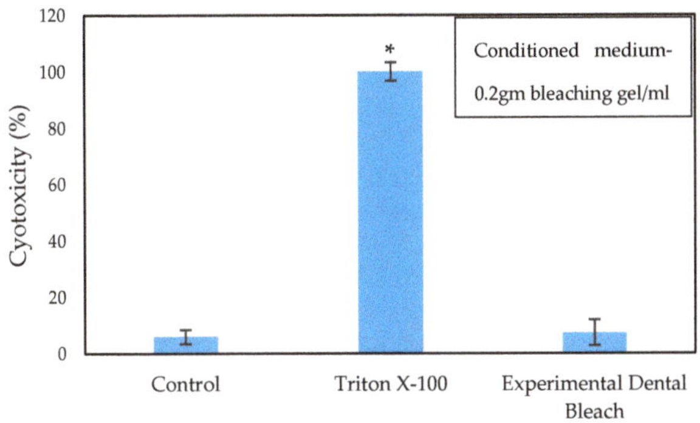

Figure 6. Cytotoxicity of the cells cultured in dental bleach-conditioned medium ($n = 6$), * $p < 0.05$ compared to the control group by student *t*-test.

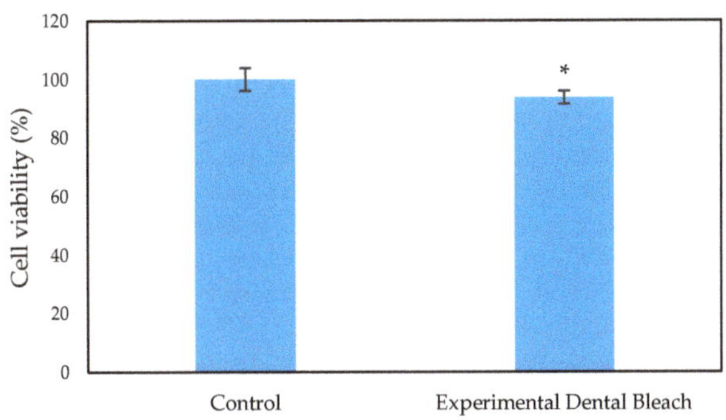

Figure 7. WST-1 assay for 3T3 cell viability evaluation on exposure to dental bleach-conditioned medium ($n = 6$), * $p < 0.05$ compared to the control group by student *t*-test.

3. Discussion

The study focused on the bleaching efficiency and cytotoxicity of the formulated dental bleaching gel using a discolored bovine tooth model. In view of the existing literature, there have been several in vitro models appointed for evaluating the efficacy of the dental bleaching agent, the important ones being the human or bovine tooth model, either cut or whole [26–28]. In this current study, bovine teeth were used as an in vitro model because of its similarity with human teeth in terms of physical and chemical properties, such as hardness, permeability, and density of the dentin tubule. Moreover, the collection of extracted human teeth is difficult for experiments with large sample sizes, and the consistency in the results may be affected by various factors, such as shade, age, thickness of enamel, and mineralization of the extracted human tooth surface [29].

In this study, discoloration of the bovine teeth was achieved by tea and coffee staining solutions, respectively, due to their cost effectiveness, easy preparation, and regular use by the majority of the population [30,31]. Shade change of the teeth after bleaching can be evaluated using several methods, such as visual comparison of the tooth surface with the standard tooth shade card or parametric analysis using a chroma meter. In this study, a chroma meter was used to evaluate the tooth shade as it can provide quantitative analysis in color matching, so that statistical analysis can be easily achieved and is more accurate than the visual analysis using a tooth shade guide.

The main and effective attraction of in-office dental bleaching products is the use of H_2O_2 as an active ingredient, but due to the diverse side effects of using high-concentration H_2O_2 as mentioned before, many researchers have now directed their research towards finding alternatives to control high-concentration H_2O_2 to provide effective bleaching without harm to the dental tissues. Therefore, CaO_2, used as a bleaching agent in this study, is expected to be a potential alternative as compared to the other commercial agents with high concentrations of H_2O_2 (20–40%). H_2O_2 released from CaO_2 can be controlled with the alterations in temperature and pH. Therefore, the release rate of H_2O_2 was controlled by adjusting the pH of the bleaching gel to 7. Moreover, the neutral pH of the dental bleaching products has been reported to be safe on enamel because it does not alter the surface roughness of the tooth even after several applications [10]. CaO_2 is also capable of reacting for a longer duration in the medium, making its functionality even more efficient.

In previous studies, it was demonstrated that the efficiency of H_2O_2-based bleaching products was increased with the addition of visible light-activating TiO_2 when exposed to visible light [32,33]. Therefore, in this study, N-TiO_2 was used as a photocatalyst to increase the efficiency of CaO_2-based bleaching gel. The nitrogen doping in the TiO_2 lattice results in the formation of a new electronic state above the valance band, which leads to the shift in the absorption spectrum of N-TiO_2 from UV light to the visible light region as analyzed by the UV-vis spectrophotometer (Figure 4). The formation of a new electronic band led to a decrease in the band gap energy value and was calculated by the Tauc plot method. The band energy value for pure TiO_2 was found to be 3.2 eV while on the other hand, the nitrogen-incorporated TiO_2 lattice exhibited a decrease in the band gap value from 3.23 eV to 2.85 eV. Moreover, to investigate the effect of nitrogen doping on the crystal phase of TiO_2, XRD was performed. The phase composition of N-TiO_2 contains major intensity of the anatase phase and minor rutile phase, similar to that of pure TiO_2 (Figure 1). In view of the existing literature, the anatase phase is a better photocatalyst than the rutile phase [34]. Therefore, in the present study, it was important to retain the intensity of the anatase phase in N-TiO_2 in order to fabricate a strong photocatalyst to make a stronger dental bleach. Additionally, the results of XPS analysis (Figure 2b) confirmed the existence of N in the TiO_2 lattice along with the elements Ti, O, and C. The observed N1s peak with binding energy at 400.1 eV is assigned to N-O-Ti linkage, i.e., nitrogen bonded to oxygen sites (interstitial doping). The high nitrogen content results in an effective visible light-activating photocatalyst even though there is slight particle aggregation at high temperatures due to the particle growth process.

The bleaching efficiency of the experimental bleaching gel was determined by comparing the color values at each bleaching time. The important indicator for the bleaching evaluation is the determination of the L value; the higher the L value, the brighter or whiter the teeth appear. In the present study, the value of L increased with each bleaching time in both the groups, indicating the brightness or whiteness was upgraded with each consecutive bleaching. On the other hand, the total color difference, i.e., ΔE, gradually increased during each bleaching treatment. ΔE is categorized into six ranks: 0.5 or less ΔE value, wherein macroscopically no color difference was seen; between 0.5 and 1.5, a small difference was observed with effort to the naked eye; between 1.5 and 3.0, a slight difference was clearly noticed; 3.0–6.0 values, when a substantial difference was seen; between 6.0 and 12.0, a marked difference was observed; and 12.0 or higher, when a different color line was noticed [35]. In this study, ΔE after the last post-treatment was in the range of 6.0–12.0, representing a significant difference (Figure 5).

Successful bleaching treatment brightens the teeth effectively without compromising safety. Therefore, a cytotoxicity test was conducted in order to evaluate the safety and biologic properties of the bleaching gel. We found that the bleaching components were non-toxic to 3T3 cells (Figures 6 and 7). No significant difference was observed between the experimental group and control group. Therefore, the biocompatibility of the experimental bleaching gel was good when tested on 3T3 cell lines and should be safe in clinical use. Future study will focus on the quantification of released H_2O_2 from CaO_2 and a more delicate mechanism under different conditions.

4. Materials and Methods

4.1. Preparation of N-TiO_2

An organic nitrogen source urea was used for nitrogen doping. Briefly, 3M urea solution was prepared, and titanium (IV) oxide (21-nm primary particle size, Sigma-Aldrich, St. Louis, MI, USA) was dispersed in it at a ratio of 0.1 g/mL. The mixture was stirred for 24 h and later dried in the oven at 80 °C, followed by calcination in a furnace at 350 °C for 2 h.

4.2. Characterization

The phase identification of pure TiO_2 and N-TiO_2 was carried out using an X-ray diffractometer (XRD) (TTRAX 3, Rigaku, Tokyo, Japan). The surface composition and electron binding energy of TiO_2 and N-TiO_2 was measured using an X-ray photoelectron spectroscopy (XPS) (Theta probe, Thermo Scientific, Waltham, MA, USA). The morphology was observed using a scanning electron microscope (SEM) (JSM6510, JEOL, Tokyo, Japan). The UV-Vis spectroscopy (CARY 300nc, Agilent, Santa Clara, CA, USA) was used to record the absorption spectra of the samples.

4.3. Preparation of Stained Teeth

Twelve freshly extracted bovine incisors purchased from a local meat market in Taipei, Taiwan were used in this study. After extraction, the teeth were dipped in hot boiling water for 20 s, and the soft tissues were removed using a scalpel. The enamel surface was polished with ascending grit silicon carbide papers starting from #100 up to #1000 under running water, removing 300 μm from the enamel surface to create a smooth and flat surface. The teeth were then stored in water at 4 °C.

Black tea and coffee stain solutions were prepared. Briefly, two black tea bags (Earl Grey Twinings) each weighing 2 g were immersed in 100 mL of boiling water for 5 min to prepare a black tea stain solution while the coffee stain solution was prepared by adding 4 g of ground coffee powder (Nescafe) in 100 mL of boiling water. The teeth were then divided into two groups ($n = 6$) and immersed into the respective above-mentioned staining solutions for a week in the incubator at 37 °C. The solutions were renewed after every 3 days and stirred once a day to avoid sedimentation.

4.4. Color Analysis

After the teeth were immersed in the staining solutions for a week, they were rinsed under tap water to remove excess coffee and tea from the surface, respectively, followed by drying using kimwipes. Prior to bleaching, the CIELAB values of the stained enamel surface were recorded as a baseline value using a dental chroma meter (VITA EasyShade Compact, Vident, Model # DEASYCBU, Yorba Linda, CA, USA).

4.5. Tooth Bleaching

The experimental bleaching gel was composed of 10% calcium peroxide, 1% N-TiO_2,, and 1.5% methylcellulose as a thickener. The pH of the gel was adjusted to 7.0 before the application onto the test tooth surface.

A thin layer of the experimental bleaching gel was applied on the test tooth surface with a brush and irradiated for 1 min with a light emitting diode (LED) light curing device (LITEX 696, Dentamerica Asia Inc., Taipei, Taiwan). Following which, the bleaching agent was left for 5 min on the tooth surface, and the bleaching gel was then rinsed with fresh water and dried. A VITA EasyShade Compact chroma meter was used to measure the L, a, and b values of the bleached teeth after rinsing and drying of teeth, where L* represents the lightness of the sample from black (0) to white (100), a* represents the green-red coordinate, and b* represents the blue-yellow coordinate. The bleaching and color measurements were repeated thrice per tooth. The difference between baseline and each bleaching time for L, a, and b values were denoted as ΔL, Δa, and Δb, respectively. The color difference (ΔE) was calculated according to the equation below:

$$\Delta E = [(\Delta L)^2 + (\Delta a)^2 + (\Delta b)^2]^{1/2}$$

4.6. Biological Assays

4.6.1. Cell Culture

The 3T3 cell line (mouse embryonic fibroblasts) (Bioresource Collection and Research Center, Taiwan) was used as a cell source in this study. Dulbecco's modified Eagle's medium (DMEM) (Sigma, USA) supplemented with 10% fetal calf serum (FBS) (Gibco, Gaithersburg, MD, USA) and 1% antibody (Gibco, USA) was used for the cell culture. The cells were incubated at 37 °C in a 5% CO_2-containing atmosphere.

4.6.2. Cell Viability

Cells were seeded in 96-well plates at a cell density of 1×10^4 cells/well and incubated for 24 h at 37 °C. After 24h, the used DMEM was aspirated, washed with PBS, and then cultured with the conditioned medium for 24 h at 37 °C. The conditioned medium was obtained according to the ISO standard 10993-12:2012. Briefly, 5 mL of the conditioned medium was prepared (0.2 g bleaching gel per 1 mL of DMEM culture medium incubated for 30 min at room temperature). The conditioned medium was filtered through a sterile filter and subsequently used for cell culture.

Cell viability was evaluated using the WST-1 assay kit (Takara, Japan) according to the manufacturer's instruction. The conditioned medium was replaced with WST-1 working solution for 2 h, and the absorbance of formazan, a colored dye produced by viable cells using WST-1, was measured at 450 nm by a microplate reader (Spectramax plus 384 microplate reader, Molecular Devices, CA, USA). Untreated cells, cultured with DMEM medium, were used as controls.

4.6.3. Cell Cytotoxicity

The cytotoxicity of the prepared dental bleach was quantified using the LDH assay kit (Takara, Japan) according to the manufacturer's instructions. The presence of LDH in the medium is an indicator of cellular toxicity. The conditioned medium after incubation with cells for 24 h was transferred and mixed with LDH working solution in a 1:1 ratio. The absorbance of LDH was measured at 450 nm, after incubating for 30 min in the dark using a microplate reader (Spectramax plus 384 microplate reader, Molecular Devices, CA, USA). The cells cultured with DMEM medium were used as controls, while the cells treated with medium containing 0.1% Triton X-100 were used as the positive control.

4.7. Statistical Analysis

After the calculation of ΔE data, it was subjected to statistical analysis and analysis of variance test (ANOVA) was performed, following which post hoc tukey's test was used for comparison between the groups. A probability (p) value of 0.05 was considered statistically significant. The statistical difference between the control and experimental group in the WST-1 and LDH assay was evaluated by student's t-test using GraphPad (Prism for Mac, GraphPaD software, San Diego, CA, USA).

5. Conclusions

In this work, nitrogen-doped titanium dioxide was successfully prepared and characterized. The synthesized N-TiO_2 calcinated at 350 °C contained majorly the photocatalytic anatase phase and the absorbed nitrogen in the TiO_2 lattice caused the band gap narrowing, resulting in visible light-activating N-TiO_2. The dental bleaching gel comprising 10% CaO_2 as an active ingredient and N-TiO_2 as a photocatalyst demonstrated an efficient bleaching effect with a gradual increase in ΔL and ΔE on a coffee- and black tea-stained bovine tooth model in vitro, while the cell viability and cytotoxicity data provided an affirmation of the safety of bleaching gel. The overall findings of this study suggest the CaO_2-based bleaching gel with N-TiO_2 not only results in effective bleaching but can also decrease the potential side effects that are usually caused by a high-concentrated hydrogen peroxide-based dental bleaching procedure.

Supplementary Materials: The following are available online at https://www.mdpi.com/article/10.3390/ijms22073759/s1, Figure S1. Representative image of tooth bleaching in black tea stained group; Figure S2. Representative image of tooth bleaching in coffee stained group.

Author Contributions: Conceptualization, F.-H.L.; Methodology, M.T. and F.-H.L.; Validation, C.-P.L.; Formal Analysis, M.T.; Investigation, M.T., Y.-N.C. and C.-P.L.; Writing, M.T.; Supervision, F.-H.L. and C.-P.L. All authors have read and agreed to the published version of the manuscript.

Funding: This research received no external funding.

Institutional Review Board Statement: Not applicable.

Informed Consent Statement: Not applicable.

Data Availability Statement: Not applicable.

Acknowledgments: The authors thank the members of the School of Dentistry and College of Medicine, National Taiwan University for their assistance and technical advice.

Conflicts of Interest: The authors declare no conflict of interest.

References

1. Barry, T.; Bailey, C.; Ashcraft-Olmscheid, D.; Vandewalle, K. Effect of a new bleaching gel on tooth whitening. *Oper. Dent.* **2017**, *42*, 559–566. [CrossRef]
2. Kurzmann, C.; Verheyen, J.; Coto, M.; Kumar, R.V.; Divitini, G.; Shokoohi-Tabrizi, H.A.; Verheyen, P.; De Moor, R.J.G.; Moritz, A.; Agis, H. In vitro evaluation of experimental light activated gels for tooth bleaching. *Photochem. Photobiol. Sci.* **2019**, *18*, 1009–1019. [CrossRef] [PubMed]
3. Carey, C.M. Tooth whitening: What we now know. *J. Évid. Based Dent. Pract.* **2014**, *14*, 70–76. [CrossRef]
4. Manuel, S.T.; Abhishek, P.; Kundabala, M. Etiology of tooth discoloration—A review. *Etiol. Tooth Discoloration Rev.* **2010**, *18*. [CrossRef]
5. Bizhang, M.; Domin, J.; Danesh, G.; Zimmer, S. Effectiveness of a new non-hydrogen peroxide bleaching agent after single use—A double-blind placebo-controlled short-term study. *J. Appl. Oral Sci.* **2017**, *25*, 575–584. [CrossRef] [PubMed]
6. Kwon, S.R.; Wertz, P.W. Review of the mechanism of tooth whitening. *J. Esthet. Restor. Dent.* **2015**, *27*, 240–257. [CrossRef]
7. Kihn, P.W. Vital tooth whitening. *Dent. Clin. N. Am.* **2007**, *51*, 319–331. [CrossRef]
8. Alqahtani, M.Q. Tooth-bleaching procedures and their controversial effects: A literature review. *Saudi Dent. J.* **2014**, *26*, 33–46. [CrossRef]
9. Abdullah, A.O.; Muhammed, F.K.; Zheng, B.; Liu, Y. An overview of extrinsic tooth bleaching and its impact on oral restorative materials. *World J. Dent.* **2017**, *8*, 503–510. [CrossRef]
10. Marson, F.; Gonçalves, R.; Silva, C.; Cintra, L.; Pascotto, R.; Dos Santos, P.H.; Briso, A.; Silva, C.; Cintra, L. Penetration of hydrogen peroxide and degradation rate of different bleaching products. *Oper. Dent.* **2015**, *40*, 72–79. [CrossRef]
11. Sundfeld, R.; Neto, D.; Machado, L.; De Oliveira, F.; De Alexandre, R.; Palo, R.; Sundefeld, M.L.M.; Rh, S.; Ds, N.; Ls, M.; et al. Dental bleaching with a 10% hydrogen peroxide product: A six-month clinical observation. *Indian J. Dent. Res.* **2014**, *25*, 4. [CrossRef] [PubMed]
12. European Commission. Scientific Committee on Consumer Products SCCP Opinion on Hydrogen Peroxide in Tooth Whitening Products. 2005. Available online: https://www.google.com.hk/url?sa=t&rct=j&q=&esrc=s&source=web&cd=&cad=rja&uact=8&ved=2ahUKEwjtn5yoh9rvAhWa_7sIHc_OB50QFjAAegQIAxAD&url=http%3A%2F%2Fec.europa.eu%2Fhealth%2Fph_risk%2Fcommittees%2F04_sccp%2Fdocs%2Fsccp_cons_01_en.pdf&usg=AOvVaw3CMBtpf_qa6iSdvIB9c_xd (accessed on 31 March 2021).
13. Lilaj, B.; Dauti, R.; Agis, H.; Schmid-Schwap, M.; Franz, A.; Kanz, F.; Moritz, A.; Schedle, A.; Cvikl, B. Comparison of bleaching products with up to 6% and with more than 6% hydrogen peroxide: Whitening efficacy using BI and WID and side effects—An in vitro study. *Front. Physiol.* **2019**, *10*, 919. [CrossRef] [PubMed]

14. Dahl, J.; Pallesen, U. Tooth bleaching—A critical review of the biological aspects. *Crit. Rev. Oral Biol. Med.* **2003**, *14*, 292–304. [CrossRef]
15. Duque, C.C.D.O.; Soares, D.G.; Basso, F.G.; Hebling, J.; Costa, C.A.D.S. Bleaching effectiveness, hydrogen peroxide diffusion, and cytotoxicity of a chemically activated bleaching gel. *Clin. Oral Investig.* **2013**, *18*. [CrossRef]
16. Dantas, C.M.G.; Vivan, C.L.; Ferreira, L.S.; De Freitas, P.M.; Marques, M.M. In vitro effect of low intensity laser on the cytotoxicity produced by substances released by bleaching gel. *Braz. Oral Res.* **2010**, *24*, 460–466. [CrossRef] [PubMed]
17. Northup, A.; Cassidy, D. Calcium peroxide (CaO_2) for use in modified Fenton chemistry. *J. Hazard. Mater.* **2008**, *152*, 1164–1170. [CrossRef]
18. Lu, S.; Zhang, X.; Xue, Y. Application of calcium peroxide in water and soil treatment: A review. *J. Hazard. Mater.* **2017**, *337*, 163–177. [CrossRef]
19. Wang, H.; Zhao, Y.; Li, T.; Chen, Z.; Wang, Y.; Qin, C. Properties of calcium peroxide for release of hydrogen peroxide and oxygen: A kinetics study. *Chem. Eng. J.* **2016**, *303*, 450–457. [CrossRef]
20. Ziemba, S.L.; Felix, H.; Macdonald, J.; Ward, M. Clinical evaluation of a novel dental whitening lamp and light-catalyzed peroxide gel. *J. Clin. Dent.* **2005**, *16*, 123–127.
21. Bruzell, E.M.; Johnsen, B.; Aalerud, T.N.; Dahl, J.E.; Christensen, T. In vitro efficacy and risk for adverse effects of light-assisted tooth bleaching. *Photochem. Photobiol. Sci.* **2009**, *8*, 377–385. [CrossRef]
22. Lin, Y.-T.; Weng, C.-H.; Hsu, H.-J.; Lin, Y.-H.; Shiesh, C.-C. The synergistic effect of nitrogen dopant and calcination temperature on the visible-light-induced photoactivity of N-doped TiO_2. *Int. J. Photoenergy* **2013**, *2013*, 268723. [CrossRef] [PubMed]
23. Ansari, S.A.; Khan, M.M.; Ansari, M.O.; Cho, M.H. Nitrogen-doped titanium dioxide (N-doped TiO2) for visible light photocatalysis. *New J. Chem.* **2016**, *40*, 3000–3009. [CrossRef]
24. Nosaka, Y.; Matsushita, M.; Nishino, J.; Nosaka, A.Y. Nitrogen-doped titanium dioxide photocatalysts for visible response prepared by using organic compounds. *Sci. Technol. Adv. Mater.* **2005**, *6*, 143–148. [CrossRef]
25. Chainarong, S.; Sikong, L.; Pavasupree, S.; Niyomwas, S. Synthesis and characterization of nitrogen-doped TiO_2 nanomaterials for photocatalytic activities under visible light. *Energy Procedia* **2011**, *9*, 418–427. [CrossRef]
26. Joiner, A.; Philpotts, C.J.; Alonso, C.; Ashcroft, A.T.; Sygrove, N.J. A novel optical approach to achieving tooth whitening. *J. Dent.* **2008**, *36*, 8–14. [CrossRef]
27. Yui, K.C.K.; Rodrigues, J.R.; Mancini, M.N.G.; Balducci, I.; Gonçalves, S.E.P. Ex vivo evaluation of the effectiveness of bleaching agents on the shade alteration of blood-stained teeth. *Int. Endod. J.* **2008**, *41*, 485–492. [CrossRef] [PubMed]
28. Kishi, A.; Otsuki, M.; Sadr, A.; Ikeda, M.; Tagami, J. Effect of light units on tooth bleaching with visible-light activating titanium dioxide photocatalyst. *Dent. Mater. J.* **2011**, *30*, 723–729. [CrossRef]
29. Tano, E.; Otsuki, M.; Kato, J.; Sadr, A.; Ikeda, M.; Tagami, J. Effects of 405 nm diode laser on titanium oxide bleaching activation. *Photomed. Laser Surg.* **2012**, *30*, 648–654. [CrossRef]
30. Sulieman, M.; Addy, M.; Rees, J. Development and evaluation of a method in vitro to study the effectiveness of tooth bleaching. *J. Dent.* **2003**, *31*, 415–422. [CrossRef]
31. Penha, K.-V.-D.F.; Sousa, A.-C.-S.; Oliveira, C.-A.; De Andrade, R.-S.-B.; Vasconcelos, D.-F.-P. A swift, easy and cheap protocol to evaluate the tooth bleaching in vitro. *J. Clin. Exp. Dent.* **2018**, *10*, e579–e584. [CrossRef]
32. Suyama, Y.; Otsuki, M.; Ogisu, S.; Kishikawa, R.; Tagami, J.; Ikeda, M.; Kurata, H.; Cho, T. Effects of light sources and visible light-activated titanium dioxide photocatalyst on bleaching. *Dent. Mater. J.* **2009**, *28*, 693–699. [CrossRef] [PubMed]
33. Suemori, T.; Kato, J.; Nakazawa, T.; Akashi, G.; Igarashi, A.; Hirai, Y.; Kumagai, Y.; Kurata, H. Effects of light irradiation on bleaching by a 3.5% hydrogen peroxide solution containing titanium dioxide. *Laser Phys. Lett.* **2008**, *5*, 379–383. [CrossRef]
34. Zhang, J.; Zhou, P.; Liu, J.; Yu, J. New understanding of the difference of photocatalytic activity among anatase, rutile and brookite TiO2. *Phys. Chem. Chem. Phys.* **2014**, *16*, 20382–20386. [CrossRef] [PubMed]
35. Sakai, K.; Kato, J.; Nakazawa, T.; Hirai, Y. Bleaching effect of a 405-nm diode laser irradiation used with titanium dioxide and 3.5% hydrogen peroxide. *Laser Phys.* **2007**, *17*, 1166–1170. [CrossRef]

International Journal of
Molecular Sciences

MDPI

Article

Role of Phage Capsid in the Resistance to UV-C Radiations

Laura Maria De Plano [1], Domenico Franco [1,*], Maria Giovanna Rizzo [1], Vincenzo Zammuto [1], Concetta Gugliandolo [1], Letteria Silipigni [2], Lorenzo Torrisi [2] and Salvatore P. P. Guglielmino [1]

[1] Department of Chemical, Biological, Pharmaceutical and Environmental Sciences, University of Messina, 98166 Messina, Italy; ldeplano@unime.it (L.M.D.P.); mgrizzo@unime.it (M.G.R.); vzammuto@unime.it (V.Z.); cgugliandolo@unime.it (C.G.); sguglielm@unime.it (S.P.P.G.)

[2] Department of Mathematical and Computational Sciences, Physical Sciences and Earth Sciences, University of Messina, 98166 Messina, Italy; lsilipigni@unime.it (L.S.); ltorrisi@unime.it (L.T.)

* Correspondence: dfranco@unime.it

Abstract: The conformational variation of the viral capsid structure plays an essential role both for the environmental resistance and acid nuclear release during cellular infection. The aim of this study was to evaluate how capsid rearrangement in engineered phages of M13 protects viral DNA and peptide bonds from damage induced by UV-C radiation. From in silico 3D modelling analysis, two M13 engineered phage clones, namely P9b and 12III1, were chosen for (i) chemical features of amino acids sequences, (ii) rearrangements in the secondary structure of their pVIII proteins and (iii) in turn the interactions involved in phage capsid. Then, their resistance to UV-C radiation and hydrogen peroxide (H_2O_2) was compared to M13 wild-type vector (pC89) without peptide insert. Results showed that both the phage clones acquired an advantage against direct radiation damage, due to a reorganization of interactions in the capsid for an increase of H-bond and steric interactions. However, only P9b had an increase in resistance against H_2O_2. These results could help to understand the molecular mechanisms involved in the stability of new virus variants, also providing quick and necessary information to develop effective protocols in the virus inactivation for human activities, such as safety foods and animal-derived materials.

Keywords: M13 engineered phage; UV-C; ionizing direct and indirect damage; virus stability

Citation: Plano, L.M.D.; Franco, D.; Rizzo, M.G.; Zammuto, V.; Gugliandolo, C.; Silipigni, L.; Torrisi, L.; Guglielmino, S.P.P. Role of Phage Capsid in the Resistance to UV-C Radiations. *Int. J. Mol. Sci.* **2021**, *22*, 3408. https://doi.org/10.3390/ijms22073408

Academic Editor: Victor Ladero

Received: 1 March 2021
Accepted: 23 March 2021
Published: 26 March 2021

1. Introduction

Viruses are biological entities, consisting of a single element of nucleic acid (RNA or DNA, in either single- or double-stranded forms) enclosed in a protective protein shell, or capsid, and are only able to replicate inside living cells. The virus's resistance to environmental stresses (temperature, pH, interaction with host cells, etc.) is closely related to the protein composition of the external structures of the virion. This is generally made up of repeated protein sub-structures, named capsomers, organized in polyhedral or helical symmetry [1]. These simple geometries can be formed with a limited number of capsid protein subunits, increasing their volume easily by repeatedly using the same proteins and consequently limiting the number of genes required for the synthesis of the capsid. The arrangement of capsid proteins in helical symmetry is a highly effective way to form tridimensional structures with repetitive subunits that, following the symmetry of the nucleic acid, have the same identical interactions that capsomeres have with each other. Therefore, modifications of capsid proteins, even of a single amino acid, may change the stability and/or metastability of the virion particles [2]. This eventuality is critical for the risk of viral infection, only reduced by several control techniques, such as heat sterilization, chemical disinfectants, air filtration, and ultraviolet (UV) irradiation. In particular, disinfection using UV radiation is a fast-growing because it is chemical-free technology and with low risk of damage to materials (especially heat-labile ones) [3]. For this reason, main parameters should be considered for the elaboration of protocols to guarantee the efficacy of the UV sterilization process, mainly in the latest period strongly marked by COVID-19 pandemic, where UV disinfection of surfaces and air has attracted tremendous attention [4].

Bacteriophages (or simply phage) are viruses that infect bacteria and have several biotechnological applications, including their use as models for human viruses. Specifically, M13 phage is well known for the development of the phage display technique, which allows at foreign peptides to be exposed on the virion capsid [5]. The capsid of M13 phage is characterized by a helical symmetry structure, almost exclusively constituted by the major coat protein pVIII (responsible for phage integrity), and four other minor coat proteins, namely pIII, pV, pVII and pIX, which together enclose the circular single-stranded DNA molecule [6]. In the phage display application, the engineering with foreign peptides in the pVIII protein (representing about 98% of capsid proteins) permits the formation of landscape phage libraries of billions of clones, wherein each individual phage displays a random peptide on its surface. Phage libraries based on vector systems (88 or 8 + 8) result in a hybrid structure with mixture of recombinant and wild-type pVIII molecules on the phage capsid [7]. These random phage libraries are used in screening processes against desired targets to isolate specific ligands in the marker discovery, in vitro or in vivo [8–10], or to functionalize surfaces for biosensors [11–14], as specific scaffolds for tissue-regenerating [15], for immunization [16], for drug-targeting [17–19], for hybrid-materials [20–22], for quantum dots, and for semi-conducting or magnetic nanowire devices [23,24]. It is known that the capsid structure has been suggested to be responsible for the great resistance of M13 to various physical and chemical stresses (i.e., heat, both acidic or alkaline organic solvents, and variations in pH values) [25,26]. About the susceptibility to the UV-C exposure, genome type and organization can play an important role against UV inactivation of phage. Phages with circular ssDNA, such as PhiX174, are most sensitive respect to that with linear ssRNA, such as MS2; while phage with linear dsDNA, such as PRD1, revealed the highest photoreactivation after UV exposure [27,28]. In contrast to UV inactivation, MS2 showed a greater susceptibility to gamma or electron beam irradiation compared to PhiX174, whose inactivation revealed to have a clear dose rate effect [29]. Since phages have a limited ability to repair DNA, the shielding of their genomes by capsids may play a key role in their resistance to radiation. Studies on phage T7 would seem to highlight a greater sensitivity towards UV inactivation of intraphage DNA compared to the isolated one, indicating the role of phage proteins in the DNA damage [30]. Moreover, it has been seen that fd virus (with single-stranded nucleic acid) shows a UV sensibility midway between characteristic for the single-stranded and double-stranded nucleic acid viruses [31]. These findings had been attributed to the capsid structure, consisting in a helical structure, with a 60 A pitch and six repeating units per turn that protect the phage nucleic acid from tertiary structural changes induced by the UV-C irradiation [31,32]. On the other hand, effects on biological components from UV-C exposure also involve protein, other than nucleic acids (RNA or DNA) [3]. Although many studies have been focused on the influence of nucleic acid on the UV-C radiation response, at our knowledge, no study has been reported on the role of physical shape and protein on the virus resistance to UV-C or other qualities and quantities of radiations.

The structural model of the M13 capsid, analogue of pC89 wild-type vector, has been used to derive the newly rearrangement and electrostatic interactions due to the addition or modification of amino acids in the capsid structure [6]. In this way, Passaretti et al. investigated the surface charge and the surface area of M13 to predict the mechanisms behind of interactions in bio-nanocomponents, finding that the presence of ionizable groups in foreign peptides can drastically change the chemical characteristics of the engineered virus compared to wild-type, playing a role in the resistance to several stresses [33].

The aim of this study was to evaluate how the capsid rearrangement protects DNA and peptide bonds from damage after exposure to UV-C radiations. By speculative in silico 3D models, two engineered phages, previously selected against *Pseudomonas aeruginosa* (P9b) and antibodies binding β-amyloid structures (12III1), have been chosen for (i) chemical features of amino acids sequences, (ii) rearrangements in the secondary structure of their pVIII proteins and (iii) in turn the interactions involved in phage capsid. Then, their resistances to UV-C radiations were investigated and compared with M13 wild-type vector (pC89) without peptide insert. In addition, the resistance to hydrogen peroxide (H_2O_2), one of the most common products generated by the radiolysis of water, has been also evaluated for indirect damage due to the radiation exposure.

2. Results

2.1. In Silico 3D Models

Engineered phage shows a mosaic of wild-type and recombinant pVIIIs proteins on the capsid structure. The 3D in-silico models provided the constructions of recombinant pVIII proteins with in-frame the amino acids of foreign peptides by Modeller. From 3D models of pVIII proteins, we deduced that the presence of the foreign peptide induced a linear extension of the N-terminal end or a circular form in the same position. In Figure S1 is showed the rearrangement of the secondary structure of the recombinant PVIII of P9b (amino acid sequence of the foreign peptide QRKLAAKLT) and 12III1 (RWPPHFEWHFDD). The interactions between the PVIIIs of the hybrid phage structure, consisting in a mixture of recombinant and wild-type pVIII molecules, also resulted in a new supramolecular rearrangement with new H-bond and steric interactions (Figure 1).

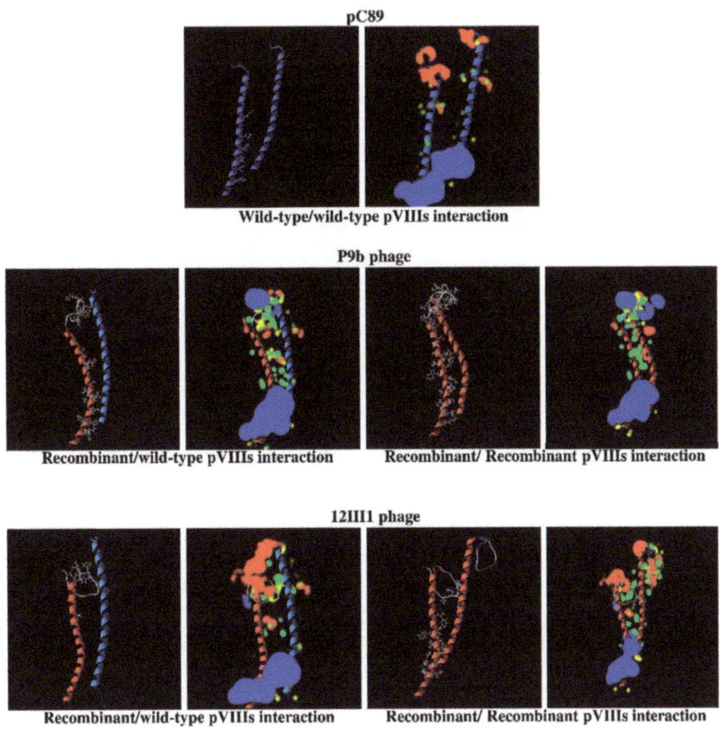

Figure 1. Speculative in silico 3D models of interactions between wild-type (chain blue) and recombinant (chain red) pVIIIs in pC89 (upper panel), P9b (central panel) and 12III1 (lower panel). For each phage, the two pVIII proteins are displayed in the same workspace in backbones style, with the amino-acids of the chains involved in the interaction in wireframe style and coupled to energy map-colored clouds. For energy map, legend colors: in green, the amino acids with stearic favorable bond; in red, a nearby negative electro-static charge of amino acids; in blue, a nearby positive charge of amino acids; in yellow, hydrogens donator favorable; in light blue, hydrogens acceptor favorable.

The electrostatic charges highlighted by the colored clouds around the models indicate that the aggregation levels of the chains were modified in all the sequence in P9b and 12III1 phages respect to pC89. 3D in-silico models also allow to derive the amino-acids, including their position number in the primary sequence, which participate in H-bond and steric interactions between the PVIII proteins (in Table 1 and Figure S2).

Table 1. Position number in the primary sequence of amino acids involved in the interactions between PVIII proteins in pC89, P9b and 12III1 phages, deduced by in silico 3D modeling analysis (Figure S2).

	pC89	**P9b (QRKLAAKLT)**		**12III1 (RWPPHFEWHFDD)**	
	Wild-Type/ Wild-Type pVIIIs	**Recombinant/ Wild-Type pVIIIs**	**Recombinant/ Recombinant pVIIIs**	**Recombinant/ Wild-Type pVIIIs**	**Recombinant/ Recombinant pVIIIs**
H-bond	43	42–43, 45–46, 49	1–2, 4–5, 7–10, 22–23, 31, 42, 45–46, 53	14	12–13, 40–42, 45–47, 49–54
Steric interactions	21, 28, 32, 35–36, 39	1, 4–5, 7–8, 26–27, 30–31, 34, 37–46, 48–50, 52–53	1–12, 14–16, 18–20, 22–24, 26–28, 30–32, 34, 38, 41–43, 45–50, 53	13–17, 20, 31	9–14, 34, 38, 40–54

pC89 showed the lowest number of amino acids involved in the interaction between wild-type pVIIIs (1 and 6 for H-bond and steric interaction, respectively). Otherwise, in P9b phage, amino acids involved in recombinant/wild-type pVIIIs interaction are 5 (H-bond) and 25 (steric interaction), while recombinant/recombinant pVIIIs interaction are 15 (H-bond) and 39 (steric interaction). Finally, in 12III1 phage, amino acids involved in recombinant/wild-type pVIIIs interaction are 1 (H-bond) and 7 (steric interaction), while recombinant/recombinant pVIIIs interaction are 14 (h-bond) and 23 (steric interaction).

2.2. UV-C Resistance Test

To assess whether the capsid rearrangement, due to the foreign peptide in PVIII protein, could preserve the infectivity of engineered phages (P9b and 12III1) from effects of radiations, phages were subjected to different UV-C radiations. Before comparative tests between phages, the radio-resistance of pC89 alone was tested up to the lethal dose able to completely reduce the infecting phage load. The last was quantified as Transducing Unit per milliliter (TU/mL), that means number of active virus particles able to infect the bacterial host.

A reduction of about eight orders of magnitude was already found after 1540 J/m^2, corresponding to 70 s of radiation exposure, while no TU were observed after 3080 J/m^2 (that means 140 s of radiation exposure).

Based on these findings, the radio-resistances of the engineered phage, P9b (amino acid sequence QRKLAAKLT) and 12III1 (RWPPHFEWHFDD) were compared to that of pC89 up to 3080 J/m^2 (Table 2).

Table 2. Active virus particles of pC89, P9b and 12III1, expressed as Transducing Unit per milliliter of infected *E. coli* (TU/mL), after exposure to UV-C.

Phage	**Exposed**	**385 J/m^2**	**770 J/m^2**	**1540 J/m^2**	**3080 J/m^2**
pC89	$(4 \pm 0.9) \times 10^9$	$(3 \pm 1.7) \times 10^7$	$(2.2 \pm 1.2) \times 10^5$	$(3.1 \pm 1.3) \times 10$	No TU
P9b (QRKLAAKLT)	$(3.9 \pm 0.2) \times 10^9$	$(1.1 \pm 0.6) \times 10^9$	$(2.6 \pm 0.7) \times 10^7$	$(3.2 \pm 1.3) \times 10^5$	$(1.4 \pm 0.7) \times 10^4$
12III1 (RWPPHFEWHFDD)	$(3.9 \pm 0.9) \times 10^9$	$(1.8 \pm 0.9) \times 10^8$	$(5.7 \pm 1.4) \times 10^6$	$(1.2 \pm 1) \times 10^4$	$(1.7 \pm 0.9) \times 10^2$

After UV-C exposure to a 1540 J/m^2 of fluence the 12III1 and P9b infectivity decreased by about six and four orders of magnitude, respectively. Both engineered phages' resistance further decreased by about one order of magnitude after the exposure to a 3080 J/m^2 fluence.

UV-C inactivation rates and related UV-C damage in pC89 wild-type, P9b and 12III1 phages are reported in Figure 2.

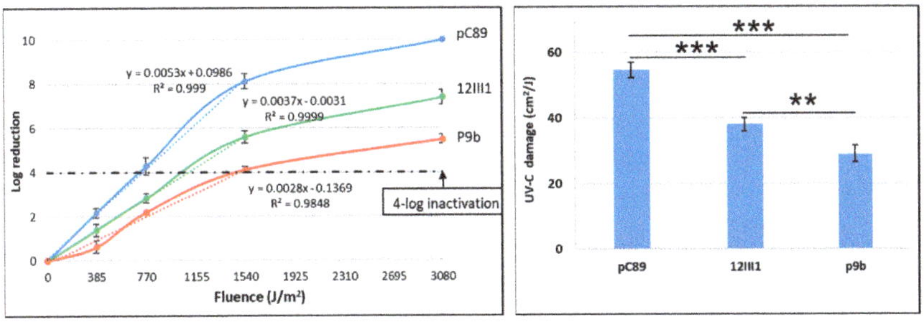

Figure 2. UV inactivation rates (**left** panel) and related UV-C damage (**right** panel) of pC89, 12III1 and P9b phage. The black dotted line marks the 4-log inactivation level to deduce the UV-C damage by linear trendline to the exponential phase of log-inactivation for each phage. For ANOVA test from Tukey's multiple comparisons test, two (**) and three (***) asterisks identify *p*-value < 0.001 and 0.0001, respectively. For each point, the mean and standard deviation were derived from 5 experimental data.

For all phages, UV-C inactivation rate showed a logarithmic trend, consisting in a first dose-dependent exponential phase (up to 1540 J/m^2). Specifically, the UV-C damage for each phage was determined at the threshold value of four-log reduction. pC89 showed a significantly higher value than the two engineered phages (54.5 ± 2.3 J/cm^2) than the two engineered phages (38 ± 2 and 29 ± 2.5 J/cm^2 for 12III1 and P9b, respectively). The difference in resistance to UV-C exposure of the phages was also maintained at 3080 J/m^2 (maximum dose tested in our experiments). In addition, a significant difference in susceptibility to UV-C exposure was also observed between the two engineered phages (Figure 2, left panel). In fact, P9b showed the lowest UV-C damage value, that was 24% lower than 12III1 (Figure 2, right panel).

2.3. Hydrogen Peroxide Resistance Test

The infectivity of pC89, P9b and 12III1 phages (TU/mL) after treatment with 1.5 M H_2O_2 for several incubation times is reported in Table 3.

Table 3. Active virus particles of pC89, P9b and 12III1, expressed as Transducing Unit per milliliter of infected *E. coli* (TU/mL), after exposure to H_2O_2 (1.5 M).

Phage	Exposed	30 Min	60 Min	90 Min
pC89	(1.2 ± 0.2) × 10^{10}	(5 ± 0.4) × 10^9	(2.5 ± 0.3) × 10^9	(1.2 ± 0.2) × 10^9
P9b (QRKLAAKLT)	(4.0 ± 0.3) × 10^{10}	(3.2 ± 0.4) × 10^{10}	(3 ± 0.3) × 10^{10}	(2.8 ± 0.1) × 10^{10}
12III1 (RWPPHFEWHFDD)	(3.4 ± 0.2) × 10^{10}	(1.7 ± 0.2) × 10^9	(1.1 ± 0.2) × 10^9	(5.7 ± 0.6) × 10^9

The pC89 and 12III1 infectivity decreased by about 0.5 orders of magnitude after 30 and 60 min, and one order of magnitude after 90 min, whereas P9b was the most resistant by a significant margin, since TU maintained viability throughout all treatment durations (Figure 3).

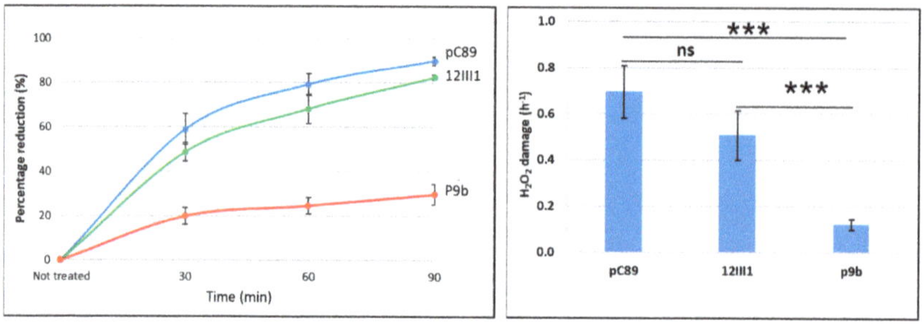

Figure 3. H_2O_2 inactivation rates (**left** panel) and related H_2O_2 damage (**right** panel) of pC89, 12III1 and P9b phage. For ANOVA test from Tukey's multiple comparisons test, three (***) asterisks identify p-value < 0.0001. ns = not significant. For each point, the mean and standard deviation were derived from 5 experimental data.

Additionally in this case, inactivation of phages has a logarithmic trend, consisting in a first dose-dependent exponential phase in the first 30 min. In addition, P9b maintained the highest resistance to treatment, as confirmed by the lowest H_2O_2 damage value, that was 82% and 75% lower than pC89 and 12III1, respectively. However, respect to UV-inactivation, no significant difference between pC89 and 12III was observed, while as also confirmed by H_2O_2 damage (Figure 3, right panel).

3. Discussion

Some studies have reported virus inactivation after exposure to different types of radiation (i.e., electron beam, gamma irradiation, lethal ultraviolet), particularly applied to water and food safety [34] or to phage therapy to plants [35]. However, the extraordinary diversity of viruses does not allow defining universal parameters about their susceptibility to the radiation exposure. Indeed, opposing conclusions have been reported about the inactivation of bacteriophages by electron beam and gamma irradiation [29].

Since phages have limited ability to repair DNA, viruses' resistance is closely related to the protein composition of the external structures of the virion. Although structural studies have been carried out by mutational inductions abrogating/increasing interactions within viral structures, they generally do not consider the effects of radiations on viral structures, as well as the molecular mechanisms responsible for more or less resistance to infectivity inactivation. Moreover, these studies cannot apply to engineered phages, for which the addition of amino acid sequences in viral capsid structure derive from screening/selection of engineered phages able to detect specific target, key step of research projects for the development of new diagnostic/therapeutic systems.

In this work, we investigated whether the foreign peptides in the N-terminal pVIII of the P9b and 12III1 phages, could influence the resistance to UV-C.

First of all, speculative 3D models have been used to screen the engineered phages (available in our laboratories) to employ in the biological tests. Based on the hydrogen (H) and steric bonds that could occur on the hybrid capsid structure, two engineered phage clones were chosen, because of more (P9b) and less (12III1) interactions due to their foreign amino acids. Specifically, in P9b the foreign peptide (amino acid sequence QRKLAAKLT) led to linear extension of engineered pVIIIs, respect to the pVIII wild type counterpart, with a significant increase of interactions in the hybrid capsid structure, both recombinant/wild-type and recombinant/recombinant pVIIIs (Figure 1; Table 1). Otherwise, in 12III1 the foreign peptide (amino acid sequence RWPPHFEWHFDD) induced a curvature of the engineered pVIIIs, with an increase of interaction numbers only between the recombinant pVIIIs (Figure 1; Table 1). Because of the strong interaction between capsomers and nucleic acids, the capsid rearrangement of engineered phages could protect the viral DNA from tertiary structural changes induced by the surrounding environmental stresses, such as UV-C radiation.

Results from the experimental design of radiation exposure revealed that the presence of the foreign peptide had a greater effect on phage infectivity. In fact, pC89 was almost totally inhibited after the exposure to UV-C for 1540 J/m^2 fluence, according to the data reported previously [32]. In contrast, P9b and 12III1 phages maintained their viability under the UV-C exposure until 3080 J/m^2 ($1.4 \pm 0.7 \times 10^4$ and $1.7 \pm 0.9 \times 10^2$ TU/mL for P9b and 12III, respectively). These results suggest that both the phage clones had acquired an advantage against direct radiation damage from UV-C exposure, respect to pC89. Moreover, the increase in UV-C resistance was in agreement with the increase of the interactions (predicted by the model) in the capsids of the engineered phages, as evidenced by the higher resistance of P9B compared to 12III1. H-bond and steric interactions favorably contribute to protein stability and could reduce the susceptibility of the hybrid phage structure to the modifications induced by UV-C, as also suggested by the experimental results. In fact, the number increasing of H-bond and steric interactions (pC89 < 12III1 < P9b) was positively correlated with the phage resistance to UV-C.

UV-C radiation, usually used in the disinfection of wastewater, is predominantly absorbed by nucleic acids, causing damage by the formation of photoproducts [36]. Formation of dimers (cyclobutane pyrimidine and pyrimidine) and photoproducts (pyrimidine) has been reported to increase in proportion to the UV-C dose, modifying the DNA structure. On the other hand, radiations can induce an indirect damage, due to OH^- and H^+ products resulting from the water radiolysis [37]. Since the latter mainly involves protein structures, including individual amino acids, the radiolysis of water could play a role in the different behavior of the two engineered phage clones to ionizing radiation. At this purpose, the infectivity of the tested phage has also been evaluated after treatment with 1.5 M H_2O_2. Surprisingly, only P9b did not lose its viability, while infectivity of pC89 and 12III1 were decreased of 1 and 0.5 orders of magnitude, respectively, after 90 min of treatment (Table 3 and Figure 3). In this prospective, the amino acid composition of the foreign peptide could alter the degree of oxidative modifications in the capsid proteins, due to H_2O_2, which can be considered the most common indirect damage from radiation exposure. Specifically, peroxyl radicals can induce the formation of oxidation products by radical mechanisms, leading to fragmentation of the protein backbone. It is known that aromatic amino acids are significantly more reactive with the dominant reaction pathway being OH^- addition to the aromatic ring [38]. In this work, 12III1 phage has four aromatic amino acids in its foreign peptide, while P9b none. Consequently, hybrid phage structure of 12III1 would not have an advantage in H_2O_2 resistance, when compared to pC89. The remaining amino acids of the foreign peptide of 12III1 (namely arginine, proline, histidine, glutamic acid and aspartic acid) could also negatively affect the oxidation resistance of the phage clone. In fact, the radiation-generated ROS modifies protein by carbonylation to the amino-acid side chains, particularly occurring on Lys, His, Thr, Pro, Glu, Asp, and Arg residues [38]. On the other hand, amino acids sequence of the P9b foreign peptide, with only one amino acid favorable to carbonylation (namely arginine) makes the phage clone more resistant to H_2O_2 respect to other phage, subject of this study.

4. Conclusions

Phage display technique allows to screen and select peptides with several biotechnological functions. These foreign peptides are exposed in-frame inside the capsid protein pIII or pVIII. Consequently, the entire capsid structure may be affected by the onset of new interactions. In order to understand the incidence of capsid organization on resistance to environmental stresses, in this work the resistance to UV-C radiation of M13 wild-type vector (pC89) and two engineered phage clones in pVIII, expressing 9 or 12 additional amino acids in N-terminal end of pVIII capsid protein, was evaluated.

Our data indicate that the presence of the foreign peptide increase the number of interactions between the pVIII in engineered phages, bringing advantages in resistance to UV-C radiation. Moreover, among the engineered phages under study, the increase in UV-C resistance was directly linked to the increase in the number of interactions, which in turn depended on the peptide sequence of the exogenous peptide. Phages engineered to express known peptides on their capsid could represent models extremely useful to understand the molecular mechanisms involved in the stability of new virus variants, also providing quick and necessary information to develop effective protocols in the virus inactivation for human activities, such as safety foods and animal-derived materials.

5. Materials and Methods

5.1. Bacteriophages

Phages used in this work were obtained from two M13 phage display libraries (kind gift of Prof. F. Felici). These libraries consist of filamentous phage particles displaying random 9- or 12-mer peptides fused to the major coat protein (pVIII). The libraries were constructed in the vector pC89 [39], by cloning a random DNA insert between the third and fifth codon of the mature pVIII-encoding segments of gene VIII [40]. pC89 (M13 wild-type vector without peptide insert) and two engineered phage clones with different amino acid sequences of the foreign peptide were used. Specifically, the P9b phage clone displayed the sequence QRKLAAKLT [41], while the 12III1 the sequence RWPPHFEWHFDD [42]. Propagation of pC89 and engineered phage clones was carried out in *Escherichia coli* TG1 bacterial host (Lac-Z deleted).

5.2. Phages Production

E. coli strain TG1 (Kan-, Amp-, lacZ-) broth culture (OD_{600} = 0.7) was infected with pC89 or engineered phage clone (Amp+), then incubated at 37 °C in static condition for 15 min, followed by shaking (250 rpm) for 20 min. After incubation, suitable aliquots of culture were plated onto Luria–Bertani agar (agar 20 g/L) plates containing ampicillin (50 μg/mL) and incubated at 37 °C in static condition. One colony of *E. coli* strain TG1, containing phage, was inoculated into 10 mL of LB medium containing ampicillin (50 μg/mL) and incubated at 37 °C with shaking (250 rpm) until reaching OD_{600} = 0.2. Then, the culture was added with isopropylthio-β-galactoside (IPTG, 40 μg/mL) and helper phage M13K07 (Kan+) (10^9 TU/mL), incubated at 37 °C in static condition for 30 min, and gently shaken for 30 min. The cells were harvested by centrifugation at 8000× *g*, transferred to 500 mL of LB medium containing ampicillin (50 μg/mL) and kanamycin (10 μg/mL), and incubated overnight with shaking at 37 °C. The infected culture was centrifuged 8000× *g* for 20 min at 25 °C, the supernatant was then mixed with 25% (*v*/*v*) of PEG/NaCl solution, cooled in ice for 4 h, and precipitated by centrifugation at 15,000× *g* for 45 min at 4 °C. The pellet was resuspended in 10% (*v*/*v*) of TBS, mixed again with 25% (*v*/*v*) of PEG/NaCl, cooled in ice for 4 h, and the solution was centrifuged as above. The pellet, containing phage particles, was suspended in 10% (*v*/*v*) of TBS, filtered through 0.22 μm-pore size membrane, and stored at 4 °C.

5.3. In Silico 3D Modelling Analysis

Engineered pVIII Protein Structure Modeling was performed using the MODELLER9.20 (https://salilab.org/modeller, (accessed on 15 March 2021).) software for the comparative protein structure modeling. The models were built as reported previously [43]. Briefly, the structure of pVIII protein (PDB ID: 2mjz) was obtained. In the context of the whole virus particle of 2mjz, two pVIII proteins, Chain [1a] and Chain [1e], have been chosen for modeling the pVIII engineered proteins. Since these proteins are located at the same position between pentamer rings 1° and 3°, they allow identification of the contributing protein-protein interactions in the capsid structure [6]. The two pVIII proteins were kept like the single pVIII protein, in a PDB-format, namely chain A and chain E, and used as templates. Amino acid sequences of engineered pVIII proteins of P9b and 12III1 clones, with a foreign peptide inside of fourth/fifth amino acids of the M13 wild-type pVIII (plus two extra residues of Phenylalanine and Glutamine, encoded by the EcoRI site as reported in Felici et al. [39] were written on the following FASTA format to derive pVIII-engineered models (in bold the foreign amino acid sequence):

1-AEGDDPAKAAFNSLQASATEYIGYAWAMVVVIVGATIGIKLFKKFTSKAS-50 (pC89);
1-AEGEFQRKLAAKLTDPAKAAFNSLQASATEYIGYAWAMVVVIVGATIGIKLFKKFTSKAS-60 (P9b);
1-AEGEFRWPPHFEWHFDDGDPAKAAFNSLQASATEYIGYAWAMVVVIVGATIGIKLFKKFTSKAS-64 (12III1).

Several models were calculated for the same target. The best 3D model was selected, according the lowest value of DOPE, which indicates the construction energy. For each engineered pVIII protein, two models were obtained, one from the chain A and one from the chain E of the capsid complex.

5.4. Analysis of Amino Acids Involved in the Structure Capsid Interactions

The models performed in PDB format were opened and processed using Molegro Molecular Viewer v1.2.0 (http://www.molegro.com (accessed on 20 February 2020)). The models were processed as ligand-protein in "Energy Map" and "Ligand Map" to identify the amino acids involved in the interactions. ProtParam (https://web.expasy.org/protparam (accessed on 20 February 2020)) bioinformatics tool was used to predict the following characteristics: molecular weight, theoretical pI, amino acid composition, atomic composition, extinction coefficient, estimated half-life, instability index, aliphatic index, and grand average of hydropathicity (GRAVY) of the amino-acid sequence of the peptide [44].

5.5. Virus Irradiation

Due to the extremely low penetrative ability of the UV-C radiation, were separately placed in uncovered petri dishes (sample exposure surface and thickness approximately 0.8 cm^2 and 0.15 cm, respectively) and exposed to UV-C radiation (G15T8/OF, OSRAM germicidal, puritec HNS 15WG13, operating with the emission line at the 253.7 nm wavelength) for different time exposure at a distance of 150 mm. Specifically, the germicidal lamp has a UV output power of 4.9 W and an intensity of 50 $\mu W/cm^2$ at a 1-m distance. Since the intensity decreases with the square of the distance between the lamp and the target, at a 15 cm distance from our operation, the lamp intensity is 2.2 mW/cm^2. All irradiations were carried out in the biological hood at 25 °C. After the treatment, the infective ability of recovered phages was evaluated by titration. All tests were performed in quintuplicate.

5.6. Hydrogen Peroxide Resistance Assay

Hydrogen peroxide (H_2O_2) was evaluated at a sub-lethal dose, according to Eisenstark et al. [45]. In detail, 30% stock solution of H_2O_2 was diluted with sterile double distilled water immediately prior to each experiment. Each phage stock (1×10^{11} phage/mL) in TBS (100 µL) was treated with 1.5 M H_2O_2 (final concentration) for 30, 60, and 90 min. At each time, 10 µL catalase (2 mg/mL) were added to stop the reaction. All tests were carried out in the biological hood at room temperature. After the treatments, the infective ability of the recovered bacteriophages was evaluated by titration, as described below. All tests were performed in quintuplicate.

5.7. Phage Titration (TU/mL)

Tenfold serial dilutions of pC89 or engineered phage clones were prepared, and 10 µL of each phage sample were dispensed into sterile micro-centrifuge tubes containing 90 µL of *E. coli* TG1 culture (OD_{600} = 0.7). Tubes were first incubated at 37 °C for 15 min in static conditions and then for 20 min in shaking (250 rpm) conditions. 100 µL of pC89/*E. coli* TG1 or engineered phage/*E. coli* TG1 suspension was spotted onto LB agar plates containing ampicillin (50 µg/mL). All plates were incubated at 37 °C overnight. Colonies from plates (between 30–300) were counted and the number of active virus particles was determined as Transducing Unit per milliliter (TU/mL) according to the following Equation (1)

$$TU = \frac{\text{number of colonies}}{\text{volume (0.1 mL)} \times \text{diluition factor}} \quad (1)$$

All assays were performed in quintuplicate.

Using the TU/mL value, the damage from UV-C or H_2O_2 per fluence (J/m^2) or time (h) was derived from the following Equation (2)

$$\text{UVC or } H_2O_2 \text{ damage} = \frac{\log_{10} Ns(r) - \log_{10} Ns(s)}{\text{fluence}\left(\frac{J}{cm^2}\right) \text{ or time (h)}} \quad (2)$$

where Ns(r) and Ns(s) represent the number of active virus particles (TU/mL) not irradiated and irradiated, respectively. The significant differences in the UVC or H_2O_2 damage values were determined by analysis of variance (ANOVA) using adjusted p-values at 0.01 and 99% for family-wise significance and confidence level, respectively.

Supplementary Materials: The following are available online at https://www.mdpi.com/article/10.3390/ijms22073408/s1, Figure S1: 3D models of the pVIII protein chain A in pC89 (A), P9b (B) and 12III1 (C), with the amino-acids of the foreign peptide in wireframe style. (PDB ID: 2mjz, Morag et al., 2015). Figure S2: Ligand map from 3D models of the interactions in pC89, P9b and 12III1 phage, including abbreviation and position of the amino acids involved in H-bond and steric interactions.

Author Contributions: L.M.D.P., D.F., C.G., L.T. and S.P.P.G. planned the experiments. L.M.D.P., D.F., M.G.R., V.Z., C.G. and S.P.P.G. performed, acquired and interpreted biological experiments. L.M.D.P. performed bioinformatics and statistical analyses. L.T. and L.S. performed, acquired and interpreted radiation experiments. L.M.D.P., D.F., C.G., L.T. and S.P.P.G. analyzed the results and prepared the manuscript. All authors have read and agreed to the published version of the manuscript.

Funding: This research received no external funding.

Institutional Review Board Statement: Not applicable.

Informed Consent Statement: Not applicable.

Acknowledgments: The authors wish to thank Franco Felici for the kind gift of phage display M13 libraries.

Conflicts of Interest: The authors declare no conflict of interest.

References

1. Prasad, B.V.; Schmid, M.F. Principles of virus structural organization. *Adv. Exp. Med. Biol.* **2012**, *726*, 17–47. [CrossRef] [PubMed]
2. Mateu, M.G. Virus engineering: Functionalization and stabilization. *Protein Eng. Des. Sel.* **2011**, *24*, 53–63. [CrossRef] [PubMed]
3. Raeiszadeh, M.; Adeli, B. A Critical Review on Ultraviolet Disinfection Systems against COVID-19 Outbreak: Applicability, Validation, and Safety Considerations. *ACS Photonics* **2020**, *7*, 2941–2951. [CrossRef]
4. Shining a light on COVID-19. *Nat. Photonics* **2020**, *14*, 337. [CrossRef]
5. Petrenko, V.A.; Smith, G.P. Phages from landscape libraries as substitute antibodies. *Protein Eng.* **2000**, *13*, 589–592. [CrossRef]
6. Morag, O.; Sgourakis, N.G.; Baker, D.; Goldbourt, A. The NMR-Rosetta capsid model of M13 bacteriophage reveals a quadrupled hydrophobic packing epitope. *Proc. Natl. Acad. Sci. USA* **2015**, *112*, 971–976. [CrossRef] [PubMed]
7. Smith, G.P.; Petrenko, V.A. Phage display. *Chem. Rev.* **1997**, *97*, 391–410. [CrossRef]
8. Wu, C.H.; Liu, I.J.; Lu, R.M.; Wu, H.C. Advancement and applications of peptide phage display technology in biomedical science. *J. Biomed. Sci.* **2016**, *23*, 8. [CrossRef]
9. Rakonjac, J.; Bennett, N.J.; Spagnuolo, J.; Gagic, D.; Russel, M. Filamentous bacteriophage: Biology, phage display and nanotechnology applications. *Curr. Issues Mol. Biol.* **2011**, *13*, 51–76. [PubMed]
10. Gillespie, J.W.; Yang, L.; De Plano, L.M.; Stackhouse, M.A.; Petrenko, V.A. Evolution of a landscape phage library in a mouse xenograft model of human breast cancer. *Viruses* **2019**, *11*, 988. [CrossRef] [PubMed]
11. Han, L.; Liu, P.; Petrenko, V.A.; Liu, A.H. A Label-Free Electrochemical Impedance Cytosensor Based on Specific Peptide-Fused Phage Selected from Landscape Phage Library. *Sci. Rep.* **2016**, *6*, 10. [CrossRef]

12. Han, L.; Wang, D.; Yan, L.; Petrenko, V.A.; Liu, A. Specific phages-based electrochemical impedimetric immunosensors for label-free and ultrasensitive detection of dual prostate-specific antigens. *Sens. Actuators B Chem.* **2019**, *297*, 126727. [CrossRef]
13. Rizzo, M.G.; Carnazza, S.; De Plano, L.M.; Franco, D.; Nicolò, M.S.; Zammuto, V.; Petralia, S.; Calabrese, G.; Gugliandolo, C.; Conoci, S.; et al. Rapid detection of bacterial pathogens in blood through engineered phages-beads and integrated Real-Time PCR into MicroChip. *Sens. Actuators B Chem.* **2021**, *329*, 129227. [CrossRef]
14. De Plano, L.M.; Fazio, E.; Rizzo, M.G.; Franco, D.; Carnazza, S.; Trusso, S.; Neri, F.; Guglielmino, S.P.P. Phage-based assay for rapid detection of bacterial pathogens in blood by Raman spectroscopy. *J. Immunol. Methods* **2019**, *465*, 45–52. [CrossRef] [PubMed]
15. Yoo, S.Y.; Merzlyak, A.; Lee, S.W. Synthetic phage for tissue regeneration. *Mediat. Inflamm.* **2014**, 192790. [CrossRef] [PubMed]
16. Bazan, J.; Całkosiński, I.; Gamian, A. Phage display-a powerful technique for immunotherapy: 1. Introduction and potential of therapeutic applications. *Hum. Vaccines Immunother.* **2012**, *8*, 1817–1828. [CrossRef] [PubMed]
17. Petrenko, V.A.; Gillespie, J.W.; Xu, H.; O'Dell, T.; De Plano, L.M. Combinatorial avidity selection of mosaic landscape phages targeted at breast cancer cells-an alternative mechanism of directed molecular evolution. *Viruses* **2019**, *11*, 785. [CrossRef] [PubMed]
18. Brown, K.C. Peptidic tumor targeting agents: The road from phage display peptide selections to clinical applications. *Curr. Pharm. Des.* **2010**, *16*, 1040–1054. [CrossRef]
19. Hartman, E.C.; Jakobson, C.M.; Favor, A.H.; Lobba, M.J.; Álvarez-Benedicto, E.; Francis, M.B.; Tullman-Ercek, D. Quantitative characterization of all single amino acid variants of a viral capsid-based drug delivery vehicle. *Nat. Commun.* **2018**, *9*, 1385. [CrossRef]
20. Han, L.; Shi, J.; Liu, A. Novel biotemplated MnO2 1D nanozyme with controllable peroxidase-like activity and unique catalytic mechanism and its application for glucose sensing. *Sens. Actuators B Chem.* **2017**, *252*, 919–926. [CrossRef]
21. Han, L.; Shao, C.; Liang, B.; Liu, A. Genetically engineered phage-templated MnO_2 nanowires: Synthesis and their application in electrochemical glucose biosensor operated at neutral pH condition. *ACS Appl. Mater. Interfaces* **2016**, *8*, 13768–13776. [CrossRef]
22. Franco, D.; De Plano, L.M.; Rizzo, M.G.; Scibilia, S.; Lentini, G.; Fazio, E.; Neri, F.; Guglielmino, S.P.P.; Mezzasalma, A.M. Bio-hybrid gold nanoparticles as SERS probe for rapid bacteria cell identification. *Spectrochim. Acta Part A Mol. Biomol. Spectrosc.* **2020**, *224*, 117394. [CrossRef]
23. De Plano, L.M.; Scibilia, S.; Rizzo, M.G.; Crea, S.; Franco, D.; Mezzasalma, A.M.; Guglielmino, S.P.P. One-step production of phage–silicon nanoparticles by PLAL as fluorescent nanoprobes for cell identification. *Appl. Phys. A* **2018**, *124*, 222. [CrossRef]
24. Henry, K.A.; Arbabi-Ghahroudi, M.; Scott, J.K. Beyond phage display: Non-traditional applications of the filamentous bacteriophage as a vaccine carrier, therapeutic biologic, and bioconjugation scaffold. *Front. Microbiol.* **2015**, *6*, 755. [CrossRef]
25. Mao, C.; Solis, D.J.; Reiss, B.D.; Kottmann, S.T.; Sweeney, R.Y.; Hayhurst, A.; Georgiou, G.; Iverson, B.; Belcher, A.M. Virus-based toolkit for the directed synthesis of magnetic and semiconducting nanowires. *Science* **2004**, *303*, 213–217. [CrossRef] [PubMed]
26. Royston, E.; Lee, S.Y.; Culver, J.N.; Harris, M. Characterization of silica-coated tobacco mosaic virus. *J. Colloid Interface Sci.* **2006**, *298*, 706–712. [CrossRef]
27. Rodriguez, R.A.; Bounty, S.; Beck, S.; Chan, C.; McGuire, C.; Linden, K.G. Photoreactivation of bacteriophages after UV disinfection: Role of genome structure and impacts of UV source. *Water Res.* **2014**, *55*, 143–149. [CrossRef]
28. Bae, K.S.; Shin, G.A. Inactivation of various bacteriophages by different ultraviolet technologies: Development of a reliable virus indicator system for water reuse. *Environ. Eng. Res.* **2016**, *21*, 350–354. [CrossRef]
29. Gehringer, P.; Eschweiler, H.; Leth, H.; Pribil, W.; Pfleger, S.; Cabaj, A.; Haider, T.; Sommer, R. Bacteriophages as viral indicators for radiation processing of water: A chemical approach. *Appl. Radiat. Isot.* **2003**, *58*, 651–656. [CrossRef]
30. Fekete, A.; Kovács, G.; Hegedüs, M.; Módos, K.; Lammer, H. Biological responses to the simulated Martian UV radiation of bacteriophages and isolated DNA. *J. Photochem. Photobiol. B* **2008**, *92*, 110–116. [CrossRef] [PubMed]
31. Rauth, A.M. The physical state of viral nucleic acid and the sensitivity of viruses to ultraviolet light. *Biophys. J.* **1965**, *5*, 257–273. [CrossRef]
32. Kurosaki, Y.; Abe, H.; Morioka, H.; Hirayama, J.; Ikebuchi, K.; Kamo, N.; Nikaido, O.; Azuma, H.; Ikeda, H. Pyrimidine dimer formation and oxidative damage in M13 bacteriophage inactivation by ultraviolet C irradiation. *Photochem. Photobiol.* **2003**, *78*, 349–354. [CrossRef]
33. Passaretti, P.; Sun, Y.; Dafforn, T.R.M.; Oppenheimer, P.G. Determination and characterisation of the surface charge properties of the bacteriophage M13 to assist bio-nanoengineering. *RSC Adv.* **2020**, *10*, 25385. [CrossRef]
34. Nims, R.W.; Plavsics, M. Efficacy of electron beam for viral inactivation. *J. Microb. Biochem. Technol.* **2015**, *7*, 173–176. [CrossRef]
35. Tom, E.F.; Molineux, I.J.; Paff, M.L.; Bull, J.J. Experimental evolution of UV resistance in a phage. *Peer J.* **2018**, *6*, e5190. [CrossRef] [PubMed]
36. Sommer, R.; Pribil, W.; Appelt, S.; Gehringer, P.; Eschweiler, H.; Leth, H.; Cabaj, A.; Haider, T. Inactivation of bacteriophages in water by means of non-ionizing (UV-253.7 nm) and ionizing (gamma) radiation: A comparative approach. *Water Res.* **2001**, *35*, 3109–3116. [CrossRef]
37. Horneck, G.; Klaus, D.M.; Mancinelli, R.L. Space microbiology. *Microbiol. Mol. Biol. Rev.* **2010**, *74*, 121–156. [CrossRef]
38. Reisz, J.A.; Bansal, N.; Qian, J.; Zhao, W.; Furdui, C.M. Effects of ionizing radiation on biological molecules–mechanisms of damage and emerging methods of detection. *Antioxid. Redox Signal.* **2014**, *21*, 260–292. [CrossRef]
39. Felici, F.; Castagnoli, L.; Musacchio, A.; Jappelli, R.; Cesareni, G. Selection of antibody ligands from a large library of oligopeptides express on a multivalente exposition vector. *J. Mol. Biol.* **2019**, *222*, 301–310. [CrossRef]

40. Luzzago, A.; Felici, F. Construction of disulfide-constrained random peptide libraries displayed on phage coat protein VIII. *Methods Mol. Biol.* **1998**, *87*, 155–164. [CrossRef]
41. Carnazza, S.; Foti, C.; Gioffrè, G.; Felici, F.; Guglielmino, S.P.P. Specific and selective probes for *Pseudomonas aeruginosa* from phage-displayed random peptide libraries. *Biosens. Bioelectron.* **2008**, *23*, 1137–1144. [CrossRef] [PubMed]
42. De Plano, L.M.; Carnazza, S.; Franco, D.; Rizzo, M.G.; Conoci, S.; Petralia, S.; Nicoletti, A.; Zappia, M.; Campolo, M.; Esposito, E.; et al. Innovative IgG biomarkers based on phage display microbial amyloid mimotope for state and stage diagnosis in Alzheimer's disease. *ACS Chem. Neurosci.* **2020**, *11*, 1013–1026. [CrossRef] [PubMed]
43. Webb, B.; Sali, A. Comparative protein structure modeling using MODELLER. *Curr. Protoc. Bioinform.* **2014**, *54*, 5.6.1–5.6.37. [CrossRef]
44. Gasteiger, E.; Hoogland, C.; Gattiker, A.; Duvaud, S.; Wilkins, M.R.; Appel, R.D.; Bairoch, A. Protein identification and analysis rools on the ExPASy server. In *The Proteomics Protocols Handbook*; Walker, J.M., Ed.; Springer Protocols Handbooks; Humana Press: Totowa, NJ, USA, 2005; pp. 571–607.
45. Eisenstark, A.; Buzard, R.L.; Hartman, P.S. Inactivation of phage by near-ultraviolet radiation and hydrogen peroxide. *Photochem. Photobiol.* **1986**, *44*, 603–606. [CrossRef] [PubMed]

International Journal of *Molecular Sciences*

MDPI

Article

In Vitro Photodynamic Effects of the Inclusion Nanocomplexes of Glucan and Chlorin e6 on Atherogenic Foam Cells

Jae Won Ahn †, Jin Hyuk Kim † and Kyeongsoon Park *

Department of Systems Biotechnology, Chung-Ang University, Anseong, Gyeonggi 17546, Korea; pinkymonkey1@naver.com (J.W.A.); wlsgur0524@naver.com (J.H.K.)
* Correspondence: kspark1223@cau.ac.kr; Tel.: +82-031-670-3357
† These authors contributed equally to this work.

Abstract: Macrophage-derived foam cells play critical roles in the initiation and progression of atherosclerosis. Activated macrophages and foam cells are important biomarkers for targeted imaging and inflammatory disease therapy. Macrophages also express the dectin-1 receptor, which specifically recognizes β-glucan (Glu). Here, we prepared photoactivatable nanoagents (termed Glu/Ce6 nanocomplexes) by encapsulating hydrophobic chlorin e6 (Ce6) within the triple-helix structure of Glu in aqueous condition. Glu/Ce6 nanocomplexes generate singlet oxygen upon laser irradiation. The Glu/Ce6 nanocomplexes were internalized into foam cells and delivered Ce6 molecules into the cytoplasm of foam cells. Upon laser irradiation, they induced significant membrane damage and apoptosis of foam cells. These results suggest that Glu/Ce6 nanocomplexes can be a photoactivatable material for treating atherogenic foam cells.

Keywords: β-glucan; chlorin e6; dectin-1; photodynamic therapy; foam cells

Citation: Ahn, J.W.; Kim, J.H.; Park, K. In Vitro Photodynamic Effects of the Inclusion Nanocomplexes of Glucan and Chlorin e6 on Atherogenic Foam Cells. *Int. J. Mol. Sci.* **2021**, *22*, 177. https://dx.doi.org/10.3390/ijms22010177

Received: 1 December 2020
Accepted: 23 December 2020
Published: 26 December 2020

1. Introduction

Atherosclerosis is a chronic inflammatory disease characterized by the accumulation of lipids in the large arteries. Various types of cells, such as endothelial cells, smooth muscle cells, and inflammatory cells are involved in the formation of atherosclerosis. Endothelial dysfunction initiates the entry of lipids and inflammatory cells into the large arteries [1–4]. In the artery, monocytes differentiate into macrophages, and the macrophages that take up the lipid become foam cells. These macrophage-derived foam cells are formed at early stages in the atherosclerotic lesions and play critical roles in the development of atherosclerosis [5,6]. Macrophages express a variety of surface recognition receptors that serve as target biomarkers for imaging and therapy [7–12]. Dectin-1, a specific receptor for β-glucan, is one such receptor expressed on macrophages [13], including plaque macrophages of atherosclerotic mice [14]. Therefore, the specific targeting of the dectin-1 receptor on foam cells is an alternative approach for imaging and treatment of atherosclerosis.

Photodynamic therapy (PDT) is an effective therapeutic strategy for cancer. The use of photosensitizers and a specific wavelength of light generate reactive oxygen species (ROS), such as singlet oxygen (SO), hydroxyl radicals, and superoxide that subsequently induce membrane damage in the target cells, leading to cell death [15,16]. PDT is also used as an alternative therapy for atherosclerosis [17,18]. Previous studies have shown that PDT promotes plaque stabilization in atherosclerotic lesions and inhibits plaque progression by reducing the macrophage content [19–23]. Among various photosensitizers, chlorin e6 (Ce6) is widely used for PDT because it has high photosensitizing effect, low dark toxicity, long absorption band (663 nm), deep tissue penetration, and higher production of singlet oxygen [24]. However, despite the success of these PSs for PDT in in vitro and in vivo studies, some issues such as its insolubility in water and low selectivity to target cells or tissues still need to be overcome [25,26].

β-Glucan is a natural polysaccharide comprising linear β-1,3-linked D-glucose molecules, and possessing anti-inflammatory, anti-oxidant, and antifungal activities [13,27,28]. Glu acts as a ligand for the dectin-1 receptor and complement receptor 3 [27] on macrophages and dendritic cells in the immune system. As the uptake of Glu molecules is dectin-1-dependent in vitro [29], the selective targeting of dectin-1 makes the Glu nanoparticles an attractive drug carrier for the delivery of proteins [29], nucleic acids such as DNA [30] and siRNA [31], and small drug molecules [32].

In this study, we developed new photoactivatable nanoagents (termed Glu/Ce6 nanocomplexes) to deliver Ce6 into foam cells and enhance the effects of PDT on foam cells after laser irradiation. These nanoagents were prepared by encapsulating hydrophobic Ce6 within the triple-helix structure of Glu molecules. We analyzed their physicochemical and optical properties, as well as their ability to generate SO after laser treatment. We evaluated their intracellular uptake into foam cells to check whether Glu/Ce6 nanocomplexes can deliver Ce6 into the foam cells. We also confirmed their effects on foam cells after laser irradiation.

2. Results and Discussion

2.1. Synthesis of Inclusion Nanocomplexes of Glucan and Chlorin e6 (Glu/Ce6)

Typically, Glu has triple-helix structures in aqueous solution, which change into single strands upon solubilization at temperatures over 150 °C, or with NaOH or dimethyl sulfoxide (DMSO). Interestingly, they can re-assemble the triple-helix structure when exposed to an aqueous environment again [33,34]. Based on this fact, we prepared photoactivatable Glu/Ce6 nanocomplexes. For the synthesis of Glu/Ce6 nanocomplexes, Glu was dissolved in DMSO with heating at 60 °C to release single spiral structures from triple-spiral structures due to the loss of hydrophilicity and hydrogen bonds [35]. Then, hydrophobic Ce6 was added to the Glu solution and stirred for 3 h. The mixture was then added into a large amount of water. On exposure to an aqueous environment, the Glu triple structures may be reformed with the Ce6 molecules loaded into them to form Glu/Ce6 inclusion nanocomplexes (Figure 1).

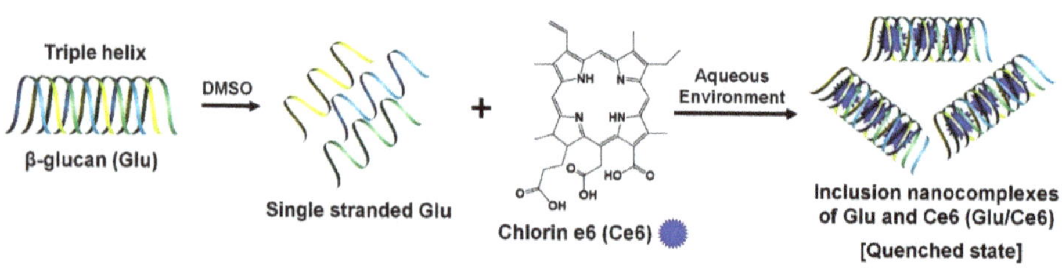

Figure 1. Schematic illustration of the preparation procedures of the Glu/Ce6 nanocomplexes.

2.2. Characterizations of Glu/Ce6 Nanocomplexes

After synthesis and purification of the Glu/Ce6 nanocomplexes, the loading amount of Ce6 within the nanocomplexes was determined by performing ultraviolet/visible (UV/Vis) analysis. Based on the standard curve of Ce6, it was determined that 1 mL of Glu/Ce6 nanocomplex solution contained 1.652 mM of Ce6. The synthesis of Glu/Ce6 nanocomplexes was further confirmed by Fourier-transform infrared (FTIR) analysis (Figure 2a). After loading of Ce6 into Glu, two new peak bands were observed at 1698 and 1570 cm^{-1} for the C = O and C = C stretching vibrations of Ce6 molecules, respectively.

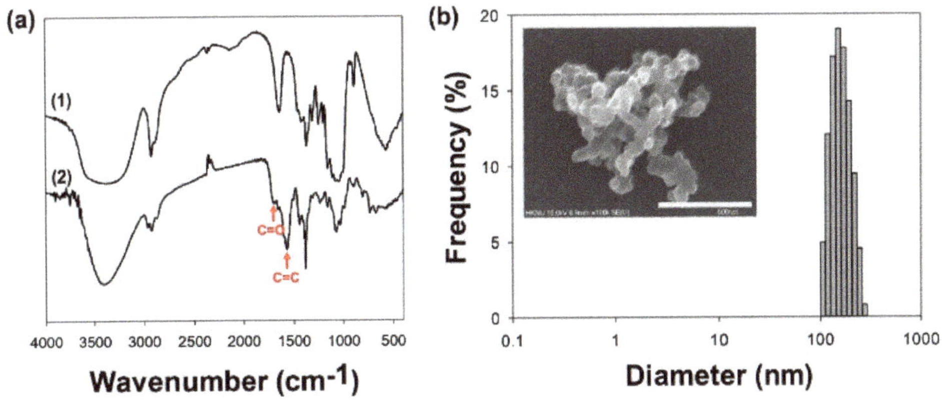

Figure 2. Characterizations of Glu/Ce6 nanocomplexes. (**a**) FTIR spectra of (1) Glu and (2) Glu/Ce6. (**b**) Particle size distribution of Glu/Ce6. Inset SEM image: the morphology of Glu/Ce6. Scale bar: 500 nm.

The dissolved Glu in DMSO has a single stranded structure, but it is converted into the triple-helix structure (known as a renaturation process) when exposed to water (at a ratio of 20 or higher) or when granulated to a size of ~ 800 nm [35]. In this study, the determined particle mean sizes of granulated Glu were 488 ± 35.6 nm (Supplementary Materials Figure S1), but polydispersity index (PDI) and Z-average sizes were not detected, suggesting that granulated Glu has the multi-disperse particle structures. After loading of Ce6 into Glu molecules, the mean sizes of the Glu/Ce6 nanocomplexes were 153.8 ± 35.6 nm, their Z-average sizes were 169.0 nm, with a PDI of 0.343 and a spherical shape (Figure 2b). SEM image showed that Glu/Ce6 nanocomplexes formed the multimeric or aggregated structures in dried state. Compared to the sizes of granulated Glu, Glu/Ce6 nanocomplexes were much smaller, due to stronger intermolecular hydrophobic interactions between Ce6 and Glu within the nanocomplexes.

Optical properties were examined by using UV/Vis and fluorescence spectrophotometers to demonstrate whether Glu/Ce6 re-assembled its triple-helix structure in aqueous solution. The solubilized Ce6 and Glu/Ce6 nanocomplexes in PBS (pH 7.4) solution with 1% Tween 20 (surfactant) showed a strong and sharp Soret band at 405 nm and Q-bands ranging from 480 to 700 nm. However, these peaks were relatively lower and broadened when Glu/Ce6 nanocomplexes were exposed to PBS (pH 7.4) without Tween 20 due to the interference of light transmittance (Figure 3a). In line with this result, fluorescence analyses demonstrated that the solubilized Ce6 and Glu/Ce6 in PBS (pH 7.4) with 1% Tween 20 showed strong fluorescence intensities, whereas Glu/Ce6 nanocomplexes dispersed in PBS without Tween 20 exhibited weak fluorescence intensities owing to the fluorescence quenching mechanism between Ce6 molecules within the Glu/Ce6 nanocomplexes (Figure 3b). These results suggested that the fabricated Glu/Ce6 re-assembled its triple-helix structure by a renaturation process due to intermolecular hydrophobic interactions between Ce6 and Glu molecules.

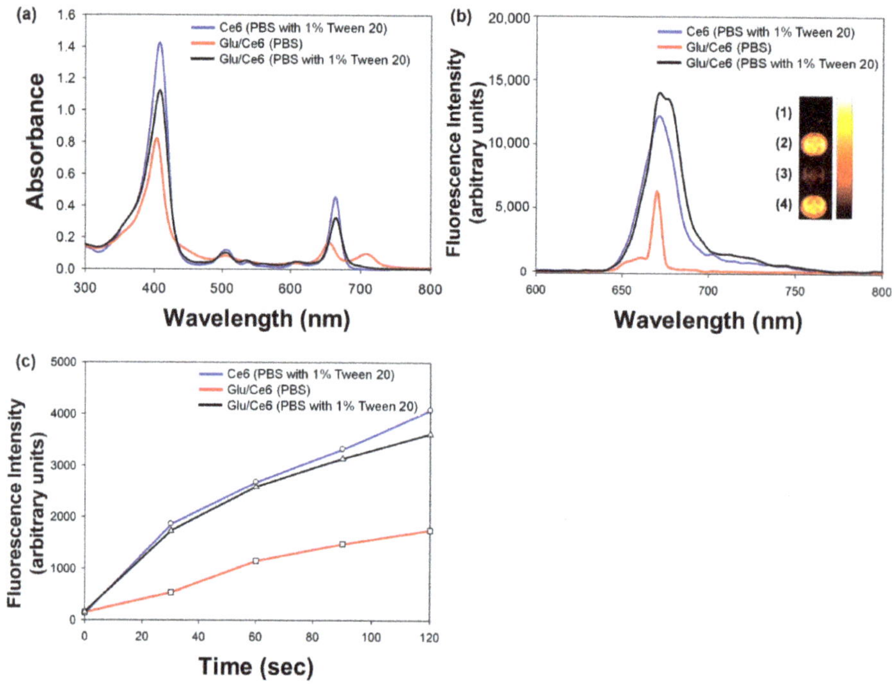

Figure 3. Optical characterizations of Glu/Ce6 and singlet oxygen generation by Glu/Ce6 upon laser irradiation. (**a**) UV/Vis and (**b**) fluorescence spectra of Ce6 (PBS with 1% Tween 20), Glu/Ce6 (PBS only or PBS with 1% Tween 20). Inset: Fluorescence image of (1) PBS only, (2) Ce6 (PBS with 1% Tween 20), (3) Glu/Ce6 (PBS), and (4) Glu/Ce6 (PBS with 1% Tween 20). (**c**) Singlet oxygen generation from Ce6 (PBS with 1% Tween 20), Glu/Ce6 (PBS), and Glu/Ce6 (PBS with 1% Tween 20), obtained by monitoring the fluorescence intensity (FI) of SOSG at 525 nm at different times of exposure to 670 nm laser.

2.3. Singlet Oxygen (SO) Generation Study

Photoactivatable nanomaterials should generate SO upon laser irradiation for PDT applications. To demonstrate whether Glu/Ce6 nanocomplexes can generate SO upon laser irradiation, a singlet oxygen sensor green (SOSG) study was performed because it is highly selective for $^1O_2$ [36]. Figure 3c shows the changes in SOSG fluorescence intensity with laser irradiation time. Ce6 and Glu/Ce6 solubilized in PBS (pH 7.4) containing 1% Tween 20 increased SOSG fluorescence signals with increasing irradiation time, whereas Glu/Ce6 nanocomplexes dispersed in PBS without a surfactant showed comparatively decreased fluorescence signals. As seen in UV/Vis and fluorescence analyses, the nanoparticle state (quenched state) of Glu/Ce6 has a relatively low absorbance and fluorescence intensity. Based on these results, we hypothesized that Glu/Ce6 nanocomplexes dispersed in PBS without a surfactant generate relatively lower levels of SO upon laser irradiation compared to the solubilized Ce6 and Glu/Ce6.

2.4. In Vitro Cytotoxic Study of Glu/Ce6 Nanocomplexes on Macrophages and Foam Cells

In vitro cytotoxicity of the Glu/Ce6 nanocomplexes was evaluated in normal macrophages and foam cells. Various concentrations of Glu/Ce6 (equivalent of 1, 2, 5, and 10 μM Ce6) were allowed to react with macrophages and foam cells in the dark. Figure 4a,b shows that macrophages and foam cells maintained their cell viabilities above 95% at all tested concentrations, suggesting that Glu/Ce6 nanocomplexes did not induce any cellular toxicity.

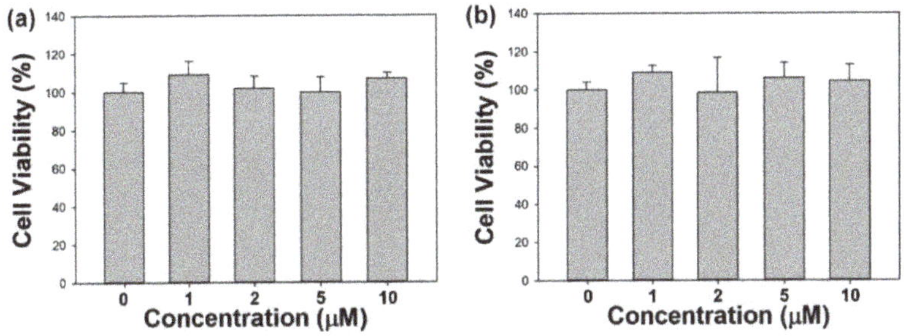

Figure 4. In vitro cytotoxic effects of Glu/Ce6 on (**a**) normal macrophages and (**b**) foam cells.

2.5. Intracellular Uptake of Glu/Ce6 Nanocomplexes by Foam Cells

Glu is a specific ligand for the dectin-1 receptor, which is expressed on macrophages [13]. We investigated whether Glu/Ce6 nanocomplexes are taken up by both macrophages and foam cells, using a customized multi-channel confocal laser scanning microscope (CLSM). Figure 5a shows that the intracellular uptake of Glu/Ce6 nanocomplexes by foam cells increased with increasing doses and was approximately nine-fold higher compared to macrophages (Figure 5b), suggesting that Glu/Ce6 nanocomplexes can deliver Ce6 molecules into foam cells with a better efficacy than normal macrophages.

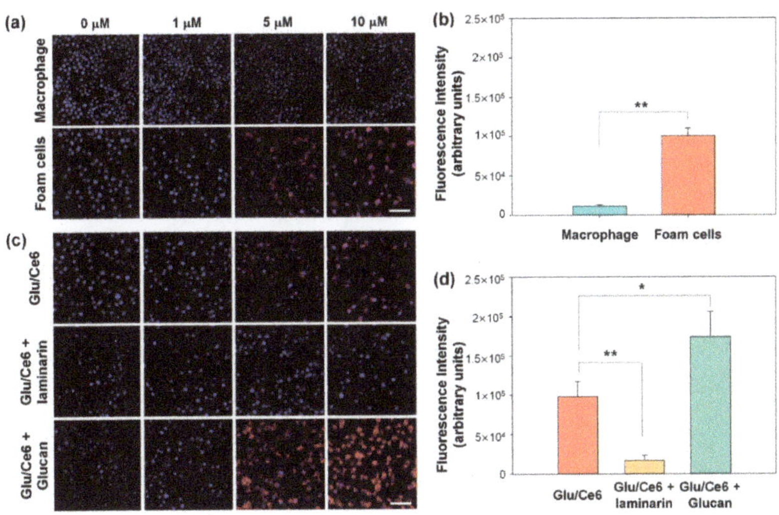

Figure 5. In vitro cellular internalization of Glu/Ce6 into macrophages and foam cells. (**a**) Confocal microscopy images of Glu/Ce6 internalization into macrophages and foam cells. Scale bar: 50 μm. (**b**) Comparison of cellular internalization of Glu/Ce6 (equivalent 5 μM Ce6) into normal macrophages and foam cells. (**c**) Cellular uptake of Glu/Ce6 against foam cells with or without pretreatment with laminarin or β-glucan. Scale bar: 50 μm. (**d**) Comparison of Glu/Ce6 (equivalent 5 μM Ce6) cellular uptake against foam cells with or without laminarin or β-glucan. ** $p < 0.01$, * $p < 0.05$.

Next, we performed blocking experiments by pretreating the foam cells with free laminarin (soluble β-glucan for dectin-1 ligand) and free Glu to confirm whether the cellular uptake of Glu/Ce6 nanocomplexes occurs via dectin-1 receptor-mediated endocytosis. The pretreatment of foam cells with free laminarin led to a significant decrease (of up to 7.15-fold) in the cellular uptake of Glu/Ce6 nanocomplexes because laminarin blocks the binding of foam cells to dectin-1 of Glu/Ce6 nanocomplexes (Figure 5c,d) [37,38]. In contrast, pretreatment with free Glu increased the cellular uptake of Glu/Ce6 nanocomplexes up to approximately 1.75-fold (Figure 5c,d). This increase might be associated with the increase of dectin-1 protein in macrophages treated with β-glucan [39]. Although we have limitation for dectin-1 binding affinity of Glu/Ce6 nanocomplexes on foam cells, these results suggest that Glu/Ce6 nanocomplexes deliver Ce6 molecules into foam cells with a better efficacy compared to normal macrophages.

2.6. In Vitro Phototoxic Effects of Glu/Ce6 Nanocomplexes

In vitro phototoxic effects of Glu/Ce6 nanocomplexes on macrophages or foam cells were first evaluated by the cell counting kit-8 (CCK-8) assay. Cell viability was normalized to control cells (no drug treatment, but laser irradiation). Macrophages or foam cells were not damaged by Glu/Ce6 nanocomplexes under dark conditions, as shown in Figure 4. In addition, after these cells were treated with free Ce6 (2 μM), their cell viabilities were above 95% and almost all cells were viable after NIR laser irradiation (Figure 6a,b). However, the cells treated with Glu/Ce6 nanocomplexes (equivalent 2 μM Ce6) were photodamaged after laser irradiation, and the viabilities of macrophages and foam cells were 73.1% and 51.8%, respectively.

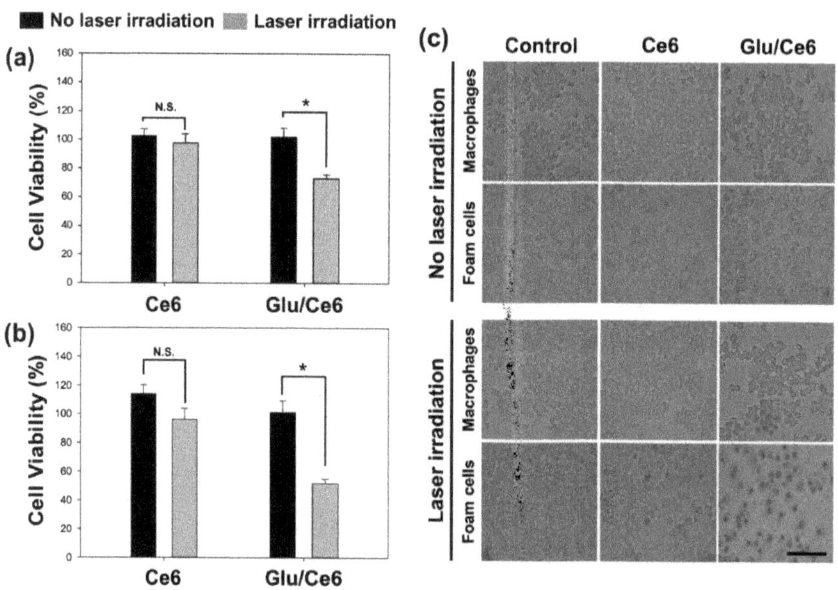

Figure 6. In vitro phototoxic effects of Ce6 and Glu/Ce6 on macrophages and foam cells. Comparison of in vitro phototoxic effects of Ce6 (2 μM) and Glu/Ce6 (equivalent 2 μM Ce6) on (**a**) macrophages and (**b**) foam cells before and after laser irradiation (670 nm, 1 min irradiation at 50 mW/cm^2). N.S.: Not Significant. * $p < 0.05$. (**c**) Trypan blue staining of macrophages and foam cells treated with Ce6 (2 μM Ce6) and Glu/Ce6 (equivalent 2 μM Ce6) before and after laser irradiation. Scale bar: 100 μm.

To confirm the in vitro PDT effects of Glu/Ce6 nanocomplexes, we performed a trypan blue dye exclusion test, which is a widely used test to identify dead cells. The live cells with intact membranes remain unstained, whereas dead cells with compromised membranes become stained [40]. As shown in Figure 6c, macrophages and foam cells treated with free Ce6 or Glu/Ce6 without laser irradiation were not stained, implying that their membranes were intact. Additionally, although some of the foam cells treated with both free Ce6 and NIR laser were stained, most of the cell membranes were not photo-damaged. However, when both macrophages and foam cells were treated with Glu/Ce6, as well as laser irradiation, the number of trypan blue-stained cells increased, indicating that the membranes of foam cells were more photo-damaged, as compared to those of macrophages.

Photoactivatable nanoagents induce cell apoptosis after laser treatment and then exert PDT effects. To demonstrate whether Glu/Ce6 nanocomplexes induce the apoptosis of foam cells compared to free Ce6, an annexin V (AV)/propidium iodide (PI) stain experiment was performed. The AV/PI staining method is useful for monitoring the progression of cell apoptosis because the early apoptotic cells are stained with AV but not PI, and late apoptotic cells are positively stained with both AV and PI [41]. Without NIR laser application, the control cells, free Ce6-treated and Glu/Ce6-treated foam cells did not show fluorescence signals (Figure 7a). After laser irradiation, both free Ce6-treated and Glu/Ce6-treated foam cells displayed similar AV-fluorescence signals in the representative fluorescence images (Figure 7a) and the quantitative comparison result of AV-fluorescence signals in both groups supported the fluorescence images (Figure 7b). However, Glu/Ce6-treated foam cells had significantly higher PI fluorescence intensities than free Ce6-treated foam cells (Figure 7a,c), suggesting that Glu/Ce6 nanocomplexes exhibit stronger PDT effects on foam cells compared to free Ce6. This enhanced PDT effect of Glu/Ce6 nanocomplexes on foam cells is associated with higher intracellular internalization compared to free Ce6. Indeed, Glu/Ce6 nanocomplexes showed strong fluorescence signals in the cytoplasm of foam cells and their intracellular uptake increased up to 2.6-fold, compared to free Ce6 (Supplementary Materials Figure S2), which in turn led to enhanced PDT effects. Thus, the delivery of Ce6 molecules into foam cells, using Glu/Ce6 nanocomplexes improves the light-triggered PDT effects.

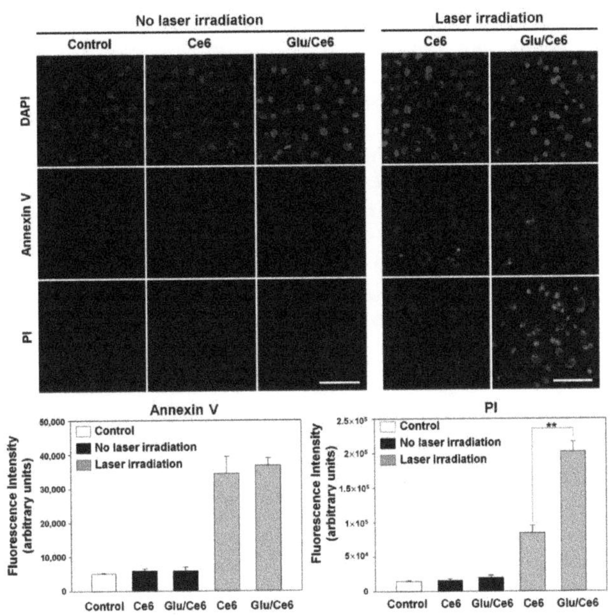

Figure 7. In vitro phototoxic effects of Ce6 and Glu/Ce6 on foam cells before and after laser irradiation (670 nm, 1 min irradiation at 50 mW/cm^2). (**a**) Annexin V and propidium iodide (PI) stainings were done to evaluate PDT effects of Ce6 and Glu/Ce6 on foam cells. Scale bar: 30 μm. Quantitative fluorescence intensity (FI) of (**b**) annexin V (AV) and (**c**) propidium iodide (PI) for each experimental group. ** $p < 0.01$.

3. Materials and Methods

3.1. Materials

Glu (β-1,3-Glucan from *Euglena gracilis*), potassium bromide (KBr), and LPS were purchased from Sigma-Aldrich (St. Louis, MO, USA). DMSO was obtained from Duksan (Ansan, Korea). A dialysis membrane (MWCO: 12–14 kDa) was purchased from Spectrum Laboratories (Rancho Dominguez, CA, USA). Ce6 and SOSG were procured from Frontier Scientific (Logan, UT, USA) and Invitrogen (Carlsbad, CA, USA), respectively.

3.2. Fabrication of Glu and Ce6 (Glu/Ce6) Inclusion Nanocomplexes

To fabricate the Glu/Ce6 nanocomplexes, 40 mg of Glu was added to 10 mL of DMSO and dissolved by heating at 90 °C overnight. Then, 20 mg of Ce6 dissolved in DMSO (1 mL) was added to the Glu solution and stirred for 3 h. Afterwards, the mixture was added to 200 mL of ultrapure water at 2 mL/min, using a syringe pump (NE-300, New Era Pump Systems Inc., Farmingdale, NY, USA) and stirred for 3 h. The resulting mixture was then dialyzed against deionized (DI) water for 3 days, using a dialysis membrane (MWCO: 12–14 kDa). Finally, the dialyzed mixture was centrifuged at 879× *g* for 10 min to remove large particles, and the supernatant was carefully collected and concentrated, using a rotary evaporator (N-1200BS, EYELA, Bohemia, NY, USA). Concentrated Glu/Ce6 nanocomplexes in DI water were used for the in vitro experiments. For FTIR analysis, a powder of Glu/Ce6 was obtained by freeze-drying the concentrated Glu/Ce6 nanocomplexes for 3 days.

3.3. Characterizations of Glu/Ce6 Nanocomplexes

To determine inclusion efficiency (loading efficiency) of Ce6 in Glu/Ce6, the concentrated Glu/Ce6 (1 mL) was added to PBS (1 mL) containing 1% Tween 20, and the mixed solution was sequentially diluted, using 1% Tween-containing PBS to measure the absorbance of Ce6. The amount of Ce6 in Glu/Ce6 was calculated, using the following equation, which was obtained from the standard curve of Ce6 ($Y = 243.46x + 0.0018$ ($R^2 = 1$)) by measuring the absorbance at 407 nm with a UV/Vis spectrophotometer (Neo-S450, Neogen, Korea).

The obtained Glu/Ce6 powder was further characterized by Fourier transform infrared (FTIR, Shimadzu 8400S, Kyoto, Japan) spectroscopy. The FTIR spectrum was acquired by using the KBr pellet method at a resolution of 4000–400 cm^{-1}.

The particle sizes and distribution of Glu/Ce6 were determined as follows: A 10-fold diluted Glu/Ce6 solution was prepared by using DI water. Then, the particle size, Z-average size, and polydispersity index were determined at a scattering angle of 90° with a particle size analyzer (SZ-100, HORIBA, Kyoto, Japan). To observe the morphology, a drop of a 10-fold diluted Glu/Ce6 solution was added to a cover slip and air-dried. Then, after platinum coating, the morphology of the dried Glu/Ce6 nanocomplexes was analyzed by using a field-emission SEM (FE-SEM, Baltec S-4700, Hitachi, Tokyo, Japan).

To confirm whether the prepared Glu/Ce6 forms nanosized particles, UV/Vis absorbance (from 300 to 800 nm) and fluorescence spectra (from 600 to 800 nm) of Glu/Ce6 (equivalent 5 μM Ce6) diluted in PBS (pH 7.4) were recorded with a UV/Vis spectrophotometer (Neo-S450, Neogen, Sejong, Korea) and fluorescence spectrometer (FS-2, Scinco, Korea), respectively. For comparison, free Ce6 (5 μM) and Glu/Ce6 (equivalent 5 μM Ce6) solutions were prepared, using 1% Tween 20-containing PBS, and their UV/Vis absorbance and fluorescence spectra were also monitored. In addition, the fluorescence images of these three solutions were acquired by using a small animal imaging equipment (In Vivo Smart-LF, VIEWORKS, Anyang, Korea).

3.4. Singlet-Oxygen-Generation Study

The SO generation potency of Glu/Ce6 was investigated by monitoring the changes in SOSG fluorescence intensity upon laser irradiation (670 nm, irradiation dose rate: 50 mW/cm^2). Oxygen-saturated PBS (pH 7.4) was prepared by bubbling oxygen gas for 30 min. Three solutions –Glu/Ce6 (equivalent 5 μM Ce6) in O_2-saturated PBS (pH 7.4) without Tween 20, Glu/Ce6 (equivalent 5 μM Ce6) in O_2-saturated PBS containing 1% Tween 20, and Ce6 (5 μM) in O_2-saturated PBS containing 1% Tween 20, were prepared. These three solutions contained 1 μM of the SOSG reagent. Under different exposure times from 0 to 120 sec with 670 nm laser irradiation, the fluorescence intensities of SOSG (Ex/Em: 504 nm/525 nm) were recorded with a fluorescence spectrometer (FS-2, Scinco, Korea).

3.5. Cells and Culture Condition

Murine macrophages (RAW 264.7) were obtained from the Korean Cell Line Bank (KCLB, Seoul, Korea). Cells were maintained in Dulbecco's Modified Eagle Medium (DMEM, Welgene, Seoul, Korea) containing 10% heat-inactivated FBS and 1% penicillin-streptomycin at 37 °C in a humidified 5% CO_2 incubator.

3.6. Cytocompatibility Study

In vitro cytocompatibility of the Glu/Ce6 on macrophages and atherogenic foam cells was evaluated by using the CCK-8 assay kit (Dojindo, Kumamoto, Japan) according to the manufacturer's instructions. Macrophages (1×10^4 cells/well) were seeded into 96-well plates and incubated for 24 h. To prepare atherogenic foam cells, the cultured macrophages were treated with LDL (100 μg/mL) and LPS (200 ng/mL) for 24 h. The cells were then treated with various concentrations of Glu/Ce6 (equivalent of 1, 2, 5, and 10 μM Ce6) for 24 h. After carefully washing the cells with culture medium, CCK-8 reagent solution was added to each well and further incubated for 2 h. The absorbance was measured at 450 nm, using Multiskan Go (Thermo Fisher Scientific, MA, USA). Cell viability was expressed as the percentage of viable cells compared to the survival of the non-treated control group.

3.7. Intracellular Uptake of Glu/Ce6

To compare the intracellular internalization ability of free Ce6 and Glu/Ce6 into normal macrophages and atherogenic foam cells, macrophages (1×10^5 cells/mL) incubated into 4-well chamber slides. Atherogenic foam cells were prepared by treating the prepared normal macrophages with LPS and LDL for 24 h. The two kinds of cells were treated with various concentrations of free Ce6 (1, 5, and 10 μM Ce6) and Glu/Ce6 (equivalent of 1, 5, and 10 μM Ce6) for 2 h. The cells were then carefully washed with PBS (pH 7.0) and fixed with formalin for 30 min. The nuclei of the cells were counterstained, using fluoromount solution containing 4′, 6′-diamidino-2-phenylindole hydrochloride (DAPI) (SouthernBiotech, AL, USA). To further evaluate whether Glu/Ce6 nanocomplexes are internalized into atherogenic foam cells, the foam cells were prepared by using the same protocols as described above. After pretreating the cells with free laminarin (1 mg/mL) as a soluble dectin-1 ligand or free β-glucan (1 mg/mL) as a dectin-1 receptor activator, the cells were treated with various concentrations of Glu/Ce6 (equivalent of 1, 5, and 10 μM Ce6) for 2 h. Afterwards, the cells were washed and fixed with formalin. The cell nuclei were counterstained with DAPI. The intracellular uptake of Glu/Ce6 into the cells was observed with a customized multi-channel confocal laser scanning fluorescence microscope.

Moreover, to quantitatively compare the intracellular uptake of free Ce6 and Glu/Ce6 in the absence and presence of laminarin or Glu, the average fluorescence intensity of a single cell was quantified by using Image J software (Ver 1.53a, National Institutes of Health, MD, USA). This was done by dividing the number of selected cells (approximately 50 cells) by the total fluorescence intensity of the selected cells from the confocal images.

3.8. In Vitro Phototoxic Effects of Glu/Ce6

To investigate whether Glu/Ce6 has phototoxic effects on normal macrophages or atherogenic foam cells under laser irradiation, CCK-8 assay, trypan blue dye exclusion, and AV/PI staining were performed. For the CCK-8 assay, the prepared normal macrophages and foam cells were treated with Ce6 (2 μM) or Glu/Ce6 (equivalent 2 μM Ce6). After 2 h of incubation, the culture medium was freshly changed and the cells were irradiated with the NIR laser (670 nm CW laser, 50 mW/cm^2) for 1 min. The laser power density and irradiation time were chosen based on our previous study [21]. They were then treated with CCK-8 solution (10 μL) and further incubated for 1 h. Phototoxic effects of each group were analyzed by measuring the absorbance at 450 nm, using a microplate reader (Thermo Fisher Scientific, Waltham, MA, USA), and the cell viability was represented as the percentage of viable cells compared to that of non-treated control cells.

For the trypan blue dye exclusion test, normal macrophages or foam cells were treated with free Ce6 or Glu/Ce6 at the same concentration as described above for 2 h. After the cells were carefully washed with fresh culture medium, they were irradiated with an NIR laser (670 nm CW laser, 50 mW/cm^2) for 1 min. Next, the cells were treated with trypan blue solution for 30 min and the excess trypan blue solution was then removed. After carefully adding PBS (pH 7.4) into each cell, optical cell images were acquired under a light microscope at 20× magnification, using LEICA DMi1 (Leica Microsystems, Wetzlar, Germany).

To evaluate apoptosis induction after laser irradiation, macrophages (1×10^5 cells/mL) were seeded into a 4-well chamber and allowed to adhere for 24 h. Atherogenic foam cells were prepared by treating LPS and LDL for 24 h as described above. The cells were exposed to free Ce6 (5 μM) or Glu/Ce6 (equivalent of 5 μM Ce6) for 2 h. After carefully washing with PBS, they were irradiated with or without an NIR laser (670 nm, 50 mW/cm^2) for 1 min. Then, the cells were stained with AV/PI for 5 min according to the manufacturer's instructions. Next, the cells were washed with the 1× binding buffer, fixed with paraformaldehyde (4%) for 30 min, and washed with PBS. The nuclei of the cells were counterstained with a mounting solution containing DAPI. Apoptosis induction was observed under a confocal microscope (LSM 700, Carl Zeiss, Germany).

3.9. Statistical Analysis

Statistical analysis was performed by using the R software (v3.4.3, Boston, MA, USA). Data are expressed as the mean ± SD. Statistical comparisons between two groups were carried out by using the Mann–Whitney test. Statistical significance was represented by p-values less than 0.01 or 0.05.

4. Conclusions

We fabricated Glu/Ce6 nanocomplexes as photoactivatable agents by encapsulating Ce6 molecules within Glu with a triple-helix structure, in aqueous conditions. The Glu/Ce6 nanocomplexes generated SO after NIR laser irradiation. Cytocompatible Glu/Ce6 nanocomplexes efficiently delivered Ce6 molecules into foam cells, as compared to normal macrophages. Without laser irradiation, the Glu/Ce6 nanocomplexes did not induce cell death. However, laser treatment significantly damaged the foam cell membranes and increased cell apoptosis. In the present study, we demonstrated the in vitro PDT effects of Glu/Ce6 nanocomplexes on foam cells. Unfortunately, the storage stability issue of Glu/Ce6 nanocomplexes should be improved because the prepared Glu/Ce6 nanocomplexes formed much bigger aggregates during the storage. In the near future, we will improve the colloidal stability of Glu/Ce6 nanocomplexes by modifying them and perform further PDT study to demonstrate whether Glu/Ce6 nanocomplexes are effective in regressing or stabilizing atherosclerotic plaques in vivo.

Supplementary Materials: The following are available online at https://www.mdpi.com/1422-0067/22/1/177/s1, Figure S1: Particle size distribution of granulated Glu, Figure S2: Intracellular uptake of free Ce6 and Glu/Ce6 against foam cells.

Author Contributions: Conceptualization, K.P.; methodology, J.W.A. and J.H.K.; formal analysis, K.P., J.W.A., and J.H.K.; writing—original draft preparation, J.W.A. and J.H.K.; writing—review and editing, K.P.; funding acquisition, K.P. All authors have read and agreed to the published version of the manuscript.

Funding: This work was supported by grants from the National Research Foundation of Korea (NRF-2018R1A2B3002001 and 2019M3A9E2066883).

Conflicts of Interest: The authors declare no conflict of interest.

References

1. Lusis, A.J. Atherosclerosis. *Nature* **2000**, *407*, 233–241. [CrossRef] [PubMed]
2. Hansson, G.K. Inflammation, atherosclerosis, and coronary artery disease. *N. Engl. J. Med.* **2005**, *352*, 1685–1695. [CrossRef] [PubMed]
3. Mantovani, A.; Garlanda, C.; Locati, M. Macrophage diversity and polarization in atherosclerosis: A question of balance. *Arterioscler. Thromb. Vasc. Biol.* **2009**, *29*, 1419–1423. [CrossRef] [PubMed]
4. Chinetti-Gbaguidi, G.; Colin, S.; Staels, B. Macrophage subsets in atherosclerosis. *Nat. Rev. Cardiol.* **2015**, *12*, 10–17. [CrossRef]
5. Beckers, L.; Heeneman, S.; Wang, L.; Burkly, L.C.; Rousch, M.M.; Davidson, N.O.; Gijbels, M.J.; de Winther, M.P.; Daemen, M.J.; Lutgens, E. Disruption of hedgehog signalling in ApoE -/- mice reduces plasma lipid levels, but increases atherosclerosis due to enhanced lipid uptake by macrophages. *J. Pathol.* **2007**, *212*, 420–428. [CrossRef]
6. Rader, D.J.; Pure, E. Lipoproteins, macrophage function, and atherosclerosis: Beyond the foam cell? *Cell Metab.* **2005**, *1*, 223–230. [CrossRef]
7. Chellat, F.; Merhi, Y.; Moreau, A.; Yahia, L. Therapeutic potential of nanoparticulate systems for macrophage targeting. *Biomaterials* **2005**, *26*, 7260–7275. [CrossRef]
8. Weissleder, R.; Nahrendorf, M.; Pittet, M.J. Imaging macrophages with nanoparticles. *Nat. Mater.* **2014**, *13*, 125–138. [CrossRef]
9. Amirbekian, V.; Lipinski, M.J.; Briley-Saebo, K.C.; Amirbekian, S.; Aguinaldo, J.G.; Weinreb, D.B.; Vucic, E.; Frias, J.C.; Hyafil, F.; Mani, V.; et al. Detecting and assessing macrophages in vivo to evaluate atherosclerosis noninvasively using molecular MRI. *Proc. Natl. Acad. Sci. USA* **2007**, *104*, 961–966. [CrossRef]
10. Quillard, T.; Libby, P. Molecular imaging of atherosclerosis for improving diagnostic and therapeutic development. *Circ. Res.* **2012**, *111*, 231–244. [CrossRef]
11. Jaffer, F.A.; Libby, P.; Weissleder, R. Molecular imaging of cardiovascular disease. *Circulation* **2007**, *116*, 1052–1061. [CrossRef] [PubMed]
12. McCarthy, J.R. Nanomedicine and Cardiovascular Disease. *Curr. Cardiovasc. Imaging Rep.* **2010**, *3*, 42–49. [CrossRef] [PubMed]
13. Brown, G.D.; Taylor, P.R.; Reid, D.M.; Willment, J.A.; Williams, D.L.; Martinez-Pomares, L.; Wong, S.Y.; Gordon, S. Dectin-1 is a major beta-glucan receptor on macrophages. *J. Exp. Med.* **2002**, *196*, 407–412. [CrossRef] [PubMed]
14. Szilagyi, K.; Gijbels, M.J.; van der Velden, S.; Heinsbroek, S.E.; Kraal, G.; de Winther, M.P.; van den Berg, T.K. Dectin-1 deficiency does not affect atherosclerosis development in mice. *Atherosclerosis* **2015**, *239*, 318–321. [CrossRef] [PubMed]
15. Sharma, S.K.; Chiang, L.Y.; Hamblin, M.R. Photodynamic therapy with fullerenes in vivo: Reality or a dream? *Nanomedicine (Lond.)* **2011**, *6*, 1813–1825. [CrossRef] [PubMed]
16. Dolmans, D.E.; Fukumura, D.; Jain, R.K. Photodynamic therapy for cancer. *Nat. Rev. Cancer* **2003**, *3*, 380–387. [CrossRef] [PubMed]
17. Chen, Y.C.; Huang, A.L.; Kyaw, T.S.; Bobik, A.; Peter, K. Atherosclerotic Plaque Rupture: Identifying the Straw That Breaks the Camel's Back. *Arterioscler. Thromb. Vasc. Biol.* **2016**, *36*, e63–e72. [CrossRef]
18. Houthoofd, S.; Vuylsteke, M.; Mordon, S.; Fourneau, I. Photodynamic therapy for atherosclerosis. The potential of indocyanine green. *Photodiagnosis Photodyn. Ther.* **2020**, *29*, 101568. [CrossRef]
19. Kossodo, S.; LaMuraglia, G.M. Clinical potential of photodynamic therapy in cardiovascular disorders. *Am. J. Cardiovasc. Drugs* **2001**, *1*, 15–21. [CrossRef]
20. Peng, C.; Li, Y.; Liang, H.; Cheng, J.; Li, Q.; Sun, X.; Li, Z.; Wang, F.; Guo, Y.; Tian, Z.; et al. Detection and photodynamic therapy of inflamed atherosclerotic plaques in the carotid artery of rabbits. *J. Photochem. Photobiol. B* **2011**, *102*, 26–31. [CrossRef]
21. Yi, B.G.; Park, O.K.; Jeong, M.S.; Kwon, S.H.; Jung, J.I.; Lee, S.; Ryoo, S.; Kim, S.E.; Kim, J.W.; Moon, W.J.; et al. In vitro photodynamic effects of scavenger receptor targeted-photoactivatable nanoagents on activated macrophages. *Int. J. Biol. Macromol.* **2017**, *97*, 181–189. [CrossRef] [PubMed]
22. Waksman, R.; McEwan, P.E.; Moore, T.I.; Pakala, R.; Kolodgie, F.D.; Hellinga, D.G.; Seabron, R.C.; Rychnovsky, S.J.; Vasek, J.; Scott, R.W.; et al. PhotoPoint photodynamic therapy promotes stabilization of atherosclerotic plaques and inhibits plaque progression. *J. Am. Coll. Cardiol.* **2008**, *52*, 1024–1032. [CrossRef] [PubMed]
23. Kim, H.; Kim, Y.; Kim, I.H.; Kim, K.; Choi, Y. ROS-responsive activatable photosensitizing agent for imaging and photodynamic therapy of activated macrophages. *Theranostics* **2013**, *4*, 1–11. [CrossRef] [PubMed]
24. Juzeniene, A. Chlorin e6-based photosensitizers for photodynamic therapy and photodiagnosis. *Photodiagnosis Photodyn. Ther.* **2009**, *6*, 94–96. [CrossRef]
25. Hsiang, Y.; Stonefield, M.; Bower, R.D.; Fragoso, M.; Tsang, V.; Crespo, M.T.; Lundkvist, A. Assessing Photofrin uptake in atherosclerosis with a fluorescent probe: Comparison with photography and tissue measurements. *Lasers Surg. Med.* **1993**, *13*, 271–278. [CrossRef]
26. Kereiakes, D.J.; Szyniszewski, A.M.; Wahr, D.; Herrmann, H.C.; Simon, D.I.; Rogers, C.; Kramer, P.; Shear, W.; Yeung, A.C.; Shunk, K.A.; et al. Phase I drug and light dose-escalation trial of motexafin lutetium and far red light activation (phototherapy) in subjects with coronary artery disease undergoing percutaneous coronary intervention and stent deployment: Procedural and long-term results. *Circulation* **2003**, *108*, 1310–1315. [CrossRef]
27. Brown, G.D.; Gordon, S. Immune recognition. A new receptor for beta-glucans. *Nature* **2001**, *413*, 36–37. [CrossRef]

28. Drummond, R.A.; Brown, G.D. The role of Dectin-1 in the host defence against fungal infections. *Curr. Opin. Microbiol.* **2011**, *14*, 392–399. [CrossRef]
29. Huang, H.; Ostroff, G.R.; Lee, C.K.; Wang, J.P.; Specht, C.A.; Levitz, S.M. Distinct patterns of dendritic cell cytokine release stimulated by fungal beta-glucans and toll-like receptor agonists. *Infect. Immun.* **2009**, *77*, 1774–1781. [CrossRef]
30. Soto, E.R.; Ostroff, G.R. Characterization of multilayered nanoparticles encapsulated in yeast cell wall particles for DNA delivery. *Bioconjug. Chem.* **2008**, *19*, 840–848. [CrossRef]
31. Aouadi, M.; Tesz, G.J.; Nicoloro, S.M.; Wang, M.; Chouinard, M.; Soto, E.; Ostroff, G.R.; Czech, M.P. Orally delivered siRNA targeting macrophage Map4k4 suppresses systemic inflammation. *Nature* **2009**, *458*, 1180–1184. [CrossRef] [PubMed]
32. Soto, E.; Kim, Y.S.; Lee, J.; Kornfeld, H.; Ostroff, G. Glucan Particle Encapsulated Rifampicin for Targeted Delivery to Macrophages. *Polymers* **2010**, *2*, 681–689. [CrossRef]
33. Nasrollahi, Z.; Mohammadi, S.R.; Mollarazi, E.; Yadegari, M.H.; Hassan, Z.M.; Talaei, F.; Dinarvand, R.; Akbari, H.; Atyabi, F. Functionalized nanoscale beta-1,3-glucan to improve Her2+ breast cancer therapy: In vitro and in vivo study. *J. Control. Release* **2015**, *202*, 49–56. [CrossRef] [PubMed]
34. Xu, X.; Wang, X.; Cai, F.; Zhang, L. Renaturation of triple helical polysaccharide lentinan in water-diluted dimethylsulfoxide solution. *Carbohydr. Res.* **2010**, *345*, 419–424. [CrossRef] [PubMed]
35. Sletmoen, M.; Stokke, B.T. Higher order structure of (1,3)-beta-D-glucans and its influence on their biological activities and complexation abilities. *Biopolymers* **2008**, *89*, 310–321. [CrossRef]
36. Rota, C.; Chignell, C.F.; Mason, R.P. Evidence for free radical formation during the oxidation of 2′-7′-dichlorofluorescin to the fluorescent dye 2′-7′-dichlorofluorescein by horseradish peroxidase: Possible implications for oxidative stress measurements. *Free Radic. Biol. Med.* **1999**, *27*, 873–881. [CrossRef]
37. Goodridge, H.S.; Reyes, C.N.; Becker, C.A.; Katsumoto, T.R.; Ma, J.; Wolf, A.J.; Bose, N.; Chan, A.S.; Magee, A.S.; Danielson, M.E.; et al. Activation of the innate immune receptor Dectin-1 upon formation of a 'phagocytic synapse'. *Nature* **2011**, *472*, 471–475. [CrossRef]
38. Rothfuchs, A.G.; Bafica, A.; Feng, C.G.; Egen, J.G.; Williams, D.L.; Brown, G.D.; Sher, A. Dectin-1 interaction with Mycobacterium tuberculosis leads to enhanced IL-12p40 production by splenic dendritic cells. *J. Immunol.* **2007**, *179*, 3463–3471. [CrossRef]
39. Li, X.; Luo, H.; Ye, Y.; Chen, X.; Zou, Y.; Duan, J.; Xiang, D. betaglucan, a dectin1 ligand, promotes macrophage M1 polarization via NFkappaB/autophagy pathway. *Int. J. Oncol.* **2019**, *54*, 271–282. [CrossRef]
40. Crescenzi, E.; Varriale, L.; Iovino, M.; Chiaviello, A.; Veneziani, B.M.; Palumbo, G. Photodynamic therapy with indocyanine green complements and enhances low-dose cisplatin cytotoxicity in MCF-7 breast cancer cells. *Mol. Cancer Ther.* **2004**, *3*, 537–544.
41. Wlodkowic, D.; Telford, W.; Skommer, J.; Darzynkiewicz, Z. Apoptosis and beyond: Cytometry in studies of programmed cell death. *Methods Cell Biol.* **2011**, *103*, 55–98. [CrossRef] [PubMed]

MDPI
St. Alban-Anlage 66
4052 Basel
Switzerland
www.mdpi.com

International Journal of Molecular Sciences Editorial Office
E-mail: ijms@mdpi.com
www.mdpi.com/journal/ijms

www.ingramcontent.com/pod-product-compliance
Lightning Source LLC
Chambersburg PA
CBHW061617220926
51850CB00005B/532